T0338246

BIODIVERSITY AND INSECT PESTS

BIODIVERSITY AND INSECT PESTS

KEY ISSUES FOR SUSTAINABLE MANAGEMENT

Edited by

Geoff M. Gurr

EH Graham Centre for Agricultural Innovation
(Charles Sturt University and NSW Department of Primary Industries)
Orange, New South Wales, Australia

Steve D. Wratten

Bio-Protection Research Centre
Lincoln University
Canterbury, New Zealand

William E. Snyder

Department of Entomology
Washington State University
Pullman, Washington, USA

With

Donna M.Y. Read

Charles Sturt University
Orange, New South Wales, Australia

A John Wiley & Sons, Ltd., Publication

This edition first published 2012 © 2012 by John Wiley & Sons, Ltd.

Wiley-Blackwell is an imprint of John Wiley & Sons, formed by the merger of Wiley's global Scientific, Technical and Medical business with Blackwell Publishing.

Registered office: John Wiley & Sons, Ltd, The Atrium, Southern Gate, Chichester, West Sussex, PO19 8SQ, UK

Editorial offices: 9600 Garsington Road, Oxford, OX4 2DQ, UK
　　　　　　　　　The Atrium, Southern Gate, Chichester, West Sussex, PO19 8SQ, UK
　　　　　　　　　111 River Street, Hoboken, NJ 07030-5774, USA

For details of our global editorial offices, for customer services and for information about how to apply for permission to reuse the copyright material in this book please see our website at www.wiley.com/wiley-blackwell.

Library of Congress Cataloging-in-Publication Data

Biodiversity and pests : key issues for sustainable management / edited by Geoff M. Gurr, Steve D. Wratten,
William E. Snyder ; with Donna M.Y. Read.
　　　p. cm.
　Includes bibliographical references and index.
　ISBN 978-0-470-65686-0 (cloth)
　1. Agricultural pests–Control.　2. Insect pests–Control.　3. Agrobiodiversity.　4. Biodiversity.　5. Sustainable
agriculture.　6. Sustainability.　I. Gurr, Geoff.　II. Wratten, Stephen D.　III. Snyder, William E.　IV. Read, Donna M. Y.
　SB950.B47 2012
　363.7'8–dc23
　　　　　　　　　　　　　2011046054

A catalogue record for this book is available from the British Library.

Wiley also publishes its books in a variety of electronic formats. Some content that appears in print may not be available in electronic books.

Set in 9/11 pt PhotinaMT by Toppan Best-set Premedia Limited
Printed and bound in Malaysia by Vivar Printing Sdn Bhd

1   2012

CONTENTS

This book has a companion website

www.wiley.com/go/gurr/biodiversity

with Figures and Tables from the book for downloading.

PREFACE

Agriculture accounts for approximately 40% of the land area on planet Earth and has been a major factor in global biodiversity decline. It is ironic, then, that agriculture is now showing conspicuous signs of faltering because of a breakdown in the services provided by nature. Pest control, soil fertility and nutrient cycling are amongst the most important of these. Industrialised agriculture, in striving for greater levels of productivity, uses inputs such as millions of tons of pesticides and fertilisers to replace natural processes. Reliance on technologies based on non-renewable resources has widely acknowledged problems including pollution, human safety and – in the case of pesticides – reduced efficacy as a result of resistance developing in pest populations.

It is time to consider how agriculture worked in such a sustainable manner for thousands of years before the rise of industrialised agriculture. Much is to be learned from traditional practices of diverse crop systems in which biodiversity is maintained. But if agriculture is also to meet the future needs of an increasing human population, ecological science must rise to the challenge of providing more than theoretical understanding and ingenious new research methods. Practicable methods are also required that will permit highly productive farming systems which, by virtue of their ecological foundation, are more sustainable.

The chapters in this book address this challenge from the perspective of insect pest management. Insect pests continue to cause severe crop losses worldwide but novel pesticides and genetically modified plants are not the only technologies available for their control. This book explores ways in which biodiversity can be harnessed to achieve sustainable pest management. Vegetation diversification at scales ranging from the field up to the landscape can reduce pests either directly or by enhancing the activity of predators and parasites. Biodiversity is also a source of genes for better crop varieties and of compounds that can be used as botanical insecticides or that work by more subtle chemical ecology mechanisms.

The role of biodiversity in pest management is a burgeoning area of research and novel pest management strategies are now being implemented successfully in many countries. Forms of ecologically based pest suppression are important examples of the ecosystem services that can be provided by biodiversity. Moreover, pest suppression can be achieved concurrently with providing other benefits such as pollinator enhancement, wildlife conservation, dual crop production and even carbon sequestration.

Our aim as editors as we planned this book in 2010, the United Nations International Year of Biodiversity, was to achieve a comprehensive synthesis of this exciting and important field of applied science. To this end we recruited authors who include leading researchers and practitioners and combined their wide experience with that of carefully selected younger scientists with innovative thinking. With wide international coverage including Africa, America, Asia, Australasia and Europe, our treatment of the subject is significantly broader that that available from mainstream, English-language journals.

We have strived to make the material in this book accessible to advanced undergraduates and newcomers to the field, with plenty of illustrative features, while still offering the specialist reader a current synthesis and stimulating new ideas. Chapters are arranged under a series of headings (Introduction, Fundamentals, Methods, Application and Synthesis), but these should not be viewed too rigidly. Many of the chapters include a blend of material; especially when stressing the link between aspects of theory and the success of real-world use. Ultimately, we hope that the book will prove useful in placing pest management on a more sustainable footing.

We thank the chapter authors for their generous contribution of ideas, attention to detail and (nearly always) keeping to schedule. Many also served as reviewers for other chapters. We very much appreciate the assistance of the many colleagues who acted as reviewers for chapters: Helmut van Emden, Daniel Karp, Myron Zalucki, Sarah Wheeler, Debbie Finke, Gary Chang, Cory Straub, Deborah Letourneau, Stephen Duke, James Hagler, Wopke van der Werf, Nuria Agusti, Matt Greenstone, Mark Jervis, Jana Lee, Marcel Dicke, Liu Shu Sheng, A. Raman, Bob Bugg, Samantha Cook, Norman Arancon and Katja Poveda. This book would not have been possible without Donna Read, whose input went way beyond proofreading and formatting.

Geoff M. Gurr
Steve D. Wratten
William E. Snyder
August 2011

FOREWORD

Agriculture has been practised for several thousand years but it is only in the past few generations that the traditional practices that sustained agriculture have come to be replaced by modern and largely industrialised systems. Despite dramatic increases in food production, it is now recognised that agriculture can negatively affect the environment through overuse of natural resources as inputs or through their use as a sink for waste and pollution. Such effects are called negative externalities because they impose costs that are not reflected in market prices. What has also become clear in recent years is that the apparent success of some modern agricultural systems has masked significant negative externalities, with environmental and health problems widely documented. These environmental costs shift conclusions about which agricultural systems are the most efficient, and suggest that alternative practices and systems which reduce negative and increase positive externalities should be sought.

The growing human population and rapidly changing consumption patterns will bring increasing demands for food, fuel and fibre. It is estimated that world population will reach some nine billion by the middle of the twenty-first century. This will require agricultural production to increase by at least two-thirds; perhaps doubled if those in developing countries are to approach levels of animal protein intake that are taken for granted in industrialised countries. The scale of this challenge is daunting but studies of agricultural sustainability in developing countries suggest overall yield increases of 80–100% are possible in many countries and systems. One analysis of 286 projects in 57 countries showed improvements had been made by 12 million farmers on 37 million hectares of farmland (Pretty et al., 2006, *Environmental Science and Technology*, 4, 1114–1119); a recent study of African agriculture found that 10 million farmers and their families had more than doubled yields on another 13 million hectares (Pretty et al., 2011, *International Journal of Agricultural Sustainability*, 9, 5–24). Food outputs by such sustainable intensification have been multiplicative – by which yields per hectare increased by combinations of the use of new and improved varieties and new agronomic-agroecological management, and additive – by which diversification resulted in the emergence of a range of new crops, livestock or fish that added to the existing staples or vegetables already being cultivated.

Realising the promise of ecologically based agriculture will require a massive and coordinated effort. A key component is the role of science to both provide a better understanding of the natural resource base and develop new technologies. The significance of this book is that it amply demonstrates the power of biodiversity to combat one of the major causes of crop loss: insect pests. Methods such as growing secondary crops on the embankments around rice fields, incorporating agroforestry into farming systems, using locally appropriate crop varieties and adopting integrated pest management were widely used in these agricultural sustainability studies. Much of what is now happening on farms has drawn from the work of the authors and editors of this book. Each of these methods, and many other biodiversity-based approaches, are detailed in chapters that span the full spectrum from underlying theory, to methods for research and implementation and, ultimately, to cases of successful application and use. This book compellingly shows that biodiversity on farms and across landscapes can provide a range of benefits to humans at the same time as contributing to suppressing pests.

Understanding, protecting and harnessing biodiversity is a key to the agricultural and food challenge before us.

Professor Jules Pretty OBE, University of Essex
August, 2011

CONTRIBUTORS

ALTIERI, MIGUEL A.: Department of Environmental Science, Policy and Management, University of California, Berkeley, USA

BRUCE, TOBY J.A.: Rothamsted Research, Harpenden, Hertfordshire AL5 2JQ, UK

CATINDIG, J.: Crop and Environmental Sciences Division, International Rice Research Institute, DAPO Box 7777 Metro Manila, Philippines

CHENG, J.A.: Institute for Insect Sciences, Zhejiang University, 268 Kaixuan Road, Zhejiang Province, China, 310029

EKBOM, BARBARA: Department of Ecology, Swedish University of Agricultural Sciences, Box 7044, 75007 Uppsala, Sweden

ESCALADA, M.M.: Department of Development Communication, Visayas State University, Baybay, Leyte, Philippines

GÁMEZ-VIRUÉS, SAGRARIO: EH Graham Centre for Agricultural Innovation (Industry and Innovation NSW and Charles Sturt University), PO Box 883 Orange, NSW 2800, Australia

GARDINER, MARY M.: Department of Entomology, The Ohio State University, Ohio Agricultural Research and Development Center, Wooster, OH, USA

GILLESPIE, MARK: Bio-Protection Research Centre, PO Box 84, Lincoln University, Lincoln 7647, New Zealand and Institute of Integrative and Comparative Biology, University of Leeds, Leeds LS2 9JT, UK

GURR, GEOFF M.: EH Graham Centre for Agricultural Innovation (Charles Sturt University and NSW Department of Primary Industries), PO Box 883 Orange, NSW 2800, Australia

HARWOOD, J.D.: Department of Entomology, University of Kentucky, Lexington, Kentucky 40546, USA

HEONG, K.L.: Crop and Environmental Sciences Division, International Rice Research Institute, DAPO Box 7777 Metro Manila, Philippines

HOLLAND, J.M.: Game & Wildlife Conservation Trust, Burgate Manor, Fordingbridge, Hampshire SP6 1EF, UK

HORGAN, FINBARR G.: Crop and Environmental Sciences Division, International Rice Research Institute, DAPO Box 7777, Metro Manila, Philippines

JAMES, DAVID G.: Department of Entomology, Washington State University, Irrigated Agriculture Research and Extension Center, 24106 N. Bunn Road, Prosser, Washington 99350, USA

JONSSON, MATTIAS: Department of Ecology, Swedish University of Agricultural Sciences, Box 7044, 75007 Uppsala, Sweden

KHAN, ZEYAUR R.: International Centre of Insect Physiology and Ecology, PO Box 30772, Nairobi, Kenya

KOUL, OPENDER: Insect Biopesticide Research Centre, 30 Parkash Nagar, Jalandhar-144003, India

LANDIS, DOUGLAS A.: Department of Entomology and Great Lakes Bioenergy Research Center, Michigan State University, East Lansing, MI, USA

LAVANDERO, BLAS: Instituto de Biologia Vegetal y Biotecnolgia, University of Talca, 2 Norte 685, Talca, Chile, Fax : 56-71-200-276, Fono : 56-71-200-280, 200-Talca, Chile

LEATHER, SIMON R.: Division of Biology, Imperial College London, Silwood Park Campus, Ascot, SL5 7PY, UK

LU ZHONGXIAN: Institute of Plant Protection and Microbiology, Zhejiang Academy of Agricultural Sciences, Hangzhou 310021, China

MEYER, KATRIN M.: Department of Ecosystem Modelling, Büsgen-Institut, Georg-August-University of Göttingen, Büsgenweg 4, 37077 Göttingen, Germany

MIDEGA, CHARLES A.O.: International Centre of Insect Physiology and Ecology, PO Box 30772, Nairobi, Kenya

NICHOLLS, CLARA I.: Department of Environmental Science, Policy and Management, University of California, Berkeley, USA

ORRE-GORDON, SOFIA: Bio-Protection Research Centre, PO Box 84, Lincoln University 7647, Canterbury, New Zealand

PEROVIC, DAVID: EH Graham Centre for Agricultural Innovation (NSW Department of Primary Industries and Charles Sturt University), PO Box 883 Orange, NSW 2800, Australia.

PFANNENSTIEL, R.S.: Beneficial Insects Research Unit, USDA-ARS, Weslaco, TX 78599, USA

PICKETT, JOHN A.: Rothamsted Research, Harpenden, Hertfordshire AL5 2JQ, UK

PITTCHAR, JIMMY: International Centre of Insect Physiology and Ecology, PO Box 30772, Nairobi, Kenya

PONTI, LUIGI: Laboratorio Gestione Sostenibile degli Agro-Ecosistemi (UTAGRI-ECO), Agenzia nazionale per le nuove tecnologie, l'energia e lo sviluppo economico sostenibile (ENEA), Centro Ricerche Casaccia, Via Anguillarese 301, 00123 Roma, Italy and Center for the Analysis of Sustainable Agricultural Systems (CASAS), Kensington, CA 94707, USA

READ, DONNA M.Y.: Charles Sturt University, PO Box 883 Orange, New South Wales, Australia

REYNOLDS (NÉE KVEDARAS), OLIVIA L.: EH Graham Centre for Agricultural Innovation (Charles Sturt University and Industry & Investment NSW), Elizabeth Macarthur Agricultural Institute, Woodbridge Road, Menangle, New South Wales 2568, Australia

SCHERBER, CHRISTOPH: Georg-August-University Goettingen, Department of Crop Science, Agroecology, Grisebachstr. 6, D-37077 Goettingen, Germany

SHREWSBURY, PAULA M.: Department of Entomology, University of Maryland, College Park, Maryland 20742, USA

SIMPSON, MARJA: Charles Sturt University, PO Box 883, Leeds Parade, Orange, New South Wales 2800, Australia

SMITH, H.A.: University of Florida, Gulf Coast Research and Education Center, 14625 CR 672, Wimauma, Florida, 33598, USA.

SNYDER, WILLIAM E.: Department of Entomology, Washington State University, Pullman, Washington 99163, USA

SYMONDSON, WILLIAM O.C.: Cardiff School of Biosciences, Biomedical Sciences Building, Museum Avenue, Cardiff, CF10 3AX, UK

TILLMAN, P.G.: USDA, ARS, Crop Protection & Management Research Laboratory, PO Box 748, Tifton, Georgia, 31793, USA

TOMPKINS, JEAN: Bio-Protection Research Centre, Lincoln University, Canterbury, New Zealand.

TSCHARNTKE, TEJA: Department of Crop Sciences, Georg-August-University, Grisebachstr. 6, D-37077 Göttingen, Germany

TYLIANAKIS, JASON M.: Biological Sciences, University of Canterbury, Christchurch 8140, New Zealand

VAN RIJN, PAUL C.J.: Institute for Biodiversity and Ecosystem Dynamics (IBED), University of Amsterdam, The Netherlands

VISSER, UTE: Georg-August-University, Grisebachstr. 6, D-37077 Göttingen, Germany

WÄCKERS FELIX L.: Lancaster University, LEC, Centre for Sustainable Agriculture, Lancaster, UK and Biobest, Ilse Velden 18, Westerlo, Belgium

WELCH, K.D.: Department of Entomology, University of Kentucky, Lexington, Kentucky 40546, USA

WIEGAND, KERSTIN: University of Goettingen, Büsgenweg 4, 37077, Göttingen, Germany.

WRATTEN, STEVE D.: Bio-Protection Research Centre, PO Box 84, Lincoln University, Lincoln 7647, New Zealand

YANG PUYUN: National Agro-Technical Extension and Service Centre, Ministry of Agriculture, Beijing 100125, China

YANG YAJUN: Institute of Plant Protection and Microbiology, Zhejiang Academy of Agricultural Sciences, Hangzhou 310021, China

ZHAO ZHONGHUA: National Agro-Technical Extension and Service Centre, Ministry of Agriculture, Beijing 100125, China

Introduction

Chapter 1

BIODIVERSITY AND INSECT PESTS

Geoff M. Gurr, Steve D. Wratten and William E. Snyder

Biodiversity and Insect Pests: Key Issues for Sustainable Management, First Edition. Edited by Geoff M. Gurr, Steve D. Wratten, William E. Snyder, Donna M.Y. Read.

INTRODUCTION: INSECTS, PLANTS AND HUMANS

This book is essentially about interactions between the three most important life forms on planet Earth: insects, plants and humans, and the ways in which they are affected by biodiversity, the complex web of life. Over a million species of insect have been formally described (20 times the number of all vertebrates), with just one insect order, the beetles (Coleoptera), representing 25% of all described species of all forms of life (Hunt *et al.*, 2007). It has been estimated that the biomass of insects in temperate terrestrial ecosystems is 10 times that of the usually more conspicuous vertebrates, and that for each human there are 1,000,000,000,000,000,000 living insects (Meyer, 2009).

Insect and plant biodiversity are tightly linked, and it is generally accepted that the rise of angiosperm plants during the Cretaceous period (145–65 million years ago) was accompanied by the development of many intricate coadaptations between plants and insects. These included pollination and seed dispersal (Ehrlich and Raven, 1964; Scriber, 2010), such that many insects benefit plants. However, many other insect species are herbivores harmful to plants, and there is compelling evidence for coevolution between plant defences and the ability of insect herbivores to overcome them. An example of great relevance to agricultural pest management is the phenomenon of 'resistance breakdown'. This occurs when a pest population responds to the resistance genes bred into into a widely used crop variety by the development of increased virulence over successive generations of the adapting pest (e.g. McMenemy *et al.*, 2009). This renders the host plant's resistance mechanism(s) ineffective.

In contrast to the two 'mega taxa' sketched out above, *Homo sapiens* is an evolutionary newcomer, as anatomically modern humans have existed for much less than a million years. Of course it is only in the last few centuries that technological advances have allowed numbers of this single species to escalate, approaching seven billion as of June 2011 (US Census Bureau, 2011). The impacts of this rise are such that we are now said to be living in the Anthropocene era (Crutzen, 2006), characterised by very high rates of species extinctions, pollution (including elevated atmospheric carbon dioxide levels) affecting every corner of the globe, destruction of natural ecosystems and high levels of land use for urban and agricultural purposes. Amongst the most important technological advances that have allowed this dramatic success ('success' at least in terms of the population size of *H. sapiens*) is agriculture.

The concept of 'pests' has arisen out of human agricultural practice and the desire to preserve food security by protecting crops from ubiquitous insects. Some, such as the locust (most likely desert locust, *Schistocerca gregaria* Forsk. (Orthoptera, Acridiidae)), are mentioned in the Bible and in other early written works (Nevo, 1996). For many centuries, farmers combated pests with cultural techniques ranging from hand removal of pests to the use of crop rotations. Saving the best seeds from each year's crop to sow in the following season led to the development of many landraces (locally adapted varieties) of major crop species, some of which persist to the present day (Thomson *et al.*, 2009). These landraces often had useful levels of broadly based resistance to various pests to which they were exposed for hundreds of generations. More recently, other technologies were brought to bear against pests including chemistry to produce ever more sophisticated insecticides (Casida and Quistad, 1998), radiation technology to allow the development of the sterile insect technique (Dyck *et al.*, 2005) and molecular biology to support plant breeding efforts (Sanchis and Bourguet, 2008). Many pest management technologies, however, are beset with problems of a technical nature (e.g. pollution, resistance breakdown, cost, etc. (van Emden and Peakall, 1996)) or a social nature (e.g. public acceptance of biotechnology in agriculture, deregistration of insecticides because of safety concerns (Cullen *et al.*, 2008; Lemaux, 2009)).

Although the term 'pest' is a human construct, and pest management involves humans modifying natural processes, there is much to be learned from nature. For more than 100 million years plants have been developing strategies to defend themselves from insect herbivores. In addition to familiar morphological adaptations such as hairs and thickened cuticles, plants have also evolved a powerful arsenal of chemical defences. Insecticide scientists and plant breeders are learning much from nature about new compounds that might be used in future insecticides (Isman, 2006) and about plant genetics that might be manipulated through molecular biology (Yencho *et al.*, 2000). The value of plant biodiversity as a resource from which botanical insecticides may be discovered is another important field, and is covered in chapter 6 of this book. Biological control

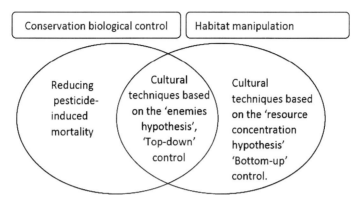

Figure 1.1 The relationship between conservation biological control and habitat manipulation approaches to arthropod pest management. Enemies hypothesis and resource concentration hypotheses are as described by Root (1973).

workers, too, have developed a very active interest in plant defences. Morphology such as glandular trichomes can directly impede natural enemies (Simmons and Gurr, 2005) and chemical defences can be exploited to make plants more attractive to predators and parasitoids (Kvedaras et al., 2010; Simpson et al., 2011). Chapter 11 of this volume explores the latter aspect, offering scope to manipulate or mimic the chemical ecology of plants to rapidly recruit natural enemies from nearby source habitats.

Aside from the various direct interactions that occur between plant and herbivore species, interactions involving other trophic levels are increasingly understood to be important in determining the magnitude of herbivore impact on plants and offer promise to pest management scientists. Indeed, for well over 100 years the action of predators, parasites and pathogens on pests has been exploited to provide biological control for pest management (Gurr et al., 2000). Pesticides, too, have a long history, although for much of this time they were broad-spectrum and used in a manner that was dangerous to non-target species including biological control agents. The human health and environmental negatives of widespread use of broad-spectrum pesticides are well documented. In the 1950s, Californian entomologists created the 'integrated control concept' (Stern et al., 1959), which included pesticides applied on the basis of crop scouting rather than prophylactic calendar spraying. This concept acknowledged that 'background' populations and communities of natural enemies had a key role in suppressing pests.

This landmark work paved the way for modern integrated pest management (IPM).

There are three common approaches to biological control – conservation, classical and inundative – all of which might be harnessed to improve natural pest control. In the early days of IPM, there was little explicit emphasis on conservation biological control, achieved by enhancing food, shelter and other resources needed by natural enemies (Figure 1.1) (Barbosa, 1998). IPM practices initially focused on enemies imported from overseas, ideally to target one particular pest species (classical biological control). These enemies were also more likely to be effective under a regime of insecticides that were target-specific for pests but again the ecological needs of these enemies were not researched. Inundative releases of natural enemies reared in very large numbers pre-dates conservation biological control, having taken place since the 1930s, especially in commercial glasshouse crops (van Lenteren and Woets, 1988; Albajes et al., 2000).

Classical biological control of arthropods by arthropods has been practised worldwide since the 1880s but this approach has had at least three problems associated with it. The first is that for a period of at least 100 years, successful suppression of the target species remained at around 10% (Gurr et al., 2000). Failure to establish biological control agents was the major cause of this low success rate. Secondly, some introduced classical biological control agents attacked arthropods other than the 'target' species (Howarth, 1991) and work on how to manage this risk has become an important strand in the biological control literature (Barratt *et al.*, 2010). Thirdly, the introduced agent may become a pest in its own right. For example, the cane toad (*Bufo marinus* L.) has devoured and poisoned non-target native species and caused other adverse ecological effects in Australia (Shine, 2010). Despite its problems, however, classical biological control is considered by most practitioners as 'risky but necessary' (Thomas and Willis, 1998).

We have briefly reviewed so far patterns of insect and plant biodiversity, the growing global impacts of agriculture, and the development of IPM and biological control; but how might biodiversity itself be exploited to improve pest management? Indeed, why might Schoonhoven *et al.* (2005), at the end of a detailed treatise on insect–plant interactions, conclude that 'diversification holds the clue to control of pestiferous insects'? In seeking to answer these questions it is necessary to explore the nature of biodiversity.

BIODIVERSITY

The much-used term 'biodiversity' is a contraction of 'biological diversity'. In popular usage is often taken to refer casually to the plants and animals that humans cannot directly eat or otherwise use and, for often poorly defined reasons, 'good stuff' that needs to be valued and protected. Various technical definitions have been proposed and a significant volume of literature exists on this subject (Gaston, 1996). Generally, definitions refer to biodiversity encompassing the variety of life on Earth at organisational scales ranging from genes, through species, to entire ecosystems. Genetic diversity covers the genes found within a given population of a single species, and the pattern of variation across different populations of that species. For example, genes in rice might provide useful traits such as insect resistance and salt tolerance. Species diversity is the more familiar level of biodiversity, referring to the assemblage of species in a given area. An example is the insects present in a cotton crop. At the higher organisational level, ecosystem diversity is the variety of habitats that occur within a region, or the mosaic of patches found within a landscape. This might include the crops, woodland, built environment, aquatic and wetland habitats found on a farm.

Species diversity and its relationship to ecological functions and their provision of ecosystem services has been the subject of intense study, reflecting its signifi-cance in ecology and the future of man's management of the environment. Of particular importance is the distinction between alpha, beta and gamma diversity (Whittaker, 1972). This is best explained by a hypothetical example addressing the issue of whether woodland vegetation is valuable in conserving spider species that might colonise wheat fields via a network of hedgerows (Table 1.1). Spiders can be captured by pitfall traps or vacuum sampling and identified to species. Alpha diversity is the resulting measure of species diversity for each of the above three habitats: 10 in the woodland, 7 in hedgerows and 3 in the wheat fields. Beta diversity, in contrast, is a comparison of habitats that provides an index of the number of species that are not common to both habitats. Thus, there is a different beta diversity statistic for each of the permutations of two-way habitat comparison. In this case, the beta diversity value for woodland versus hedgerows is 7, a relatively low value (given the alpha diversity in each) because many of the species are common to both habitats. In contrast, the woodland to field beta diversity value is 13 because none of the species is common to both habitats. Finally, the gamma diversity value of 14 is an index of spider species richness over all of the farm's three habitats. As is evident from this hypothetical example, species richness is amongst the aspects of biodiversity of direct relevance to pest management and the landscape-level effects are particularly important. Reflecting this, much of the remainder of this chapter discusses how natural enemy diversity can suppress pest populations, and how it might be managed to improve these benefits. For example, intensification of the landscape that comprises crop fields only (each with associated high levels of disturbance) (Figure 1.2, left) or agricultural areas in a naturally inhospitable matrix (Figure 1.2, right) can deprive natural enemies of refuges and important non-crop resources such as plant foods. Reflecting the importance of landscape-scale effects and associated gamma diversity, one chapter in this volume explores this area from a theoretical perspective (Gamez-Virues *et al.*, chapter 7) and one explores it from a methodological perspective (Scherber *et al.*, chapter 8).

Manipulating plant biodiversity to control pests

The notion that plant biodiversity could help suppress pests has origins dating back to the polycultures that

Table 1.1 Example of alpha, beta and gamma diversity of spider species in adjacent habitats of a farm landscape (based on the hypothetical example given by Meffe *et al.* (2002)).

Spider species	Woodland	Hedgerow	Wheat field
1	present		
2	present		
3	present		
4	present		
5	present		
6	present	present	
7	present	present	
8	present	present	
9	present	present	
10	present	present	
11		present	
12		present	present
13			present
14			present
Alpha diversity	10	7	3
Beta diversity	Woodland vs. hedgerow: 7	Hedgerow vs. field: 8	Woodland vs. field: 13
Gamma diversity		14	

Figure 1.2 Challenging habitats for natural enemies: landscape composed entirely of arable fields and towns in Western Europe (left) and isolated patches of irrigated agriculture in the arid landscape of the US Midwest (right) (photos by G.M. Gurr).

were the norm in pre-industrialised agriculture and persist in the concept termed 'companion planting'. This practice recommends, for example, that aromatic plants such as basil (*Ocimum basilicum* L.) or *Allium* spp. be inter-sown with pest-prone vegetables. These aro-matic plants supposedly repel pests or interfere with their location of a suitable host plant (Cunningham, 1998), but rigorous testing of the approach does not always yield encouraging findings (Held *et al.*, 2003). More sound ecological support for the significance of non-crop vegetation came from early work suggesting the importance of nectar availability to predatory insects such as parasitoid adults (Thorpe and Caudle, 1938). In that study, newly emerged *Pimpla ruficollis* Gravenhorst, an ichneumonid parasitoid of the pine shoot moth (*Rhyacionia (Evetria) buoliana* Schiff. (Eucosmidae)) demonstrated repellency to the pine oil volatiles from *Pinus sylvestris* L. trees. This led young adult parasitoids to leave areas with trees where the dense shade was likely to mean an absence of an under-storey. Outside the forest they were presumed to feed on nectar, including that of plants in the family Apiaceae, returning to the trees 3–4 weeks later when suitable larval hosts were available.

Another important early example, and one that demonstrates a separate ecological mechanism by which plant biodiversity may benefit natural enemies, is the study by Cate (1975) on the ecology of the western grape leafhopper *Erythroneura elegantula* Osborn (Homoptera: Cicadellidae). That, and subsequent studies, showed that the presence of blackberry bushes (*Rubus* spp.) in riparian habitats close to vineyards could improve biological control of this pest by the parasitoid *Anagrus epos* Girault. The mechanism for this is that the blackberry bushes fill the temporal absence of *E. elegantula* eggs which are the host of the parasitoids. The leafhopper overwinters as adults but these are unsuitable as hosts because the parasitoids can overwinter only as eggs inside host eggs. Clearly, parasitoids are unable to overwinter within the vineyard itself. The presence of a suitable overwintering host, the blackberry leafhopper *Dikrella californica* Osborn, on non-crop vegetation throughout the year allows populations of *A. epos* to persist in the region. If these overwintering sites are close to vineyards the parasitoid is better able to colonise those vineyards and help check development of pest leafhoppers (Murphy *et al.*, 1998).

Pivotal work by Root (1973) suggested two ways that greater plant diversity within crops might improve pest suppression. The first was the 'enemies hypothesis', which postulated that diverse plantings would encourage greater prey, nectar and pollen resources for natural enemies, building their densities and encouraging stronger impacts on pests. The second was the 'resource concentration hypothesis', which holds that herbivorous insects (at least specialists) should more easily find, and choose to remain within, large monoculture plantings of suitable host plants. Testing the relative importance of these two hypotheses has remained a research-rich challenge amongst insect ecologists (e.g. Grez and González, 1995) and has led to the concept of 'top-down, bottom-up' trophic effects. The former refers to the action of predators and other natural-enemy species in the third trophic level, while the latter emphasises plant defences and benefits of plant biodiversity such as disruption of herbivore visual and olfactory cues (as well as other mechanisms reviewed below). The complexity of analysing and separating these effects was discussed by Lawton and McNeill (1979) under the compelling title 'Between the devil and the deep blue sea: on the problems of being a herbivore'.

BIOTIC FORCES SHAPING PESTS: BETWEEN THE DEVIL AND THE DEEP BLUE SEA REPRISE

Just as the design of a coin is derived from pressure to each face, so too may a pest population be viewed as taking shape by pressure from opposing forces (Figure 1.3). First, plants are far from passive players in the game of herbivory. Millions of years of evolution have given plants a formidable arsenal of defences to which the animals seeking to feed upon them have had to adapt. Plant defences include conspicuous morphological features such as spines, hairs (including trichomes that poison and entrap pests (Figure 1.3, bottom insert), thickened cuticles and protected growing points as well as sophisticated metabolic defences that give constitutive and induced defences designed to poison or otherwise impede herbivores (Wu and Baldwin, 2010). In addition to this 'bottom-up' pressure from the first trophic level, herbivorous arthropods also have to contend with the action of 'top-down' forces from the third trophic level. Predators and parasitoids have forced insect herbivores to evolve adaptations ranging from morphological (e.g. hairs (Figure 1.3, centre)), physiological (e.g. encapsulation of parasitoid eggs (Namba *et al.*, 2008)) to behavioural (e.g caterpillars dropping from plants when sensing a predator (Steffan and Snyder, 2010)).

Of course, factors other than top-down and bottom-up forces will also shape pest adaptations. Competition and the abiotic environment are two of the most important. In agriculture, however, the importance of competition is reduced by the usual super-abundance of food resources for pests of the relevant crop. Aspects of the physical environment that are of particular

Figure 1.3 The pest as a coin: shaped by pressures from top-down trophic force of natural enemies (e.g. tiger beetle) and the bottom-up force of plant defences (e.g. glandular trichomes) (centre photo by J. Liu, other photos by G.M. Gurr).

importance in agricultural systems include the weather (frosts, flood, etc.) but often these effects are ameliorated for the sake of efficient crop production, by protected cropping (greenhouses, cloches, etc.), site selection or by a carefully selected sowing date. Then, human imposed disturbance becomes the most important form of abiotic mortality factor for pests (e.g. irrigation, harvest, tillage).

Bottom-up trophic effects of biodiversity on pests

Host accessibility for herbivores is unrestricted in large monocultures (assuming the host is suitable for the herbivore in question) and two ecological mechanisms can be at play. First, insect herbivores tend to locate suitable hosts and remain upon them more readily

in monocultures (Root, 1973). Baliddawa (1985) reviewed 36 papers and found that 24 provided evidence that suitable hosts were less apparent in polycultures. Cases where this applied included herbivores that located hosts by random landings and which were not directed by host cues. These insects – such as wind-dispersed aphids – may have limited opportunity to leave unsuitable host plants or patches of plants so are unable to make repeated attempts to land on a host. Second, herbivores can be expected to be more numerous in large patches of suitable habitat (Kareiva, 1983). An example of how such an effect may operate is provided by bark beetles (*Ips* spp.). Generally, these herbivores are repelled by plant defences so they are usually unable to overcome the defences of a healthy tree. Therefore their fitness is greater on stressed hosts with weakened defences. Normally these poorly defended trees are scarce and widely dispersed so the herbivore population increase is prevented. Only after a storm event that weakens sufficient host trees will the pest population build up to high enough numbers to successfully attack and overwhelm the defences of healthy trees (Speight and Wainhouse, 1989).

For both of the above cases, any spatial or temporal break in availability of susceptible hosts can reduce pest build-up (Jactel *et al.*, 2005). Temporal barriers may result in cases where the herbivore is able to feed only on a certain phenological stage of the host plant; on young leaves, for example (Wratten, 1974). In such cases egg hatch must coincide with bud burst. In a polyculture forest system, bud burst is staggered across tree species so many trees, although potentially suitable hosts, will not be available to neonate larvae, thus restricting food resources available to the pest population. Plant diversity might also lead to physical barriers that protect crop plants from herbivores. For example, understorey plants may be protected by the presence of an overstorey that impedes host plant detection. Chemical barriers, too, are important because many insect species use the volatiles produced by plants as host location cues. Mixed species vegetation will provide a more complex chemical environment in which it is more difficult for a specialist herbivore to locate and settle on suitable plants. Chapter 19 of this volume, on cover crops, provides examples where the close proximity of the primary crop to the secondary (cover) crop can evoke bottom-up effects.

The other way in which insect pests may be suppressed in a plant stand with more than two species is through 'trap cropping' (Rea *et al.*, 2002). At its simplest, one plant might be a preferred site for egg laying so might be sown alongside the main crop to divert pests. There are cases of such 'trap crops' being attractive to egg-laying pest females but providing poor support for the development of their larvae (Khan *et al.*, 2006). In a further example, females of the cerambycid stalk boring beetle *Dectes texanus* LeConte prefer to oviposit on sunflower (*Helianthus annuus* L.) over soybean (*Glycine max* L. Merr.), to the extent that an individual host plant may accumulate multiple eggs. Larvae subsequently fight, typically leading to the death of all but one individual per plant. Even where such biological mechanisms do not operate to kill pests, 'trap crops' may be established and methods such as targeted insecticide application or mechanical destruction used to prevent pest development. A particularly elegant form of trap cropping is the 'push–pull' strategy (Cook *et al.*, 2007) whereby a synergistic behavioural manipulation of pests is brought about. This usually uses non-host volatiles, anti-aggregation or alarm pheromones, oviposition deterrents or antifeedants on the focal crop to 'push' the pest away from it. Visual distractions might also be involved. Simultaneously, pests are 'pulled' to a trap crop using visual oviposition or gustatory stimulants, pheromones or host volatiles. The most successful example of the push–pull strategy is for control of stem borers in African maize and sorghum using the trap plants Napier grass, *Pennisetum purpureum* Schumach and Sudan grass, *Sorghum sudanensis* Stapf (Khan *et al.*, 2000). Chapter 16 of this volume provides an analysis of reasons for the outstanding success of this push–pull approach.

Meta-analyses of plant-biodiversity benefits for pest control

As well as the mechanisms and associated pest management approaches summarised above, several meta-analyses have been conducted over many experimental studies on the effects of plant diversity on herbivores. For example, such an analysis of 21 studies of the effects of diversified crops on insects pests found a 60% reduction in mean insect density in diverse compared with simple crop situations (Tonhasca and Byrne, 1994). In a non-agricultural context, Hillebrand and Cardinale (2004) examined effects of grazers on the biomass of periphytic algae and found

a robust trend whereby the impact of grazing tended to decrease as the diversity of algae increased. Similarly, in a particularly comprehensive meta-analysis of the biodiversity effects on ecosystem functioning and services, Balvanera *et al.* (2006) found evidence for positive effects of biodiversity on pest control whereby higher plant diversity was associated with reduced plant damage. Allied to this there was also evidence of benefit against an important category of pests: invasive species. Under conditions of higher plant diversity, invader abundance, survival, fertility and diversity were all reduced. Most recently, Jactel and Brockerhoff (2007) also found that diverse plant communities were less affected by pests. Their meta-analysis covered 119 forest-related studies of 47 different tree:pest systems and found overall a significant reduction of herbivory in more diverse forests. Importantly, however, in terms of formulating any specific recommendations, the response varied with host specificity of the pest species. In diverse forests, herbivory by oligophagous species was generally reduced but the response of polyphagous pests varied. An important effect that explains instances of oligophagous species sometimes being favoured by tree diversity is 'associational susceptibility'. This operates when a herbivore develops high population densities on a palatable host and then spills over to the other, less preferred plant species. An example is the gypsy moth (*Lymantria dispar* L.) which feeds on conifers once it has defoliated its preferred broadleaved hosts, such that white pine (*Pinus strobes* (L.) growing in mixed stands with oaks (*Quercus* spp.) is more likely to be attacked than when in monoculture (Brown *et al.*, 1988). Another finding from the meta-study by Jactel and Brockerhoff (2007) was that, as might be expected, the effects on herbivory were greater when the diverse tree species were more distantly related. The authors claimed that this finding lends support for the action of bottom-up effects based on the notion that trees from taxonomically distant groups would be more likely to have dissimilar volatiles, so impeding host location by pests. The same trend could, however, result from natural enemy activity if the diversity of trees (e.g. nectar-producing angiosperms with conifers) enhanced the top-down effects. Indeed, it is very likely that the effects in many of the publications covered by the foregoing meta-studies include a mixture of bottom-up and top-down effects, even where the original authors did not specifically seek evidence of enhanced natural enemy activity.

Top-down trophic effects

The second suite of hypotheses that may account for the suppressive effects of biodiversity on pests involves the third trophic level: natural enemies attacking herbivores. Pest control by natural enemies is now widely acknowledged as an important ecosystem service with annual values estimated at US$2, $23 and $24 per hectare in forests, grassland and cropland, respectively (Costanza *et al.*, 1997). More recent work using in-field experimental approaches, has put the value of 'background' biological control of pests at over US$100/ha/year even though the effects of only one pest were explored (Sandhu *et al.*, 2008). Partly because of these recent results, the influence of natural enemies on pests has emerged as an important aspect of the wider field of biodiversity and ecosystem function (Wilby and Thomas, 2002). Cardinale *et al.* (2006) performed a meta-analysis of 111 field, greenhouse and laboratory studies that manipulated species diversity to examine its effect on ecosystem function in a range of trophic groups and ecosystems. On average, decreasing species richness led to a decrease in the abundance or biomass of the relevant trophic group and reduced ecological process rate (e.g. predation).

BIODIVERSITY AND ECOSYSTEM FUNCTION

An intuitive view is that a more diverse community of natural enemies should yield higher consumption rates across the entire community of natural enemies (Wilby and Thomas, 2002). This would be expected when different species occupy different feeding niches, so that more unique niches are filled when more species are present (e.g. Finke and Snyder, 2008). Yet increasing amounts of empirical research and modelling (Casula *et al.*, 2006) indicate that this relationship is more complex than a simple additive one where each new enemy species provides incrementally more ecosystem function. Indeed, the addition of more enemy species can lead to an overall reduced consumption of pests when predator species interfere strongly with one another (e.g. Finke and Denno, 2004). On the other hand, the addition of enemy species may lead to the opposite effect: synergy (e.g. Cardinale *et al.*, 2006). Synergy among natural enemies occurs when one predator species enhances prey capture by another (e.g. Losey and Denno, 1998). In still other cases, pred-

ator species fill similar niches (that is, are functionally redundant) such that adding new species to a community is neither beneficial nor harmful to pest control (e.g. Straub and Snyder, 2006). Thus, positive, negative, and neutral enemy-diversity effects can result from niche or functional complementarity, predator interference, and functional redundancy, respectively (Straub *et al.*, 2008).

A recent review of the effects of natural enemy biodiversity on suppression of arthropod herbivores in terrestrial systems (Letourneau *et al.*, 2009) is important in distilling the now considerable volume of experimental work in this field. The meta-analysis of 62 published studies covering 266 comparisons of herbivore and natural enemy communities revealed a significant overall strengthening of herbivore suppression with greater natural enemy species richness (Plate 1.1). The analysis of these comparisons revealed herbivore suppression from increased enemy richness in 185 cases, one instance of no effect and 80 where herbivores were favoured by enemy richness. The overall significant effect of natural enemy richness on herbivores was consistent for studies conducted in tropical and temperate agriculture. Indicative of the robustness of predator biodiversity's benefit, this effect was also significant across both of the common approaches for conducting such work: cages with artificially manipulated arthropod community structure and insect numbers and open-field investigations of systems where natural enemy communities differed in response to an aspect of the local environment. The overall finding of Letourneau *et al.* (2009) is consistent with an earlier meta-analysis of predator removal studies (Halaj and Wise, 2001) which concluded that herbivore abundance increased as predation pressure decreased in 77% of cases, with the opposite occurring in only 20% of studies. These meta-analyses are powerful evidence for the influence of natural enemy diversity on pests but it is clear that the outcome of enhancing the enemy community in any particular system is still unpredictable.

An important complement to biodiversity:ecosystem function (BEF) studies that has undergone rapid advances in the past decade is the use of molecular techniques to analyse the diet of predators. These approaches have the potential to firmly define feeding-niche overlap among predator species, and the frequency with which predators feed upon one another (intraguild predation). Chapter 10 of this volume provides a state-of-the-art view of how newly available methods can be used to move from a general understanding of the effect of predators on pests to a quantified understanding of 'who eats whom'. At this level, however, assessing the effects of predators on prey populations, using other methods, may still be needed. The relatively new technique of pyrosequencing can help in this regard, as shown by work in New Zealand by Boyer and Wratten (2004).

Specialist and generalist natural enemies: the importance of partitioning

A good generalisation that helps understand the results of the meta-analyses by Letourneau *et al.* (2009) and Halaj and Wise (2001), is that suppression of pests by enemies is reduced when intraguild predation takes place (Finke and Denno 2003). Conversely, pest suppression is enhanced when enemy species are able to partition prey by life stage, size or microhabitat use effects (Wilby *et al.*, 2005). This partitioning might also result from enemy species having some kind of synergy such as 'predator facilitation' (Charnov *et al.*, 1976) whereby prey is more readily captured by one predator after being disturbed by another. Although the classical example of this phenomenon (Soluk and Collins, 1988) concerns trout and stoneflies there is also evidence for predator facilitation effects amongst arthropod natural enemies of pests (Losey and Denno, 1998). Certainly there is an important difference between the way that generalist and specialist enemies interact to drive effects on pest populations. In the study by Finke and Snyder (2008), a model system with radish (*Raphanus sativus* L.), aphids (green peach aphid (*Myzus persicae* Sulzer)), cabbage aphid (*Brevicoryne brassicae* L.) and turnip aphid (*Lipaphis erysimi* Kaltenbach) and parasitoids (*Diaeretiella rapae* McIntosh, *Aphidius colemani* Viereck, *and Aphidius matricariae* Haliday (Braconidae)) was used to tease apart the relative effects of resource partitioning and diversity *per se*. That study exploited the phenonenom of natal fidelity whereby a given wasp individual will prefer to attack a host of the same species from which it emerged. This is despite the fact that each of the three wasp species is potentially able to parasitise all three of the aphid species. The experimentation involved rearing batches of each wasp species on each aphid species (nine permutations). This then allowed arenas to be set up in which parasitoids were confined with aphids such that the wasps either fully partitioned the available hosts or

were generalists that overlapped in terms of resource use. At the same time, parasitoid species richness was varied from one to three species. Increasing the number of resource partitioning parasitoids from one to three species markedly increased the parasitism rate and reduced aphid abundance. In contrast, when the parasitoids were effectively generalists (i.e. they were competing for hosts rather than each searching for one species in a specialist manner) there was no effect of increasing species diversity. The increase in aphid use by specialist parasitoids but not generalists demonstrated that the extent to which enemies partitioned the resource was the dominant factor.

Remarkably, such positive influences of enemy diversity can be mediated even independently of actual predation events. This is because, rather than simply staying put and waiting to be killed, herbivores often deploy a wide range of chemical, physical and behavioural defences. These defences often are energetically costly, however, such that herbivores bear a cost in their deployment. For example, in work with lepidopteran pests of *Brassica oleracea* L., Steffan and Snyder (2010) examined the effects of predator diversity. The pest in that system, *Plutella xylostella* L., drops from the host plant when disturbed by a predator but remains suspended by a silken thread, presumably to avoid falling to the ground where it is likely to be vulnerable to soil-associated natural enemies (such as those covered by Altieri *et al.*; see chapter 5 of this volume). Only after some minutes does the larva return to the leaf and resume feeding, so the defence strategy carries an opportunity (i.e. feeding) cost. The manipulative experiment replaced caterpillars predated by *Diadegma* and *Hippodamia* enemies to ascertain the effects of the different enemy communities on pests via behavioural mechanisms independent of the actual predation. Another treatment was predator-free but caterpillars were carefully removed to simulate predation free of the induction of larval defence reactions. This study demonstrated that plant production was increased by enemy diversity-induced anti-predation behaviour by the caterpillars in the absence of any actual predation. These 'predation-free' effects are considered in more detail in chapter 2 of this volume.

Functional redundancy and complementarity

Several factors influence the relationship between the number of natural enemy species in a system and the

Box 1.1 Examples of functional redundancy and functional complementarity of natural enemies

Several egg parasitoid species of a pest species that forage in the same microhabitat and season exhibit *functional redundancy*. This means that the loss of one species is unlikely to result in a pest population growth.

Several spider species that attack the eggs, small nymphs and adults of a pest species with different hunting strategies in different microhabitats exhibit *functional complementarity*. This means that the loss of a single species is more likely to result in pest population growth.

resulting rate of prey consumption. An important aspect is the distinction between functional redundancy and functional complementarity (Box 1.1; Rosenfeld, 2002). Species of natural enemies that exhibit functional redundancy are similar to one another in terms of the life state of the pest attacked, the microhabitat used, the season of the year in which they are active and so on. In contrast, enemies with functional complementarity differ markedly in terms of their niche characteristics (Bográn *et al.*, 2002).

Although the characteristics of a given species will be profoundly influenced by its genotype – a spider is unable to parasitise a pest egg, for example – phenotypic plasticity can also allow the members of an enemy species to respond to the availability of prey. For example, Tahir and Butt (2009) showed in a study of spiders of Pakistani rice systems that Diptera were the dominant prey early in the season. Only later, when planthopper numbers in the crop began to increase, did this prey become dominant in the diet. Such plasticity – in this case responding to temporal shifts in prey availability – has clear importance in pest suppression. The availability of dipteran prey early in the season allowed spider numbers to increase to high levels and thereby provided effective control of pest planthoppers. In a situation where diet plasticity was not exhibited by the predator this would not be possible. Clearly the early season build-up of the spider community also depends on the availability of prey species and this in turn is dependent largely on the use of inputs of

organic matter such as animal manure (Settle *et al.*, 1996). A similar idea comes from the use of nectar by natural enemies. Lacewings (Hemerobiidae in this case) use and benefit from nectar when aphid prey numbers are low, but nectar does not contribute significantly to their fitness at high prey densities (Robinson *et al.*, 2008).

NATURAL ENEMY EVENNESS

Considerable research attention has been given to increasing overall numbers of natural enemies or the numbers of species (species richness). However, recent work has shown that relative evenness of the numbers of individuals across the species in an enemy community is also important (Crowder *et al.*, 2010). In that study, field enclosures were used to test the effect of relatively even versus less even communities of enemies. Pest population reduction and plant productivity were higher when enemy evenness was high; an effect that was independent of which enemy species was numerically dominant.

Unevenness can leave niches under-exploited and the common enemies are likely to be competing for prey as a result of low levels of resource partitioning. This difference between species diversity and species evenness can be important but, perhaps surprisingly, many studies of biodiversity in agricultural systems measure diversity alone and ignore evenness (Bengtsson *et al.*, 2005).

From theory to practice: exploiting top-down effects with agri-environmental schemes, 'SNAP' and ecological engineering

As awareness of the potential of natural enemies as biological control agents increased, a great deal of modelling work was undertaken to understand the ecological mechanisms that would lead to density-dependent population regulation and therefore persistence of the parasitoid–host relationship (Nicholson and Bailey, 1935). This work brought about a realisation that density-dependent regulation was not required for population reduction to take place, and that parasitoid– and predator–host communities do not exist as single, homogeneous units. Rather, such communities exist in patchy environments (Hassell *et al.*, 1991) that require models to take into account meta-population effects (Hanski and Simberloff, 1997). Further, although concepts such as the area of discovery (a) and instantaneous attack rate (a') of the natural enemy were crucial components of these models, there was no recognition that these two key parameters could change substantially if non-host/prey resources such as nectar were part of the system. Kean et al. (2003) showed the profound effect on these parameters of nectar provision by using a development of the models produced by Hassell et al. (1991). The awareness that biological control effectiveness in monocultures is almost always operating at a sub-optimal level grew along with knowledge of the value of uncultivated land as a refuge and as a source of non-prey food (van Emden, 1965) and the emergence of conservation biological control began to develop as a science in its own right (Barbosa, 1998). This awareness began to be reflected in farm environmental policies within the European Union and elsewhere by the use of approaches such as unsprayed crop strips ('conservation headlands'), areas taken out of crop production ('set aside'), and the broader 'Countryside Stewardship Scheme' set up in 1991 and now replaced by the 'Environmental Stewardship Scheme' (Natural England, undated). These agri-environmental schemes have a broad public good and environmental protection and remediation mission rather than being focused on the management of farm biodiversity for any specific ecosystem service such as pest control (Wade et al., 2008a). Accordingly, they are not informed by appropriate ecological research addressing aspects such as which plant species are best to sow or conserve, the optimal layout of non-crop features such as 'weed strips', the nature of effects on pest and natural enemy species and the ecological mechanisms at play. Notwithstanding these potential problems, some land use practices such as cover crops to enhance breakdown of prunings or tree strips to shelter crops and livestock offer scope to promote natural enemy biodiversity (Plate 1.2). A caution was provided, however, by an assessment of agri-environment schemes in the Netherlands that showed no positive effects on plant and bird species diversity (Kleijn et al., 2001). Nevertheless, of some relevance to pest management, the hoverfly (Syrphidae) fauna was slightly more diverse (Kleijn et al., 2001). This is, however, a minor gain from the major funding allocated to these schemes.

As the science of conservation biological control grew, the freshwater ecology concept of 'resource subsidies' (inputs from external habitats to support the food web in a focal habitat) (Takimoto et al., 2002) became increasingly used to stress the significance to natural enemies of external habitats and the resources available therein (Tylianakis et al., 2004). The role of plant-provided foods for predators and parasitoids is now well understood and actively exploited for pest management (Wäckers et al., 2007). A simple acronym, 'SNAP', is used to summarise the ways in which non-crop resources can help natural enemies. The letters stand for shelter, nectar, alternative prey and pollen. The value of shelter is apparent in a British research programme that led to the development of grassy, overwintering strips ('beetle banks', Plate 1.2) in arable farmland (Thomas et al., 1991; 1992; 2001). These raised earth banks, sown with cocksfoot grass (orchard grass), Dactylis glomerata L., are established across fields. Large numbers of predatory carabid and staphylinid beetles as well as spiders overwinter in the shelter provided. Many of these emigrate into the crop in spring, leading to reductions in aphid pest numbers (Collins et al., 2002). Subsequently, other ecosystem services have been demonstrated for these refuges, including breeding populations of the harvest mouse (Micromys minutus Pallas) (Bence et al., 2003), a species of conservation relevance, and gamebirds such as the grey partridge (Perdix perdix L.), a species of cultural and economic significance for recreational shooters. The extent to which features such as beetle banks can support wildlife is greatly influenced by the level of use of native as opposed to exotic plant species; a subject explored in chapter 17 of this volume. Beetle banks are effectively a 'service providing unit' (Kontogianni et al., 2010) in that the protocol for improved ecosystem services is clear and emphasises to farmers how, where and why these enhancements should be made. The use of biodiversity as a pest management tool is explored from the perspective of ecological economics in chapter 4 of this volume.

Other examples of research and uptake of conservation biological control have concerned the other three components of SNAP, especially the provision of nectar (N) and pollen (P) for natural enemies such as parasitoid wasps, hoverflies, lacewings, ladybirds (Wäckers et al., 2007). Nectar provides the carbohydrates for energy, as well as amino acids and minerals, while pollen provides much of the protein required by these insects for egg maturation. Plant species commonly used include buckwheat (Fagopyron esculentum

Moench) (e.g. Berndt *et al.*, 2002), phacelia (*Phacelia tanacetifolia* Benth.) (e.g. Hickman and Wratten, 1996), alyssum (*Lobularia maritima* L.) (Begum *et al.*, 2004) and, sometimes plants in the Apiaceae (e.g. Idris and Grafius, 1995). The majority of studies tend to focus on nectar use by hymenopteran parasitoids but the Hickman and Wratten (1996) study is an example of one on pollen use by adult hoverflies.

As pointed out by Wade *et al.* (2008b), a hierarchy of effects of floral provision is usually expected. This hierarchy is:

1. Natural enemies aggregate on the flowers
2. The ecological fitness of natural enemies increases
3. Searching behaviour of the insects changes
4. The proportion of pests killed increases
5. Pest populations are reduced
6. Pest populations are brought below the economic threshold.

For conservation biological control practitioners, achieving the effects in the above hierarchy becomes more difficult as the steps in the hierarchy are progressed. One way in which the science has risen to this challenge of increasing efficacy whilst avoiding possible negative effects has become known as 'ecological engineering' for pest management (Gurr *et al.*, 2004). Essentially, ecological engineering aimed to place conservation biological control on a more rigorous theoretical foundation with an experimental framework informing decisions such as the choice of nectar plant species. This was an advance because many previous attempts at CBC were not well targeted, consisting of seed mixes for example. The species in these were not generally tested for efficacy for any particular natural enemies, or to deny benefit to pests (e.g. moths taking nectar (Lavandero *et al.*, 2006)). In tropical rice, in particular, the need for ecological engineering has been stressed (Settele *et al.* 2008) and is now the focus of significant research in Asia (Gurr *et al.*, 2011). The broader state of research on the use of biodiversity to increase availability to natural enemies of important plant foods is explored in chapter 9 of this volume.

The least actively researched aspect of the SNAP acronym is provision of alternative hosts and prey. The pioneering work by Cate (1975) which identified non-crop plants that support alternative hosts of *E. elegantula* parasitoids has led to only sporadic work to look at similar relationships in other systems (e.g. in pome (pip) fruit (Pfannenstiel *et al.*, 2010)). The availability

of alternative prey has received more research attention by virtue of a growing interest in the importance of generalist natural enemies. Chapter 3 of this volume explores this phenomenon whilst chapter 13 provides a detailed example of how detritivores can be enhanced in rice systems as alternative prey to support early season build-up of generalist predators.

CONCLUSION: BIODIVERSITY FOR PEST MANAGEMENT

Irrespective of whether bottom-up or top-down ecological effects are being exploited, and whichever aspect of the SNAP acronym is targeted, an advantage of manipulating biodiversity for pest management is that it can be initiated and carried out by individual landowners. This is in direct contrast to classical biological control where phytosanitary-related quarantine regulations restrict the introduction of exotic agents, making it the realm of government agencies. Only these and large research providers are able to conduct the necessary host specificity testing to clear regulatory hurdles. In contrast conservation biological control is much more in the hands of the individual farmer. Further, the implementation tends to lead to intensely visual improvements in landscape features, making it easier for the growers to demonstrate that they are making a tangible, biodiversity-based attempt to improve pest suppression on their land. Demonstrating this can have benefits in terms of farm tourism and sales of 'branded' products as well as qualifying for government payments under agri-environmental schemes. Several reviews of this biological control approach analyse the ecology and utility of this method (Landis *et al.*, 2000; Zehnder *et al.*, 2007; Jonsson *et al.*, 2008 and other papers in that special issue of *Biological Control*). Uptake of biodiversity-based strategies by farmers and other land managers is, however, contingent on effectively communicating to them the need to implement new approaches, and the means by which such approaches can be implemented. This is an area often overlooked in research on pest, disease and weed management and several chapters in this volume seek to redress this. Chapter 12 examines the sociological dimension of effective communication with farmers, drawing on a successful project that is persuading Asian rice farmers to reduce dependence on insecticides. Also, chapter 14 examines the importance of policy in driving change in pest management, reporting on a national 'Green Plant Protection' initiative being implemented in China. Finally, chapter 18 considers the use of biodiversity-based strategies in the urban environment; an important arena given that an increasingly large proportion of the world's population live in cities and are potentially exposed to the detrimental effects of insecticide use in ornamental and amenity areas.

It is critical for the successful use of biodiversity in pest management that farmers and other practitioners, as well as policy-makers responsible for incorporating this ecosystem service into agri-environmental schemes, are well served by the research community. Readers of this book will need to rise to this challenge and ensure that research and conventional 'outputs' (e.g. scientific publications) are converted into 'outcomes' (i.e. changed practices) that enhance the three pillars of sustainability: economy, society and the environment. However, too great a role for Gross Domestic Product as an indicator of improvement is unwise, as GDP is increasingly recognised as a poor measure of human wellbeing (Costanza et al., 2009). This will demand ongoing work to more completely understand the ecology of the mechanisms that drive the effects of biodiversity on pests and their natural enemies. Important also is the need to convert such knowledge into practicable technologies that are compatible with modern and future farming systems. Fortunately for this mission, it seems likely that future farmers will significantly broaden their enterprises beyond food, fibre and fuel production. The expansion of agri-environmental schemes will increasingly provide revenue streams to farmers for providing 'public good' services such as conserving biodiversity, and for practices that help capture atmospheric carbon dioxide. These practices will include planting farm trees, perennial forages and green manure crops to increase soil carbon. With careful planning these practices might simultaneously harness the power of biodiversity to reduce the impact of pests (i.e. multiple ecosystem services on farmland), and developing a structure to pay for them must be tackled in future agricultural policies.

ACKNOWLEDGEMENTS

Donna Read provided invaluable help in the production of this chapter.

REFERENCES

Albajes, R., Gullino, M.L., Lenteren, J.C. and van Elad, Y. (eds) (2000) *Integrated pest and disease management in greenhouse crops*, Kluwer Academic Publishers, Dordrecht.

Baliddawa, C.W. (1985) Plant species diversity and crop pest control: an analytical review. *Insect Science and its Application*, 6, 479–487.

Balvanera, P., Pfisterer, A.B., Buchmann, N., He, J-S., Nakashizuka, T., Raffaelli, D. and Schmid, B. (2006) Quantifying the evidence for biodiversity effects on ecosystem functioning and services. *Ecology Letters*, 9, 1146–1156.

Barbosa, P. (ed.) (1998) *Conservation biological control*, Academic Press, San Diego.

Barratt, B.I.P., Howarth, F.G., Withers, T.M., Kean, J.M. and Ridley, G.S. (2010) Progress in risk assessment for classical biological control. *Biological Control*, 52, 245–254.

Begum, M., Gurr, G.M. and Wratten, S.D. (2004) Flower colour affects tri-trophic biocontrol interactions. *Biological Control*, 30, 584–590.

Bence, S.L., Stander, K., and Griffiths, M. (2003) Habitat characteristics of harvest mouse nests on arable farmland. *Agriculture, Ecosystems and Environment*, 99, 179–186.

Bengtsson, J., Ahnström, J. and Weibull, A-C. (2005) The effects of organic agriculture on biodiversity and abundance: a meta-analysis. *Journal of Applied Ecology*, 42, 261–269.

Berndt, L.A., Wratten, S.D. and Hassan, P.G. (2002) Effects of buckwheat flowers on leafroller (Lepidoptera: Tortricidae) parasitoids in a New Zealand vineyard. *Agricultural and Forest Entomology*, 4, 30–45.

Bográn, C.E., Heinz, K.M. and Ciomperlik, M.A. (2002) Interspecific competition among insect parasitoids: field experiments with whiteflies as hosts in cotton. *Ecology*, 83, 653–668.

Boyer, S. and Wratten, S.D. (2004) *Using molecular tools to identify New Zealand endemic earthworms in a mine restoration project (Oligochaeta: Acanthodrilidae, Lumbricidae, Megascolecidae)*. Proceedings of the 4th International Oligochaeta Taxonomy Meeting, Diyarbakir, Turkey.

Brown, J.H., Cruickshank, V.B., Gould, W.P. and Husband, T.P. (1988) Impact of gypsy moth defoliation in stands containing white pine. *Northern Journal of Applied Forestry*, 5, 108–111.

Cardinale, B.J., Srivastava, D., Duffy, J.E. et al. (2006) Effects of biodiversity on the functioning of trophic groups and ecosystems. *Nature*, 443, 989–992.

Casida, J.E. and Quistad, G.B. (1998) Golden age of insecticide research: past, present, or future? *Annual Review of Entomology*, 43, 1–16.

Casula, P., Wilby, A. and Thomas, M.B. (2006) Understanding biodiversity effects on prey in multi-enemy systems. *Ecology Letters*, 9, 995–1004.

Cate, J.R. (1975) Ecology of Erythroneura elegantula Osborn (Homoptera: Cicadellidae) in grape agroecosystems in California. Doctor of Philosophy thesis, University of California, Berkeley.

Charnov, E.L., Orians, G.H. and Hyatt, K. (1976) Ecological implications of resource depression. *American Naturalist*, 110, 247–259.

Collins, K.L., Boatman, N.D., Wilcox, A., Holland, J.M. and Chaney, K. (2002) Influence of beetle banks on cereal, aphid predation in winter wheat. *Agriculture, Ecosystems and Environment*, 93, 337–350.

Cook, S.M., Khan, Z.R. and Pickett, J.A. (2007) The use of push–pull strategies in integrated pest management. *Annual Review of Entomology*, 52, 375–400.

Costanza, R., d'Arge, R., de Groot, R. *et al.* (1997) The value of the world's ecosystem services and natural capital. *Nature*, 387, 253–260.

Costanza R., Hart M., Posner, S. and Talberth, J. (2009) *Beyond GDP: the need for new measures of progress*, The Pardee Papers No. 4, January, Boston University Press.

Crowder, D.W., Northfield, T.D., Strand, M.R. and Snyder, W.E. (2010) Organic agriculture promotes evenness and natural pest control. *Nature*, 466, 109–112.

Crutzen, P.J. (2006) The Anthropocene, in *Earth system science in the Anthropocene: emerging issues and problems* (eds E. Ehlers and K. Krafft), Springer, Berlin, pp. 13–18.

Cullen, R., Warner, K.D., Jonsson, M. and Wratten, S.D. (2008) Economics and adoption of conservation biological control. *Biological Control*, 45, 272–280.

Cunningham, S.J. (1998) *Great garden companions: a companion planting system for a beautiful, chemical-free vegetable garden*, Rodale Press, Emmaus, PA.

Dyck, V.A., Hendrichs, J. and Robinson, A.S. (eds) (2005) Sterile insect technique: principles and practice in area-wide integrated pest management, Springer, Dordrecht.

Ehrlich, P.R. and Raven, P.H. (1964) Butterflies and plants: a study on coevolution. *Evolution*, 18, 586–608.

Finke, D.L. and Denno, R.F. (2003) Intra-guild predation relaxes natural enemy impacts on herbivore populations. *Ecological Entomology*, 28, 67–73.

Finke, D.L. and Denno, R.F. (2004) Predator diversity dampens trophic cascades. *Nature*, 429, 407–410.

Finke, D.L. and Snyder, W.E. (2008) Niche partitioning increases resource exploitation by diverse communities. *Science*, 321, 1488–1490.

Gaston, K.J. (1996) What is biodiversity? in *Biodiversity: a biology of numbers and difference* (ed. K.J. Gaston), Blackwell, Oxford, pp. 1–9.

Grez, A.A. and González, R.H. (1995) Resource concentration hypothesis: effect of host plant patch size on density of herbivorous insects. *Oecologia*, 103, 471–474.

Gurr, G.M., Barlow, N., Memmott, J., Wratten, S.D. and Greathead, D.J. (2000) A history of methodological, theoretical and empirical approaches to biological control, in *Biological control: Measures of success* (eds G.M. Gurr and S.D. Wratten), Kluwer, Dordrecht, pp. 3–37.

Gurr, G.M., Scarratt, S.L., Wratten, S.D., Berndt, L. and Irvin, N (2004) Ecological engineering, habitat manipulation and pest management, in *Ecological engineering for pest management: Advances in habitat manipulation for arthropods* (eds G.M. Gurr, S.D. Wratten and M.A. Altieri), CSIRO Publishing, Collingwood, Victoria, pp. 1–12.

Gurr, G.M., Liu, J., Read, D.M.Y., Catindig, J.L.A., Cheng, J.A., Lan, L.P. and Heong, K.L. (2011) Parasitoids of Asian rice planthopper (Hemiptera: Delphacidae) pests and prospects for enhancing biological control. *Annals of Applied Biology*, 158, 149–176.

Halaj, J. and Wise, D.H. (2001) Terrestrial trophic cascades: how much do they trickle? *American Naturalist*, 157, 262–281.

Hanski, I. and Simberloff, D. (1997) The meta-population approach, its history, conceptual domain, and application to conservation, in *Metapopulation biology: Ecology, genetics, and evolution* (eds I. Hanski and M.E Gilpin), Academic Press, San Diego, pp. 5–26.

Hassell, M.P., Comins, H.N. and May, R.M. (1991) Spatial structure and chaos in insect population dynamics. *Nature*, 353, 255–258.

Held, D.W., Gonsiska, P. and Potter, D.A. (2003) Evaluating companion planting and non-host masking odors for protecting roses from the Japanese beetle (Coleoptera: Scarabaeidae). *Journal of Economic Entomology*, 96, 81–87.

Hickman, J.M., and Wratten, S.D. (1996) Use of *Phacelia tanacetifolia* strips to enhance biological control of aphids by hoverfly larvae in cereal fields. *Journal of Economic Entomology*, 89, 832–840.

Hillebrand, H. and Cardinale, B.J. (2004) Consumer effects decline with prey diversity. *Ecology Letters*, 7, 192–201.

Howarth, F.G. (1991) Environmental impacts of classical biological control. *Annual Review of Entomology*, 36, 485–509.

Hunt, T., Bergsten, J., Levkanicova, Z. *et al.* (2007) A comprehensive phylogeny of beetles reveals the evolutionary origins of a superradiation. *Science*, 318, 913–1916.

Idris, A.B. and Grafius, E. (1995) Wildflowers as nectar sources for *Diadegma insulare* (Hymenoptera: Ichneumonidae), a parasitoid of diamondback moth (Lepidoptera: Yponomeutidae). *Environmental Entomology*, 24, 1726–1735.

Isman, M.B. (2006) Botanical insecticides, deterrents, and repellents in modern agriculture and an increasingly regulated world. *Annual Review of Entomology*, 51, 45–66.

Jactel, H. and Brockerhoff, E.G. (2007) Tree diversity reduces herbivory by forest insects. *Ecology Letters*, 10, 835–848.

Jactel, H., Brockerhoff, E. and Duelli, P. (2005) A test of the biodiversity-stability theory: meta-analysis of tree species diversity effects on insect pest infestations, and re-examination of responsible factors. *Ecological Studies*, 176, 235–262.

Jonsson, M., Wratten, S.D., Landis, D.A. and Gurr, G.M. (2008) Recent advances in conservation biological control of arthropods by arthropods. *Biological Control*, 45, 172–175.

Kareiva, P. (1983) Influence of vegetation texture on herbivore populations: resource concentration and herbivore movement, in *Variable plants and herbivores in natural and managed systems* (eds R. Denno and M. McClure), Academic Press, New York, pp. 259–289.

Kean, J., Wratten, S.D., Tylianakis, J. and Barlow, N. (2003) The population consequences of natural enemy enhancement, and implications for conservation biological control. *Ecology Letters*, 6, 1–9.

Khan, Z.R., Pickett, J.A., van den Berg, J., Wadhams, L.J. and Woodcock, C.M. (2000) Exploiting chemical ecology and species diversity: stem borer and striga control for maize and sorghum in Africa. *Pest Management Science*, 56, 957–962.

Khan, Z.R., Midega, C.A.O., Hutter, N.J., Wilkins, R.M. and Wadhams, L.J. (2006) Assessment of the potential of Napier grass (*Pennisetum purpureum*) varieties as trap plants for management of *Chilo partellus*. *Entomologia Experimentalis et Applicata*, 119, 15–22.

Kleijn, D., Berendse, F., Smit, R. and Gilissen, N. (2001) Agri-environment schemes do not effectively protect biodiversity in Dutch agricultural landscapes. *Nature*, 413, 723–725.

Kontogianni, A., Luck, G.W. and Skourtos, M. (2010) Valuing ecosystem services on the basis of service-providing units: a potential approach to address the 'endpoint problem' and improve stated preference methods. *Ecological Economics*, 69, 1479–1487.

Kvedaras, O.L., An, M., Choi, Y.S. and Gurr, G.M. (2010) Silicon enhances natural enemy attraction and biological control through induced plant defences. *Bulletin of Entomological Research*, 100, 367–371.

Landis, D., Wratten, S.D. and Gurr, G.M. (2000) Habitat management to conserve natural enemies of arthropod pests in agriculture. *Annual Review of Entomology*, 45, 175–201.

Lavandero, B., Wratten, S., Didham, R. and Gurr, G. (2006) Increasing floral diversity for selective enhancement of biological control agents: A double-edged sward? *Basic and Applied Ecology*, 7, 236–243.

Lawton, J.H. and McNeill, S. (1979) Between the devil and the deep blue sea: on the problems of being a herbivore, in *Population dynamics* (eds R.M Anderson, B.D. Turner and L.R. Taylor), Blackwell, Oxford, pp. 223–244.

Lemaux, P.G. (2009) Genetically engineered plants and foods: a scientist's analysis of the issues (Part II). *Annual Review of Plant Biology*, 60, 511–559.

Letourneau, D.K., Jedlicka, J.A., Bothwell, S.G. and Moreno, C.O. (2009) Effects of natural enemy biodiversity on the suppression of arthropod herbivores in terrestrial ecosystems. *Annual Review of Ecology, Evolution and Systematics*, 40, 573–592.

Losey, J.E. and Denno, R.F. (1998) Positive predator–predator interactions: enhanced predation rates and synergistic suppression of aphid populations. *Ecology*, 79, 2143–2152.

McMenemy, L.S., Mitchell, C. and Johnson, S.N. (2009) Biology of the European large raspberry aphid (*Amphorophora idaei*): its role in virus transmission and resistance breakdown in red raspberry. *Agricultural and Forest Entomology*, 11, 61–71.

Meffe, G.K., Nielsen, L.A., Knight, R.L. and Schenborn, D.A. (2002) *Ecosystem management: adaptive, community-based conservation*, Island Press, Washington.

Meyer, J. (2009) *A class of distinction*. North Carolina State University, URL: http://www.cals.ncsu.edu/course/ent425/library/tutorials/importance_of_insects/class_of_distinction.html

Murphy, B.C., Rosenheim, J.A., Dowell, R.V. and Granett, J. (1998) Habitat diversification tactic for improving biological control: parasitism of the western grape leafhopper. *Entomologia Experimentalis et Applicata*, 87, 225–235.

Namba, O., Nakamatsu, Y., Miura, K. and Tanaka, T. (2008) *Autographa nigrisigna* looper (Lepidoptera: Noctuidae) excludes parasitoid egg using cuticular encystment induced by parasitoid ovarian fluid. *Applied Entomology and Zoology*, 43, 359–367.

Natural England (undated) *Countryside stewardship scheme (CSS)*. Natural England. URL: http://www.naturalengland.org.uk/ourwork/farming/funding/closedschemes/css/default.aspx

Nevo, D. (1996) The desert locust, *Schistocerca gregaria*, and its control in the land of Israel and the near east in antiquity, with some reflections on its appearance in Israel in modern times. *Phytoparasitica*, 24, 7–32.

Nicholson, A.J. and Bailey, V.A. (1935) The balance of animal populations. *Proceedings of the Zoological Society of London*, 1, 551–598.

Pfannenstiel, R.S., Unruh, T.R. and Brunner, J.F. (2010) Overwintering hosts for the exotic leafroller parasitoid, *Colpoclypeus florus*: implications for habitat manipulation to augment biological control of leafrollers in pome fruits. *Journal of Insect Science*, 10:75, available online: insectscience.org/10.75.

Rea, J.H., Wratten, S.D., Sedcole, J., Cameron, P.J. and Chapman, R.B. (2002) Trap cropping to manage green vegetable bug *Nezara viridula* (L.) (Heteroptera: Pentatomidae) in sweet corn in New Zealand, *Agricultural and Forest Entomology*, 4, 101–108.

Robinson, K.A., Jonsson, M., Wratten, S.D., Wade, M.R. and Buckley, H. (2008) Implications of floral resources for predation by an omnivorous lacewing. *Basic and Applied Ecology*, 9, 172–181.

Root, R.B. (1973) Organization of a plant–arthropod association in simple and diverse habitats: the fauna of collards (*Brassica oleracea*). *Ecological Monographs*, 43, 95–124.

Rosenfeld, J.S. (2002) Functional redundancy in ecology and conservation. *Oikos*, 98,156–162.

Sanchis, V. and Bourguet, D. (2008) Bacillus thuringiensis: applications in agriculture and insect resistance management. A review. *Agronomy for Sustainable Development*, 28, 11–20.

Sandhu, H., Wratten, S.D., Cullen, R. and Case, B. (2008) The future of farming: The value of ecosystem services in conventional and organic arable land. An experimental approach. *Ecological Economics*, 64, 835–848.

Schoonhoven, L.M., van Loon, J.J.A. and Dicke, M. (2005) *Insect–Plant Biology*, 2nd edn, Oxford University Press, Oxford.

Scriber, J.M. (2010) Integrating ancient patterns and current dynamics of insect–plant interactions: Taxonomic and geographic variation in herbivore specialization. *Insect Science*, 17, 471–507.

Settele, J., Biesmeijer, J. and Bommarco, R. (2008) Switch to ecological engineering would aid independence. *Nature*, 456–570.

Settle, W., Ariawa, H., Astuti, E.T. *et al.* (1996) Managing tropical rice pests through conservation of generalist natural enemies and alternative prey, *Ecology*, 77, 1795–1988.

Shine, R. (2010) The ecological impact of invasive cane toads (*Bufo marinus*) in Australia. *Quarterly Review of Biology*, 85, 253–291.

Simmons, A.T. and Gurr, G.M. (2005) Trichomes of Lycopersicon species and their hybrids: effects on pests and natural enemies. *Agricultural and Forest Entomology*, 7, 265–276.

Simpson, M., Gurr, G.M., Simmons, A.T. *et al.* (2011) Insect attraction to synthetic herbivore-induced plant volatile treated field crops. *Agriculture and Forest Entomology*, 13, 45–57.

Soluk, D.A. and Collins, N.C. (1988) Synergistic interactions between fish and stoneflies: facilitation and interference among stream predators. *Oikos*, 52, 94–100.

Speight, M.R. and Wainhouse, D. (1989) *Ecology and management of forest insects*, Clarendon Press, Oxford.

Steffan, S.A. and Snyder, W.E. (2010) Cascading diversity effects transmitted exclusively by behavioral interactions. *Ecology*, 91, 2242–2252.

Stern, V.M., Smith, R.F., van den Bosch, R. and Hagen, K.S. (1959) The integrated control concept. *Hilgardia*, 29, 81–101.

Straub, C.S. and Snyder, W.E. (2006) Species identity dominates the relationship between predator biodiversity and herbivore suppression. *Ecology*, 87, 277–282.

Straub, C.S., Finke, D.L. and Snyder, W.E. (2008) Are the conservation of natural enemy biodiversity and biological control compatible goals? *Biological Control*, 45, 225–237.

Tahir, H.M. and Butt, A. (2009) Predatory potential of three hunting spiders inhabiting the rice ecosystems. *Journal of Pest Science*, 82, 217–225.

Takimoto, G., Iwata, .T and Murakami, M. (2002) Seasonal subsidy stabilizes food web dynamics: Balance in a heterogeneous landscape. *Ecological Research*, 17, 433–439.

Thomas, M.B. and Willis, A.J. (1998) Biocontrol – risky but necessary? *Trends in Ecology and Evolution*, 13, 325–329.

Thomas, M.B., Wratten, S.D. and Sotherton, N.W. (1991) Creation of island habitats in farmland to manipulate populations of beneficial arthropods – predator densities and emigration. *Journal of Applied Ecology*, 28, 906–917.

Thomas, M.B., Wratten, S.D. and Sotherton, N.W. (1992) Creation of island habitats in farmland to manipulate populations of beneficial arthropods – predator densities and species composition. *Journal of Applied Ecology*, 29, 524–531.

Thomas, S.R., Goulson, D. and Holland, J.M. (2001) Resource provision for farmland gamebirds: the value of beetle banks. *Annals of Applied Biology*, 139, 111–118.

Thomson, M.J., Polato, N.R., Prasetiyono, J., Trijatmiko, K.R., Silitonga, T.S. and McCouch, S.R. (2009) Genetic diversity of isolated populations of Indonesian landraces of rice (*Oryza sativa* L.) collected in East Kalimantan on the island of Borneo. *Rice*, 2, 80–92.

Thorpe, W.H. and Caudle, H.B. (1938) A study of the olfactory responses of insect parasites to the food plant of their host. *Parasitology*, 30, 523–528.

Tonhasca, A. and Byrne, D.N. (1994) The effects of crop diversification on herbivorous insects – a metaanalysis approach. *Ecological Entomology*, 19, 239–244.

Tylianakis, J.M., Didham, R.K. and Wratten, S.D. (2004) Improved fitness of aphid parasitoids receiving resource subsidies. *Ecology*, 85, 658–666.

US Census Bureau (2011) *World POPClock projection*. International Programs Center, URL: http://www.census.gov/ipc/www/popclockworld.html

Van Emden, H.F. (1965) The role of uncultivated land in the biology of crop pests and beneficial insects. *Scientific Horticulture*, 17, 121–136.

Van Emden, H.F. and Peakall, D.B. (eds) (1996) *Beyond silent spring: integrated pest management and chemical safety*. Chapman & Hall, London.

Van Lenteren, J.C. and Woets, J. (1988) Biological and integrated pest control in greenhouses. *Annual Review of Entomology*, 33, 239–269.

Wäckers, F.L., Romeis, J. and van Rijn, P. (2007) Nectar and pollen feeding by insect herbivores and implications for multitrophic interactions. *Annual Review of Entomology*, 52, 301–323.

Wade, M.R., Gurr, G.M. and Wratten, S.D. (2008a) Ecological restoration of farmland: progress and prospects. *Philosophical Transactions of the Royal Society B*, 363, 831–847.

Wade, M.R., Zalucki, M.P., Wratten, S.D. and Robinson, K.A. (2008b) Conservation biological control of arthropods using artificial food sprays: current status and future challenges. *Biological Control*, 45, 185–199.

Whittaker, R.H. (1972) Evolution and measurement of species diversity. *Taxon*, 21, 213–251.

Wilby, A. and Thomas, M.B. (2002) Natural enemy diversity and pest control: patterns of pest emergence with agricultural intensification. *Ecology Letters*, 5, 353–360.

Wilby, A., Villareal, S.C., Lan, L.P., Heong, K.L. and Thomas, M.B. (2005) Functional benefits of predator species diversity depend on prey identity. *Ecological Entomology*, 30, 497–501.

Wratten, S.D. (1974) Aggregation in the birch aphid *Euceraphis punctipennis* (Zett.) in relation to food quality. *Journal of Animal Ecology*, 43, 191–198.

Wu, J. and Baldwin, I.T. (2010) New insights into plant responses to the attack from insect herbivores. *Annual Review of Genetics*, 44, 1–24.

Yencho, G.C., Cohen, M.B. and Byrne, P.F. (2000) Applications of tagging and mapping insect resistance loci in plants. *Annual Review of Entomology*, 45, 393–422.

Zehnder, G., Gurr, G.M., Kühne, S., Wade, M.R., Wratten, S.D. and Wyss, E. (2007) Arthropod pest management in organic crops. *Annual Review of Entomology*, 52, 57–80.

Fundamentals

Chapter 2

THE ECOLOGY OF BIODIVERSITY– BIOCONTROL RELATIONSHIPS

William E. Snyder and Jason M. Tylianakis

Biodiversity and Insect Pests: Key Issues for Sustainable Management, First Edition. Edited by Geoff M. Gurr, Steve D. Wratten, William E. Snyder, Donna M.Y. Read.

TROPHIC CASCADES

Hairston *et al.* (1960) first suggested that the Earth's land masses are (largely) green because plants are indirectly protected by their herbivores' predators, pathogens and parasites. Naturally, this has come to be known as the 'green world hypothesis' (GWH). These authors argued that ecosystems consistently collapse into three trophic levels – plants, herbivores, and natural enemies – with natural enemies driving herbivores to densities too low to substantially damage plants (in a process called a 'trophic cascade'; Figure 2.1a). It is important to note that the GWH was not devised within a biodiversity framework, and that its authors did not address how species richness would affect trophic cascade strength. However, we take some liberties here in extrapolating the GWH to consider such issues, as biodiversity considerations were key concerns of the GWH's critics (as discussed below). Within the GWH framework, because all species within a trophic level function similarly, we suppose that biodiversity within the predator trophic level would be relatively un-influential (Figure 2.2). Oksanen *et al.* (1981) expanded the GWH by pointing out that different communities have different numbers of trophic levels depending on their overall productivity. These authors suggest that systems with very low productivity can only support plant and herbivore trophic levels, with unregulated herbivores devastating plants (Figure

2.1b). Likewise, exceptionally productive communities support four trophic levels, with top predators freeing herbivores from control by intermediate predators (Figure 2.1b). This again leads to heavily damaged plants. Thus, while trophic cascades occur in all cases,

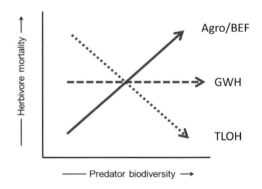

Figure 2.2 Various theories encompass all possible relationships between predator biodiversity and the strength of biological control. Green world hypothesis (GWH) proponents expect consistently strong trophic cascades regardless of intra-trophic level biodiversity; the trophic-level omnivory hypothesis (TLOH) predicts a weakening of trophic cascades as biodiversity increases; and agroecologists and biodiversity–ecosystem function researchers (Agro/BEF) propose stronger herbivore suppression as predator biodiversity increases.

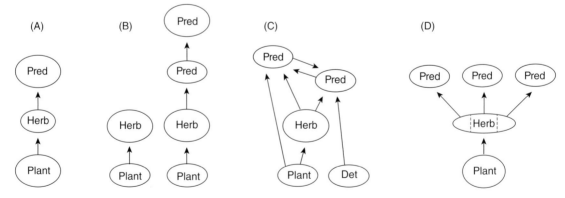

Figure 2.1 Diagrams of the food web structures envisioned by (A) proponents of the green world hypothesis and (B) its extension by Oksanen *et al.* (1981), (C) advocates of the trophic-level omnivory hypothesis, and (D) agroecologists and biodiversity–ecosystem function researchers interested in complementarity. The circles variously indicate plant (Plant), herbivore (Herb), detritivore (Det) and predator (Pred) trophic levels. In the figures circles are scaled to show relative biomass at that trophic level, and arrows denote the direction of energy flow and so point from resource to consumer.

herbivores are controlled by their natural enemies in communities with an odd number of trophic levels, but freed from control in communities with an even number of trophic levels. And again, because species can be lumped into coherent trophic levels, biodiversity within a trophic level would not be expected to have a major effect on the strength of trophic cascades (Figure 2.2).

While influential, the simple trophic-level cascade model described above is not universally accepted. Among food web ecologists, there has been scepticism that this simple trophic-level structure is common (or present at all) in real ecosystems (Polis, 1991; Strong, 1992; Polis and Strong, 1996). This is because communities typically include many species of generalist predator, which often feed not only on herbivores but also upon detritus-feeders and even other predators (Polis et al., 1989). Such interactions have the potential to entirely blur the designation of simple, distinct trophic levels (Figure 2.1c). Of course, if distinct trophic levels cannot be delineated, then the simple trophic cascades envisioned by Hairston et al. (1960) cannot occur. This point of view is known as the 'trophic-level omnivory hypothesis' (TLOH), and suggests that greater species richness leads to more reticulate connections among species and weak (or no) trophic cascades (Figure 2.2). The TLOH has been fiercely rejected by GWH supporters (Hairston and Hairston, 1993; 1997).

A third view of the relationship between biodiversity and biocontrol is presented in the agroecology literature (Figure 2.1d). Early agroecologists noted that, whereas highly diverse natural systems rarely experience devastating herbivore outbreaks, such outbreaks are common in simplified agricultural systems (Pimentel, 1961). Indeed, movement to huge monocultures within modern agriculture appears to correlate both with increasingly species-poor ecological communities (Tylianakis et al., 2005) and increasingly intense pest problems (Matson et al., 1997). This led to the obvious suggestion that restoring greater biodiversity to agricultural systems would restore natural 'balance' between pests and their enemies, reducing the frequency and intensity of pest outbreaks (Root, 1973; van Emden and Williams, 1974; Altieri and Whitcomb, 1979). That is, greater biodiversity leads to stronger pest suppression (Figure 2.2). Such a reduction in pests could occur through alteration to the way pests locate their host plant (the crop) amongst a diverse assemblage of plants (the so-called 'resource concentration hypothesis' (Root, 1973)), which forms the ecological basis for cultural pest control practices such as trap-cropping, intercropping or living ('green') mulches. Plant diversity may also provide resources to natural enemies that allow them to better control the herbivore pest (Landis et al., 2000; Tylianakis et al., 2004). Finally, diversity of natural enemies themselves may potentially benefit biocontrol. This latter precept is consistent with findings from the emerging field of 'biodiversity–ecosystem function' (BEF) research. BEF research originally focused on the effects of the ongoing mass extinction event worldwide, triggered by the growing human population and associated environmental degradation, on the ability of ecosystems to function effectively (Naeem et al., 1995; Hooper and Vitousek, 1997; Hooper et al., 2005). Initial experimental work generally focused on prairie-plant communities, and deployed a common experimental approach. Typically, the number of plant species present was manipulated in large field plots, and such community attributes as total biomass and resource use were tracked through time (e.g. Tilman et al., 1997; Hector et al., 1999). The consistent result from these studies was that greater species richness nearly always improved the ecological functioning of communities (Hooper et al., 2005; Cardinale et al., 2006). If these results hold true for natural enemy communities, one would again expect a positive relationship between natural enemy biodiversity and the intensity of pest suppression (Figure 2.2; Naylor and Ehrlich, 1997; Duffy, 2003). It has been suggested that the positive relationships typical of most BEF experiments follow a variety of different forms, reflecting an equally broad variety of possible underlying mechanisms (Box 2.1).

Thus, across the GWH, TLOH, agroecology and biodiversity–ecosystem function perspectives, all possible biodiversity–biocontrol relationships – neutral, negative and positive – have been suggested. Frustratingly, empirical studies have revealed each of these patterns within particular model predator–prey systems (Sih et al., 1998; Ives et al., 2005; Straub et al., 2008; Bruno and Cardinale, 2008; Finke and Snyder, 2010). However, across these many studies, a few clear patterns have begun to emerge within research undertaken from both a trophic cascade and a biodiversity–ecosystem function perspective.

First, it is clear that trophic cascades commonly occur across a wide array of terrestrial and aquatic

Box 2.1 Relationship between predator biodiversity and herbivore suppression

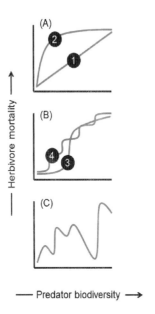

Ecologists have envisioned a wide variety of forms for the relationship between predator biodiversity and herbivore suppression (Naeem *et al.*, 2002; Snyder *et al.*, 2005). (A) When all predators attack an entirely unique subset of the prey population (line 1), herbivore suppression will increase linearly with increasing predator biodiversity; when predators partially overlap in prey taken, herbivore suppression will eventually plateau as species become redundant (line 2). (B) Sometimes just a few predator species make particular, unique contributions to pest suppression. When one predator has a particularly unique effect this is called a 'keystone' species and suppression suddenly jumps when that species joins the community (line 3); where there are several such important species this is called the 'rivet effect' and suppression jumps each time one of these unique species is added (line 4). (C) When predator communities include species with widely varying positive and negative effects, the biodiversity–biocontrol relationship is said to be 'idiosyncratic' and no overriding trend is obvious.

communities (Schmitz *et al.*, 2000; Shurin *et al.*, 2002). Thus, the worst fears of TLOH proponents appear not to be realised, although there is some evidence that as biodiversity increases the strength of trophic cascades is somewhat ameliorated, as the TLOH suggests (Halaj and Wise, 2001). For work undertaken from a BEF perspective, there clearly are a few case studies where greater predator biodiversity only serves to disrupt herbivore suppression (e.g. Finke and Denno, 2004). Nonetheless, neutral and positive predator-diversity effects appear to be predominant (Ives *et al.*, 2005; Cardinale *et al.*, 2006; Straub *et al.*, 2008; Snyder, 2009; Finke and Snyder, 2010). Thus, the preponderance of evidence appears to be falling in favour of the long-standing view of agroecologists that increasing biodiversity can be a particularly effective way to diffuse pest problems. It is puzzling, then, why practitioners of classical biological control (where introduced natural enemies are imported to attack introduced pests) fail to find a relationship between the number of species introduced and the resulting success of biocontrol (Denoth *et al.*, 2002; Stiling and Cornelissen, 2005). This may be in part because of the inherently unpredictable nature of planned species introductions, with the vagaries of establishment success being more important for eventual control than any advantages of natural enemy biodiversity itself (Pedersen and Mills, 2004). Alternatively, it may be that certain characteristics of the pest species or crop type predispose them to being more or less susceptible to diverse natural enemy assemblages (Tylianakis and Romo, 2010).

In summary, work to date has revealed biodiversity–biocontrol relationships consistent with each of the predicted relationships shown in Figure 2.2. Nonetheless, our growing understanding of the mechanisms underlying biodiversity effects of various types breeds optimism that general, predictable patterns do occur in nature. We next review the mechanisms thought to underlie the complex array of biodiversity–biocontrol relationships found across these many studies.

MECHANISMS UNDERLYING BIODIVERSITY–BIOCONTROL RELATIONSHIPS

As reviewed above, various studies have recorded neutral, positive and negative relationships between natural enemy biodiversity and the resulting strength of biological control.

Neutral biodiversity–biocontrol relationships: no change in pest suppression with increasing predator richness

Sometimes, the combined effects of multiple natural enemy species can be precisely predicted by summing or averaging (depending on the experimental design; Box 2.2) the impact each species has when on its own (e.g. Sokol-Hessner and Schmitz, 2002; Straub and Snyder, 2006). In this case, enemy biodiversity has no relationship with herbivore suppression, as long as there is no relationship between predator diversity and overall predator abundance (Cardinale *et al.*, 2003). Thus, predator species identity effects dominate. Diverse predator communities may nonetheless exert stronger pest control than particular single natural enemy species, when diverse predator communities including particularly effective natural enemy species are compared to relatively ineffective single species (Straub and Snyder, 2006). This is called the 'sampling

---

Box 2.2 Determining impacts of multiple predator species

 (A) Multi-species additive

 (B) Multi-species substitutive

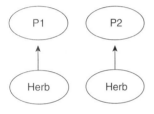

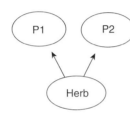

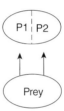

Single-species treatments are the same under both designs

Ecologists have used two very different experimental designs to examine the impacts of multiple predator species on their prey (Sih *et al.*, 1998; Straub *et al.*, 2008; Finke and Snyder, 2010). Early studies generally deployed an additive design (A). In these experiments the density of each predator species is held constant across diversity levels. This means that total predator densities increase as more predator species are added within diverse communities. The advantage of this design is that, because densities of each predator species are held constant, the intensity of interactions among members of the same predator species might also stay the same across diversity levels. This isolates the impacts of interactions between predator species. A problem with additive designs is that, because prey densities are also held constant across all treatments, overall competition for prey increases as predator species richness increases; thus, there may be more negative predator–predator interactions at higher diversity levels purely because competition for prey is more intense (rather than because diverse communities necessarily inspire predators to have more conflicts). In recent years, more and more predator diversity studies have used a substitutive design (B). In these experiments the number of predator individuals is kept constant across all treatments. So, the number of individuals of a particular predator species is cut in half when two predator species are present together, but by two-thirds when three predator species are present, and so on. The advantage of this design is that overall predator–prey ratios are kept constant across predator diversity levels. A problem with this design is that, because densities of each predator species decline as predator species richness increases, any negative interactions within species also weaken. This means that diverse predator communities may do better purely because negative interactions within species are relaxed, rather than because of any positive interactions among different predator species. The ideal design would include both additive and substitutive manipulations of predator richness at once, but such large experiments can be very difficult to carry out (Northfield *et al.*, 2010). In the figure above, the prey is a herbivore (Herb) that is fed upon by predator species 1 (P1) and predator species 2 (P2); arrows point from resource (the prey) to consumer (the predators).

effect' in the BEF literature (Hooper *et al.*, 2005). Whenever a natural enemy species exerts particularly strong effects on a pest, from a biological control perspective it would be best to focus efforts on preserving and augmenting densities of that key natural enemy species. Neutral diversity effects can be recorded when predator species act largely independently from one another, and do not strongly interact; when prey are so abundant that any predator species-specific differences in prey use are not revealed; and when positive predator–predator interactions and negative predator–predator interactions perfectly counteract one another (Finke and Snyder, 2010).

Positive biodiversity–biocontrol relationships: pest suppression grows stronger as more predator species are added to a community

Often the combined impact of a collection of natural enemy species exceeds that which even the single most effective species can achieve on its own (e.g. Snyder *et al.*, 2006). These cases provide evidence for 'emergent' biodiversity effects, where in essence the whole exceeds the sum of the parts. This scenario is consistent with the world-view championed by agroecologists and BEF researchers, who see greater biodiversity as a means to dampen pest outbreaks (Figure 2.2). Generally speaking, emergent biodiversity effects are attributed to one of two mechanisms: complementarity and facilitation (Hooper *et al.*, 2005; Ives *et al.*, 2005). Each of these mechanisms is discussed in turn below.

Predator–predator 'complementarity' occurs when natural enemy species differ from one another in some ecologically significant way, such that different natural enemies attack different pest species or different subsets of the same pest species (Hooper *et al.*, 2005; Ives *et al.*, 2005). That is, predator species occupy unique and complementary feeding niches (Finke and Snyder, 2008). Various studies have revealed, or suggested, that predator–predator complementarity can assume a wide array of forms (Snyder, 2009). Most obvious is the case where different natural enemy species feed on entirely different pest species. In this case, predator impacts across the entire pest complex can be achieved only by pairing many different natural enemy species, so that each pest faces attack by at least one enemy species. For example, a series of exotic pests have invaded citrus plantings in Florida, USA. Each pest has required, in turn, the introduction of a different bio-

logical control agent that specialises on that pest (Michaud, 2002). In the absence of the complete community of biological control agents, any single pest, unregulated by natural enemies, is capable of devastating citrus crops (Michaud, 2002). In such cases, predator–predator complementarity is complete with no overlap in function between species, although complementarity can still occur when there is some sharing of prey species among predators (e.g. Tamaki and Weeks, 1972). The completeness of functional overlap will depend not only on the fundamental niche of the different predator species, but also on the availability of varied niches within the particular habitat, such that a heterogeneous prey base may increase the realisation of predator complementarity (Tylianakis *et al.*, 2008; Tylianakis and Romo, 2010).

Complementarity can also occur when different natural enemy species attack the same pest species, but at least partially subdivide the prey population along a spatial or temporal axis (Wilby and Thomas, 2002; Casula *et al.*, 2006). Many pests exhibit complex life cycles that bring different life stages into contact with different natural enemy species, which provides an opportunity for enemy complementarity. For example, Wilby *et al.* (2005) found that a diverse community of predators was necessary to maximise mortality of a moth pest of rice, apparently because different enemy species focused their attacks on different moth life stages and so were complementary. In contrast, for a planthopper pest of rice with morphologically similar life stages there was no predator partitioning among pest life stages and so no benefit to greater predator diversity (Wilby *et al.*, 2005). Nonetheless, it now is clear from other studies that positive predator-diversity effects can occur even when prey species have simple development (e.g. Cardinale *et al.*, 2003; Snyder *et al.*, 2006; Finke and Snyder, 2008; Straub and Snyder, 2008). Similarly, predators that are active at different times of the day or year can exert complementary impacts on shared prey species. For example, Pfannenstiel and Yeargan (2002) found that predation of a lepidopteran pest of corn was maximised through the combined effects of two natural enemies: a lady beetle that foraged primarily during the day, and a predatory bug active primarily at night. Only with both enemies present were the moths deprived of a daily refuge from predation (Pfannenstiel and Yeargan, 2002). Operating on a longer time scale, Neuenschwander *et al.* (1975) found seasonal complementarity within a community of predators attacking pea aphids on alfalfa: several lady beetles provided strong impacts during relatively mild weather earlier in the season, while predatory bugs were most active during the hottest parts of summer. Thus, only a diverse predator community provided attacks on aphids throughout the growing season. Finally, natural enemy species have been shown to complement one another across space. For example, Straub and Snyder (2008) found that aphid pests of *Brassica oleracea* (L.) plants were most effectively controlled by a diverse group of natural enemy species, because some predators foraged mostly on leaf edges while others also accessed aphids at the centre of leaves. Thus, only a diverse predator community took away all spatial refuges that the aphids might otherwise exploit (Straub and Snyder, 2008). Of course, predator species that differ in where and/or when they hunt are less likely to encounter one another, and so are also less likely to prey upon or otherwise interfere with one another (Musser and Shelton, 2003; Schmitz, 2007). This further encourages positive, rather than negative, biodiversity–biocontrol relationships.

Finally, natural enemy species often differ in their hunting styles, and it has been suggested that these differences might also lead to complementarity. For example, generalist predators can exist in a crop independent of the density of any particular prey species, and so are present to attack the first colonising pests. But once pest densities begin to grow, generalists are generally unable to increase quickly in response (Symondson *et al.*, 2002). Thus, generalists tend to exert constant, but density-independent, mortality (Hassell, 1980; Hassell and May, 1986). In contrast, specialists cannot exist in a crop before the pest arrives, but can often 'keep up' numerically with pests as pest density increases (Hassell, 1980; Hassell and May, 1986). Thus, the two natural enemies may complement one another in a dynamic sense, such that overall control is most effective when the constant mortality of generalists is paired with the density-dependent (but delayed) response of specialists (e.g. Snyder and Ives, 2003). Combinations of predator hunting styles differing in other ways, such as deploying a sit-and-wait versus an active hunting style, likewise might lead to complementarity (Snyder *et al.*, 2006). However, predator–predator interference may erase any benefits of such pairings when differing hunting styles heighten the risk of one predator species falling victim to another (Rosenheim *et al.*, 2004; Schmitz, 2007).

Predator–predator 'facilitation' occurs when the presence of one natural enemy species improves the capture success of another enemy species. Many exam-

ples of this phenomenon come from what might be called the 'between a rock and a hard place' literature, wherein pests fleeing from a predator in one habitat instead fall victim to a second predator species lurking in the would-be refuge. Perhaps the best-known example of predator–predator facilitation in a biocontrol setting comes from the work of Losey and Denno (1998). These authors examined single and combined impacts of lady beetles and ground beetles attacking pea aphids (*Acyrthosiphon pisum* (Harris)) on alfalfa plants. The lady beetles hunt in the foliage where pea aphids feed, whereas ground beetles forage on the soil surface where aphids usually fear to tread. However, when accosted by lady beetles, pea aphids attempt to escape by dropping from plants onto the soil surface, (inadvertently) bringing themselves into contact with voracious ground beetles. Thus, the two natural enemies together exert combined impacts greater than either enemy alone, because only in the presence of lady beetles do ground beetles regularly have the chance to encounter and kill aphids (Losey and Denno, 1998). Sometimes, facilitation among natural enemies can trace an even more convoluted path (Box 2.3). Whereas some ecologists suggest that facilitation is a relatively important and common interaction among predators (Sih *et al.*, 1998; Ives *et al.*, 2005), Schmitz (2007) suggests that predator–predator facilitation is exceedingly rare. Indeed, Schmitz (2007) argues that facilitation can occur only when two predators forage in the same way and in the same location, and prey forage in a narrow subset of the two predators' foraging range.

The best means to manage agricultural systems to maximise the benefits of predator biodiversity may differ when complementarity, versus facilitation, is the underlying mechanism. When complementarity among predator species is key, it will be necessary to preserve and augment as broad a range of different species as possible, in order to maximise the total number of niches that are filled by one predator or another. When facilitation is key, it will be important to focus conservation efforts on those species that foster one another's prey capture. A key gap in the literature to date, however, is that the benefits of predator biodiversity for pest suppression are not often followed to see if crop yields also improve (but see Cardinale *et al.*, 2003; Snyder *et al.*, 2006; Steffan and Snyder, 2010; Crowder *et al.*, 2010). Thus, it is not entirely clear that greater predator biodiversity consistently leads to stronger trophic cascades that benefit plants.

Box 2.3 Facilitation between predators and natural enemies of the potato beetle: a convoluted path

Ramirez and Snyder (2009) examined how biodiversity among predator and pathogen natural enemies impacts biological control of herbivorous Colorado potato beetles (*Leptinotarsa decemlineata* (Say)). Potato beetles have a complex life history, with the eggs and larvae feeding on plant foliage above ground before burrowing into the soil to pupate. The beetles face entirely different natural enemies in these two habitats. Predatory *Hippodamia convergens* (Guérin-Méneville) (Hc) and *Pterostichus melanarius* (Illiger) (Pmel) beetles, and *Nabis alternatus* (Parshley) (Nabis) bugs, dominate above ground. Insect-attacking entomopathogenic *Steinernema carpocapsae* (Weiser) (Scarp) and *Heterorhabditis marelatus* (Lui & Berry) (Hmar) nematodes, and *Beauveria bassiana* (Balsamo) (Bbass) fungi, dominate below ground. So, predators and pathogens attack different life stages of the beetles in different habitats, and so complement one another through both space and time. Indeed, when the authors experimentally manipulated species richness among predators and pathogens they found that greater natural enemy biodiversity substantially increased beetle suppression. However, the underlying mechanism was complex. Subsequent experiments revealed that potato beetles that faced predators in early stages, but were not killed, nonetheless were more susceptible to pathogen infection once they entered the soil. The potato beetles engage in a variety of energetically costly anti-predator behaviours (Ramirez *et al.*, 2010), and the authors propose that frequent deployment of these defences earlier in life drains resources needed later to fight off pathogens. Thus, predators and pathogens partitioned resources internal to the herbivores themselves: an unusual mode of complementary resource use among species (Ramirez and Snyder, 2009).

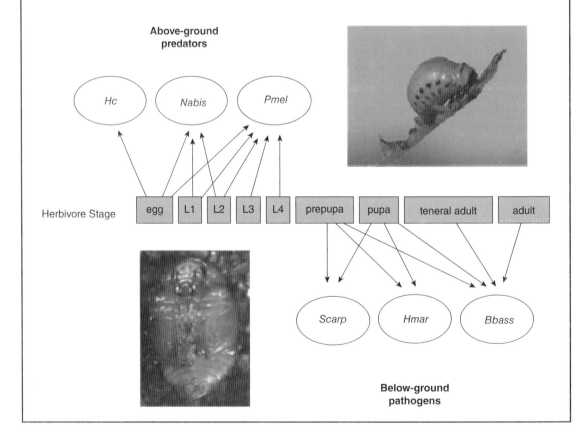

Negative predator diversity effects: pest control weakens as more species are added to a community

As suggested by proponents of the TLOH, generalist predators often feed on one another (Polis *et al.*, 1989). This interaction, known as 'intraguild predation', can be strongly disruptive of herbivore suppression within species-rich predator communities (Rosenheim, 1998). Disruptive effects of intraguild predation are most common when a very large predator species that feeds infrequently on the focal herbivore is paired with a vulnerable smaller predator that is the herbivore's key regulator (Ives *et al.*, 2005). A good example of such a scenario is provided by Finke and Denno (2004), who examined predator diversity effects among a community of spider, beetle and bug predators attacking planthoppers on *Spartina* marsh grass. They found that a relatively specialised predator of planthopper eggs, the bug *Tytthus*, was a particularly effective regulator of planthopper densities. Indeed, *Tytthus* alone was best able to suppress planthoppers and increase plant growth (Figure 2.3A). In contrast, more diverse predator communities including the large wolf spider *Hogna* caused planthopper numbers to explode and plants to be stunted (Figure 2.3B). This counterintuitive finding of more herbivores and smaller plants when more predators were present resulted from intraguild predation, with wolf spiders killing off *Tytthus* egg-predators and freeing the herbivore from its key mortality source (Finke and Denno, 2004). Biological control will be most effective when conservation and augmentation schemes focus on the predator species that interact well together, and discourage the predators that feed primarily upon other natural enemies.

NON-TROPHIC ENEMY BIODIVERSITY EFFECTS

It is obvious how plants might benefit when predators kill herbivores. But surprisingly, there is growing evidence that predators can protect plants even when the herbivore gets away! This is because herbivores (and other prey) have often evolved strategies to escape predator attack. Because these defences often come at the cost of lost feeding opportunities for the herbivore, the plant is nonetheless protected. Indeed, these non-trophic predator effects, mediated through changing herbivore behaviour, are often as beneficial to plants as

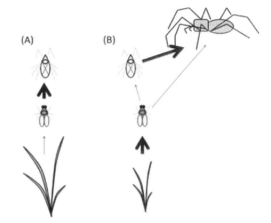

Figure 2.3 In salt marshes along the coast of New Jersey, USA (A) the specialist predatory bug *Tytthus* (top left) by itself strongly controls planthoppers (middle left), protecting *Spartina* grass (bottom left) from herbivory. (B) Surprisingly, a more diverse predator community that also includes the large wolf spider *Hogna* exhibits herbivore outbreaks and stunted plants, because the wolf spider feeds little on planthoppers but heavily on *Tytthus* (Finke and Denno, 2004). Arrows point from resource to consumer, and are scaled to reflect the magnitude of the interaction.

actual predation (Werner and Peacor, 2003; Schmitz *et al.*, 2004; Preisser *et al.*, 2005). For example, Nelson *et al.* (2004) found that *Nabis* predators of pea aphids substantially reduced aphid population growth, even when the predators' mouthparts were snipped such that actual predation was impossible (see also Beckerman *et al.*, 1997). This is because the predators disrupted aphid feeding as the aphids attempted to escape the unknowingly impotent predators. Similarly, predators can disrupt one another's feeding even in the absence of actual intraguild predation. A good example of this comes from the work of Moran and Hurd (1994), who found non-trophic disruption between praying mantid and wolf spider predators in old fields. In this system, praying mantids would capture and eat wolf spiders if given the chance. Apparently recognising this threat, wolf spiders simply 'ran away' from plots housing mantids. Thus, praying mantids acted primarily to scare off wolf spiders, rather than kill them, reducing spider impacts on prey through non-lethal means.

Non-trophic predator effects have rarely been examined as drivers of predator-diversity effects, but the limited work to date suggests that they could be important. For example, Steffan and Snyder (2010) recorded a positive predator-diversity effect that was mediated entirely through behavioural means. Their study system was a community of predators and parasitoids attacking diamondback moth caterpillars on *B. oleracea* plants. The caterpillars drop onto short silk threads when threatened by predators, and of course cannot feed while suspended in mid-air; a non-trophic effect that is readily observed and quantified (Plate 2.1). These authors independently manipulated predator species richness, predator trophic effects, and predator non-trophic effects, through a series of complicated experimental procedures. Trophic predator effects were isolated by removing caterpillars, at a rate typical of that at which caterpillars were killed by real predators, from one series of field cages that did not actually contain any predators (such that prey were 'killed' without ever being 'scared'). Non-trophic predator effects were isolated, in turn, by replacing caterpillars killed by predators in a separate set of cages (such that prey were 'scared' but 'resurrected' once killed). They found that diverse predator communities were more effective at protecting *B. oleracea* plants than any single enemy species, but that this diversity effect was entirely mediated by non-trophic means. This occurred because the two predator species most likely to induce silk-drop behaviour by caterpillars, a parasitoid and a lady beetle, spent more time on plants when within diverse rather than single-species predator communities. Thus, predator diversity reduced behavioural interference among predator species and so incited more frequent escape behaviours among the herbivore – a cascading series of non-trophic diversity effects (Steffan and Snyder, 2010). This study raises the possibility that many diversity cascades attributed to predator feeding may instead (or additionally) be attributable to behavioural effects of predators on their prey, as it is rare to attempt to isolate trophic from non-trophic effects in such studies.

THE UNDERAPPRECIATED ROLE OF THE SECOND COMPONENT OF BIODIVERSITY: EVENNESS

All of the predator biodiversity work discussed thus far equates greater biodiversity with an increase in the number of species present (i.e. the 'species richness' of the natural enemy community). However, biodiversity is generally envisioned to encompass a second component as well: the relative abundance of species (i.e. 'species evenness'). Communities with more equitably abundant species are generally considered to be healthiest, whereas communities dominated by one or just a few very abundant species are considered relatively unhealthy (Hillebrand *et al.*, 2008). However, relative to the large body of experimental studies recording natural enemy richness effects and their underlying mechanisms (Bruno and Cardinale, 2008), very few studies have examined the importance of having evenly abundant natural enemy communities (and almost no consideration has been given to the mechanisms underlying any enemy evenness effects). The small body of work to date suggests that natural enemy evenness could be as important as species richness for biological control. For example, Crowder *et al.* (2010) examined richness and evenness among a community of predatory insects and insect pathogens, both of which attack Colorado potato beetle (*Leptinotarsa decemlineata* (Say)) on potato (*Solanum tuberosum* (L.)) crops in Washington, USA. Potato crops in this region are grown using either conventional methods, where applications of broad-spectrum pesticides are frequent, or organic methods, where pesticide use is more selective. Presumably conventional management is most disruptive to communities of biological control agents. These authors found that predator and pathogen richness differed little among potato fields and was not affected by the choice of pest management approach. In stark contrast, evenness of both predators and pathogens was significantly higher in organic than conventional fields. This means that predator and pathogen communities in conventionally managed fields tended to be dominated by single enemy species, whereas in organic fields many enemy species were approximately equal in number. When even versus uneven predator communities were reproduced in field cages, it was found that the even enemy communities were far more effective at killing potato beetles and encouraging larger plants. Crowder *et al.* (2010) then re-examined a large number of comparisons of the natural enemy communities in organic versus conventional crops, from many different crops and world regions, and found that greater natural enemy evenness was a general feature of organically farmed fields. This suggests that organic agriculture often promotes the benefits of enemy evenness for pest control. The specific mechanisms at work in this case study were not examined, but the authors suggest that evenness and richness effects may be related. This could occur if the loss of species from a community is the end result of sensitive species becoming increasingly uncommon and 'weedy' species increasingly common as disturbance intensity increases. Thus, unevenness grows with increasing disruption, leading only gradually to the loss of species. This growing unevenness would, in turn, render the complementary ecological roles of vulnerable species increasingly unfilled even before extinction occurs (Crowder *et al.*, 2010). Unfortunately, these ideas have never been experimentally examined. Predator evenness likely reflects evenness within prey communities, although this relationship also is poorly understood. If predator evenness does indeed reflect greater evenness in prey and non-prey foods, then conservation methods that diversify the prey base may be the best way to harness the benefits of natural enemy evenness (see chapter 1 of this volume).

EFFECTS OF PREY DIVERSITY ON PREDATOR IMPACTS

One commonly attempted approach to conserving natural enemies is to manage agricultural fields to provide predators with non-pest prey (or other foods) in addition to target pests (Altieri and Whitcomb, 1979; Landis *et al.*, 2000; Wäckers, 2004). Here we provide a brief overview of these issues, which are covered in greater detail in chapters 3, 9 and 17 of this volume. This approach is successful in some cases, but is fraught with ecological risks (Heimpel and Jervis, 2005). This is partly because non-target prey can serve to either heighten or weaken predator impacts on a particular pest, depending on at least two factors: (1) the time scale under consideration, and (2) the relative preferences of predators for the pest versus non-pest prey. In the short term, very tasty non-pest prey may serve primarily to distract predators from feeding on a target pest, drawing away predator attacks and partially sheltering the target from biological control (e.g. Koss and Snyder, 2005). However, these effects may be reversed when a longer time scale is considered, as alternative prey supplement predator diets, building predator densities and eventually increasing overall predator impacts on pests (Holt, 1977; Holt and Kotler, 1987; Polis and Hurd, 1995; Chaneton and Bonsall, 2000). A good example of how short- versus long-term effects of alternative foods might differ is presented by Eubanks and Denno (2000a; 2000b). These authors examined predation of pest caterpillars on bean plants by predatory big-eyed bugs (*Geocoris punctipes* (Say)). Big-eyed bugs are omnivores that do some plant feeding in addition to attacking herbivores, and are particularly fond of bean pods. In simple feeding-choice trials in the laboratory, the presence of bean pods substantially reduced the likelihood that big-eyed bugs would attack caterpillars. This suggested that bean pods would only serve to distract big-eyed bugs from feeding on pests (Figure 2.4A). However, in an open-field trial the result was the opposite: predation rates on pests were highest where bean pods were present. This was because the presence of bean pods drew in substantially higher overall densities of big-eyed bugs, such that their net effect was greater even if per-capita predator impacts fell (Figure 2.4B; Eubanks and Denno, 2000a; 2000b).

Likewise, the relative preferences of predators for pest versus non-pest foods can determine their benefit or harm to biological control. For example, Halaj and Wise (2002) attempted to build populations of predatory ground beetles and wolf spiders in cucumber plots by adding supplemental detritus (composted straw and horse manure) to the system. The detritus provided additional food for springtails (Collembola) and other detritivores, which are important non-pest prey for the predators. The hope was that the extra detritivores would fill out predator diets and thus build predator densities, and indeed this was the case: densities of both springtails and wolf spiders increased approximately threefold in plots receiving supplemental detritus. However, despite this dramatic increase in predator densities, no increase was recorded in predation on the cucumber beetles that were the target pests. Apparently, the predators preferred detritivores to cucumber beetles, and never switched their attention to attacking the target pest. This problem can be overcome by providing non-pest prey that conveniently disappear around the time that the pest colonises the crop. For example, Settle *et al.* (1996) found a dramatic improvement in biocontrol of rice pests when early-season insecticide applications were avoided. This is because cancelling these sprays allowed detritivores to build up in rice plots early in the growing cycle, with these non-pest prey providing early-season food for spiders and other generalists. Detritivore densities declined naturally just as herbivorous pests began to colonise the rice, freeing the predators to concentrate their attacks on the suddenly abundant target pests (Settle *et al.*,

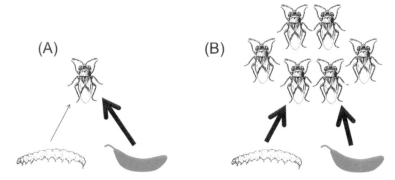

Figure 2.4 In bean fields in Maryland, USA, omnivorous bugs (*Geocoris punctipes* (Say), top left) feed on both pest caterpillars (bottom left) and bean pods (second from bottom left). (A) Individual *Geocoris* reduce their feeding on caterpillars when beans are available, reducing their per-capita contribution to biological control (B). However, bean pods attract large numbers of *Geocoris*, such that their net impact on biological control is a beneficial heightening of overall caterpillar predation (Eubanks and Denno, 2000a; 2000b). Arrows point from resource to consumer, and are scaled to reflect the magnitude of the interaction.

1996). The application of this and related approaches in rice pest management is explored in chapter 13 of this volume.

A perhaps understudied, but potentially disruptive, effect of non-pest foods is that these resources may also be exploited by intraguild predators. For example, Jonsson *et al.* (2009) found that floral resources provided as supplemental foods for lacewings were also heavily exploited by a specialist parasitoid of the lacewings. For this reason, any benefit to lacewings of the floral resources was more than counteracted by harm due to greater lacewing-parasitoid attack (Jonsson *et al.*, 2009). Similarly, Prasad and Snyder (2006) found that non-target aphid prey disrupted biocontrol of root maggot pests by ground beetles. This was because the aphids incited greater foraging activity by a dangerous intraguild predator, the relatively large ground beetle *Pterostichus melanarius* (F.); when the large beetles were foraging the smaller ground beetles, which were the most effective root maggot predators, hid in their burrows to avoid falling victim to *P. melanarius* and so fed far less often on root maggot pests (Prasad and Snyder, 2006).

BIODIVERSITY AND STABILITY

Obviously, in a pest management context it is important to kill a large number of pest individuals, thus high average control rates (as discussed above) are a crucial management objective, and benefit to biological control. However, in addition to high average control rates, stability of control is also important, and before this can be discussed, we must define what we mean by 'stability'. Most people have an intuitive understanding of what stability means (i.e. something roughly synonymous with 'unchanging'). However, this broad concept can have very different meanings when looking at ecological communities. For example, Pimm (1984) identified three key definitions of stability: resilience (how fast a system returns to equilibrium following perturbation), resistance (the extent to which a system is altered by a perturbation), and variability (variance in the system through time and/or space, often measured by the coefficient of variation of a population density or amplitude of fluctuations in ecosystem properties). All of these have obvious implications for pest management, particularly when disturbances such as crop harvest or press-perturbations such as climate change can affect pest and enemy dynamics. Most important from a pest control perspective, though, is variability in pest abundance. Even short periods of high pest abundance can cause significant damage to crops (e.g. feeding damage may reduce food quality, or virus transmission by pest vectors can continue to damage crops after the vector is eliminated). Thus, it is often necessary to maintain pests consistently below a damage threshold, preventing outbreaks (high-amplitude population fluctuations), and making stability a key aspect of pest control.

A review by Ives and Carpenter (2007) extended the ideas of Pimm, by differentiating between different kinds of perturbations and adding several more definitions of stability to this list, though these tended to be based on aggregate measures of ecosystems, rather than densities of individual populations. For example, the number of alternative stable states in which a system can exist is one definition of stability (more stable systems being those with fewer states), as is the predisposition to switch between such states (i.e. stable systems recover quickly after disturbance (Holling's resilience (Holling, 1973)) and do not switch states easily). Alternatively, in a dynamic context, chaotic non-point attractors can be seen as unstable, and resistance of the system to the addition (invasion) or removal (extinction) of species has also received considerable attention (McCann, 2000; Hooper et al., 2005; Ives and Carpenter, 2007). From a pest control perspective, the latter definition of stability is the most worthy of attention (invasion by new pests or extinction of natural enemies), though chaotic non-point attractors may lead to high amplitude fluctuations in pest abundance, as discussed above.

Not surprisingly, given these rather different definitions of stability, there has been considerable contention over whether or how biodiversity may influence the stability of ecological systems. In fact, the question of whether diversity/complexity begets stability has been one of the major research questions in community ecology over the past few decades (e.g. May, 1973; McNaughton, 1978; Givnish, 1994; Hanski, 1997; McGrady-Steed et al., 1997; Hughes and Roughgarden, 2000; McCann, 2000; Worm and Duffy 2003; Hooper et al., 2005, Ives and Carpenter, 2007). Nevertheless, the evolution of thinking that has accompanied this debate has not only shaped our current answers to the question, but even the nature of the questions being asked.

The classic paper on ecological succession by Clements (1936) defined how communities change over time towards a stable climax. Along this line, ecologists such as Charles Elton and Eugene Odum argued that diverse, complex communities such as forests were stable, whereas simplified ecosystems such as agricultural fields were characterised by extreme fluctuations in densities of organisms, and frequent invasions by new species (Elton, 1958; Odum, 1969). The idea that diverse communities are, other things (e.g. propagule

pressure) being equal, more difficult to invade, is now relatively well supported (Hooper et al., 2005), and may be relevant for reducing pest invasions, following similar reasoning to the resource concentration hypothesis (Root, 1973) discussed above. Further, it is also widely accepted that diverse systems tend to show less extreme fluctuations in ecosystem functioning than do simplified systems, through time or in response to environmental perturbations (Hooper et al., 2005). By occupying distinct temporal niches or showing different responses to perturbation, a statistical averaging effect can reduce temporal variance of ecosystem functions (Petersen et al., 1998; Yachi and Loreau, 1999; Elmqvist et al., 2003) such as biocontrol. The strength of this buffering or 'insurance' effect increases with asynchronicity in the responses of individual predator species to environmental fluctuations, and also depends on the specific nature of their responses (i.e. response diversity; Yachi and Loreau, 1999; Elmqvist et al., 2003).

These insurance effects are basically an extension of the temporal complementarity effects discussed above, and comprise one of the major potential benefits of biodiversity to sustained biological control. Similarly, in patchy environments biodiversity can provide spatial insurance effects (Loreau et al., 2003), whereby species that occupy multiple habitats can move between patches, controlling pests in certain crops and/or at certain times (e.g. following disturbance or harvest of adjacent crops; Lundberg and Moberg, 2003), even though they may be less important at other times (Srivastava and Vellend, 2005). Furthermore, turnover of species among habitats (i.e. 'beta diversity') can add to the spatial insurance effect of having different biological control agents in different habitats (Tscharntke et al., 2007).

FOOD WEBS: PUTTING TROPHIC INTERACTIONS IN THEIR PLACE

Much of the research on biological control has necessarily focused on a target pest species and one or a few species of natural enemies. However, even simplified agricultural systems contain numerous herbivore and enemy species, and each of these species comprises one element (node) within a network of feeding interactions ('food web'). Current food web research has been enriched by the addition of concepts from other types of networks (e.g. social, computer, business networks), and it has long been known that the structure of food webs makes them more than the sum of their parts (Bascompte, 2009). In particular, there has been considerable research (and debate) examining whether complex (i.e. highly connected) food webs are more stable (e.g. MacArthur, 1955; May, 1972; McCann, 2000; Montoya *et al.*, 2006; Thebault and Fontaine, 2010), and as with diversity discussed above, the outcome can depend considerably on the definition of stability employed.

Food web structure can be closely tied to biodiversity, with certain structural attributes such as 'connectance' (the proportion of potential interactions between species that are actually realised) being generally related to the number of species in the food web (McCann 2000; de Ruiter *et al.*, 2005a; 2005b; Banašek-Richter *et al.*, 2009). Thus, agricultural management may affect food web structure indirectly by altering local biodiversity. In addition, management intensity may also directly alter food web structure beyond what might be expected due to changing biodiversity. Tylianakis *et al.* (2007) examined quantitative parasitoid–host food webs from 48 sites comprising a gradient of land-use intensification in coastal Ecuador. They found that, although species diversity did not change significantly, there was a strong change in the structure of the food webs, which became highly dominated by one or two interactions in the most highly modified habitats (rice and pasture). Food webs are static representations of a system, compiled by sampling interactions through space and time, and it has been suggested that this may obscure the true dynamic variability of the system (e.g. de Ruiter *et al.*, 2005a; 2005b). By breaking down the dataset from Ecuador into its individual spatial and temporal subsamples, Laliberté and Tylianakis (2010) were able to show that not only did land-use intensification alter the average structure of parasitoid–host food webs, but it also homogenised them (reduced their structural variability) in space and time. This occurred because parasitoids could more easily find hosts in more open, simplified habitats, so a greater proportion of the potential hosts were actually utilised. Therefore, there is likely to be some context-dependency in the relationship between diversity and food web complexity, and this relationship may be moderated by land-use intensity or by the physical structure of the crop itself. In terms of biological control, this suggests that any influences (functional or stabilising) of food web structure may be highly context-dependent, and it is too early to give any kind of prescriptive advice for application to biocontrol.

Despite a long history of theoretical research showing that food web structure can affect system stability, and recent findings that human changes to the environment may alter this structure, the link between web structure and functions such as biological control remains less clear (Tylianakis *et al.*, 2010). However, attributes of food web architecture that promote resistance to changing abundance of species (e.g MacArthur, 1955) should make the system resistant to outbreaks of a pest species, and this will be an important avenue for future research from a biocontrol perspective. Furthermore, if, as outlined above, diverse natural enemy assemblages provide greater levels of prey suppression, then food webs with a large average number of predators attacking each prey species should be associated with high attack rates. In food web terminology, the average number of predator species per prey species is called 'vulnerability', and Tylianakis *et al.* (2007) showed that high food web vulnerability was in fact associated with high parasitism rates in their sites. In contrast, Macfadyen *et al.* (2009) simulated the invasion of a new pest in organic and conventional farms and found no differences in attack rates of the introduced leafminer 'pest', despite differences in quantitative food web vulnerability. Thus, despite the potential importance of food web structure for ecosystem properties such as stability, the importance of this structure for biological control requires further investigation.

CONCLUSIONS

In summary, there is growing evidence that greater species richness and evenness among natural enemies improves pest control (Crowder *et al.*, 2010; Finke and Snyder, 2010). In some instances this leads to 'predator diversity cascades', where the stronger pest suppression that diverse enemy communities engender leads to bigger plants and higher yields (Cardinale *et al.*, 2003; Snyder *et al.*, 2006; Ramirez and Snyder, 2009; Crowder *et al.*, 2010; Steffan and Snyder, 2010). Therefore, natural pest control can generally be assumed to be strengthened when natural enemy diversity is encouraged through the provisioning of a diverse prey base, and a reduction in harmful insecticide sprays (e.g. Landis *et al.*, 2000). Unfortunately, intensification of agriculture threatens natural enemy biodiversity and its many benefits for pest control

(Tylianakis *et al.*, 2007). Harnessing the benefits of biodiversity for biocontrol will therefore require radical transformation of modern farming practices.

REFERENCES

Altieri, M.A. and Whitcomb, W.H. (1979) The potential use of weeds in the manipulation of beneficial insects. *Horticultural Science*, 14, 12–18.

Banašek-Richter, C., Bersier, L.F., Cattin, M.F. *et al.* (2009) Complexity in quantitative food webs. *Ecology*, 90, 1470–1477.

Bascompte, J. (2009) Disentangling the web of life. *Science*, 325, 416–419.

Beckerman, A.P., Uriarte, M. and Schmitz, O.J. (1997) Experimental evidence for a behavior-mediated trophic cascade in a terrestrial food chain. *Proceedings of the National Academy of Sciences USA*, 94, 10735–10738.

Bruno, J.F. and Cardinale, B.J. (2008) Cascading effects of predator richness. *Frontiers in Ecology and the Environment*, 6, 539–546.

Cardinale, B.J., Harvey, C.T., Gross, K. and Ives, A.R. (2003) Biodiversity and biocontrol: emergent impacts of a multi-enemy assemblage on pest suppression and crop yield in an agroecosystem. *Ecology Letters*, 6, 857–865.

Cardinale B.J., Srivastava D.S., Duffy J.E. *et al.* (2006) Effects of biodiversity on the functioning of trophic groups and ecosystems. *Nature*, 443, 989–992.

Casula, P., Wilby, A. and Thomas, M.B. (2006) Understanding biodiversity effects on prey in multi-enemy systems. *Ecology Letters*, 9, 995–1004.

Chaneton, E.J. and Bonsall, M.B. (2000) Enemy-mediated apparent competition: empirical patterns and the evidence. *Oikos*, 88, 380–394.

Clements, F.E. (1936) Nature and structure of the climax. *The Journal of Ecology*, 24, 252–284.

Crowder, D.W., Northfield, T.D., Strand, M.R. and Snyder, W.E. (2010) Organic agriculture promotes evenness and natural pest control. *Nature*, 466, 109–112.

De Ruiter, P.C., Wolters, V. and Moore, J.C. (eds) (2005a) *Dynamic food webs: multispecies assemblages, ecosystem development and environmental change*, Elsevier, Amsterdam.

De Ruiter, P.C., Wolters, V., Moore, J.C. and Winemiller, K.O. (2005b) Food web ecology: playing Jenga and beyond. *Science*, 309, 68–71.

Denoth, M., Frid, L. and Myers, J.H. (2002) Multiple agents in biological control: improving the odds? *Biological Control*, 24, 20–30.

Duffy, J.E. (2003) Biodiversity loss, trophic skew and ecosystem functioning. *Ecology Letters*, 6, 680–687.

Elmqvist, T., Folke, C., Nystrom, M. *et al.* (2003) Response diversity, ecosystem change, and resilience. *Frontiers in Ecology and the Environment*, 1, 488–494.

Elton, C.S. (1958) *Ecology of invasions by animals and plants*, Chapman & Hall, London.

Eubanks, M.D. and Denno, R.F. (2000a) Host plants mediate ominivore–herbivore interactions and influence prey suppression. *Ecology*, 81, 936–947.

Eubanks, M.D. and Denno, R.F. (2000b) Health food versus fast food: the effects of prey quality and mobility on prey selection by a generalist predator and indirect interactions among prey species. *Ecological Entomology*, 25, 140–146.

Finke, D.L. and Denno, R.F. (2004) Predator diversity dampens trophic cascades. *Nature*, 429, 407–410.

Finke, D.L. and Snyder, W.E. (2008) Niche partitioning increases resource exploitation by diverse communities. *Science*, 321, 1488–1490.

Finke, D.L. and Snyder, W.E. (2010) Conserving the benefits of predator biodiversity. *Biological Conservation*, 143, 2260–2269.

Givnish, T.J. (1994) Does diversity beget stability? *Nature*, 371, 113–114.

Hairston, N.G. Jr. and Hairston, N.G. Sr. (1993) Cause–effect relationships in energy flow, trophic structure, and interspecies interactions. *American Naturalist*, 142, 379–411.

Hairston, N.G. Jr. and Hairston, N.G. Sr. (1997) Does food-web complexity eliminate trophic-level dynamics? *American Naturalist*, 149, 1001–1007.

Hairston, N.G., Smith, F.E. and Slobodkin, L.B. (1960) Community structure, population control and competition. *American Naturalist*, 94, 421–425.

Halaj, J. and Wise, D.H. (2001) Terrestrial trophic cascades: how much do they trickle? *American Naturalist*, 157, 262–281.

Halaj, J. and Wise, D.H. (2002) Impact of a detrital subsidy on trophic cascades in a terrestrial grazing food web. *Ecology*, 83, 3141–3151.

Hanski, I. (1997) Be diverse, be predictable. *Nature*, 390, 440–441.

Hassell, M.P. (1980) Foraging strategies, population models and biological control – a case-study. *Journal of Animal Ecology*, 49, 603–628.

Hassell, M.P. and May, R.M. (1986) Generalist and specialist natural enemies in insect predator prey interactions. *Journal of Animal Ecology*, 55, 923–940.

Hector, A., Schmid, B., Beierkuhnlein, C. *et al.* (1999) Plant diversity and productivity experiments in European grasslands. *Science*, 286, 1123–1127.

Heimpel, G.E. and Jervis, M.A. (2005) Does floral nectar improve biological control by parasitoids? in *Plant-provided food and herbivore–carnivore interactions* (eds F.L. Wacker, P.C.J. van Rijn and J. Bruin), Cambridge University Press, New York, pp. 267–304.

Hillebrand, H., Bennett, D.M. and Cadotte, M.W. (2008) Consequences of dominance: a review of evenness effects on local and regional ecosystem processes. *Ecology*, 89, 1510–1520.

Holling, C.S. (1973) Resilience and stability of ecological systems. *Annual Review of Ecology and Systematics*, 4, 1–23.

Holt, R.D. (1977) Predation, apparent competition, and the structure of prey communities. *Theoretical Population Biology*, 12, 197–229.

Holt, R.D. and Kotler, B.P. (1987) Short-term apparent competition. *American Naturalist*, 130, 412–430.

Hooper, D.U. and Vitousek, P.M. (1997) The effects of plant composition and diversity on ecosystem processes. *Science*, 277, 1302–1305.

Hooper, D.U., Chapin, F.S., Ewel, J.J. *et al.* (2005) Effects of biodiversity on ecosystem functioning: A consensus of current knowledge. *Ecological Monographs*, 75, 3–35.

Hughes, J.B. and Roughgarden, J. (2000) Species diversity and biomass stability. *American Naturalist*, 155, 618–627.

Ives, A.R. and Carpenter, S.R. (2007) Stability and diversity of ecosystems. *Science*, 317, 58–62.

Ives, A.R., Cardinale, B.J. and Snyder, W.E. (2005) A synthesis of subdisciplines: predator–prey interactions, and biodiversity and ecosystem functioning. *Ecology Letters*, 8, 102–116.

Jonsson, M., Wratten, S.D., Robinson, K.A. and Sam, S.A. (2009) The impact of floral resources and omnivory on a four trophic level food web. *Bulletin of Entomological Research*, 99, 275–285.

Koss, A.M. and Snyder, W.E. (2005) Alternative prey disrupt biocontrol by a guild of generalist predators. *Biological Control*, 32, 243–251.

Laliberté, E. and Tylianakis, J.M. (2010) Deforestation homogenizes tropical parasitoid–host networks. *Ecology*, 91, 1740–1747.

Landis, D.A., Wratten, S.D. and Gurr, G.M. (2000) Habitat management to conserve natural enemies of arthropod pests in agriculture. *Annual Review of Entomology*, 45, 175–201.

Loreau, M., Mouquet, N. and Gonzalez, A. (2003) Biodiversity as spatial insurance in heterogeneous landscapes. *Proceedings of the National Academy of Sciences USA*, 100, 12765–12770.

Losey, J.E. and Denno, R.F. (1998) Positive predator–predator interactions: enhanced predation rates and synergistic suppression of aphid populations. *Ecology*, 79, 2143–2152.

Lundberg, J. and Moberg, F. (2003) Mobile link organisms and ecosystem functioning: implications for ecosystem resilience and management. *Ecosystems*, 6, 87–98.

MacArthur, R. (1955) Fluctuations of animal populations, and a measure of community stability. *Ecology*, 36, 533–536.

Macfadyen, S., Gibson, R., Polaszek, A. *et al.* (2009) Do differences in food web structure between organic and conventional farms affect the ecosystem service of pest control? *Ecology Letters*, 12, 229–238.

Matson, P.A., Parton, W.J., Power, A.G. and Swift, M.J. (1997) Agricultural intensification and ecosystem properties. *Science*, 277, 504–509.

May, R.M. (1972) Will a large complex system be stable? *Nature*, 238, 413–414.

May, R.M. (1973) *Stability and complexity in model ecosystems*, Princeton University Press, Princeton.

McCann, K.S. (2000) The diversity–stability debate. *Nature*, 405, 228–233.

McGrady-Steed, J., Harris, P.M. and Morin, P.J. (1997) Biodiversity regulates ecosystem predictability. *Nature*, 390, 162–165.

McNaughton, S.J. (1978) Stability and diversity of ecological communities. *Nature*, 274, 251–253.

Michaud, J.P. (2002) Classical biological control: a critical review of recent programs against citrus pests in Florida. *Annals of the Entomological Society of America*, 95, 531–540.

Montoya, J.M., Pimm, S.L. and Sole, R.V. (2006) Ecological networks and their fragility. *Nature*, 442, 259–264.

Moran, M.D. and Hurd, L.E. (1994) Short-term responses to elevated predator densities: Noncompetitive intraguild interactions and behaviour. *Oecologia*, 98, 269–273.

Musser, F.R. and Shelton, A.M. (2003) Factors altering the temporal and within-plant distribution of coccinellids in corn and their impact on potential intra-guild predation. *Environmental Entomology*, 32, 575–583.

Naeem, S., Loreau, M. and Inchausti, P. (2002) Biodiversity and ecosystem functioning: the emergence of a synthetic ecological framework, in *Biodiversity and ecosystem functioning: synthesis and perspectives* (eds M. Loreau, S. Naeem and P. Inchausti), Oxford University Press, New York, pp. 3–11.

Naeem, S., Thompson, L.J., Lawler, S.P., Lawton, J.H. and Woodfin, R.M. (1995) Empirical evidence that declining species diversity may alter the performance of terrestrial ecosystems. *Philosophical Transactions of the Royal Society B*, 347, 249–262.

Naylor, R.L. and Ehrlich, P.R. (1997) Natural pest control services and agriculture, in *Nature's Services: societal dependence on natural ecosystems* (ed. G.C. Daily), Island Press, Washington, DC, pp. 151–174.

Nelson, E.H., Matthews, C.E. and Rosenheim, J.A. (2004) Predator reduce prey population growth by inducing changes in prey behavior. *Ecology*, 85, 1853–1858.

Neuenschwander, P., Hagen, K.S. and Smith, R.F. (1975) Predation of aphids in California's alfalfa fields. *Hilgardia*, 43, 53–78.

Northfield, T.D., Snyder, G.B., Ives, A.R. and Snyder, W.E. (2010) Niche saturation reveals resource partitioning among consumers. *Ecology Letters*, 13, 338–348.

Odum, E.P. (1969) The strategy of ecosystem development. *Science*, 164, 262–269.

Oksanen, L., Fretwell, S.D., Arruda, J. and Niemala, P. (1981) Exploitation ecosystems in gradients of primary productivity. *American Naturalist*, 118, 240–261.

Pedersen, B.S. and Mills, N.J. (2004) Single vs. multiple introduction in biological control: the roles of parasitoid efficiency, antagonism and niche overlap. *Journal of Applied Ecology*, 41, 973–984.

Petersen, G., Allen, C. and Holling, C. (1998) Ecological resilience, biodiversity, and scale. *Ecosystems*, 1, 6–18.

Pfannenstiel, R.S. and Yeargan, K.V. (2002) Identification and diel activity patterns of predators attacking *Helicoverpa zea* (Lepidoptera: Noctuidae) eggs in soybean and sweet corn. *Environmental Entomology*, 31, 232–241.

Pimentel, D. (1961) Species diversity and insect population outbreaks. *Annals of the Entomological Society of America*, 54, 76–86.

Pimm, S.L. (1984) The complexity and stability of ecosystems. *Nature*, 307, 321–326.

Polis, G.A. (1991) Complex trophic interactions in deserts: an empirical critique of food-web theory. *American Naturalist*, 138, 123–155.

Polis, G.A. and Hurd, S.D. (1995) Extraordinarily high spider densities on islands: flow of energy from the marine to terrestrial food webs and the absence of predation. *Proceedings of the National Academy of Sciences USA*, 92, 4382–4386.

Polis, G.A. and Strong, D.R. (1996) Food web complexity and community dynamics. *American Naturalist*, 147, 813–846.

Polis, G.A., Myers, C.A. and Holt, R.D. (1989) The ecology and evolution of intraguild predation: potential competitors that eat each other. *Annual Review of Ecology and Systematics*, 20, 297–330.

Prasad, R.P. and Snyder, W.E. (2006) Polyphagy complicates conservation biological control that targets generalist predators. *Journal of Applied Ecology*, 43, 343–352.

Preisser, E.L., Bolnick, D.I. and Benard, M.F. (2005) Scared to death? The effects of intimidation and consumption in predator–prey interactions. *Ecology*, 86, 501–509.

Ramirez, R.A. and Snyder, W.E. (2009) Scared sick? Predator–pathogen facilitation enhances exploitation of a shared resource. *Ecology*, 90, 2832–2839.

Ramirez, R.A., Crowder, D.W., Snyder, G.B., Strand, M.R. and Snyder, W.E. (2010) Antipredator behavior of Colorado potato beetle larvae differs by instar and attacking predator. *Biological Control*, 53, 230–237.

Root, R.B. (1973) Organization of a plant–arthropod association in simple and diverse habitats: the fauna of collards. *Ecological Monographs*, 43, 95–124.

Rosenheim, J.A. (1998) Higher-order predators and the regulation of insect herbivore populations. *Annual Review of Entomology*, 43, 421–447.

Rosenheim, J.A., Glik, T.E., Goeriz, R.E. and Ramert, B. (2004) Linking a predator's foraging behavior with its effects on herbivore population suppression. *Ecology*, 85, 3362–3372.

Schmitz, O.J. (2007) Predator diversity and trophic interactions. *Ecology*, 88, 2415–2426.

Schmitz, O.J., Hambäck, P.A. and Beckerman, A.P. (2000) Trophic cascades in terrestrial systems: A review of the effects of carnivore removals on plants. *American Naturalist*, 155, 141–153.

Schmitz, O.J., Krivan, V. and Ovadia, O. (2004) Trophic cascades: the primacy of trait-mediated indirect interactions. *Ecology Letters*, 7, 153–163.

Settle, W.H., Ariawan, H., Astuti, E.T. *et al.* (1996) Managing tropical rice pests through conservation of generalist natural enemies and alternative prey. *Ecology*, 77, 1975–1988.

Shurin, J.B., Borer, E.T., Seabloom, E.W. *et al.* (2002) A cross-ecosystem comparison of the strength of trophic cascades. *Ecology Letters*, 5, 785–791.

Sih, A., Englund, G. and Wooster, D. (1998) Emergent impacts of multiple predators on prey. *Trends in Ecology and Evolution*, 13, 350–355.

Snyder, W.E. (2009) Coccinellids in diverse communities: which niche fits? *Biological Control*, 51, 323–335.

Snyder, W.E. and Ives, A.R. (2003) Interactions between specialist and generalist natural enemies: Parasitoids, predators, and pea aphid biocontrol. *Ecology*, 84, 91–107.

Snyder, W.E., Chang, G.C. and Prasad, R.P. (2005) Conservation biological control: biodiversity influences the effectiveness of predators, in *Ecology of Predator–Prey Interactions* (eds P. Barbosa and I. Castellanos), Oxford University Press, London, pp. 324–343.

Snyder, W.E., Snyder, G.B., Finke, D.L. and Straub, C.S. (2006) Predator biodiversity strengthens herbivore suppression. *Ecology Letters*, 9, 789–796.

Sokol-Hessner, L. and Schmitz, O.J. (2002) Aggregate effects of multiple predator species on a shared prey. *Ecology*, 83, 2367–2372.

Srivastava, D.S. and Vellend, M. (2005) Biodiversity–ecosystem function research: is it relevant to conservation? *Annual Review of Ecology, Evolution and Systematics*, 36, 267–294.

Steffan, S.A. and Snyder, W.E. (2010) Cascading diversity effects transmitted exclusively by behavioral interactions. *Ecology*, 91, 2242–2252.

Stiling P. and Cornelissen T. (2005) What makes a successful biocontrol agent? A meta-analysis of biological control agent performance. *Biological Control*, 34, 236–246.

Straub, C.S. and Snyder, W.E. (2006) Species identity dominates the relationship between predator biodiversity and herbivore suppression. *Ecology*, 87, 277–282.

Straub, C.S. and Snyder, W.E. (2008) Increasing enemy biodiversity strengthens herbivore suppression on two plant species. *Ecology*, 89, 1605–1615.

Straub, C.S., Finke, D.L. and Snyder, W.E. (2008) Are the conservation of natural enemy biodiversity and biological control compatible goals? *Biological Control*, 45, 225–237.

Strong, D.R. (1992) Are trophic cascades all wet? Differentiation and donor-control in speciose ecosystems. *Ecology*, 73, 747–754.

Symondson, W.O.C., Sunderland, K.D. and Greenstone, M.H. (2002) Can generalist predators be effective biocontrol agents? *Annual Review of Entomology*, 47, 561–594.

Tamaki, G. and Weeks, R.E. (1972) Efficiency of three predators, *Geocoris bullatus, Nabis americoferus,* and *Coccinella transversoguttata*, used alone or in combination against three insect prey species, *Myzus persicae, Ceramica picta*, and *Mamestra configurata*, in a greenhouse study. *Environmental Entomology*, 1, 258–263.

Thebault, E. and Fontaine, C. (2010) Stability of ecological communities and the architecture of mutualistic and trophic networks. *Science*, 329, 853–856.

Tilman, D., Knops, J., Wedin, D., Reich, P., Ritchie, M. and Siemann, E. (1997) The influence of functional diversity and composition on ecosystem processes. *Science*, 277, 1300–1302.

Tscharntke, T., Bommarco, R., Clough, Y. *et al.* (2007) Conservation biological control and enemy diversity on a landscape scale. *Biological Control*, 43, 294–309.

Tylianakis, J.M. and Romo, C. (2010) Natural enemy diversity and biological control: making sense of the context-dependency. *Basic and Applied Ecology*, 11, 657–668.

Tylianakis, J.M., Didham, R.K. and Wratten, S.D. (2004) Improved fitness of aphid parasitoids receiving resource subsidies. *Ecology*, 85, 658–666.

Tylianakis, J.M., Klein, A.M. and Tscharntke, T. (2005) Spatio-temporal variation in the effects of a tropical habitat gradient on Hymenoptera diversity. *Ecology*, 86, 3296–3302.

Tylianakis, J.M., Tscharntke, T. and Lewis, O.T. (2007) Habitat modification alters the structure of tropical host–parasitoid food webs. *Nature*, 455, 202–205.

Tylianakis, J.M., Rand, T.A., Kahmen, A., Klein, A.M., Buchmann, N., Perner, J. and Tscharntke, T. (2008) Resource heterogeneity moderates the biodiversity-function relationship in real world ecosystems. *PLoS Biology*, 6, 947–956.

Tylianakis, J.M., Laliberté, E., Nielsen, A. and Bascompte, J. (2010) Conservation of species interaction networks. *Biological Conservation*, 143, 2270–2279.

Van Emden, H.F. and Williams, G.F. (1974) Insect stability and diversity in agro-ecosystems. *Annual Review of Entomology*, 19, 455–475.

Wäckers, F.L. (2004) Assessing the suitability of flowering herbs as parasitoid food sources: flower attractiveness and nectar accessibility. *Biological Control*, 29, 307–314.

Werner, E.E. and Peacor, S.D. (2003) A review of trait-mediated indirect interactions in ecological communities. *Ecology*, 84, 1083–1100.

Wilby, A. and Thomas, M.B. (2002) Natural enemy diversity and pest control: patterns of pest emergence with agricultural intensification. *Ecology Letters*, 5, 353–360.

Wilby, A., Villareal, S.C., Lan, L.P., Heong, K.L. and Thomas, M.B. (2005) Functional benefits of predator species diversity depend on prey identity. *Ecological Entomology*, 30, 497–501.

Worm, B. and Duffy, J.E. (2003) Biodiversity, productivity and stability in real food webs. *Trends in Ecology and Evolution*, 18, 628–632.

Yachi, S. and Loreau, M. (1999) Biodiversity and ecosystem productivity in a fluctuating environment: the insurance hypothesis. *Proceedings of the National Academy of Sciences of the United States of America*, 96, 1463–1468.

Chapter 3

THE ROLE OF GENERALIST PREDATORS IN TERRESTRIAL FOOD WEBS: LESSONS FOR AGRICULTURAL PEST MANAGEMENT

K.D. Welch, R.S. Pfannenstiel and J.D. Harwood

Biodiversity and Insect Pests: Key Issues for Sustainable Management, First Edition. Edited by Geoff M. Gurr, Steve D. Wratten, William E. Snyder, Donna M.Y. Read.

INTRODUCTION

Agroecosystems, although subjected to monocultural practices that typically suppress arthropod diversity, are surprisingly rich in invertebrate fauna. Temperate agricultural production systems, for example, contain at least 1,500–3,000 species (Nentwig, 1987) and a very large number of these can be given the somewhat arbitrary classification of 'generalist predators'. Thus, these rather homogeneous crop environments offer resources to arthropod communities that not only allow survival, but foster growth and development of individual populations. Even within apparently uniform landscapes closer inspection reveals high levels of temporal heterogeneity, not just in terms of structural complexity of the crop or habitat but also in terms of food availability. As will become apparent later in this chapter, this is not solely limited to invertebrate 'prey' but is rather more diverse than traditionally thought. Furthermore, the predisposition of many predators to remain in place despite limited resource availability, driven by the inherent risks associated with site or habitat abandonment, facilitates their capacity for pest suppression, potentially impacting prey populations during colonisation ('early-season control' is discussed below). There is an extensive literature available that focuses on aspects of conservation biological control through habitat management approaches (reviewed by Landis et al., 2000), but this topic is beyond the scope of this chapter. Such approaches to agricultural management have certainly proven successful in a variety of agro-environment schemes, especially where government subsidies provide economic compensation for areas set aside from crop production to conservation.

It therefore becomes increasingly evident that generalist predators, with their diverse foraging habits, rapid colonisation ability, high levels of site tenacity and, in many cases, voracious appetite for pest species are at the forefront of many conservation biological control schemes. This is demonstrated convincingly in the review by Symondson et al. (2002), which highlighted how predators are frequently among the most important mortality factors for arthropod pests. Indeed, in 75% of studies, the presence of generalist predators resulted in significant reduction of pest populations (Symondson et al., 2002), although teasing apart economically viable reductions from statistically significant ones poses many additional complications. This chapter therefore focuses on the roles generalist predators play in terrestrial food webs and emphasises their capacity for biological control.

WHAT IS A PREDATOR?

This question may seem obtuse; yet an accurate, unambiguous definition of 'predator' is surprisingly difficult to find. At its most basic, a predator can be defined as 'an animal that naturally preys on others' (Oxford English Dictionary, 2008). However, upon closer examination, a large number of complications can significantly blur this definition. For example, the distinction between 'predator' and 'parasite' is not always clear, and therefore somewhat arbitrary. The natural world reveals a gradient of organisms ranging from consummate parasites to consummate predators, with a large portion of the intermediate types often being classified as 'parasitoids'. A parasitoid is an organism that behaviourally and developmentally resembles a parasite – its larval form develops on or inside a single host organism – but ecologically resembles a predator as the host organism is consumed and killed by the parasitoid. Table 3.1 provides several char-

Table 3.1 A comparison of various attributes of different types of natural enemies.

	Parasite	Parasitoid	Predator
Diet range	Usually stenophagous or monophagous	Usually stenophagous or monophagous	Usually oligophagous or polyphagous
Development	On or inside a single host	On or inside a single host	Free-living
Number of pests attacked	One	One	Multiple
Effect on host/prey	Usually does not kill host	Kills host	Kills prey
Body size	Very small	Smaller than host	Often larger than prey
Foraging	Entire life cycle in a host	Parent usually locates the offspring's host	Immatures usually forage for their own prey

acteristics that can be useful in differentiating between these three categories of organisms. However, even with recognition of parasitoids as an intermediate category, the distinctions between categories are still not clear. For example, organisms that develop inside an egg-mass, rather than inside a single host organism, and consume all the eggs within the egg-mass (e.g. mantisflies, Neuroptera: Mantispidae), bear similarities to both predators and parasitoids. Additionally, the larvae of spider wasps (Hymenoptera: Pompilidae) develop on a single host like a parasitoid; however, the 'host' is captured and killed in a predatory fashion by the adult wasp, and provisioned to the larva (Endo and Endo, 1994). Organisms such as these are difficult to classify definitively as 'predators' or 'parasitoids'. This chapter focuses primarily on predators, which are generally defined as free-living organisms that kill and consume multiple prey items throughout their lifetime (see Table 3.1). However, much of the information presented can be applied equally well to parasitoids and other natural enemies.

GENERALISTS VERSUS SPECIALISTS

Predators have traditionally been categorised by diet breadth (Figure 3.1). On a coarse scale, predators can be considered either trophic specialists or generalists;

although, again, a range of intermediary types render the distinction ambiguous. Basically, a specialist feeds on a narrow range of prey, while a generalist feeds on a broader range of prey. An alternative set of terminology used to describe herbivores has also been adopted by predator ecologists. In this terminology, a *monophagous* predator consumes only one type of prey, while a *polyphagous* predator has the capacity to consume many types of prey (or other food – see below for a discussion of non-prey foods). While these terms broadly mirror the terms 'specialist' and 'generalist', they are more precise: a specialist need not be truly monophagous, but need only show preference for or special adaptation to a specific prey item, and may also utilise other prey items to a lesser extent. This ambiguity leads to complications in determining whether a given predator qualifies as a generalist or specialist (Huseynov *et al.*, 2005; 2008; Polidori *et al.*, 2010); while monophagy and polyphagy are considerably less ambiguous. Additionally, the terms *stenophagous* (referring to a narrow diet breadth) and *oligophagous* (referring to a moderate diet breadth) are sometimes used to further partition the intermediate range of the diet-breadth spectrum (Figure 3.1). In truth, nearly all predators are polyphagous to some extent, and most can be appropriately considered generalists; while truly monophagous predators are very rare. The situation is reversed for parasitoids: although generalist

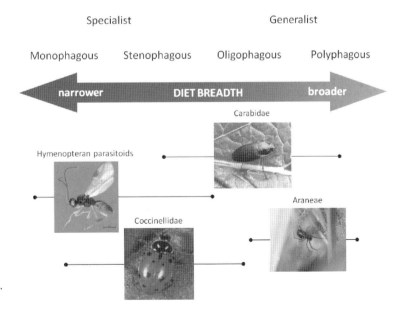

Figure 3.1 Generalised ranges of diet breadths for four common groups of natural enemies, with two schemes of categorising diet ranges shown above (photos: Kacie Johansen (parasitoid), Ric Bessin (coccinellid), Blake Newton (carabid), Kelton Welch (Araneae)).

parasitoids are not uncommon, parasitoids frequently have narrow host ranges. In fact, monophagous parasitoids are often seen as the archetypal specialists (e.g. Sheehan, 1986), and some works have even implicitly synonymised the comparison between 'specialist' and 'generalist' with the comparison between 'parasitoid' and 'predator' (e.g. Sabelis, 1992; see discussion in Symondson *et al.*, 2002).

Both specialist and generalist natural enemies can play important roles in biological control. While specialist natural enemies have the advantage of close interrelationships with and specific adaptations to the pest, they have the weakness of inflexibility: adaptation to a specific pest prey often entails adaptations to a specific habitat, life cycle or other conditions that maximise the ability to exploit that specific pest as a prey item. In contrast, generalist predators can more readily adjust to the conditions that the environment provides them with, and can take advantage of whatever prey or food resources are available. These differences in ecological requirements undoubtedly have consequences for 'opportunistic' biological control. Agronomic practices are not always conducive to the rigid ecological requirements of specialist predators, underlining the potential importance of generalist predators, whose versatility is a definite advantage in ephemeral habitats such as annual crop fields (Wissinger, 1997). This weakness of specialists has not always been as well appreciated as it is currently. For much of the history of biological control research, study focused strongly on importation of specialist predators, due in part to the belief that the adaptations of generalist natural enemies to a diverse diet prevented them from efficiently suppressing pest populations (Wardle and Buckle, 1923; Symondson *et al.*, 2002). Furthermore, dramatic successes in the field built a strong case for imported specialist predators as biological control agents. As the evidence continued to accumulate, however, it became clear that there was a characteristic profile for cases in which specialist predators were successful (Southwood, 1977; Wissinger, 1997; Hawkins *et al.*, 1999). Most successes with single specialist natural enemies were accomplished in simple systems with relatively low levels of disturbance, such as perennial crops (Wissinger, 1997), and with the natural enemy acting against an exotic pest (Hawkins *et al.*, 1999); while situations involving more complex landscapes and native pests are more suited to pest suppression by assemblages of generalist natural enemies. In fact, because real landscapes are often sur-

prisingly complex, pest control should be considered with a 'community-level' approach (Sunderland *et al.*, 1997) whereby synergistic and additive interactions among organisms promote the service of biological control offered by the organisms.

The tendency of generalists to utilise multiple types of prey results in less dramatic and fewer direct interrelationships with other organisms. The diffuse nature of the interactions of generalist predators with their environment and the large number of reticulations in their interaction webs make simple models less applicable to anything but very simple real world settings. Of greater relevance is that much of predator–prey theory addresses regulation of prey populations at some stable cyclic level that is likely not desirable (or even achievable) in agricultural fields. However, there is an acute need for biological control to be guided by theory, and the basic models of predator–prey interactions have much to offer for biological control research (see Box 3.1 and Box 3.2).

IMPLICATIONS OF ECOLOGICAL THEORY FOR BIOLOGICAL CONTROL BY GENERALISTS

The foraging ecology of predators is described by several theoretical models that provide insights into the dynamics of natural communities and food webs. One of the most influential theories is optimal foraging theory (OFT), which models the fundamental aspects of the decision-making processes of animals as they forage (Box 3.1). In its basic form, this theory addresses static decisions made by predators, although it has since been expanded to account for more dynamic patterns of behaviour over time (Box 3.2). An understanding of these theories is important for an understanding of the roles of generalist predators in agroecological food webs.

Optimal foraging theory provides a framework of viewing predator foraging ecology through the lens of economics, based on the principle that natural selection favours organisms that balance and optimise resource usage (see Box 3.1 for a discussion of OFT). From this theory, it is predicted that predators will selectively forage on abundant or high-quality prey and ignore rare or lower-quality prey. As the relative abundance of different prey types fluctuates over time, however, preferences for different prey types can also shift, leading predators to alter their proportional

Box 3.1 Optimal foraging theory

MacArthur and Pianka (1966) pioneered a system of modelling the decision-making processes of foraging organisms through the lens of economics. Called 'optimal foraging theory' (OFT), this system includes two basic models: the prey model and the patch model, which address decisions made before and after (respectively) attacking a prey item or entering a patch of resources. The **prey model** addresses the decision of whether to attack an encountered prey item or to forage in an encountered patch through an economic comparison of the availability, nutritional quality and handling time of various prey types. The **patch model** addresses the decision of how long to continue feeding on a captured prey item or foraging in a patch of resources based on the rate of nutritional gain from feeding or foraging and the potential rate of gain from searching for a new prey item or patch (Charnov, 1976).

OFT predicts that a predator will rank prey types based on their quality. Prey types will be attacked only if the gain from attacking the prey type is enough to offset the cost of losing the opportunity to encounter a prey type of higher quality. Although these conclusions rest on a number of unrealistic assumptions, OFT has dominated the thinking of theoretical ecologists since its introduction. The conceptual premise that foraging behaviour can be viewed as an economic decision-making process has been borne out in numerous studies (e.g. Viswanathan *et al*., 1999; Sayers *et al*., 2010), although the strict view that all animals' diets are optimised has long been controversial (e.g. Pierce and Ollason, 1987). Despite its shortcomings, OFT has provided important insights for biological control and natural enemy research. For example, OFT predicts that predators will reject pest prey if higher-quality, non-pest prey are abundant, which can disrupt pest suppression; a prediction that has been borne out in many real-world examples.

(For an in-depth discussion of optimal foraging theory, see Stephens and Krebs, 1986.)

these relatively simple ecological dynamics, ecologists can begin to develop an understanding of community functions in agricultural systems and the service of pest suppression offered by predators.

Generalist predators are inherently less amenable to theoretical study than specialist predators. While the dynamics of specialist predators can be modelled by accounting for a relatively simple set of ecological dynamics related to the single predator and its single prey, the breadth of a generalist predator's diet leads to a reticulate web of ecological interactions that are often very difficult to visualise and manage. Each prey item consumed by the predator possesses its own unique suite of characteristics (diet, climatic tolerances, behaviour patterns, etc.), to which the predator may respond (functionally or numerically) in very different ways, which can quickly clutter any model seeking an exhaustive understanding of the intricate dynamics of the food web. Additionally, prey may compete with or otherwise interact with one another and with the predator in a variety of ways. Fortunately, many of these complexities can be distilled into a few, generalised patterns that are informative for ecologists and can guide biological control efforts.

ALTERNATIVE PREY AND OMNIVORY

The implications of foraging on multiple prey items are an important consideration for generalist predators in ecological and biological control research. In agroecology, the assemblage of prey fed on by a generalist predator can be partitioned into two broad groups: the pest(s) of interest, and alternative prey that are not considered pests (usually referred to simply as 'alternative prey'). Of primary interest is the impact of alternative prey on the pest-feeding activity of natural enemies. A broad spectrum of possible impacts, ranging from very negative to very positive, can be envisioned, dealing both with individual-level and population-level effects.

OFT predicts that predators will reject low-quality prey in favour of higher-quality prey (Box 3.1). As some types of pest prey, such as aphids, are of poor nutritional quality for many generalist predators (Toft, 1995), it is reasonable to predict that these pests will frequently be rejected by generalist predators in favour of higher-quality alternative prey. In accordance with this prediction, the performance of generalist predators in pest suppression is diminished in many cases in which high-quality alternative prey are abundant

impacts on the populations of different prey types (for functional responses see Box 3.2). Additionally, as prey abundance changes, the number of predators that can be supported by the prey population also changes; thus, increasing prey density often increases predator density (for numerical responses see Box 3.2). From

Box 3.2 Density-dependent responses

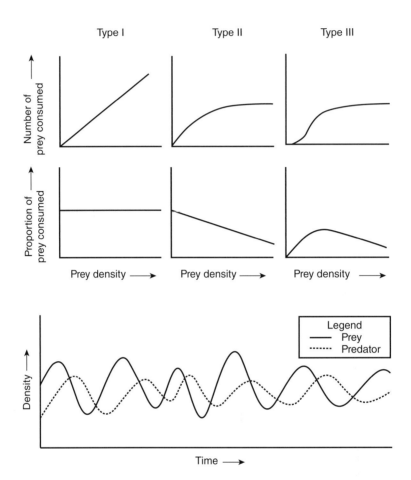

Predators may respond to changes in prey density in two basic ways: 1) change in the per-capita predation rate (a functional response); 2) change in the density of predators (a numerical response). **Functional responses** take three generalised forms (above). Only the Type III response entails a proportional increase in the mortality of the prey caused by the predator; thus, such a functional response has long been considered a crucial characteristic for generalist predators as biological control agents. However, more recent developments have questioned this notion.

Numerical responses can result from two ecological processes: 1) aggregation to areas of high prey density; 2) increased fecundity due to increased consumption of prey. Generalist predators are not inherently dependent on specific types of prey, and thus, are not typically expected to show strong numerical responses to individual prey types. While this limits their capacity to control pests on a short time scale, it allows them to maintain relatively steady populations over time, even as the populations of pests and their specialist natural enemies fluctuate (below), which can improve their ability to contribute to control of the pest (see section on early-season control).

(Harwood *et al.*, 2004; Gavish-Regev *et al.* 2009; Kuusk and Ekbom, 2010). This has long been considered a major weakness of generalist predators in biological control. However, alternative prey may also have important positive benefits for predators. Generalist predators are capable of 'prey switching', or altering preference for different prey types as their relative densities change. This allows generalist predators to subsist on alternative prey in times when pest populations are low, and switch to consuming the pest when it becomes abundant. This has led to the suggestion that the most suitable alternative prey are distributed opposite to the pest temporally: that is, the alternative prey and pest each reach their highest densities at the point in time when the other reaches its lowest densities (Settle *et al.*, 1996; Symondson *et al.*, 2002), thereby minimising interference between the two prey types and forcing the predator to switch to the pest. Prey switching often results in a type-III functional response, in which the proportional impact of the predator on the prey is actually increased (Box 3.2).

However, a direct functional response to the pest is not necessarily required. For example, a high enough predator density can result in significant levels of pest suppression even in the absence of preference for the pest (Symondson *et al.*, 2002). In this case, alternative prey can be highly beneficial to biological control, as they can sustain generalist predators during periods when pest prey are rare or absent, thereby helping maintain their populations at high densities in the absence of pest prey and promoting a 'lying in wait' strategy for pest suppression (Settle *et al.*, 1996) (described in detail below). Indeed, non-pest prey have even been shown to *increase* predator density through population growth (a numerical response), thereby maintaining high underlying populations prior to pest arrival: the ideal example of a successful 'lying in wait' strategy (Butler and O'Neil, 2007). The practical application of this strategy in rice pest management is covered by Gurr *et al.* in chapter 13 of this volume. Accordingly, numerical responses by a polyphagous predator to alternative prey items, rather than to a pest, can promote high population densities of the predator and increase its impact on the pest without altering the predator's preference for or density-dependent response to the pest (Evans and Toler, 2007). However, it should be noted that generalist predators seem unlikely to exhibit strong numerical responses to the density of individual prey types, because they are not dependent on a specific prey type

for survival (Symondson *et al.*, 2002). Thus, as one prey item declines in availability, generalist predators can switch to another prey type and maintain their densities, thereby buffering themselves against significant fluctuations in response to pest density changes (Symondson *et al.*, 2002).

It is therefore clear that alternative prey are inherently crucial to predator foraging and the temporal separation of such resources from pest species are undoubtedly important not only for sustaining natural enemies in agroecosystems, but also preventing them from diverting attention towards such food items, which are often favoured.

Early-season control

Generalist predators, with their diverse feeding habits, typically lack the tight dynamic relationship with their prey that commonly exists between specialist parasitoids and their hosts (Crawley, 1992). Thus, as pest populations accumulate exponentially, as is typical in many systems, generalist predators simply cannot keep up and pest outbreaks (leading to crop yield loss) typically occur. Despite this perceived weakness, one attribute placing generalist predators at a distinct advantage over specialist natural enemies is their aforementioned capacity to 'lie in wait'; that is, to subsist on non-pest food resources before pests arrive in the crop, allowing their populations to establish and even grow while target pest populations are still very low. Thus, the generalist predator is given a 'head start', and will be able to impact pests during the early growth phase of the pest population cycle resulting in favourable predator:pest ratios (Settle *et al.*, 1996; Ishijima *et al.* 2004; Piñol *et al.*, 2009). By comparison, a specialist natural enemy would be less able to subsist on other food sources before the pest arrives, and would thus be likely to experience unfavourable predator:pest ratios as its own population would still have to accumulate. Indeed, the specialist's dependence on the pest may cause its population density to track pest density, rather than control it (Piñol *et al.*, 2009); or lag behind pest density, in which case it may be unable to provide timely suppression of the pest. This concept is supported by mathematical modelling of predator–prey populations (Fleming, 1980; Piñol *et al.*, 2009). If pest populations have the ability to grow exponentially, they will eventually reach a point at which their density passes an economic threshold,

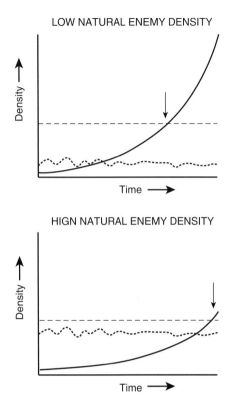

LOW NATURAL ENEMY DENSITY

Density →

Time →

HIGN NATURAL ENEMY DENSITY

Density →

Time →

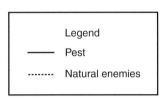

Legend

——— Pest

········ Natural enemies

Figure 3.2 High densities of natural enemies (lower graph) can slow pest population growth relative to low densities of natural enemies (higher graph). This can delay pest outbreaks, such that pests do not cross an economic threshold (dotted line) until later in the season.

and the damage they cause begins to reduce the grower's profits. At low densities of natural enemies, the pest can cross the economic threshold unimpeded (Figure 3.2, top graph). However, at higher densities of natural enemies, the pest population growth can be counteracted by mortality due to predation, and the pest does not cross the economic threshold until later in the season (Figure 3.2, bottom graph). This delay in the pest outbreak may provide enough time for the crop to grow out of a critical vulnerable stage or for specialist enemies to establish and effect biological control. For crops with short growing cycles, such as hay fields, which are harvested multiple times each season, the delay caused by natural enemies may be sufficient to prevent pest outbreaks during an entire growth cycle, making pesticides and other major control efforts unnecessary.

The role for generalist predators in early-season pest suppression as predicted by these theoretical works has also been demonstrated in several empirical field studies (Edwards *et al.*, 1979; Chiverton, 1987; Landis and van der Werf, 1997; Harwood *et al.*, 2004; 2007;

2009). Clearly, these empirical field studies confirm the ability of predators to provide significant levels of pest suppression at times in the season when pest populations are very small. The specific mechanisms underlying these phenomena are, however, unclear; it is possible that predators specifically seek out these 'rare' prey for nutritional diversification, which promotes fitness and reproduction. Greenstone (1979) suggested that generalist predators seek a mixed diet to balance their amino acid requirement while, more recently, Mayntz *et al.* (2005) reported that food selection by predators occurs primarily as a means for addressing dietary imbalances. Perhaps rare pests can provide these additional nutrient resources that could be particularly important in prey-limited agroecosystems. However, it is unclear if predators in these conditions have the ability to selectively forage given that they are widely reported as being highly prey-limited in agricultural crops (Wise, 1975; Lenski, 1984; Bilde and Toft, 1998).

Despite the lack of understanding concerning the specific mechanisms underlying early-season control, predators have the ability, both from a theoretical and applied viewpoint, to contribute to biological control before pest populations increase exponentially. The challenge is providing predators with needed resources early in the season to produce a population-scale (i.e. numerical) response in the predators without causing detriment to biological control by attenuating the predator's functional response to the pest. Diversity therefore has the confounding effect of providing natural enemies with alternative food sources to consume. Undoubtedly habitat management plays an important part in conservation biological control and promoting early-season populations of natural enemies (Landis et al., 2000). However, the underlying role of prey diversity in determining predator populations during pest colonisation requires further investigation.

NON-PREY RESOURCES

In addition to the effects of prey diversity, generalist predators can also experience significant impacts from abiotic features of the environment and from other non-trophic resources. Trophic generalists may not necessarily be habitat generalists. Predators and parasitoids alike may be constrained by the climates and habitats they can successfully inhabit (Southwood, 1977), or may experience enhanced or reduced performance on different crops or substrates (Lundgren et al., 2008; Verheggen et al., 2009), perhaps due to inhibited searching efficiency or increased prey refuge. In such cases, active manipulation and/or diversification of the habitat is often employed to provide the predator with its limiting resource and overcome the inhibitions placed on predator performance (reviewed in Landis et al., 2000). Additionally, natural enemies often require additional, non-prey food resources to maintain adequate growth and reproduction, and thus provisioning with these resources can improve the survival and overall fitness of predator populations (Symondson et al., 2006).

Habitat and abiotic features

Due to patterns of habitat preference, prey availability and other ecological factors, crops differ in the composition of the natural enemy assemblages they support (Roach, 1980; Pfannenstiel and Yeargan, 1998a). For example, Nordlund et al. (1984) reported differences in the distributions of various groups of natural enemies across corn, bean and tomato monocultures and polycultures, with crop type playing a primary role in driving these differences. Likewise, Pfannenstiel and Yeargan (1998a) demonstrated that distributions across four crop types were unique to each of several taxa of heteropteran predators. The reasons for these 'preferences' for certain crops are diverse, and depend on the life history of the natural enemy in question. Predators may favour certain habitats because of the prey assemblages they support (as discussed above), because of the availability of oviposition sites (Pfannenstiel and Yeargan, 1998b), or because of structural features of the plants, such as trichomes (hairs on plant-stems) or waxy leaf coatings, which can affect the mobility of natural enemies or their larvae (Cottrell and Yeargan, 1999). Alternatively, 'preference' may not result from behavioural factors at all, but from evolutionary processes (i.e. differential success of predator populations on different crops, resulting in higher fecundity and larger populations on 'preferred' crops) even in the absence of behavioural preference for certain habitats (Lundgren et al., 2009).

Additional studies on the effect of the habitat on natural enemy populations have examined different agronomic practices, such as tillage and harvest (Thorvilson et al., 1985; Tonhasca and Stinner, 1991; Tonhasca, 1993; 1994; Ishijima et al., 2004). Disturbances such as these can severely deplete or even completely eradicate populations of pests and predators alike, and thus present an extreme challenge for natural enemies. However, the behaviour of generalist predators increases their potential role in such environments – their opportunistic feeding habits and high dispersal ability enable them to colonise an environment rapidly after an agronomic disturbance (Öberg and Ekbom, 2006), and persist on the relatively impoverished prey assemblage available thereafter. The prevalence of such disturbances varies across crops of different types. The stability of perennial crop systems is thought to allow the development of strong, continuous predator–prey relationships, such as that between a specialist predator or parasitoid and its specific prey item (Southwood, 1977). However, as disturbances from harvesting, tilling and other management practices increase, tightly linked predator–prey relationships are strained and eventually collapse. Thus, specialist predators have been considerably less effective in highly disturbed systems, such as annual crops or forage crops that are harvested several times a year. Generalist predators have been found to be resilient to such disturbance regimes, which make them more suitable for ephemeral habitats such as annual crops and forage crops (Wissinger, 1997). Nevertheless, the arthropod communities in disturbed systems can be expected to undergo major changes across a season due to the differing life-history phenologies of different predators and prey and to the timing and nature of disturbances throughout the season.

Similar changes can also occur across a diel (daily) cycle. The study of predators in the field has occurred largely during daylight hours with little consideration for nocturnally active predators. However, a body of work is building that describes significant nocturnal predation, starting with the prescient study of aphid predation by Vickerman and Sunderland (1975). More recently, nocturnal predation has been recognised as a major component of pest suppression by natural enemies (Pfannenstiel and Yeargan, 2002; Weber and Lundgren, 2009) and an increasing number of studies have shown diel variation in predation and predator activity against a variety of prey. Thus, it is clear that consideration of the activity patterns of both pest and prey will yield important insights into the roles of predators in pest suppression. Ultimately, the use of molecular approaches enables reliable breakdown of trophic relationships in food webs where nocturnal (and therefore difficult to observe) interactions are commonplace (see chapter 10 of this volume).

Omnivory

While all generalist predators, by definition, consume multiple types of prey, they differ fundamentally from true omnivores, which consume plant material in addition to pests and alternative prey. However, it is widely acknowledged today that many, if not most, arthropods that exhibit predatory behaviour are actually omnivores (Coll and Guershon, 2002). In addition to the many examples of predators that feed on plant-based resources (Wäckers *et al.*, 2005; Lundgren, 2009; Wäckers and van Rijn (chapter 9 of this volume)), many species considered pest herbivores may also feed on prey (Coll and Guershon, 2002). These 'predator' taxa frequently take and gain significant value from non-pest, plant-based resources, such as vascular juices, foliage, seeds, pollen and nectar. Our understanding of the benefits of non-prey resources for predators is still somewhat limited, yet much of the available data provides important insights into such mechanisms of foraging (Wäckers *et al.*, 2005; Lund-

gren, 2009). In some predatory taxa, a significant portion of the predator's foraging activity is dedicated to plant feeding (Coll, 1996; Hagler *et al.*, 2004). While this is not surprising for some groups, such as the plant bugs (Heteroptera: Miridae), which include a mix of species that range from entirely predatory to entirely herbivorous (Kapadia and Puri, 1991; Rosenheim *et al.*, 2004; Hagler *et al.*, 2010), plant feeding in other groups is perhaps less expected. For example, spiders are often perceived as obligate predators; however, it has been observed that many types of spiders will routinely feed on floral and extrafloral nectar (Taylor and Pfannenstiel, 2008; 2009; Taylor and Bradley, 2009; Patt *et al.*, 2011), pollen (Ludy and Lang, 2006; Peterson *et al.*, 2010) and other plant parts (Meehan *et al.*, 2009), as well as yeast (Sunderland *et al.*, 1996; Patt *et al.*, 2011). Some spiders, as well as some predatory bugs, can develop to the next instar or even to adulthood feeding on a diet of pollen or other plant materials alone (Sunderland *et al.*, 1996; Kiman and Yeargan, 1985). Additionally, the responses of some spiders to non-prey resources (such as extrafloral nectar) have been observed to mirror responses to prey (Patt and Pfannenstiel, 2008; 2009); indicating that plant-feeding in these natural enemies is not just incidental but adaptive. Although these examples focus solely on one group of predators (spiders) they provide evidence that predators exhibit extremely diverse feeding habits.

In many insects (particularly neopterous insects), the diet of larvae and adults are very different. Some groups can be considered omnivorous if one life stage consumes prey while another feeds on plants (called 'life-history omnivores'). For example, hoverflies (Diptera: Syrphidae) feed primarily on nectar and pollen as adults whereas the larvae are predaceous. In other groups, such as lacewings (Neuroptera: Chrysopidae and Hemerobiidae), larvae can be carnivorous or omnivorous and adults can be herbivorous, carnivorous or omnivorous (Patt *et al.*, 2003; Limburg and Rosenheim, 2001; Jacometti *et al.*, 2010). For most of these groups, flowers can attract adults into a crop system, which subsequently increases numbers of larvae and thereby generates large impacts on pest populations (Hickman and Wratten, 1996; White *et al.*, 1995); however there are cases in which plant resources, like alternative prey, have been shown to disrupt pest suppression (Spellman *et al.*, 2006). Just as feeding on alternative prey can have mixed effects on pest suppression by natural enemies, true omnivory has also been documented to have both positive and negative effects. Some studies have shown that plant resources improve the fitness and survival of natural enemies during periods of low prey availability (Valicente and O'Neil, 1995; Magalhäes and Bakker, 2002; Beckman and Hurd, 2003), thus allowing them to exert greater effects on pests when they are present. Much work has been done on the usefulness of alternative non-prey foods as dietary supplements and/or attractants for natural enemies such as lacewings, hoverflies and ladybird beetles (Landis *et al.*, 2000; Lundgren, 2009).

INTERACTIONS AMONG NATURAL ENEMIES

While increased resource and vegetation diversity can increase the biological control potential of natural enemies within the system, it can also increase the potential for interactions among the natural enemies, which may have positive or negative effects on pest suppression. Increased pest suppression is often observed in systems characterised by rich and even natural enemy assemblages (Snyder *et al.*, 2006; Straub and Snyder, 2006; 2008; Crowder *et al.*, 2010). Whether this can be attributed to emergent effects of diversity, or whether it is due to the inclusion of individual species that are particularly effective biological control agents is still being debated, and apparently varies from system to system. In fact, there is evidence that synergistic, neutral and antagonistic effects of multiple-enemy assemblages can all occur in different systems (Casula *et al.*, 2006; Schmitz, 2007). This diversity of effects results from the complexity of ecological communities. However, because species identity appears to be so critical to the results of inter-specific linkages (Coll and Guershon, 2002; Snyder *et al.*, 2006), focused manipulation designed to enhance populations of key predators, not predator diversity per se, is much more likely to result in improvements in biological control.

Predators may work synergistically through utilisation of different microhabitats and niche partitioning (Snyder and Tylianakis, chapter 2 of this volume). In single-predator systems, much of the habitat not utilised by the predator can serve as a refuge for the pest. However, the addition of more predators results in a greater coverage of the habitat, and thus a smaller refuge for the pest, increasing the overall impact of the assemblage on the pest (Schmitz, 2007). Additionally, if the action of one predator increases the exposure of the pest to another predator, an emergent effect can occur, in which the impact of the entire assemblage is greater than the combined impacts of the individual species working alone (Losey and Denno, 1998). Cardinale *et al.* (2003) showed synergistic effects of three predators on their impact on the pea aphid (*Acyrthosiphon pisum* (Harris)). Such positive effects of increasing diversity are the goal of conservation biological control programmes: ideally, habitat manipulations, natural enemy attractants and cultural controls could be developed for the optimisation of synergistic effects such as these.

Intraguild predation

Interactions among predator species depend on their behavioural interactions (Björkman and Liman, 2005) and can run the gamut of positive effects (as discussed above) to negative effects. The most extreme form of negative interaction is intraguild predation (IGP), in which one natural enemy preys upon another. More formally IGP is defined as 'predation by a natural enemy on another natural enemy with which it also competes for other prey resources'. IGP has been considered an important aspect of trophic ecology (Rosenheim *et al.*, 1995). The victim of IGP is referred to as the 'intraguild prey' while the consumer is referred to as the 'intraguild predator'. Where it is directional (i.e. one species is always the intraguild predator while the other is always the intraguild prey) IGP may have critical implications in determining the structure of ecological communities (Polis *et al.*, 1989; Polis and Holt, 1992) but because of the diversity of predators a variety of effects can be expected (Sih *et al.*, 1998). Indeed, Müller and Brodeur (2002, p. 217) state:

> As various outcomes for the prey resource arise from IGP, it may depend on the identities of the species that interact; this in turn makes the outcome of this indirect interaction less predictable.

IGP is likely to be ubiquitous in both natural and agricultural systems. A frequently ignored characteristic of IGP is that it is frequently not unidirectional: it may vary in direction and intensity depending on the age/size structure of the predator population and the makeup of the predator complex at any one time (Balfour *et al.*, 2003). Therefore it is hard to predict what consistent effects IGP might have across systems.

The results of intraguild interactions will be determined by the identities of the predators and stage-specific characteristics of the predator complex, and will also vary by season and by year. Generalist predators may also enhance or interfere with the activity of specialist parasitoids via predation on parasitised hosts (Snyder and Ives, 2003).

Theoretical and empirical work has shown that IGP may: have no effect on overall predation rates (when the intraguild predator is also an efficient predator of the pest); release the pest from predation (when the intraguild predator prefers or more efficiently utilises the intraguild prey than the pest) (Finke and Denno, 2003); or increase the impact on herbivore populations (by releasing some third natural enemy). Generally, however, IGP is regarded as having a negative effect on the performance of natural enemies in pest suppression and thus practices that promote coexistence (thereby minimising the significance of IGP) are highly advantageous for promoting incorporation of the community of generalist predators in biological control. Indeed, research suggests that in simple agricultural systems, increasing diversity may improve pest control (see chapter 2 of this volume).

COMPLEX INTERACTIONS IN DIVERSE SYSTEMS

The many studies of the impacts of generalist predators on pest populations have yielded mixed results, indicating that success in biological control by generalist natural enemies is contingent on system-specific factors. Therefore, an understanding of the ecology of generalist predators, and of the ecological characteristics of crop systems, is critical for promoting pest-suppression services by generalist predators. Management techniques designed to promote or enhance pest suppression by natural enemies should be tailored to the type(s) of natural enemy that can suppress the target pest. However, this is not always easy to do in practice because teasing apart the intricate web of interactions to find the important variables is a serious challenge. Experimentation with habitat manipulations can often reveal which aspects of a predator's ecology limits its utility in pest suppression, and this can reveal what types of management practices may be best suited for the promotion of the desired natural enemies. For example, Riechert and Bishop (1990) compared the effects of abiotic factors and alternative food resources on a spider community in vegetable garden plots. Their work showed that adding mulch increased spider densities and pest suppression by the spiders, whereas adding weeds to attract alternative prey did not yield success. They attribute the success of the mulch to the favourable humidity and shelter it provided the spiders. This indicates that habitat, rather than prey, limits the pest-suppression service of this generalist predator community.

However, the intricate interaction webs of generalist predators often lead to complex dynamics that are very difficult to predict and manage. For example, Dinter (2002) studied intraguild predation by web-building spiders on lacewing larvae in potted microcosms. He discovered that spiders inflicted significant mortality on lacewing larvae via IGP but, despite the strong negative interaction, the two predators in combination were still effective at suppressing aphid populations. However, when alternative prey were added predation by spiders on aphids and lacewing larvae was reduced, with the net result of reduced aphid suppression in the presence of alternative prey. In contrast, Yasuda and Kimura (2001) found that IGP by a crab spider on ladybird larvae disrupted aphid predation. In another study, Cardinale et al. (2003) discovered that alternative hosts reduced the impact of parasitoids on pest aphids but the addition of generalist predators reduced the abundance of the alternative prey and, consequently, increased suppression of the pest aphid. Onzo et al. (2005) discovered that addition of pollen as a non-prey food item altered the directionality of IGP in favour of the superior pest suppressor, thereby improving biological control. Given that complexities such as these produce unpredictable indirect effects, it may be impractical to design biological control programmes that account for all the specific ecological dynamics of all the natural enemies involved. Rather, management at a coarser scale may be more reasonable.

CONCLUSIONS

It is worthwhile considering the question 'When can we expect a generalist predator to be an effective biological control agent?' Unfortunately the answer is not straightforward as different characteristics of generalist predators may be advantageous in different circumstances. In general, a number of characteristics can be considered to underline the success of generalist predators:

1. Ability to persist on alternative prey with an opposite temporal distribution from the pest.
2. Ability to rebuild populations rapidly following disturbances.
3. Starvation tolerance during times of low prey abundance.
4. Low ecological overlap and minimal antagonistic interactions with other co-occurring natural enemies.

Beyond this basic set of characteristics success will rely on an understanding of the specific system and of the organisms and abiotic features associated with it.

Generalist predators and other natural enemies are critical components of agroecosystems. The diversity of ecological characteristics among natural enemies leads to a surprisingly high degree of complexity in these 'simple' systems and presents many unique challenges and opportunities for agroecologists. Of particular interest to researchers and growers is the possibility of synergistic pest suppression by communities of natural enemies. Pest suppression by generalist natural enemies can result from a variety of ecological processes: behavioural or population-level responses to pest dynamics, responses to other environmental variables (alternative foods, vegetation, etc.) that incidentally increase impact on pests, or positive interactions among natural enemies. Within entire assemblages of natural enemies successful pest suppression relies on the diversity of ways in which different natural enemies impact a pest, as well as the capacity of the natural enemies to coexist and interact synergistically. While the promise for such synergistic effects is high, much work remains to be done before this potential can be fully understood and utilised.

ACKNOWLEDGEMENTS

J.D. Harwood and K.D. Welch are supported by the University of Kentucky Agricultural Experiment Station State Project KY008043. This is publication number 11-08-051 of the University of Kentucky Agricultural Experiment Station.

REFERENCES

Balfour, R.A., Buddle, C.M., Rypstra, A.L., Walker, S.E. and Marshall, S.D. (2003) Ontogenetic shifts in competitive interactions and intra-guild predation between two wolf spider species. *Ecological Entomology*, 28, 25–30.

Beckman, N. and Hurd, L.E. (2003) Pollen feeding and fitness in praying mantids: the vegetarian side of a tritrophic predator. *Environmental Entomology*, 32, 881–885.

Bilde, T. and Toft, S. (1998) Quantifying food limitation of arthropod predators in the field. *Oecologia*, 115, 54–58.

Björkman, C. and Liman, S.A. (2005) Foraging behaviour influences the outcome of predator–predator interactions. *Ecological Entomology*, 30, 164–169.

Butler, C.D. and O'Neil, R.J. (2007) Life history characteristics of *Orius insidiosus* (Say) fed diets of soybean aphid, *Aphis glycines* Matsumura and soybean thrips, *Neohydatothrips variabilis* (Beach). *Biological Control*, 40, 339–346.

Cardinale, B.J., Harvey, C.T., Gross, K. and Ives, A.R. (2003) Biodiversity and biocontrol: emergent impacts of a multi-enemy assemblage on pest suppression and crop yield in an agroecosystem. *Ecology Letters*, 6, 857–865.

Casula, P., Wilby, A. and Thomas, M.B. (2006) Understanding biodiversity effects on prey in multi-enemy systems. *Ecology Letters*, 9, 995–1004.

Charnov, E.L. (1976) Optimal foraging, marginal value theorem. *Theoretical Population Biology*, 9, 129–136.

Chiverton, P.A. (1987) Predation of *Rhopalosiphum padi* (Homoptera: Aphididae) by polyphagous predatory arthropods during the aphids' pre-peak period in spring barley. *Annals of Applied Biology*, 111, 257–269.

Coll, M. (1996) Feeding and ovipositing on plants by an omnivorous insect predator. *Oecologia*, 105, 214–220.

Coll, M. and Guershon, M. (2002) Omnivory in terrestrial arthropods: mixing plant and prey diets. *Annual Review of Entomology*, 47, 267–297.

Cottrell, T.E. and Yeargan, K.V. (1999) Factors influencing dispersal of larval *Coleomegilla maculata* from the weed *Acalypha ostryaefolia* to sweet corn. *Entomologia Experimentalis et Applicata*, 90, 313–322.

Crawley, M.J. (1992) Population dynamics of natural enemies and their prey, in *Natural enemies: the population biology of predators, parasites and diseases* (ed. M.J. Crawley), Cambridge University Press, Cambridge, pp. 40–89.

Crowder, D.W., Northfield, T.D., Strand, M.R. and Snyder, W.E. (2010) Organic agriculture promotes evenness and natural pest control. *Nature*, 466, 109–122.

Dinter, A. (2002) Microcosm studies on intraguild predation between female erigonid spiders and lacewing larvae and influence of single versus multiple predators on cereal aphids. *Journal of Applied Entomology*, 126, 249–257.

Edwards, C.A., Sunderland, K.D. and George, K.S. (1979) Studies on polyphagous predators of cereal aphids. *Journal of Applied Ecology*, 16, 811–823.

Endo, T. and Endo, A. (1994) Prey selection by a spider wasp, *Batozonellus lacerticida* (Hymenoptera, Pompilidae) – effects of seasonal variation in prey species, size and density. *Ecological Research*, 9, 225–235.

Evans, E.W. and Toler, T.R. (2007) Aggregation of polyphagous predators in response to multiple prey: ladybirds

(Coleoptera: Coccinellidae) foraging in alfalfa. *Population Ecology*, 49, 29–36.

Finke, D.L. and Denno, R.L. (2003) Intra-guild predation relaxes natural enemy impacts on herbivore populations. *Ecological Entomology*, 28, 67–73.

Fleming, R.A. (1980) The potential for control of cereal rust by natural enemies. *Theoretical Population Biology*, 18, 374–395.

Gavish-Regev, E., Rotkopf, R., Lubin, Y. and Coll, M. (2009) Consumption of aphids by spiders and the effect of additional prey: evidence from microcosm experiments. *Biocontrol*, 54, 341–350.

Greenstone, M.H. (1979) Spider feeding-behavior optimizes dietary essential amino-acid composition. *Nature*, 282, 501–503.

Hagler, J.R., Jackson, C.G., Isaacs, R. and Machtley, S.A. (2004) Foraging behavior and prey interactions by a guild of predators on various lifestages of *Bemisia tabaci*. *Journal of Insect Science*, 4, 1.

Hagler, J.R., Jackson, C.G. and Blackmer, J.L. (2010) Diet selection exhibited by juvenile and adult lifestages of the omnivores western tarnished plant bug, *Lygus hesperus* and tarnished plant bug, *Lygus lineolaris*. *Journal of Insect Science*, 10, 127.

Harwood, J.D., Sunderland, K.D. and Symondson, W.O.C. (2004) Prey selection by linyphiid spiders: molecular tracking of the effects of alternative prey on rates of aphid consumption in the field. *Molecular Ecology*, 13, 3549–3560.

Harwood, J.D., Desneux, N., Yoo, H.J.S. *et al.* (2007) Tracking the role of alternative prey in soybean aphid predation by *Orius insidiosus*: a molecular approach. *Molecular Ecology*, 16, 4390–4400.

Harwood, J.D., Yoo, H.J.S., Greenstone, M.H., Rowley, D.L. and O'Neil, R.J. (2009) Differential impact of adults and nymphs of a generalist predator on an exotic invasive pest demonstrated by molecular gut-content analysis. *Biological Invasions*, 11, 895–903.

Hawkins, B.A., Mills, N.J., Jervis, M.A. and Price, P.W. (1999) Is the biological control of insects a natural phenomenon? *Oikos*, 86, 493–506.

Hickman, J.M. and Wratten, S.D. (1996) Use of *Phacelia tanacetifolia* strips to enhance biological control of aphids by hoverfly larvae in cereal fields. *Journal of Economic Entomology*, 89, 832–840.

Huseynov, E.F., Cross, F.R. and Jackson, R.R. (2005) Natural diet and prey-choice behaviour of *Aelurillus muganicus* (Araneae: Salticidae), a myrmecophagic jumping spider from Azerbaijan. *Journal of Zoology*, 267, 159–165.

Huseynov, E.F., Jackson, R.R. and Cross, F.R. (2008) The meaning of predatory specialization as illustrated by *Aelurillus m-nigrum*, an ant-eating jumping spider (Araneae: Salticidae) from Azerbaijan. *Behavioural Processes*, 77, 389–399.

Ishijima, C., Motabayashi, T., Nakai, M. and Kunimi, Y. (2004) Impacts of tillage practices on hoppers and predatory wolf spiders (Araneae: Lycosidae) in rice paddies. *Applied Entomology and Zoology*, 39, 155–162.

Jacometti, M., Jorgensen, N. and Wratten, S. (2010) Enhancing biological control by an omnivorous lacewing: Floral resources reduce aphid numbers at low aphid densities. *Biological Control*, 55, 159–165.

Kapadia, M.N. and Puri, S.N. (1991) Biology and comparative predation efficacy of 3 heteropteran species recorded as predators of *Bemisia tabaci* in Maharashtra. *Entomophaga*, 36, 555–559.

Kiman, Z.B. and Yeargan, K.V. (1985) Development and reproduction of the predator *Orius insidiosus* (Hemiptera, Anthocoridae) reared on diets of selected plant-material and arthropod prey. *Annals of the Entomological Society of America*, 78, 464–467.

Kuusk, A.K. and Ekbom, B. (2010) Lycosid spiders and alternative food: feeding behavior and implications for biological control. *Biological Control*, 55, 20–26.

Landis, D.A and van der Werf, W. (1997) Early-season predation impacts the establishment of aphids and spread of beet yellows virus in sugar beet. *Entomophaga*, 42, 499–516.

Landis, D.A., Wratten, S.D. and Gurr, G.M. (2000) Habitat management to conserve natural enemies of arthropod pests in agriculture. *Annual Review of Entomology*, 45, 175–201.

Lenski, R.E. (1984) Food limitation and competition – a field experiment with 2 *Carabus* species. *Journal of Animal Ecology*, 53, 203–216.

Limburg, D.D. and Rosenheim, J.A. (2001) Extrafloral nectar consumption and its influence on survival and development of an omnivorous predator, larval *Chrysoperla plorabunda* (Neuroptera: Chrysopidae). *Environmental Entomology*, 30, 595–604.

Losey, J.E. and Denno, R.F. (1998) Positive predator–predator interactions: enhanced predation rates and synergistic suppression of aphid populations. *Ecology*, 79, 2143–2152.

Ludy, C. and Lang, A. (2006) *Bt* maize pollen exposure and impact on the garden spider, *Araneus diadematus*. *Entomologia Experimentalis et Applicata*, 118, 145–156.

Lundgren, J.G. (2009) Nutritional aspects of non-prey foods in the life histories of predaceous Coccinellidae. *Biological Control*, 51, 294–305.

Lundgren, J.G., Fergen, J.K. and Riedell, W.E. (2008) The influence of plant anatomy on oviposition and reproductive success of the omnivorous bug *Orius insidiosus*. *Animal Behaviour*, 75, 1495–1502.

Lundgren, J.G., Wyckhuys, K.A.G. and Desneux, N. (2009) Population responses by *Orius insidiosus* to vegetational diversity. *Biocontrol*, 54, 135–142.

MacArthur, R.H. and Pianka, E.R. (1966) On the optimal use of a patchy environment. *American Naturalist*, 100, 603–609.

Magalhäes, S. and Bakker, F.M. (2002) Plant feeding by a predatory mite inhabiting cassava. *Experimental and Applied Acarology*, 27, 27–37.

Mayntz, D., Raubenheimer, D., Salomon, M., Toft, S. and Simpson, S.J. (2005) Nutrient-specific foraging in invertebrate predators. *Science*, 301, 111–113.

Meehan, C.J., Olson, E.J., Reudink, M.W., Kyser, T.K. and Curry, R.L. (2009) Herbivory in a spider through exploitation of an ant-plant mutualism. *Current Biology*, 19, R892–R893.

Müller, C.B. and Brodeur, J. (2002) Intraguild predation in biological control and conservation biology. *Biological Control*, 25, 216–223.

Nentwig, W. (ed.) (1987) *Ecophysiology of Spiders*. Springer-Verlag Publishing Company, Berlin.

Nordlund, D.A., Chalfant, R.B. and Lewis, W.J. (1984) Arthropod populations, yield and damage in monocultures and polycultures of corn, beans and tomatoes. *Agriculture, Ecosystems & Environment*, 11, 353–367.

Öberg, S. and Ekbom, B. (2006) Recolonisation and distribution of spiders and carabids in cereal fields after spring sowing. *Annals of Applied Biology*, 149, 203–211.

Onzo, A., Hanna, R., Negloh, K., Toko, M. and Sabelis, M.W. (2005) Biological control of cassava green mite with exotic and indigenous phytoseiid predators – effects of intraguild predation and supplementary food. *Biological Control*, 33, 143–152.

Patt, J.M. and Pfannenstiel, R.S. (2008) Odor-based recognition of nectar in cursorial spiders. *Entomologia Experimentalis et Applicata*, 127, 64–71.

Patt, J.M. and Pfannenstiel, R.S. (2009) Characterization of restricted area searching behavior following consumption of prey and non-prey food in a cursorial spider, *Hibana futilis*. *Entomologia Experimentalis et Applicata*, 132, 13–20.

Patt, J.M., Wainright, S.C., Hamilton, G.C., Whittinghill, D., Bosley, K., Dietrick, J. and Lashomb, J.H. (2003) Assimilation of carbon and nitrogen from pollen and nectar by a predaceous larva and its effects on growth and development. *Ecological Entomology*, 28, 717–728.

Patt, J.M., Pfannenstiel, R.S., Meikle, W.G. and Adamczyk, J.J. (2011) Supplemental diets containing yeast, sucrose, and soy powder enhance the survivorship, growth, and development of prey-limited cursorial spiders. *Biological Control*, doi:10.1016/j.biocontrol.2011.02.004.

Peterson, J.A., Romero, S.A. and Harwood, J.D. (2010) Pollen interception by linyphiid spiders in a corn agroecosystem: implications for dietary diversification and risk-assessment. *Arthropod–Plant Interactions*, 4, 207–217.

Pfannenstiel, R.S. and Yeargan, K.V. (1998a) Association of predaceous Hemiptera with selected crops. *Environmental Entomology*, 27, 232–239.

Pfannenstiel, R.S. and Yeargan, K.V. (1998b) Ovipositional preference and distribution of eggs in selected field and vegetable crops by *Nabis roseipennis* (Hemiptera: Nabidae). *Journal of Entomological Science*, 33, 82–89.

Pfannenstiel, R.S. and Yeargan, K.V. (2002) Identification and diel activity patterns of predators attacking *Helicoverpa zea* (Lepidoptera: Noctuidae) eggs in soybean and sweet corn. *Environmental Entomology*, 31, 232–241.

Pierce, G.J. and Ollason, J.G. (1987) Eight reasons why optimal foraging theory is a complete waste of time. *Oikos*, 49, 111–118.

Polidori, C., Gobbi, M., Chatenoud, L., Santoro, D., Montani, O. and Andrietti, F. (2010) Taxon-biased diet preference in the 'generalist' beetle-hunting wasp *Cerceris rubida* provides insights on the evolution of prey specialization in apoid wasps. *Biological Journal of the Linnean Society*, 99, 544–558.

Polis, G.A. and Holt, R.D. (1992) Intraguild predation: the dynamics of complex trophic interactions. *Trends in Ecology and Evolution*, 7, 151–154.

Polis, G.A., Myers, C.A. and Holt, R.D. (1989) The ecology and evolution of intraguild predation: potential predators that eat each other. *Annual Review of Ecology and Systematics*, 20, 297–330.

Piñol, J., Espadaler, X., Perez, N. and Beven, K. (2009) Testing a new model of aphid abundance with sedentary and non-sedentary predators. *Ecological Modeling*, 220, 2469–2480.

Riechert, S.E. and Bishop, L. (1990) Prey control by an assemblage of generalist predators: spiders in garden test systems. *Ecology*, 71, 1441–1450.

Roach, S.H. (1980) Arthropod predators on cotton, corn, tobacco, and soybeans in South Carolina. *Journal of the Georgia Entomological Society*, 15, 131–138.

Rosenheim, J.A., Kaya, H.K., Ehler, L.E., Marois, J.J. and Jaffee, B.A. (1995) Intraguild predation among biological control agents: theory and evidence. *Biological Control*, 5, 303–335.

Rosenheim, J.A., Goeriz, R.E. and Thacher, E.F. (2004) Omnivore or herbivore? Field observations of foraging by *Lygus hesperus* (Hemiptera: Miridae). *Environmental Entomology*, 33, 1362–1370.

Sabelis, M.W. (1992) Predatory Arthropods, in *Natural Enemies: the population biology of predators, parasites and diseases* (ed. M.J. Crawley), Blackwell, Oxford, pp. 225–264.

Sayers, K., Norconk, M.A. and Conklin-Brittain, N.L. (2010) Optimal foraging on the roof of the world: Himalayan langurs and the classical prey model. *American Journal of Physical Anthropology*, 141, 337–357.

Schmitz, O.J. (2007) Predator diversity and trophic interactions. *Ecology*, 88, 2415–2426.

Settle, W.H., Ariawan, H., Astuti, E.T. *et al.* (1996) Managing tropical rice pests through conservation of generalist natural enemies and alternative prey. *Ecology*, 77, 1975–1988.

Sheehan, W. (1986) Response by specialist and generalist natural enemies to agroecosystem diversification: a selective review. *Environmental Entomology*, 15, 456–461.

Sih, A., Englund, G. and Wooster, D. (1998) Emergent impacts of multiple predators on prey. *Trends in Ecology and Evolution*, 13, 350–355.

Snyder, W.E., Snyder, G.B., Finke, D.L. and Straub, C.S. (2006) Predator biodiversity strengthens herbivore suppression. *Ecology Letters*, 9, 789–796.

Snyder, W.E. and Ives, A.R. (2003) Interactions between specialist and generalist natural enemies: parasitoids, predators, and pea aphid biological control. *Ecology*, 84, 91–107.

Southwood, T.R.E. (1977) Habitat, the templet for ecological strategies? *Journal of Animal Ecology*, 46, 336–365.

Spellman, B., Brown, M.W. and Mathews, C.R. (2006) Effect of floral and extrafloral resources on predation of *Aphis spiraecola* by *Harmonia axyridis* on apple. *Biocontrol*, 51, 715–724.

Stephens, D.W. and Krebs, J.R. (1986) *Foraging Theory*. Princeton University Press, Princeton.

Straub, C.S. and Snyder, W.E. (2006) Species identity dominates the relationship between predator biodiversity and herbivore suppression. *Ecology*, 87, 277–282.

Straub, C.S. and Snyder, W.E. (2008) Increasing enemy biodiversity strengthens herbivore suppression on two plant species. *Ecology*, 89, 1605–1615.

Sunderland, K.D., Topping, C.J., Ellis, S., Long, S., Van de Laak, S. and Else, M. (1996) Reproduction and survival of linyphiid spiders with special reference to *Lepthyphantes tenuis* (Blackwall). *Acta Jutlandica*, 71, 81–95.

Sunderland, K.D., Axelsen, J.A., Dromph, K. *et al.* (1997) Pest control by a community of natural enemies. *Acta Jutlandica*, 72, 271–326.

Symondson, W.O.C., Sunderland, K.D. and Greenstone, M.H. (2002) Can generalist predators be effective biocontrol agents? *Annual Review of Entomology*, 47, 561–594.

Symondson, W.O.C., Cesarini, S., Dodd, P.W. *et al.* (2006) Biodiversity vs. biocontrol: positive and negative effects of alternative prey on control of slugs by carabid beetles. *Bulletin of Entomological Research*, 96, 637–645.

Taylor, R.M. and Bradley, R.A. (2009) Plant nectar increases survival, molting, and foraging in two foliage wandering spiders. *Journal of Arachnology*, 37, 232–237.

Taylor, R.M. and Pfannenstiel, R.S. (2008) Nectar feeding by wandering spiders on cotton plants. *Environmental Entomology*, 37, 996–1002.

Taylor, R.M. and Pfannenstiel, R.S. (2009) How dietary plant nectar affects the survival, growth, and fecundity of a cursorial spider *Cheiracanthium inclusum* (Araneae: Miturgidae). *Environmental Entomology*, 38, 1379–1386.

Thorvilson, H.G., Pedigo, L.P. and Lewis, L.C. (1985) *Plathypena scabra* (F.) (Lepidoptera: Noctuidae) populations and the incidence of natural enemies in four soybean tillage systems. *Journal of Economic Entomology*, 78, 213–218.

Toft, S. (1995) Value of the aphid *Rhopalosiphum padi* as food for cereal spiders. *Journal of Applied Ecology*, 32, 552–560.

Tonhasca, A. (1993) Effects of agroecosystem diversification on natural enemies of soybean herbivores. *Entomologia Experimentalis et Applicata*, 69, 83–90.

Tonhasca, A. (1994) Response of soybean herbivores in two agronomic practices increasing agroecosystem diversity. *Agriculture Ecosystems and Environment*, 48, 57–65.

Tonhasca, A. Jr. and Stinner, B.R. (1991) Effects of strip intercropping and no-tillage on some pests and beneficial invertebrates of corn in Ohio. *Environment Entomology*, 20, 1251–1258.

Valicente, F.H. and O'Neil, R.J. (1995) Effects of host plants and feeding regimes on selected life-history characteristics of *Podisus maculiventris* (Say) (Heteroptera, Pentatomidae). *Biological control*, 5, 449–461.

Verheggen, F.J., Capella, Q., Schwartzberg, E.G., Voigt, D. and Haubruge, E. (2009) Tomato-aphid-hoverfly: a tritrophic interaction incompatible for pest management. *Arthropod–Plant Interactions*, 3, 141–149.

Vickerman, G.P. and Sunderland, K.D. (1975) Arthropods in cereal crops: nocturnal activity, vertical distribution and aphid predation. *Journal of Applied Ecology*, 12, 755–766.

Viswanathan, G.M., Buldyrev, S.V., Havlin, S., da Luz, M.G.E., Raposo, E.P. and Stanley, H.E. (1999) Optimizing the success of random searches. *Nature*, 401, 911–914.

Wäckers, F.L., van Rijn, P.C.J. and Bruin, J. (eds) (2005) *Plant-provided food for carnivorous insects: a protective mutualism and its applications*, Cambridge University Press, New York.

Wardle, R.A. and Buckle, P. (1923) Parasites and Predators, in *The principles of insect control* (eds R.A. Wardle and P. Buckle), Manchester University Press, New York, pp. 43–56.

Weber, D.C. and Lundgren, J.G. (2009) Assessing the trophic ecology of the Coccinellidae: Their roles as predators and as prey. *Biological Control*, 51, 199–214.

White, A.J., Wratten, S.D., Berry, N.A. and Weigmann, U. (1995) Habitat manipulation to enhance biological-control of *Brassica* pests by hover flies (Diptera, Syrphidae). *Journal of Economic Entomology*, 88, 1171–1176.

Wise, D.H. (1975) Food limitation of spider *Linyphia marginata* – experimental field studies. *Ecology*, 56, 637–646.

Wissinger, S.A. (1997) Cyclic colonization in predictably ephemeral habitats: a template for biological control in annual crop systems. *Biological Control*, 10, 4–15.

Yasuda, H. and Kimura, T. (2001) Interspecific interactions in a tri-trophic arthropod system: effects of a spider on the survival of larvae of three predatory ladybirds. *Entomologia Experimenatlis et Applicata*, 98, 17–25.

Chapter 4

ECOLOGICAL ECONOMICS OF BIODIVERSITY USE FOR PEST MANAGEMENT

Mark Gillespie and Steve D. Wratten

Biodiversity and Insect Pests: Key Issues for Sustainable Management, First Edition. Edited by Geoff M. Gurr, Steve D. Wratten, William E. Snyder, Donna M.Y. Read.
© 2012 John Wiley & Sons, Ltd. Published 2012 by John Wiley & Sons, Ltd.

INTRODUCTION

In chapter 1 of this volume it was demonstrated that, prior to modern agriculture, the recognition and use of biodiversity in pest management and agriculture in general was a necessity. Conversely, the contemporary loss of biodiversity and degradation of ecosystems are often associated with the use of potentially more reliable, predictable and efficient means of production (Costanza *et al.*, 1997; Robinson and Sutherland, 2002). As a result, agriculture as a sector of the global economy may be threatened by unsustainable exploitation and depreciation of its most important assets: natural resources (MEA, 2005; Pretty, 2008). Therefore the key challenge facing agriculture is maintaining its assets while meeting the increasing food demands of a rapidly expanding global population (Pretty, 2008; Godfray *et al.*, 2010). To this end, it has been widely recognised that better management of natural resources could have far-reaching human welfare benefits (Kleijn *et al.*, 2004; Kremen, 2005; Tscharntke *et al.*, 2005). However, the wise use of natural resources such as biodiversity is not a simple case of returning to old values. The adoption of sustainable pest management strategies, for instance, needs to be economically viable, particularly for the already vulnerable rural poor in developing countries (WRI, 2005).

The recognition of the socioeconomic factors involved in the dual goals of biodiversity conservation and sustained food production has led to the development of ecological economics as a discipline. This expansion of traditional economics has also brought growing awareness of the concept of 'ecosystem services' (ES; also called nature's or ecological services) to highlight the importance of nature to human well-being, and enhancing ES to meet the challenges of these two goals is one of the tenets of 'transdisciplinary' approaches of natural resource management that attempt to incorporate the value of nature into economic decision-making (Costanza, 1996). From these approaches come the economic tools to encourage the numerous stakeholders involved in natural resource management to consider conservation at scales beyond protected areas. This chapter explores some of the concepts of these approaches and their application to the use of biodiversity in pest management. However, it should be noted that the subject of ecological economics fills many textbooks and has amassed 40 years of research to date. This chapter therefore seeks only to provide a primer for interested readers and to highlight some of the complexities involved in marrying economics and ecology. Initially, a broad overview of ecological economics is given to provide a context for the consideration of how the concept of ES is viewed in the political and economic world. The complications surrounding ES and valuation are then summarised, and the chapter concludes with a review of valuation techniques and their application to sustainable pest management.

WHAT IS ECOLOGICAL ECONOMICS?

Ecological economics developed in response to the failure of traditional economic systems and policies to incorporate natural resources and consequently account for their degradation, and an equal failure of ecology to account for human cultural behaviour *within* ecosystems (Costanza and Daly, 1987; Costanza, 1996; Gomez-Baggethun *et al.*, 2010). Traditional economists tend to view environmental products as free, and the environmental consequences of economic decisions as 'externalities' (Asafu-Adjaye, 2005). Negative externalities, such as the pollution of waterways by pesticide or fertiliser runoff, are external costs not included in the market price of a good because there is no market for them and/or because they do not drive supply or demand. The market is said to 'fail' because the polluter does not pay the full cost (economic + social + environmental costs) of the production of goods, and pesticide producers do not pay for the full costs of their products (Pretty, 2008). Positive externalities can include the benefits of the environment or natural resources not paid for by producers, such as soil formation and pest regulation, and these form the basis of the concept of ES (Gomez-Baggethun *et al.*, 2010). Typically, externalities occur with a time lag between the external event and the realised cost, they can damage the interests of groups not represented in decision-making and can be difficult to identify at source, often due to a lack of data (Pretty, 2008). For example, in the UK, externalities of £1.5 billion per year related to pesticide use were recorded in the 1990s, a figure greater than net farm income (Pretty *et al.*, 2001).

In contrast to conventional economics, ecological economics attempts to incorporate or 'capture' externalities into economic decision-making by seeking ways to address market failure, by developing markets for ecosystem services to encourage their supply and by valuing the costs and benefits of natural resource management strategies (Asafu-Adjaye, 2005). Consequently, in the context of agriculture, it is broadly hoped that land managers will be encouraged to reduce environmental costs while maintaining profitability and increase natural benefits to enhance overall well-being. In general, a production system with fewer negative externalities (external costs) can be considered more sustainable (Pretty, 2008). This is a simplification of the discipline, however, as it also seeks to develop a more holistic and multidisciplinary understanding of humans and nature to enable more sustainable policies (Costanza, 1991); for instance by considering just distribution of natural resources as well as economic efficiency (Farley and Costanza, 2010). The methods for capturing externalities are also complex, varying according to the nature and characteristics of the cost, the goods and the stakeholders involved. For example, the valuation of externalities is a complicated undertaking because most benefits (ES) and costs may not be easily convertible into a common measurement unit such as price (Gomez-Baggethun et al., 2010), and therefore cannot be internalised into the value or price of a good. Ecological economists have been working towards providing the tools to address these complexities for 40 years to develop and encourage sustainable management strategies.

WHAT ARE ECOSYSTEM SERVICES?

The ecosystem services (ES) concept was introduced in the 1980s (Ehrlich and Ehrlich, 1981) and was initially pedagogic in scope, with proponents aiming to highlight how biodiversity loss affected the functions that were critical to human well-being (Gomez-Baggethun et al., 2010). The functional traits of the organisms of an ecosystem interact with each other and with abiotic factors (climate, geography) to regulate ecosystem properties and functions (Chapin et al., 1997; 2000) including ecological processes (e.g. flows of energy and matter), ecosystem goods (e.g. food, medicines) and ES (e.g. climate regulation, hydrological cycles) (Hooper et al., 2005). Biodiversity (B) is therefore positively linked to ecosystem functioning (EF), and greater levels of biodiversity can enhance the stability of ecosystems and the flow of goods and services (reviewed by Hooper et al., 2005). This is also a simplification, however, as recent work has suggested that a) the B–EF relationship is not linear (e.g. Cardinale et al., 2006; see Box 4.1), b) that species traits may be more important than number (e.g. Straub et al., 2008), and c) that habitats can also influence these relationships (Tylianakis and Romo, 2010).

Box 4.1 What are ecosystem services?

The concept of ecosystem services (ES) developed from the recognition that human well-being (obtained via anything from food and water consumption to recreation and culture) relies on fully functioning ecosystems. The ecosystems may be natural mountain idylls or extensively managed farms, but the functions satisfying many of our needs are essentially the same: pollination, soil formation, flood prevention, pest regulation and so on. Biodiversity forms the structure of ecosystems and interacts with abiotic conditions to regulate its functions (e.g. trees sequester carbon and accommodate other organisms, earthworms cycle soil nutrients). These functions become ES when they benefit humans (Fisher et al., 2009). Basic theory suggests that the more species and individuals there are in an ecosystem, the more functions that ecosystem can effectively perform (Figure 4.1 – line a). However, an ecosystem may require a large number of species before ecosystem functions increase significantly (Figure 4.1 – line b) and this may be because interactions between species enhance their efficiency. Con-

versely, a few key species may perform the majority of functions, leading to 'saturation' where other species are 'redundant' because of low abundance or efficiency or both (Figure 4.1 – line c). However, theory also suggests that the more redundancy there is in an ecosystem, the more stable that ecosystem will be to variations in abiotic factors (e.g. climate, weather events) and anthropogenic disturbance, because if one species becomes locally or temporally extinct, other species can still perform the same function. Figure 4.2 demonstrates simply the level of ES resulting from different levels of landscape complexity and disturbance. Pristine natural habitats are likely to have high biodiversity levels and therefore high function levels and ES. Some slightly disturbed habitats such as traditionally coppiced woodlands may deliver even more ES, because the disturbance creates a greater mosaic of habitat types within the ecosystem. Generally, simple and highly disturbed ecosystems are associated with low levels of biodiversity and ES.

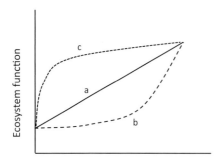

Figure 4.1 Observed relationships between ecosystem function and biodiversity (redrawn from Kremen, 2005).

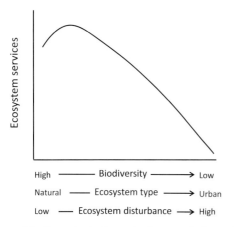

Figure 4.2 Generalised relationship between the level of ecosystem services and the level of biodiversity related to different land use intensities (redrawn from Braat and ten Brink, 2008).

Despite origins in the 1980s, the concept became important in the 1990s with broad economic valuations of global 'natural capital' and associated ES (Costanza *et al.*, 1997), which were instrumental in bringing the importance of ES and biodiversity to the forefront of science and policy (MEA, 2005). Although the valuations have since been widely criticised for their large uncertainties, extrapolation of localised, context-specific values and lack of utility in land use planning (Naylor and Ehrlich, 1997; Naidoo *et al.*, 2008), the monetary valuation of ES continues to be

viewed as vital in demonstrating the costs of inaction towards global biodiversity loss (Losey and Vaughan, 2006; EC, 2008). This financial view of ES has also increased the interest in creating economic incentives for conservation and in capturing external costs (Daily and Matson, 2008; Jack *et al.*, 2008), and research concerned with ES has increased exponentially since 1997 (Fisher *et al.*, 2009). As a result, numerous attempts have been made to develop agreed definitions, classifications and conceptual frameworks for incorporating ES into decision-making and policy (Farley and Costanza, 2010).

The most widely used definition of ES is that of the Millennium Ecosystem Assessment: 'the benefits people obtain from ecosystems' (MEA, 2005). However, use of this definition is by no means agreed upon universally (Wallace, 2007; Boyd and Banzhaf, 2007; Fisher and Turner, 2008; Fisher and Turner, 2008; Fisher *et al.*, 2008; Turner *et al.*, 2010; de Groot *et al.*, 2010), with other authors developing alternative definitions for different operational purposes (Boyd and Banzhaf, 2007; Fisher *et al.*, 2009) (Table 4.1). For example, Boyd and Banzhaf (2007) and Fisher and Turner (2008) argue that 'services' and 'benefits' of ecosystems should be distinguished to prevent double counting, and that ES should be confined to services within the sphere of ecology.

To highlight the diversity of ES and explicitly link human welfare to ES, the Millennium Ecosystem Assessment further classifies services into (a) provisioning services (the products obtained from ecosystems (e.g. food and fibre)), (b) regulating services (providing regulation of ecosystem processes (e.g. climate regulation, flood control, pest reduction)), (c) cultural services (nonmaterial benefits usually related to experiences (e.g. recreation, spiritual values)), and (d) supporting services (those that enable the production of other services (e.g. nutrient cycling, soil formation)) (MEA, 2003). However, alternative classifications have also been suggested recently, depending on the purpose of the study (Table 4.1). These include classifying ES into:
• intermediate and final services: intermediate services (pollination, pest control) contribute to final services (food production) which deliver the benefit (food for consumption) (Fisher and Turner, 2008).
• those based on excludability and rivalry: excludable services are those which people can be prevented from enjoying, and a rival service is one where a person's use of it prevents or depletes its use by another

Table 4.1 Ecosystem service definitions and classifications from some studies since the MEA (2005). The list of services are from MEA (2005) and are classified according to authors' systems.

Author	MEA (2005)	Wallace (2007)	Boyd and Banzhaf (2007)	Costanza (2008)	Fisher and Turner (2008)
Definition	'The benefits people obtain from ecosystems'	As MEA	(final ES) 'The components of nature, directly enjoyed, consumed, or used to yield human well-being'	As MEA	'Aspects of ecosystems utilised (actively or passively) to produce human well-being'
Classification scheme	• provisioning • regulating • supporting • cultural	• Adequate resources • Protection from predators, diseases, parasites • Benign physical and chemical environment • Socio-cultural fulfilment	Final and intermediate services	Spatial characteristics: e.g. global, regional, local; proximal/non-proximal to beneficiary; in situ or flow related	• abiotic inputs • intermediate services • final services • benefits
Example Service classifications					
Food	Provisioning	Adequate resources	Final service or benefit	*In situ* (point of use)	Benefit
Water	Provisioning	Adequate resources	Final service or benefit	Directional flow related: flow from point of production to point of use	Benefit
Climate regulation	Regulating	Process – not a service	Intermediate service	Global, non-proximal (does not depend on proximity)	Intermediate service
Pest and disease regulation	Regulating	Process – not a service	Intermediate service	Local, proximal (depends on proximity)	Intermediate service
Soil formation	Supporting	Process – not a service	Intermediate service	*In situ* (point of use)	Intermediate service
Recreation	Cultural	Socio-cultural fulfilment	Benefits, not services. Components of the natural landscape that provide these benefits are the services	User movement related: flow to people from unique natural features	Benefit

(Costanza, 2008). For example, most regulatory and cultural services are non-rival and non-excludable.
• those based on spatial characteristics such as scale or proximity to human populations (Costanza, 2008).

These distinctions are important because the classification and definition of ES used determines the conceptual framework used to value ES, as explored in the next section. A broad system of definition and a pluralism of classifications are considered to be necessary because of the many applications of ecosystem services (Costanza, 2008).

THE ECOLOGICAL ECONOMICS OF ECOSYSTEM SERVICES

Ecological economics and ES are tightly bound in the literature because valuing services in monetary terms is considered to be a useful way of framing their importance to non-ecologists and funding and policy managers and to aid decision-making. However, valuing ES is complex largely because there are no effective markets for most ES or ecosystem functions (Peterson *et al.*, 2010), and no one can own or have sole rights to them (Sternberg, 1996). For example, although a value for the ES of pest regulation could be ascribed to the costs avoided through reduced pesticide use (e.g. Sandhu *et al.*, 2008), or the costs of releasing commercially available natural enemies, this type of valuation may be viewed as oversimplifying the ecological resistance to perturbations of a diverse community of native natural enemies and the additional functions deriving from species interactions. The ability of natural enemies to perform their service outside the farm boundaries, demonstrating their non-rival and non-excludable classification, may also be important. Other complexities include the fact that the organisms may not even reside within farm boundaries, creating the need to consider different management scales for different ES (Zhang *et al.*, 2007), and the interconnected nature of ES, with ecosystems generating multiple services, and a single species typically performing more than one function (Diaz *et al.*, 2007). For example, hoverflies (Syrphidae) are predators of pests in their larval life stage but the adults are pollinators.

Table 4.2 presents some of the valuation methods for ES, their application and limitations. Many of these are applicable to only a handful of ES due to the relative ease of ascribing global monetary values and incorporating them into policy or market-based compensation instruments (Naidoo *et al.*, 2008). Supporting services and those with effects on the improvement of human welfare that are harder to classify and understand, such as soil formation and nutrient cycling, are more difficult to value. These difficulties can be addressed using alternative approaches to ES valuation employing tools such as mapping (Naidoo *et al.*, 2008; Raudsepp-Hearne *et al.*, 2010) and modelling (Nelson *et al.*, 2010) to evaluate the change in ES from alternative management strategies, and integrated cost–benefit analysis (Balmford *et al.*, 2002; Turner *et al.*, 2003), which attempt to expand traditional cost–benefit analysis (CBA) to incorporate non-market ES. An alternative suggestion to putting ecological functions into the same language as economics is to do the reverse, and 'externalise the internalities' by framing goods and services in terms of the energy required to produce them (Odum and Odum, 2000). Peterson *et al.* (2010) suggest that this view would refocus attention from profit to environmental protection but admit that practical application would be difficult. In the absence of alternatives, the development of valuation of ES is ongoing but requires careful consideration of the complexities of ES.

THE ECOLOGICAL ECONOMICS OF SUSTAINABLE PEST MANAGEMENT

Valuing agricultural biodiversity

The valuation of agro-biodiversity is especially difficult because of the complexity and multi-scale characteristics of biodiversity, although attempts have been made to value the ES of pest management and insects in general (Losey and Vaughan, 2006; Sandhu *et al.*, 2008; Porter *et al.*, 2009). For example, Losey and Vaughan (2006) made minimum estimates of a set of services provided by insects in the USA, including pest control, using the replacement cost method (Table 4.2). The valuations suggested a minimum value of insect pest regulation of US$4.5 billion. Broad-scale attempts at valuation have been criticised, either on the basis of the method used, or because of a lack of consideration of concepts such as double counting (Fisher and Turner, 2008; Fu *et al.*, 2011; see Box 4.2). However, valuations in these studies are developed to attract attention to the benefits of biodiversity conservation to encourage policy and landscape-scale decisions, rather than for making farm-scale decisions. For

Table 4.2 Resource economics ecosystem service valuation methods and their limitations.

Technique	Details of method	Limitations
Market prices	Values are prices of goods or services that are traded on markets (extendable to non-market goods based on effects on prices of market goods)	Not available for many services, ignores social costs
Production function	Tracing impact of changes in ES to human welfare	Inadequate data, ignores ES interdependence
Travel cost	The amount a user is willing to pay to travel to a service	Poorly reflects true value, ignores many ES
Hedonic pricing	The price of a good is reflected by services attached to it, e.g. housing located close to scenic beauty	Restricted applicability, large dataset required
Replacement cost	Cost of replacing or restoring an ES, depending on the beneficiaries' willingness to pay for the restoration	Sensitive: subtle differences in ES descriptions can lead to widely differing cost estimates
Defensive expenditure or Avoidance cost	Costs incurred in avoiding environmental damage or reduced function	Assumes substitutability, hard to disentangle multiple ES values
Contingent valuation	Stated preferences of individuals as described through questionnaires and interviews – usually assesses people's willingness to pay (WTP) for benefits or willingness to accept (WTA) costs	Potential for many sources of bias in answers, subjective, time consuming
Choice modelling or Conjoint analysis	Stated preference method where individuals choose their preferred scenario from a number of options	Potential for many sources of bias in answers, subjective, time consuming
Deliberative monetary valuation	Use of small representative groups, citizen juries and stakeholder analysis to discuss issues	Time consuming and costly, potential lack of representation in small groups, subject to dominant participants, bias potential, difficult to develop monetary values

this, more localised valuations have also been conducted. For example, the value of coffee borer control by birds in Jamaica has been estimated to be US$310 ha^{-1} (Johnson *et al.*, 2010). Similarly, estimates of the economic importance of bats for pest suppression in cotton-dominated agricultural landscapes in Texas are between US$5 to US$70 ha^{-1} (Cleveland *et al.*, 2006). By framing pest control services in this way, it is hoped that these valuations will encourage farmers and land managers to adopt biodiversity-friendly practices.

Who pays for biodiversity conservation on farmland?

It is clear that biodiversity in agriculture is potentially worth a great deal to farmers and society at large, but conserving biodiversity for pest control services may require some form of capital outlay for the farmer to convert to alternative technologies. The economic costs and benefits of biodiversity in pest management are often considered to be borne or benefited from by the grower or society as a whole (Griffiths *et al.*, 2008), although consumers may also benefit from a potentially healthier product, for example, which they may be willing to pay for.

The farmer

There are three broad types of biological control: 1) classical biological control involves the release of one or more species of appropriate natural enemies into a pest-infested area, 2) augmentation biological control aims to increase the abundance of natural enemies already present in low numbers through timely releases of predators into the cropping system, and 3)

Box 4.2 Valuing biodiversity and ES

Valuing biodiversity and ES is considered an important way to include these factors in decision-making at numerous levels. However, the recognition of common pitfalls of valuation such as double counting indicates the complexity of this task. For example, pest control may be valued using the total avoided costs of pesticides (i.e. the value of the pesticides that would have been sprayed in the absence of natural enemies). However, adding the value of this ES to other benefits derived from biodiversity could be considered as double counting under certain ES classification schemes. For example, biological pest regulation is an intermediate service contributing to the production of the final service: food (Fu et al., 2011). The value of pest regulation (and other intermediate ES) should therefore be included in the price of the food items, just as the value of labour is included in the price of a car. The true value of total ES to the farmer is the overall profit obtained and this may include savings in pesticides or consumer premiums for environmentally friendly products. Overall, under this type of classification, the true value of total ES to society is the value of final services, just as the value of final goods makes up a country's GDP (Fu et al., 2011).

In addition, large-scale valuations can sometimes ignore multi-scale effects and the non-linearity of ES. Using the economic concept of supply and demand, when biological control services are scarce, for example in a simple landscape, the value of an additional 'unit' is relatively large compared to the abundant services in complex landscapes (Figure 4.3). The additional or 'marginal' unit price could consist of the amount of new habitat needed to attract one 'unit' of natural enemies, or the amount of pesticide required to replace this unit. However, service provision is often non-linear so the addition of an extra unit of habitat will not necessarily create one unit of an ES, and this is further complicated because organisms can perform services at multiple scales. The interdependence and non-linearity of ES make them virtually impossible to separate in ways required for their valuation (Fu et al., 2011). While valuation is an important guide, inherent difficulties in definition and classification hinder attempts to successfully encourage sustainable practices.

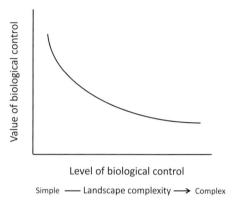

Figure 4.3 A generalised demand curve for the ecosystem service of biological control. The demand curve indicates the price individuals are willing to pay for an incremental amount of biological control (adapted from Turner et al., 2010).

conservation biological control attempts to improve populations of natural enemies through habitat management. The decision of farmers or land managers to adopt any of these techniques may be made on the basis of their perceived economic efficiency, but is often also influenced by social, cultural or environmental factors (Jackson et al., 2007) and prevailing policy and incentive support systems (Falconer and Hodge, 2000, see below). Basic and conventional cost–benefit analysis is frequently used to demonstrate the cost-effectiveness of biodiversity use in pest management (de Groote et al., 2003; de Lange and van Wilgen, 2010; see Menzler-Hokkanen, 2006 for a review). The main economic benefits of biological control to the grower consist of savings in terms of pesticide purchase, and the costs involved in applying it (Griffiths et al., 2008). For example, a key pest of vines in New Zealand is a leafroller caterpillar, *Epiphyas postvittana* (Walker), which can be managed by planting buckwheat or other nectar plants between vine rows to improve the ecological fitness of its key natural enemy, the parasitoid wasp *Dolichogenidea tasmanica* (Cameron) at a cost of NZ$2 per ha. There is no loss of productive land and annual variable costs of NZ$250 per ha can be saved (Scarratt, 2005) indicating a cost–benefit ratio of 125:1. However, this is a simple example, and a grower's decision to adopt this practice may be further influenced by any additional implementation costs and the efficacy of the control. Table 4.3 high-

Table 4.3 Obvious and less obvious costs and benefits of adopting biological control practices.

Costs	Benefits
Obvious factors	
Land use changes and loss of yield	Small savings in production costs
Loss of yield due to variability of control	Savings in pesticides and application
Costs of technique (flower seeds, attractants, natural enemies, specialised machinery)	Potential subsidies from agri-environment schemes or price premiums for organic or 'environmentally friendly' products
Labour	
Less obvious, delayed or non-monetary factors	
Training	No pesticide resistance concerns
Monitoring	Avoidance of strict pesticide regulations of taxation
Maintenance costs of new habitats	Landscape manipulation helps other ES
Augmentation of natural enemies	Pesticide spray operator health improvement
Increase in alternative pests	Clean waterways
Increase in weeds	Aesthetic value of improved landscape
Increase in diseases	Suppression of weeds
Multiple pest attacks not previously considered	Improved pollinator conservation and pollination
	Other wildlife benefits
	Benefits beyond farm boundary

lights the potential hidden costs and benefits of biological control in other situations. For example, the costs may comprise those associated with land use changes and consequent loss of yield, although removing land from production results in small savings in energy and inputs. In addition, biological control methods and chemical management may not be equally effective, and findings regarding this are equivocal (Gurr *et al.*, 2003; Griffiths *et al.*, 2008). Costs such as augmentation of natural enemies or emergency pesticide spraying may be required due to the uncertainty of biological control practices, which can be an important driver in grower adoption (Lu *et al.*, 1999).

Conversely, added to the benefit side of the equation are the more complex possibilities that may make conservation biological control more appealing; for example, the possibility that the target pest develops pesticide resistance, that pesticide regulation becomes gradually more strict and taxation of pesticides is introduced (Falconer and Hodge, 2000; Griffiths *et al.*, 2008). There are also the multiple beneficial side effects on other ES or, in economic language, the positive externalities of biological control which are typically ignored by farmers when making an economic decision (Jackson *et al.*, 2007). For example, most wildflower seed mixes available and recommended to farmers as part of agri-environment schemes in Europe

are designed to attract both pollinators and natural enemies (Haaland *et al.*, 2011), but these mixtures may also suppress weeds in field margins (Pywell *et al.*, 2005), facilitate the survival of rare plants and insects of conservation value (e.g. Marshall and Moonen, 2002), provide food to farmland birds (Boatman, 1999), enhance agroecosystem resilience (Hooper *et al.*, 2005) and provide aesthetic benefits (Forbes *et al.*, 2009).

Furthermore, as Zhang *et al.* (2007) point out, very few ES are confined to the field level, so the management actions a farmer takes are more likely to succeed if they are either farm-wide or fit into management at larger scales. This can make it particularly difficult for a farmer to implement biological control if the farm is located in a simplified degraded landscape, because landscape-scale management schemes are more effective when located in areas with some source level of biodiversity (Tscharntke *et al.*, 2005; Kremen and Chaplin-Kramer, 2007; Rundlof *et al.*, 2008; Merckx *et al.*, 2009). A key challenge in landscape-scale management of pests therefore is identifying the optimal spatial structure of connected habitats that supports farm management cost-effectively and without unnecessary loss of production (Pascual and Perrings, 2007; Gámez-Virués *et al.*, chapter 7 of this volume).

The complex list of factors in Table 4.3 that have not yet been valued but may have economic impacts on some management systems, may partly explain why less than 1% of global pest control sales are related to biocontrol methods (Griffiths *et al.*, 2008). Other reasons for this may include the prevailing policy and commercial pressures (Falconer and Hodge, 2000; Pascual and Perrings, 2007), rejection or lack of trust in cost–benefit ratios (Beck *et al.*, 1992) or lack of information from research to ground-level management (Sagoff, 2011). Incorporating complex factors into conventional CBA, or re-inventing CBA, will take some innovative research (for an example, see Box 4.3), given that few of the factors in Table 4.3 are amenable to valuation (Costanza *et al.*, 1997; de Groot *et al.*, 2002; Naidoo *et al.*, 2008). However, with growing awareness that multi-function sustainable use of farmland is more economically beneficial than single-function use (Balmford *et al.*, 2002; Turner *et al.*, 2003; de Groot *et al.*, 2010), research should also identify the fair compartmentalisation of costs and benefits accruing to the grower and to society at large.

The consumer

One way to view the internalisation of externalities into the price of goods is as a contribution by the consumer to pay for more sustainable farming practices. This occurs in organic farming, with growers earning premiums to offset what is sometimes a loss in yield due to factors such as the variable pest control of non-chemical techniques. For example, pipfruit growers in New Zealand earn a 100–150% premium for organically produced fruit with yields up to 30% lower (Walker *et al.*, 2004). Niche markets for 'biodiversity friendly' products, as opposed to organic products, could be developed to channel premiums to farmers practising biodiversity conservation (Pascual and Perrings, 2007), without requiring some of the necessary changes associated with the conversion to organic farming. For example, shade-grown coffee (also known as bird-friendly coffee) is a more traditional form of coffee production which provides habitat for a diverse flora and fauna (Perfecto *et al.*, 2005). Certified growers in Latin America earn higher prices for this type of coffee, helping them to withstand pressure from rapid expansion and overproduction by mass producers (Fleischer, 2002).

Alternatively, food producers could be taxed under 'polluter pays' principles with the intention of passing

Box 4.3 Alternative view of cost–benefit analysis and ES evaluation

Due to the complexity of ecosystems and the difficulties associated with ES valuation, a more meaningful approach may be to value the changes that occur as a result of management strategies and the effect this will have on a number of other goods and services (Fu *et al.*, 2011). For example, Farber *et al.* (2006) present a number of long-term ecological research projects to illustrate the application of an ES approach to the assessment of management strategies. They suggest considering the change in service provision from the status quo that occurs through different management strategies. For example, the Kellogg Biological Station in Michigan, USA is a long-term experimental farm site allowing the comparison of ES provision across different management practices. A service matrix reflects the changes in different ES likely to occur through three different management options compared with traditional management, and these are scored between –3 to 3. For example, biological regulation was considered to be unaffected under no tillage (0 score), and to change positively under low chemical organic farming (+1) or pasture and grazing (+2). These scores are then multiplied by a value weight (0–3), depending on their relative value to the farmer or society, and the overall service change values are aggregated to provide an overall score for each management type. The scores and value weights can be assessed using different valuation methods (Table 4.2). While the results may be a simplification of the complexity of ES, such alternative valuation schemes are more likely to be inclusive of managers and provide tools for assessing the effects of management on total ES.

the cost onto customers who choose polluting products. For example, in Sweden, a US$0.25 per kg tax on nitrogen use effectively reduced use by 10% and helped to alleviate water quality problems (OECD, 2001). Some of these ideas have been explored through product labelling in the New Zealand wine industry, for example, where growers using conservation biological control could highlight this fact on bottle labels and achieve a higher price or more sales than unlabelled wine (Cullen *et al.*, 2008). However, the creation of niche markets also depends on consumer demands for such products and their willingness to pay (WTP) premiums for biodiversity-friendly food (see Randall, 2002 for a review of WTP valuation). This has not been carried out for food produced using biological control, but a niche market for biodiversity-friendly produce may be well placed to take advantage of the growing popularity of socially responsible products (Mahe, 2010).

Society

In addition to reducing their own costs and providing pesticide-free food, farmers who adopt biological control practices reduce other externalities of intensive agriculture, including reduced runoff of toxicants into ground waters, rivers and lakes and reduced ill-health effects of operators (Griffiths et al., 2008), and perhaps should be compensated by society (Pretty, 2008). It is also in the interests of society to ensure landscape-scale coordination of farmer adoption, which can produce societal benefits through multiple effects (Tscharntke et al., 2005; de Groot et al., 2010) and aesthetic and cultural improvement when this is valued by social groups (Griffiths et al., 2008), particularly when the surrounding landscape is managed in a similar way (Rundlof et al., 2008; Merckx et al., 2009). The practical adoption of this information is slow, but a good example of an integrated agri-environment programme has been reported in China where 'agroecological engineering' has been promoted in 2,000 townships and villages in 150 counties through various subsidies, loans and tax breaks, and publicly funded marketing, technical assistance and access to research institutes (Li, 2001). Chapter 14 of this volume provides a more comprehensive update of China's 'Green Plant Protection' initiative.

The argument for compensation is further compounded by the fact that the demonstration of favourable cost–benefit ratios has not been enough to convince many farmers to conserve biodiversity; most require incentives to undertake the initial start-up costs and withstand the risks of failure of sustainable pest management techniques and possible time lags between costs and benefits (Pretty, 2008). Funding is also required for continued research, exchange of information, training and extension activities (Griffiths et al., 2008). There is much discussion in the literature over the best way to administer the compensation that is clearly needed. The most prevalent policies at the

landscape level are market-based mechanisms known as payments for ecosystem services (PES), although ecological economists suggest that a mixture of market and non-market based instruments are required for some ES to achieve this (Farley and Costanza, 2010). The literature surrounding PES is growing, and a review is beyond the scope of this chapter as they have not yet reached the world of pest control, most commonly including initiatives for improving watershed services and carbon sequestration (Corbera et al., 2007).

Of more relevance to pest control is the modern proliferation of agri-environment schemes (AES): voluntary incentive schemes that provide subsidies to farmers for adopting environmentally friendly practices, which are currently available in Europe and the US (Kleijn et al., 2006; Klimek et al., 2008; Haaland et al., 2011). Most schemes are 'action-oriented' (Klimek et al., 2008) in that farmers receive payments for meeting the schemes' requirements to compensate for loss of income. These have had mixed results and are usually positive only for the most common species (Kleijn et al., 2006). In particular, studies have highlighted the need to tailor schemes on regional scales (Whittingham, 2007), develop measurability of more precisely defined objectives (Kleijn et al., 2006) and to encourage landscape-wide adoption of sustainable practices (Whittingham, 2007; Rundlof et al., 2008; Merckx et al., 2009). Action-oriented schemes are also criticised for offering fixed payments in a situation of 'information asymmetry' where farmers know more about their opportunity costs and land characteristics than does the compensating government (Ferraro, 2008; Klimek et al., 2008). Farmers could therefore be over- or under-compensated, leading to abuse or low participation respectively. Klimek et al. (2008) have proposed a new type of 'results-oriented' approach to AES to address these concerns, whereby farmers bid in auctions for conservation contracts with the regional government and are paid for results following completion and inspection of compliance with well-defined goals. In bidding for conservation measures farmers reveal their opportunity costs and the spatial heterogeneity of costs and benefits can be accounted for by auctioning different quality levels of improvement (Klimek et al., 2008). This method provides the potential to more accurately compensate farmers for conservation measures and allows farmers to diversify their income. As with other concepts, however, research is needed to expand the system and transfer the scheme between habitat types, and this will rely on the evolution of methods for measuring and valuing ES. For any compensation scheme to be effective the incentive level has to be as accurate as possible to encourage wide adoption (Randall, 2002).

CONCLUSIONS AND FUTURE RESEARCH

The use of biodiversity in pest management can have a number of additional benefits which may not be valued by farmers. However, as the above discussion has shown, the costs and benefits of biodiversity use are complex and difficult to value. While ecological economics is a relatively young discipline, workers are continually striving to develop markets and tools for capturing externalities. To help drive this development, data are particularly required on the cascading effects of biodiversity conserving pest management on to yield (Griffiths *et al.*, 2008), quantifying the impact of biological control on other ES and their flows to and from agriculture and how this varies across time and space (Kremen, 2005), and how habitat manipulations can be modified to optimise impact (Zhang *et al.*, 2007). Such data are likely to contribute to a better understanding of the true costs and benefits of adopting biodiversity-based pest management including socio-economic factors rather than just those of technological innovation.

REFERENCES

Asafu-Adjaye, J. (2005) *Environmental economics for non-economists: techniques for sustainable development*, 2nd edn, World Scientific Publishing Co., Hackensack, NJ.

Balmford, A., Bruner, A., Cooper, P. *et al.* (2002) Ecology – economic reasons for conserving wild nature. *Science*, 297, 950–953.

Beck, N.G., Herman, T.J.B., Cameron, P.J. and New Zealand Plant Protection Society (1992) *Scouting for lepidopteran pests in commercial cabbage fields*. Proceedings of the Forty-Fifth New Zealand Plant Protection Conference, 31–34.

Boatman, N. (1999) Marginal benefits? How field edges and beetle banks contribute to game and wildlife conservation. *The Game Conservancy Trust Review of 1998*, 61–67.

Boyd, J. and Banzhaf, S. (2007) What are ecosystem services? The need for standardized environmental accounting units. *Ecological Economics*, 63, 616–626.

Braat, L. and ten Brink, P. (eds) (2008) *The cost of policy inaction: the case of not meeting the 2010 biodiversity target*. Study for the European Commission, DG Environment. Alterra report 1718, Wageningen.

Cardinale, B.J., Srivastava, D.S., Duffy, J.E. *et al.* (2006) Effects of biodiversity on the functioning of trophic groups and ecosystems. *Nature*, 443, 989–992.

Chapin, F.S., Walker, B.H., Hobbs, R.J. *et al.* (1997) Biotic control over the functioning of ecosystems. *Science*, 277, 500–504.

Chapin, F.S., Zavaleta, E.S., Eviner, V.T. *et al.* (2000) Consequences of changing biodiversity. *Nature*, 405, 234–242.

Cleveland, C.J., Betke, M., Federico, P. *et al.* (2006) Economic value of the pest control service provided by Brazilian free-tailed bats in South-central Texas. *Frontiers in Ecology and the Environment*, 4, 238–243.

Corbera, E., Kosoy, N. and Tuna, M.M. (2007) Equity implications of marketing ecosystem services in protected areas and rural communities: Case studies from Meso-America. *Global Environmental Change-Human and Policy Dimensions*, 17, 365–380.

Costanza, R. (ed.) (1991) *Ecological economics: the science and management of sustainability*, Columbia University Press, New York.

Costanza, R. (1996) Ecological economics: Reintegrating the study of humans and nature. *Ecological Applications*, 6, 978–990.

Costanza, R. (2008) Ecosystem services: Multiple classification systems are needed. *Biological Conservation*, 141, 350–352.

Costanza, R. and Daly, H.E. (1987) Toward an ecological economics. *Ecological Modelling*, 38, 1–7.

Costanza, R., dArge, R., deGroot, R. *et al.* (1997) The value of the world's ecosystem services and natural capital. *Nature*, 387, 253–260.

Cullen, R., Warner K.D., Jonsson M. and Wratten, S.D. (2008) Economics and adoption of conservation biological control. *Biological Control*, 45, 272–280.

Daily, G.C. and Matson, P.A. (2008) Ecosystem services: from theory to implementation. *Proceedings of the National Academy of Sciences of the United States of America*, 105, 9455–9456.

De Groot, R.S., Wilson, M.A. and Boumans, R.M.J. (2002) A typology for the classification, description and valuation of ecosystem functions, goods and services. *Ecological Economics*, 41, 393–408.

De Groot, R.S., Alkemade, R., Braat, L., Hein, L. and Willemen, L. (2010) Challenges in integrating the concept of ecosystem services and values in landscape planning, management and decision making. *Ecological Complexity*, 7, 260–272.

De Groote, H., Ajuonu, O., Attignon, S., Djessou, R. and Neuenschwander, P. (2003) Economic impact of biological control of water hyacinth in Southern Benin. *Ecological Economics*, 45, 105–117.

De Lange, W.J. and van Wilgen, B.W. (2010) An economic assessment of the contribution of biological control to the management of invasive alien plants and to the protection of ecosystem services in South Africa. *Biological Invasions*, 12, 4113–4124.

Diaz, S., Lavorel, S., de Bellom F., Quetier, F., Grigulis, K. and Robson, M. (2007) Incorporating plant functional diversity effects in ecosystem service assessments. *Proceedings of the National Academy of Sciences of the United States of America*, 104, 20684–20689.

EC (2008) *The economics of ecosystems and biodiversity*, European Commission, Brussels.

Ehrlich, P.R. and Ehrlich, A.H. (1981) *Extinction: the causes and consequences of the disappearance of species*, Random House, New York.

Falconer, K. and Hodge, I. (2000) Using economic incentives for pesticide usage reductions: responsiveness to input taxation and agricultural systems. *Agricultural Systems*, 63, 175–194.

Farber, S., Costanza, R., Childers, D.L. *et al.* (2006) Linking ecology and economics for ecosystem management. *Bioscience*, 56, 121–133.

Farley, J. and Costanza, R. (2010) Payments for ecosystem services: From local to global. *Ecological Economics*, 69, 2060–2068.

Ferraro, P.J. (2008) Asymmetric information and contract design for payments for environmental services. *Ecological Economics*, 65, 810–821.

Fisher, B. and Turner, R.K. (2008) Ecosystem services: Classification for valuation. *Biological Conservation*, 141, 1167–1169.

Fisher, B., Turner, K., Zylstra, M. *et al.* (2008) Ecosystem services and economic theory: integration for policy-relevant research. *Ecological Applications*, 18, 2050–2067.

Fisher, B., Turner, R.K. and Morling, P. (2009) Defining and classifying ecosystem services for decision making. *Ecological Economics*, 68, 643–653.

Fleischer, G. (2002) Toward more sustainable coffee: consumers fuel demand for more sustainable agriculture. *Agriculture Technology Notes*, 23, The World Bank (June).

Forbes, S.L., Cohen, D.A., Cullen, R., Wratten, S.D. and Fountain, J. (2009) Consumer attitudes regarding environmentally sustainable wine: an exploratory study of the New Zealand marketplace. *Journal of Cleaner Production*, 17, 1195–1199.

Fu, B.J., Su, C.H., Wei, Y.P., Willett, I.R., Lu, Y.H. and Liu, G.H. (2011) Double counting in ecosystem services valuation: causes and countermeasures. *Ecological Research*, 26, 1–14.

Godfray, H.C.J., Beddington, J.R., Crute, I.R. *et al.* (2010) Food Security: the challenge of feeding 9 billion people. *Science*, 327, 812–818.

Gomez-Baggethun, E., de Groot, R., Lomas, P.L. and Montes, C. (2010) The history of ecosystem services in economic theory and practice: from early notions to markets and payment schemes. *Ecological Economics*, 69, 1209–1218.

Griffiths, G.J.K., Holland, J.M., Bailey, A. and Thomas, M.B. (2008) Efficacy and economics of shelter habitats for conservation biological control. *Biological Control*, 45, 200–209.

Gurr, G.M., Wratten, S.D. and Luna, J.M. (2003) Multi-function agricultural biodiversity: pest management and other benefits. *Basic and Applied Ecology*, 4, 107–116.

Haaland, C., Naisbit, R.E., and Bersier, L.F. (2011) Sown wildflower strips for insect conservation: a review. *Insect Conservation and Diversity*, 4, 60–80.

Hooper, D.U., Chapin, F.S., Ewel, J.J. *et al.* (2005) Effects of biodiversity on ecosystem functioning: A consensus of current knowledge. *Ecological Monographs*, 75, 3–35.

Jack, B.K., Kousky, C. and Sims, K.R.E. (2008) Designing payments for ecosystem services: Lessons from previous experience with incentive-based mechanisms. *Proceedings of the National Academy of Sciences of the United States of America*, 105, 9465–9470.

Jackson, L.E., Pascual, U. and Hodgkin, T. (2007) Utilizing and conserving agrobiodiversity in agricultural landscapes. *Agriculture Ecosystems and Environment*, 121, 196–210.

Johnson, M.D., Kellermann, J.L. and Stercho, A.M. (2010) Pest reduction services by birds in shade and sun coffee in Jamaica. *Animal Conservation*, 13, 140–147

Kleijn, D., Berendse, F., Smit, R. *et al.* (2004) Ecological effectiveness of agri-environment schemes in different agricultural landscapes in the Netherlands. *Conservation Biology*, 18, 775–786.

Kleijn, D., Baquero, R.A., Clough, Y. *et al.* (2006) Mixed biodiversity benefits of agri-environment schemes in five European countries. *Ecology Letters*, 9, 243–254.

Klimek, S., Kemmermann, A.R., Steinmann, H.H., Freese, J. and Isselstein, J. (2008) Rewarding farmers for delivering vascular plant diversity in managed grasslands: a transdisciplinary case-study approach. *Biological Conservation*, 141, 2888–2897.

Kremen, C. (2005) Managing ecosystem services: what do we need to know about their ecology? *Ecology Letters*, 8, 468–479.

Kremen, C. and Chaplin-Kramer, R. (2007) Insects as providers of ecosystem services: crop pollination and pest control, in *Insect conservation biology* (eds A.J.A. Stewart, T.R. New and O.T. Lewis), CAB International, Wallingford, pp. 349–382.

Li, W. (2001) Agro-ecological farming systems in China. *Man and the biosphere series*, 26, UNESCO, Paris.

Losey, J.E. and Vaughan, M. (2006) The economic value of ecological services provided by insects. *Bioscience*, 56, 311–323.

Lu, Y.C., Watkins, B. and Teasdale, J. (1999) Economic analysis of sustainable agricultural cropping systems for mid-Atlantic states. *Journal of Sustainable Agriculture*, 15, 77–93.

Mahe, T. (2010) Are stated preferences confirmed by purchasing behaviours? The case of fair trade-certified bananas in Switzerland. *Journal of Business Ethics*, 92, 301–315.

Marshall, E.J.R. and Moonen, A.C. (2002) Field margins in northern Europe: their functions and interactions with agriculture. *Agriculture Ecosystems and Environment*, 89, 5–21.

MEA (2003) *Ecosystems and human well-being: a framework for assessment*, Millennium Ecosystem Assessment, Island Press, Washington, DC.

MEA (2005) *Ecosystems and human well-being: synthesis*, Millennium Ecosystem Assessment, Island Press, Washington, DC.

Menzler-Hokkanen, I. (2006) Socioeconomic significance of biological control. *Ecological and societal approach to biological control*, 2, 13–25.

Merckx, T., Feber, R.E., Riordan, P. *et al.* (2009) Optimizing the biodiversity gain from agri-environment schemes. *Agriculture Ecosystems and Environment*, 130, 177–182.

Naidoo, R., Balmford, A., Costanza, R. *et al.* (2008) Global mapping of ecosystem services and conservation priorities. *Proceedings of the National Academy of Sciences of the United States of America*, 105, 9495–9500.

Naylor, R.L. and Ehrlich, P.R. (1997) Natural pest control services and agriculture, in *Nature's services: societal dependence on natural ecosystems* (ed. G.C. Daily), Island Press, Washington, DC, pp. 151–174.

Nelson, E., Sander, H., Hawthorne, P. *et al.* (2010) Projecting global land-use change and its effect on ecosystem service provision and biodiversity with simple models. *Plos One*, 5, doi:10.1371/journal.pone.0014327.

Odum, H.T. and Odum, E.P. (2000) The energetic basis for valuation of ecosystem services. *Ecosystems*, 3, 21–23.

OECD (2001) *Multifunctionality: towards an analytical framework*, Organisation for Economic Cooperation and Development, Paris.

Pascual, U. and Perrings, C. (2007) Developing incentives and economic mechanisms for in situ biodiversity conservation in agricultural landscapes. *Agriculture Ecosystems and Environment*, 121, 256–268.

Perfecto, I., Vandermeer, J., Mas, A. and Pinto, L.S. (2005) Biodiversity, yield, and shade coffee certification. *Ecological Economics*, 54, 435–446.

Peterson, M.J., Hall, D.M., Feldpausch-Parker, A.M. and Peterson, T.R. (2010) Obscuring ecosystem function with application of the ecosystem services concept. *Conservation Biology*, 24, 113–119.

Porter, J., Costanza, R., Sandhu, H., Sigsgaard, L. and Wratten, S. (2009) The value of producing food, energy, and ecosystem services within an agro-ecosystem. *Ambio*, 38, 186–193.

Pretty, J. (2008) Agricultural sustainability: concepts, principles and evidence. *Philosophical Transactions of the Royal Society B*, 363, 447–465.

Pretty, J., Brett, C., Gee, D. *et al.* (2001) Policy challenges and priorities for internalising the externalities of agriculture. *Journal of Environmental Planning Management*, 44, 263–283.

Pywell, R.F., Warman, E.A., Carvell, C. *et al.* (2005) Providing foraging resources for bumblebees in intensively farmed landscapes. *Biological Conservation*, 121, 479–494.

Randall, A. (2002) Valuing the outputs of multifunctional agriculture. *European Review of Agricultural Economics*, 29, 289–307.

Raudsepp-Hearne, C., Peterson, G.D. and Bennett, E.M. (2010) Ecosystem service bundles for analyzing tradeoffs in diverse landscapes. *Proceedings of the National Academy of Sciences of the United States of America*, 107, 5242–5247.

Robinson, R.A. and Sutherland, W.J. (2002) Post-war changes in arable farming and biodiversity in Great Britain. *Journal of Applied Ecology*, 39, 157–176.

Rundlof, M., Bengtsson, J. and Smith, H.G. (2008) Local and landscape effects of organic farming on butterfly species richness and abundance. *Journal of Applied Ecology*, 45, 813–820.

Sagoff, M. (2011) The quantification and valuation of ecosystem services. *Ecological Economics*, 70, 497–502.

Sandhu, H.S., Wratten, S.D., Cullen, R. and Case, B. (2008) The future of farming: the value of ecosystem services in conventional and organic arable land. An experimental approach. *Ecological Economics*, 64, 835–848.

Scarratt, S.L. (2005) Enhancing the biological control of leadrollers (Lepidoptera: Tortricidae) using floral resource subsidies in an organic vineyard in Marlborough, New Zealand. Doctor of Philosophy thesis, Lincoln University, New Zealand.

Sternberg, E. (1996) Recuperating from market failure: planning for biodiversity and technological competitiveness. *Public Administration Review*, 56, 21–29.

Straub, C.S., Finke, D.L. and Snyder, W.E. (2008) Are the conservation of natural enemy biodiversity and biological control compatible goals? *Biological Control*, 45, 225–237.

Tscharntke, T., Klein, A.M., Kruess, A., Steffan-Dewenter, I. and Thies, C. (2005) Landscape perspectives on agricultural intensification and biodiversity – ecosystem service management. *Ecology Letters*, 8, 857–874.

Turner, R.K., Paavola, J., Cooper, P., Farber, S., Jessamy, V. and Georgiou, S. (2003) Valuing nature: lessons learned and future research directions. *Ecological Economics*, 46, 493–510.

Turner, R.K., Morse-Jones, S. and Fisher, B. (2010). Ecosystem valuation: a sequential decision support system and quality assessment issues. *Ecological Economics Reviews*, 1185, 79–101.

Tylianakis, J.M. and Romo, C.M. (2010) Natural enemy diversity and biological control: Making sense of the context-dependency. *Basic and Applied Ecology*, 11, 657–668.

Walker, J.T.S., Gurnsey, S., McArtney, S.J. and Wunsche, J.N. (2004) *Current issues impacting on organic apple production in New Zealand*. Presentation at the 6th International Conference on Integrated Fruit Production, September 26–30, Baselga De Piné, Italy.

Wallace, K.J. (2007) Classification of ecosystem services: problems and solutions. *Biological Conservation*, 139, 235–246.

Whittingham, M.J. (2007) Will agri-environment schemes deliver substantial biodiversity gain, and if not why not? *Journal of Applied Ecology*, 44, 1–5.

WRI (2005) *The Wealth of the Poor: Managing ecosystems to fight poverty*, World Resources Institute, Washington, DC.

Zhang, W., Ricketts, T.H., Kremen, C., Carney, K. and Swinton, S.M. (2007) Ecosystem services and dis-services to agriculture. *Ecological Economics*, 64, 253–260.

Chapter 5

SOIL FERTILITY, BIODIVERSITY AND PEST MANAGEMENT

Miguel A. Altieri, Luigi Ponti and Clara I. Nicholls

Biodiversity and Insect Pests: Key Issues for Sustainable Management, First Edition. Edited by Geoff M. Gurr, Steve D. Wratten, William E. Snyder, Donna M.Y. Read.
© 2012 John Wiley & Sons, Ltd. Published 2012 by John Wiley & Sons, Ltd.

INTRODUCTION

Optimisation of agroecosystem health is based on two pillars: habitat manipulation and soil fertility enhancement. The latter is achieved through management of organic matter and conservation of below-ground biodiversity, and is the focus of this chapter. The chapter first looks at ways in which soil fertility management can reduce plant susceptibility to pests, both directly by mediating plant health and indirectly via interactions between above-ground and below-ground biodiversity. Appropriate management of organic soil fertility may reduce crop damage by increasing plant resistance through improving the foliage's nutritional balance, or by reducing pest populations via enhancement of natural enemies. In organically fertilised systems, several insect herbivores consistently show lower abundance due to emerging synergies between plant diversity, natural enemies and soil fertility. Healthy soil is probably more important than is currently acknowledged in determining individual plant response to stresses such as pest pressure. Combining crop diversification and organic soil enhancement is a key strategy to sustainable agroecosystem management.

Traditionally entomologists have explained pest outbreaks in cropping systems as a consequence of the absence of natural enemies or the effects of insecticides, such as the development of pesticide resistance by insect pests or secondary pest outbreaks due to disruptions of biological control. Entomologists have, however, been unaware of the theory of trophobiosis offered by French scientist Francis Chaboussou (Chaboussou, 2004 (English translation of 1985 French edition)). As early as 1967, Chaboussou contended that pest problems were also linked to disturbances in the nutritional balances of crop plants and destruction of life in the soil. He explained that heavy applications of soluble nitrogen (N) fertilisers (and also certain pesticides) increase the cellular amounts of N, ammonia and amino acids, faster than the rate at which plants synthesise them for proteins. These reductions in the rate of protein synthesis result in the temporary accumulation of free N, sugars and soluble amino acids in the foliage: substances needed for growth and reproduction by insect herbivores and also plant pathogens. Chaboussou's empirical evidence led him to postulate that insect pests and diseases grow and multiply faster when plants contain more soluble free nutrients, due to the inhibition of protein synthesis. He also believed that a healthy soil life is fundamental for a balanced uptake of mineral nutrients by the plant, especially micronutrients. A lack of micronutrients also causes inhibition in protein synthesis and therefore leads to a build-up in nutrients needed by pests and pathogens.

In the last 20 years a number of research studies have corroborated Chaboussou's assertions, showing that the ability of a crop plant to resist or tolerate insect pests and diseases is tied to optimal physical, chemical and mainly biological properties of soils. Soils with high organic matter and active soil biological activity generally exhibit good soil fertility as well as complex food webs and beneficial organisms that prevent infection (Magdoff and van Es, 2000). Recent evidence suggests that the lower pest pressure observed in many organic systems, although associated with a greater use of practices that preserve beneficial insects, is also linked to enhanced soil biology and fertility (Zehnder et al., 2007). Several studies also document that farming practices which cause nutrition imbalances can lower pest resistance (Magdoff and van Es, 2000). Evidence is mounting that synthetic fertilisers can reduce plant resistance to insect pests, tend to enhance insect pest populations, and can increase the need for insecticide application (Yardlm and Edwards, 2003). Furthermore, recent research shows how biotic interactions in soil can regulate the structure and functioning of above-ground communities (Harman et al., 2004; Wardle et al., 2004), suggesting that the below-ground component of an agroecosystem can be managed through a set of agroecological practices that can exert a substantial influence on pest dynamics (Altieri and Nicholls, 2003).

Slowly agroecologists are recognising that above-ground and below-ground biodiversity components of agroecosystems cannot continue to be viewed in isolation from each other (van der Putten et al., 2009). In fact, the otherwise largely separate above-ground and below-ground components of agroecosystems are connected by the plant (Wardle et al., 2004). This recognition of the biological linkages between above-ground and below-ground biota constitutes a key step on which a truly innovative ecologically based pest management strategy can be built.

Ecologically based pest management (EBPM) considers below-ground and above-ground habitat management as equally important, because enhancing positive ecological interactions between soils and pests can provide a robust and sustainable way of optimising total agroecosystem function (Figure 5.1). The integrity of the agroecosystem relies on synergies of plant

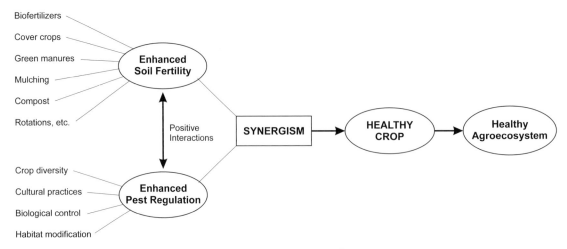

Figure 5.1 The potential synergism between soil fertility management and IPM.

diversity and the continuing function of the soil microbial community supported by a soil rich in organic matter (Altieri and Nicholls, 1990). Despite the potential links between soil fertility and crop protection, the evolution of integrated pest management (IPM) and integrated soil fertility management (ISFM) have proceeded separately (Altieri and Nicholls, 2003). Since many soil management practices are already known to influence pest management interactions, it does not make ecological sense to continue with such an atomistic approach.

The overall goal of EBPM is to create soil and above-ground conditions that promote the growth of healthy plants, while stressing pests and promoting beneficial organisms. This approach constitutes the basis of a habitat management strategy aimed at enhancing above- and below-ground biological diversity which in turn creates the conditions that are hospitable to plant roots, allowing the development of strong and healthy crops while promoting the presence of naturally occurring biological control organisms (Magdoff, 2007).

HEALTHY SOILS, HEALTHY PLANTS

One way soil fertility management can directly reduce plant susceptibility to pests is by mediating plant health (Phelan *et al.*, 1995). Many researchers and also practising farmers have observed that fertility practices that replenish and maintain high soil organic matter and that enhance the level and diversity of soil macro- and

microbiota provide an environment that through various processes enhances plant health (McGuiness, 1993). The following are a few of the suggested pest suppressive mechanisms linked to healthy soils:
• Competition: high levels and diversity of soil microbes diminish the populations or infectivity of soil-borne pathogens; this occurs because the soil microbes compete with the pathogens for food and space. Biodiverse soils also contain fungi and bacteria that consume, parasitise or are otherwise antagonistic to many soil-borne crop pathogens. Plant pathologists have known for years that a soil rich in microbiota lessens the danger of epidemic outbreaks caused by soil-borne pathogens (Campbell, 1994).
• Induced resistance: exposure to compost, compost extracts or certain microbes (both pathogenic and non-pathogenic) can induce plants to develop resistance to a broad range of soil-borne and airborne pathogens. Induced resistance is described as a broad-spectrum, long-lasting resistance and appears to be most effective against fungal pathogens (Kuć, 2001).
• Natural enemies: enriching the soil stimulates the proliferation of soil mesofauna which may serve as alternative prey for natural enemies such as carabid beetles and spiders, allowing them to develop high populations that can then respond quickly to pest outbreaks (Purvis and Curry, 1984). This effect is particularly important for generalist predators, as explored by Welch *et al.* in chapter 3 of this volume.
• Buffering of nutrient supply: humus and microbial biomass provide a more gradual and balanced release

of nutrients than is possible with synthetic fertilisers. Many insect pests and fungal pathogens are stimulated by lush growth and/or high N level in plants. As Chaboussou (2004) suggested, more balanced mineral nutrition makes crops more resistant to pests and diseases.

• Reduced stress: soils with high humus and biodiversity have improved capacity to take up and store water and thus reduce water stress. Water and other types of stress increase pest problems, possibly by restricting protein synthesis, which in turn increases soluble N in foliage making tissues more nutritious to many pests (Waring and Cobb, 1991).

Soil fertility practices can directly affect the physiological susceptibility of crop plants to insect pests by either affecting the resistance of individual plants to attack or altering plant acceptability to certain herbivores (Barker, 1975; Scriber, 1984). But the mechanisms can be more complex and include genetic and biochemical dimensions as suggested by the finding of scientists of the USDA Beltsville Agricultural Research Center in Maryland, which contributes to building a scientific basis to better understand the relationships between plant health and soil fertility (Kumar *et al.*, 2004). These researchers showed a molecular basis for delayed leaf senescence and tolerance to diseases in tomato plants cultivated in a legume (hairy vetch) mulch-based alternative agricultural system, compared to the same crop grown on a conventional black polyethylene mulch along with chemical fertiliser. Probably due to regulated release of C and N metabolites from hairy vetch decomposition, the cover-cropped tomato plants showed a distinct expression of selected genes, which ultimately led to a more efficient utilisation and mobilisation of C and N, promoting defence against disease and enhanced crop longevity. These results confirm that in intensive conventional tomato production, the use of legume cover crops offers advantages as a biological alternative to commercial fertilisers leading to disease suppression, in addition to other benefits such as minimising soil erosion and loss of nutrients, enhancing water infiltration, reducing runoff and more balanced natural control.

INTERACTIONS BETWEEN ABOVE-GROUND AND BELOW-GROUND BIODIVERSITY

Plants function in a complex multitrophic environment. However, as pointed out by van der Putten *et al.*

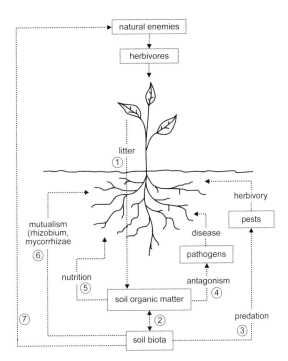

Figure 5.2 Complex ways in which above- and below-ground biodiversity interact in agroecosystems: 1) crop residues enhance soil organic matter (SOM); 2) SOM provides substratum for micro, meso, and macro soil biota; 3) soil predators reduce soil pests; 4) SOM enhances antagonists which suppress soil-borne pathogens; 5) slow mineralisation of C and N activates genes which promote disease tolerance and crop longevity as well as low free N content in foliage; 6) mutualists enhance N fixation, P uptake, water use efficiency, etc.; 7) certain invertebrates (e.g. Collembola and detritivores) serve as alternative food to natural enemies in times of pest scarcity.

(2001), most multitrophic studies have focused almost exclusively on above-ground interactions, generally neglecting the fact that above- and below-ground organisms interact in complex ways (Figure 5.2). Several studies point to the interdependence of the population dynamics of above- and below-ground herbivores and associated natural enemies as mediated through defence responses by different plant compartments (above and below ground). Because plant chemical defence pathways against herbivores and pathogens can interact, root herbivory could affect the induction of plant defence compounds in leaves. But, as argued by van der Putten *et al.* (2001), the interactions between the below- and above-ground compartments are even more complex, because the underlying mechanisms (nutrition and plant defence) are typically interlinked. In fact, the production of both direct and indirect plant defences is dependent on nutrient uptake by the roots. And the evidence in favour of such beneficial interactions is growing (Bezemer and van Dam, 2005; Erb *et al.*, 2008; Kempel *et al.*, 2010; Pineda *et al.*, 2010; Wurst, 2010; van Dam and Heil, 2011).

Below-ground organism activity can affect plant above-ground phenotype, inducing plant tolerance to herbivores and pathogens (Blouin *et al.*, 2005). In that study, an 82% decrease in nematode-infested plants was achieved when earthworms were present. Although earthworms had no direct effect on nematode population size, in their presence root biomass was not affected by nematodes and the expected inhibition of photosynthesis was suppressed. This is the first time earthworms have been shown to reduce nematode effects in infested plants. Apparently, the presence of earthworms in the rhizosphere induced systemic changes in plant gene expression, leading to increased photosynthetic activity and chlorophyll concentration in the leaves (Blouin *et al.*, 2005). Such findings indicate that soil fauna activities are probably more important than currently acknowledged in determining individual plant response to stress.

Above-ground communities are affected by both direct and indirect interactions with soil food web organisms (Wardle *et al.*, 2004). Feeding activities in the detritus food web stimulate nutrient turnover, plant nutrient acquisition and plant performance, and thereby indirectly influence above-ground herbivores. Studies in traditional Asian irrigated-rice agroecosystems showed that by increasing soil organic matter in test plots, researchers could boost populations of detritivores and plankton-feeders, which in turn significantly boosted abundance of above-ground generalist predators (Settle *et al.*, 1996). This system is explored in detail in chapter 13 of this volume. In addition, soil Collembola are regarded as important sources of alternative prey for predators such as carabid beetles when pests are scarce (Bilde *et al.*, 2000).

On the other hand, soil biota exerts direct effects on plants by feeding on roots and forming antagonistic or mutualistic relationships with their host plants (e.g. mycorrhizae). Such direct interactions with plants influence not just the performance of the host plants themselves, but also that of the herbivores and poten-

tially their predators. Vestergard *et al.* (2004) found that interactions between aphids and rhizosphere organisms were influenced by plant development and by soil nutrient status. This is one of the first agricultural reports confirming that above- and below-ground biota are able to influence each other with the plant as a mediator. In a long-term agricultural experiment, Birkhofer *et al.* (2008) found that the use of synthetic fertilisers negatively affected interactions within and between below- and above-ground agroecosystem components, with consequent reduction of internal biological cycles and pest control.

SOIL FERTILITY AND PLANT RESISTANCE TO INSECT PESTS

Plant resistance to insect pests varies with the age or growth stage of the plant (Slansky, 1990), suggesting that resistance is linked directly to the physiology of the plant. Thus any factor which affects the physiology of the plant (e.g. fertilisation) is potentially linked to changes in resistance to insect pests. In fact, fertilisation has been shown to affect all three categories of resistance proposed by Painter (1951): preference, antibiosis and tolerance. Furthermore, obvious morphological responses of crops to fertilisers, such as changes in growth rates, accelerated or delayed maturity, size of plant parts, and thickness and hardness of cuticle, can also indirectly influence the success of many pest species in utilising their host plant. For example, Adkisson (1958) reported nearly three times as many boll weevil larvae (*Anthonomus grandis* (Boheman)) from cotton receiving heavy applications of fertilisers compared to unfertilised control plants, probably due to the prolonged growing season for cotton resulting from the fertiliser amendment. Klostermeyer (1950) observed that N fertiliser increased husk extension and tightness of husks on sweet corn, which reduced corn earworm (*Heliothis zea* (Boddie)) infestation levels. Hagen and Anderson (1967) observed that zinc deficiency reduced the pubescence on corn leaves, which allowed a subsequent increase in feeding by adult western corn rootworm (*Diabrotica virgifera* (LeConte)).

Effects of soil fertility practices on pest resistance can be mediated through changes in the nutritional content of crops. At equivalent amounts of applied N (100 and 200 mg/pot), Barker (1975) found that nitrate-N concentrations in spinach leaves were higher when receiving ammonium nitrate than in plants treated with five organic fertilisers. In a comparative study of Midwestern conventional and organic farmers, Lockeretz *et al.* (1981) reported organically grown (OG) corn to have lower levels of all amino acids (except methionine) than conventionally grown (CG) corn. Eggert and Kahrmann (1984) also showed CG dry beans to have more protein than OG beans. Consistently higher N levels in the petiole tissue were also found in the CG beans. Potassium and phosphorus levels, however, were higher in the OG bean petioles than in the CG beans. In a long-term comparative study of organic and synthetic fertiliser effects on the nutritional content of four vegetables (spinach, savoy, potatoes and carrots), Schuphan (1974) reported that the OG vegetables consistently contained lower levels of nitrate and higher levels of potassium, phosphorus and iron than CG vegetables.

Fertilisation with N may decrease plant resistance to insect pests by improving the nutritional quality of host plants and reducing secondary metabolite concentrations. Jansson and Smilowitz (1986) reported that N applications increased the rate of population growth of green peach aphid on potatoes and that the growth was positively correlated with the concentrations of free amino acids in leaves. High levels of N reduced glycoalkaloid synthesis, which has an inhibitory effect on insect pests of potatoes (Fragoyiannis *et al.*, 2001). Barbour *et al.* (1991), investigating interactions between fertiliser regimes and host-plant resistance in tomatoes, showed that the survival of Colorado potato beetles to adult emergence increased with larger amounts of fertiliser, and was related to decreases in trichome- and lamellar-based beetle resistance, in response to the improved nutritional quality of the host plant. In addition to increases in the survival rates of Colorado potato beetles from the first instar to adults, larger amounts of N in tomatoes could also cause significantly faster insect development and increased pupal biomass. More recently, Hsu *et al.* (2009) found that *Pieris rapae* (L.) butterflies laid more eggs on CG than on OG cabbage, and that caterpillars then grew faster on CG cabbage due to a diet with more nutrients (N and sugar) and less allelochemicals (sinigrin, the most important and abundant glucosinolate known for its feeding deterrent and antimicrobial properties). Their findings suggest that higher biomass (dry weight) and lower pest incidence may be jointly achieved in organically vs. synthetically fertilised cropping systems.

Meyer (2000) suggests that soil nutrient availability not only affects the amount of damage that plants sustain from herbivores, but also the ability of plants to recover from herbivory. Meyer's study reported the effects of soil fertility on both the degree of defoliation and compensation for herbivory by *Brassica nigra* (L.) plants damaged by *P. rapae* caterpillars. In this study, the percentage defoliation was more than twice as great at low fertility compared to high fertility, even though plants grown at high soil fertility lost a greater absolute amount of leaf area. At both low and high soil fertility, total seed number and mean mass per seed of damaged plants were equivalent to those of undamaged plants. Apparently, soil fertility did not influence plant compensation in terms of maternal fitness.

INDIRECT EFFECTS OF SOIL NITROGEN ON CROP DAMAGE BY ARTHROPODS

Increases in N levels in plants can enhance populations of invertebrate herbivores living on them (Patriquin *et al.*, 1995). Such increases in populations of insect pests on their host plants in response to higher nitrogen levels can result from various mechanisms, depending on the insect species and host plant. Total N has been considered a critical nutritional factor mediating herbivore abundance and fitness (Mattson, 1980; Scriber, 1984; Slansky and Rodriguez, 1987; Wermelinger, 1989). Many studies report dramatic increases in aphid and mite numbers in response to increased N fertilisation rates. According to van Emden (1966), increases in fecundity and developmental rates of the green peach aphid, *Myzus persicae* (Sulzer), were highly correlated to increased levels of soluble N in leaf tissue. Changes in N content in poinsettias grown with ammonium nitrate stimulated the fecundity of the whitefly *Bemisia tabaci* (Gennadius) and attracted more individuals to oviposit on them (Bentz *et al.*, 1995).

Several other authors have also indicated increased aphid and mite populations from N fertilisation (Luna, 1988). Herbivorous insect populations associated with *Brassica* crop plants have also been reported to increase in response to increased soil N levels (Letourneau, 1988). In a two-year study, Brodbeck *et al.* (2001) found that populations of the thrips *Frankliniella occidentalis* (Pergande) were significantly higher on tomatoes that received higher rates of N fertilisation.

Other insect populations found to increase following N fertilisation include fall armyworm in maize, corn earworm on cotton, pear psylla on pear, Comstock mealybug (*Pseudococcus comstocki* (Kuwana)) on apple, and European corn borer (*Ostrinia nubilalis* (Hubner)) on field corn (Luna, 1988).

Because plants are the source of nutrients for herbivorous insects, an increase in the nutrient content of the plant may be argued to increase its acceptability as a food source to pest populations. Variations in herbivore response may be explained by differences in the feeding behaviour of the herbivores themselves (Pimentel and Warneke, 1989). For example, with increasing N concentrations in creosote bush (*Larrea tridentata* (Coville)) plants, populations of sucking insects were found to increase, but the number of chewing insects declined. It is plausible that with higher N fertilisation, the amount of nutrients in the plant increases, as well as the amount of secondary compounds that may selectively affect herbivore feeding patterns. In particular, protein digestion inhibitors that are found to accumulate in plant cell vacuoles are not consumed by sucking herbivores, but will harm chewing herbivores (Mattson, 1980). However this differential response does not seem to change the overall trend when one looks at studies on crop nutrition and pest attack (Altieri and Nicholls, 2003).

In reviewing 50 years of research relating to crop nutrition and insect attack, Scriber (1984) found 135 studies showing increased damage and/or growth of leaf-chewing insects or mites in N-fertilised crops, versus fewer than 50 studies in which herbivore damage was reduced. In aggregate, these results suggest a hypothesis with implications for fertiliser use patterns in agriculture, namely that high N inputs can result in high levels of herbivore damage in crops. As a corollary, crop plants would be expected to be less prone to insect pests and diseases if organic soil amendments are used, as these generally result in lower N concentrations in the plant tissue. However, Letourneau (1988) questions if such a 'nitrogen-damage' hypothesis, based on Scriber's review, can be extrapolated to a general warning about fertiliser inputs associated with insect pest attack in agroecosystems. Letourneau reviewed 100 studies and found that two-thirds (67) of the insect and mite studies showed an increase in growth, survival, reproductive rate, population densities or plant damage levels in response to increased N fertiliser. The remaining third of the arthropods studied showed either a decrease in damage with fertiliser N or

no significant change. The author also noted that experimental design can affect the types of responses observed.

The majority of Cakchiquel farmers responding to a survey conducted in Patzun, Guatemala, did not recognise herbivorous insects as a problem in their milpas (corn (*Zea mays*) intercropped with beans (*Phaseolus vulgaris*), fava (*Vicia faba*), and/or squash (*Cucurbita maxima, C. pepo*) (Morales *et al.*, 2001). The farmers attributed this lack of pests to preventative measures incorporated into their agricultural practices, including soil management techniques. Patzun farmers traditionally mixed ashes, kitchen scraps, crop residues, weeds, leaf litter and manure to produce compost. However, from about 1960 onward, synthetic fertilisers were introduced to the region and were rapidly adopted in the area. Today, the majority of farmers have replaced organic fertilisers with urea ($CO(NH_2)_2$), although some recognise the negative consequences of the change and complain that pest populations have increased in their milpas since the introduction of the synthetic fertilisers.

In their survey in the Guatemalan highlands, Morales *et al.* (2001) also found that corn fields treated with organic fertiliser (applied for two years) hosted fewer aphids (*Rhopalosiphum maidis* (Fitch)) than corn treated with synthetic fertiliser. This difference was attributed to a higher concentration of foliar N in corn in the synthetic fertiliser plots, although numbers of *Spodoptera frugiperda* (Smith) showed a weak negative correlation with increased N levels.

DYNAMICS OF INSECT HERBIVORES IN ORGANICALLY FERTILISED SYSTEMS

Lower abundance of several insect herbivores in low-input systems has been partly attributed to a lower N content in organically farmed crops (Lampkin, 1990). Furthermore, farming methods utilising organic soil amendments significantly promote the conservation of arthropod species in all functional groups, and enhance the abundance of natural enemies compared with conventional practices (Moreby *et al.*, 1994; Basedow, 1995; Drinkwater *et al.*, 1995; Berry *et al.*, 1996; Pfiffner and Niggli, 1996; Letourneau and Goldstein, 2001; Mäder *et al.*, 2002; Hole *et al.*, 2005). This suggests that reduced pest populations in organic systems are a consequence of both nutritional changes induced in the crop by organic fertilisation and increased natural pest control. Whatever the cause, there are many examples in which lower insect herbivore populations have been documented in low-input systems, with a variety of possible mechanisms proposed.

In Japan, the density of immigrants of the planthopper species *Sogatella furcifera* (Horváth) was significantly lower and the settling rate of female adults and survival rate of immature stages of ensuing generations were generally lower in organic than in conventional rice fields. Consequently, the density of planthopper nymphs and adults in the ensuing generations was found to decrease in organically farmed fields (Kajimura, 1995). In England, conventional winter wheat fields exhibited a larger infestation of the aphid *Metopolophium dirhodum* (Walker) than their organic counterparts. The conventionally fertilised wheat crop also had higher levels of free protein amino acids in its leaves during June, which were attributed to an N top dressing applied early in April. However, the difference in aphid infestations between crops was attributed to the aphid's response to the relative proportions of certain non-protein to protein amino acids present in the leaves at the time of aphid settling on crops (Kowalski and Visser, 1979). The authors concluded that chemically fertilised winter wheat was more palatable than its organically grown counterpart, hence the higher level of infestation.

Interesting results were found also in greenhouse experiments comparing maize grown on organic versus chemically fertilised soils collected from nearby farms (Phelan *et al.*, 1995). The researchers observed that European corn borer (*Ostrinia nubilalis* (Hübner)) females, when given a choice, laid significantly more eggs in the chemically fertilised plants versus the organically fertilised ones. But this significant variation in egg-laying between chemical and organic fertiliser treatments was present only when maize was grown on soil collected from conventionally managed farms. In contrast, egg laying was uniformly low in plants grown on soil collected from organically managed farms. Pooling results across all three farms showed that variance in egg laying was approximately 18 times higher among plants in conventionally managed soil than among plants grown under an organic regimen. The authors suggested that this difference is evidence for a form of biological buffering characteristically found more commonly in organically managed soils.

Yardlm and Edwards (2003) conducted a two-year study comparing the effects of organic (composted cow manure) and synthetic (NPK) fertilisers on pests (aphids and flea beetles) and predatory arthropods (anthocorids, coccinellids and chrysopids) associated with tomatoes. In the second year aphids exhibited significantly lower numbers on plants that received organic fertiliser than on those treated with synthetic fertilisers, suggesting that the effects of organic fertilisers in reducing pest populations may be expressed more fully in the long term. The reductions in aphid populations could not be attributed to the increases in predator populations on tomatoes in the organic fertiliser-treated plots, because the predator populations did not differ significantly between the full-rate synthetic fertiliser-treated and the organic fertiliser-treated plots. However, it seems that both synthetic and organic fertiliser inputs were able to increase flea beetle populations significantly, even when the synthetic fertiliser application rate was reduced to half. Flea beetle populations were significantly higher on plants that received the full rates of synthetic and organic fertilisers during the two years of the study, despite the significant differences between years with respect to flea beetle and other pest numbers.

Altieri *et al.* (1998) conducted a series of comparative experiments in various growing seasons between 1989 and 1996, in which broccoli was subjected to varying fertilisation regimes (conventional vs. organic). The goal was to test the effects of different N sources on the abundance of the key insect pests: cabbage aphid (*Brevicoryne brassicae* (L.)) and flea beetle (*Phyllotreta cruciferae* (Goeze)). Conventionally fertilised monocultures consistently developed a larger infestation of flea beetles and in some cases of the cabbage aphid, than the organically fertilised broccoli systems. The reduction in aphid and flea beetle infestations in the organically fertilised plots was attributed to lower levels of free N in the foliage of plants. Applications of synthetic N fertilisation to individual broccoli plants within an organic field triggered aphid densities on the treated plants but not on the surrounding organic plants (Figure 5.3). These results further support the view that insect pest preference can be moderated by alterations in the type and amount of fertiliser used.

By contrast, a study comparing the population responses of *Brassica* pests to organic versus synthetic fertilisers measured higher *Phyllotreta* flea beetle populations on sludge-amended collard (*Brassica oleracea* (L.)) plots early in the season compared to mineral fertiliser-amended and unfertilised plots (Culliney and Pimentel, 1986). However, later in the season, in these

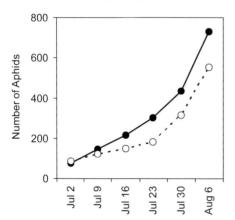

Figure 5.3 Aphid population response (cumulative numbers) to treatment of individual organic broccoli plants with chemical N fertiliser within an organically managed field in Albany, California (Altieri, unpublished data).

same plots population levels of beetle, aphid and lepidopteran pests were lowest in organic plots. This suggests that the effects of fertiliser type vary with plant growth stage and that organic fertilisers do not necessarily diminish pest populations throughout the whole season. For example, in a survey of California tomato producers, despite the pronounced differences in plant quality (N content of leaflets and shoots) both within and among tomato fields, Letourneau *et al.* (1996) found no indication that greater concentrations of tissue N in tomato plants were associated with higher levels of insect damage at harvest time.

SYNERGIES BETWEEN PLANT DIVERSITY, NATURAL ENEMIES AND SOIL FERTILITY

When examining weedy faba beans fields in Tunwath, Canada, Patriquin *et al.* (1988) found higher numbers of aphid enemies in the diverse systems but they also discovered that the reproductive rate of aphids is proportional to the supply of amino acids in the phloem. Legumes nodulate and fix gaseous N from air when the supply of mineral N in the soil is deficient; when it is

not, they preferentially take up soil N. Weeds among the faba beans take up and thereby reduce soil mineral N to a level below that which suppresses nodulation (about 5 ppm nitrate-N for faba beans at Tunwath). This causes the faba beans to nodulate more and to obtain more N from N fixation. Under those conditions, there is closer coupling of N uptake and assimilation than when mineral N predominates, and consequently accumulation of amino acids in the phloem is reduced. The reproductive rate of the aphids is restricted accordingly. When the weeds are removed, the soil N supply increases, phloem N increases, and the plants are more attractive and more nutritious to the aphids. Weedy plants also had higher yields than the plants in the weed-free plots because the benefits of increased nodulation outweighed any losses due to weeds. This beneficial interaction between weeds and the crop works only if levels of soil nitrate are relatively low (e.g. 10 ppm) to begin with. When plots were fertilised with urea, weeds overgrew the crop and greatly reduced yields. On another organic farm where soil nitrate levels were five to ten times higher than at Tunwath, weeds overgrew the faba bean crop. In spite of an abundance of natural enemies, large aphid infestations caused massive yield loss. Managing the soil N to keep it low under faba beans was thus critical for favourable crop–weed and crop–aphid interactions. At Tunwath, faba beans followed winter wheat in the rotation. The highly 'immobilising' (N robbing) wheat residues were worked into the soil following harvest. This was a deliberate strategy to lower soil nitrate levels under the faba beans and thereby stimulate nodulation and N fixation. Patriquin's data indicated that seed yield of aphid-infested plants at Tunwath Farm was not reduced and was even slightly higher than those of non-infested plants (Patriquin *et al.*, 1988).

In California, Ponti *et al.* (2007) reported that intercropping of broccoli with mustard and buckwheat significantly reduced aphid populations especially in the summer (Figure 5.4), when the proximity of flowers (i.e. polyculture with competition) significantly enhanced aphid parasitisation rates on nearby broccoli plants. Monoculture and polyculture broccoli consistently had lower aphid densities and higher parasitisation rates when fertilised with compost. In this study, synthetically fertilised broccoli produced more biomass (fresh weight), but also recruited higher pest numbers. Nevertheless, parasitism by *Diaeretiella rapae* (McIntosh) was higher in compost-fertilised plots. Intercropping and composting decreased pest abundance in

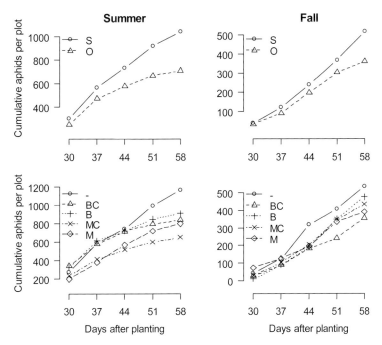

Figure 5.4 Cumulative counts of aphids on five broccoli plants per plot at the different sampling dates as influenced by cropping system levels (-, monoculture; B, buckwheat polyculture without competition; BC, buckwheat polyculture with competition; M, mustard polyculture without competition; MC, mustard polyculture with competition) and by fertiliser levels (S, synthetic fertiliser; O, organic fertiliser-compost) in two (summer and fall) experiments at Albany, California, in 2004.

broccoli cropping systems with or without interspecific competition, suggesting a synergistic relationship between plant diversity and soil organic management. In addition, depending on the intercropped plant and the growing season (summer vs. autumn), intercropping enhanced parasitism of cabbage aphid. The seasonal effectiveness of *D. rapae* was increased by composting despite lower aphid abundance in compost-fertilised broccoli.

CONCLUSIONS

Soil fertility management can have several effects on plant quality, which in turn can affect insect abundance and subsequent levels of herbivore damage. The reallocation of mineral amendments in crop plants can influence oviposition, growth rates, survival and reproduction in the insects that use these hosts (Jones, 1976). Although more research is needed, preliminary evidence suggests that fertilisation practices can influ-

ence the relative resistance of agricultural crops to insect pests. Increased soluble N levels in plant tissue were found to decrease pest resistance, although this is not a universal phenomenon (Phelan *et al.*, 1995; Staley *et al.*, 2010).

Chemical fertilisers can dramatically influence the balance of nutritional elements in plants, and it is likely that their excessive use will create nutrient imbalances, which in turn reduce resistance to insect pests. In contrast, organic farming practices promote an increase of soil organic matter and microbial activity and a gradual release of plant nutrients and should, in theory, allow plants to derive a more balanced nutrition. Thus, while the amount of N immediately available to the crop may be lower when organic fertilisers are applied, the overall nutritional status of the crop appears to be improved. Organic soil fertility practices can also provide supplies of secondary and trace elements, occasionally lacking in conventional farming systems that rely primarily on artificial sources of N, P and K. Besides nutrient concentrations, optimum fertilisation, which provides a proper balance of elements, can stimulate resistance to insect attack (Luna, 1988). Organic N sources may allow greater tolerance of vegetative damage because they release N more slowly, over the course of several years.

Phelan *et al.* (1995) stressed the need to consider other mechanisms when examining the link between fertility management and crop susceptibility to insects. Their study demonstrated that the ovipositional preference of a foliar pest can be mediated by differences in soil fertility management. Thus, the lower pest levels widely reported in organic farming systems may, in part, arise from plant–insect resistances mediated by biochemical or mineral nutrient differences in crops under such management practices. In fact, we feel such results provide interesting evidence to support the view that the long-term management of soil organic matter can lead to better plant resistance against insect pests (Birkhofer *et al.*, 2008). This view is corroborated by recent research on the relationships between above-ground and below-ground components of ecosystems, which suggests that soil biological activity is probably more important than currently acknowledged in determining individual plant response to stresses such as pest pressure (Blouin *et al.*, 2005), and that this stress response is mediated by a series of interactions outlined in Figure 5.2. These findings are enhancing our understanding of the role of biodiversity in agriculture, and the close ecological linkages between above-ground and below-ground biota. Such understanding constitutes a key step towards building a truly innovative, ecologically based pest management strategy which combines crop diversification and organic soil enhancement.

REFERENCES

Adkisson, P.L. (1958) The influence of fertiliser applications on population of *Heliothis zea* and certain insect predators. *Journal of Economic Entomology*, 51, 757–759.

Altieri, M.A. and Nicholls, C.I. (1990) Biodiversity, ecosystem function and insect pest management in agricultural systems, in *Biodiversity in agroecosystems* (eds W.W. Collins and C.O. Qualset), CRC Press, Boca Raton, pp. 69–84.

Altieri, M.A. and Nicholls, C.I. (2003) Soil fertility management and insect pests: harmonizing soil and plant health in agroecosystems. *Soil and Tillage Research*, 72, 203–211.

Altieri, M.A., Schmid, L.L. and Montalba, R. (1998) Assessing the effects of agroecological soil management practices on broccoli insect pest populations. *Biodynamics*, 218, 23–26.

Barbour, J.D., Farrar, R.R. and Kennedy, G.G. (1991) Interaction of fertilizer regime with host-plant resistance in tomato. *Entomologia Experimentalis et Applicata*, 60, 289–300.

Barker, A. (1975) Organic vs. inorganic nutrition and horticultural crop quality. *HortScience*, 10, 12–15.

Basedow, T. (1995) Insect pests: their antagonists and diversity of the arthropod fauna in fields of farms managed at different intensities over a long term – a comparative survey. *Mitteilungen der Deutschen Gesellschaft für Allgemeine und Angewandte Entomologie*, 10, 565–572.

Bentz, J.A., Reeves, J., Barbosa, P. and Francis, B. (1995) Nitrogen fertilizer effect on selection, acceptance, and suitability of *Euphorbia pulcherrima* (Euphorbiaceae) as a host plant to *Bemisia tabaci* (Homoptera: Aleyrodidae). *Environmental Entomology*, 24, 40–45.

Berry, N.A., Wratten, S.D., McErlich, A. and Frampton, C. (1996) Abundance and diversity of beneficial arthropods in conventional and organic carrot crops in New Zealand. *New Zealand Journal of Crops and Horticultural Sciences*, 24, 307–313.

Bezemer, T.M. and van Dam, N.M. (2005) Linking aboveground and belowground interactions via induced plant defenses. *Trends Ecology and Evolution*, 20, 617–624.

Bilde, T., Axelsen, J.A. and Toft, S. (2000) The value of Collembola from agricultural soils as food for a generalist predator. *Journal of Applied Ecology*, 37, 672–683.

Birkhofer, K., Bezemer, T.M., Bloem, J. *et al.* (2008) Long-term organic farming fosters below and aboveground biota: Implications for soil quality, biological control and productivity. *Soil Biology and Biochemistry*, 40, 2297–2308.

Blouin, M., Zuily-Fodil, Y., Pham-Thi, A.-T. *et al.* (2005) Belowground organism activities affect plant aboveground phenotype, inducing plant tolerance to parasites. *Ecology Letters*, 8, 202–208.

Brodbeck, B., Stavisky, J., Funderburk, J., Andersen, P. and Olson, S. (2001) Flower nitrogen status and populations of *Frankliniella occidentalis* feeding on *Lycopersicon esculentum*. *Entomologia Experimentalis et Applicata*, 99, 165–172.

Campbell, R. (1994) Biological control of soil-borne diseases: some present problems and different approaches. *Crop protection*, 13, 4–13.

Chaboussou, F. (2004) *Healthy crops: a new agricultural revolution*, Jon Carpenter Publishing, Oxford.

Culliney, T. and Pimentel, D. (1986) Ecological effects of organic agricultural practices in insect populations. *Agriculture Ecosystems and Environment*, 15, 253–266.

Drinkwater, L.E., Letourneau, D.K., Workneh, F. and van Bruggen, A.H.C. (1995) Fundamental differences between conventional and organic tomato agro-ecosystems in California. *Ecological Applications*, 5, 1098–1112.

Eggert, F.P. and Kahrmann, C.L. (1984) Responses of three vegetable crops to organic and inorganic nutrient sources. *Organic farming: current technology and its role in sustainable agriculture.* Pub. No. 46, American Society of Agronomy, Madison.

Erb, M., Ton, J., Degenhardt, J. and Turlings, T.C.J. (2008) Interactions between arthropod-induced aboveground and belowground defenses in plants. *Plant Physiology*, 146, 867–874.

Fragoyiannis, D.A., McKinlay, R.G. and D'Mello, J.P.F. (2001) Interactions of aphid herbivory and nitrogen availability on the total foliar glycoalkaloid content of potato plants. *Journal of Chemical Ecology*, 27, 1749–1762.

Hagen, A.F. and Anderson, F.N. (1967) Nutrient imbalance and leaf pubescence in corn as factors influencing leaf injury by the adult western corn rootworm. *Journal of Economic Entomology*, 60, 1071–1077.

Harman, G.E., Howell, C.R., Viterbo, A., Alexander, I.H., Grice, P.V. and Evans, A.D. (2004) Trichoderma species – opportunistic, avirulent plant symbionts. *Nature Reviews Microbiology*, 2, 43–56.

Hole, D.G., Perkins, A.J., Wilson, J.D. *et al.* (2005) Does organic farming benefit biodiversity? *Biological Conservation*, 122, 113–130.

Hsu, Y.T., Shen, T.C. and Hwang, S.Y. (2009) Soil fertility management and pest responses: a comparison of organic and synthetic fertilization. *Journal of Economic Entomology*, 102, 160–169.

Jansson, R.K. and Smilowitz, Z. (1986) Influence of nitrogen on population parameters of potato insects: Abundance, population growth, and within-plant distribution of the green peach aphid, *Myzus persicae* (Homoptera: Aphididae). *Environmental Entomology*, 15, 49–55.

Jones, F.G.W. (1976) *Pests, resistance, and fertilizers, in Fertilizer use and plant health*. Proceedings of the 12th Colloquium of the International Potash Institute, Izmir, Turkey, International Potash Institute, Bern, pp. 233–258.

Kajimura, T. (1995) Effect of organic rice farming on planthoppers: Reproduction of white backed planthopper, *Sogatella furcifera* (Homoptera: Delphacidae). *Researches on Population Ecology*, 37, 219–224.

Kempel, A., Schmidt, A.K., Brandl, R. and Schädler, M. (2010) Support from the underground: Induced plant resistance depends on arbuscular mycorrhizal fungi. *Functional Ecology*, 24, 293–300.

Klostermeyer, E.C. (1950) Effect of soil fertility on corn earworm damage. *Journal of Economic Entomology*, 43, 427–429.

Kowalski, R. and Visser, P.E. (1979) Nitrogen in a crop–pest interaction: cereal aphids, in *Nitrogen as an ecological parameter* (ed. J.A. Lee), Blackwell, Oxford, pp. 67–74.

Kuć, J. (2001) Concepts and direction of induced systemic resistance in plants and its application. *European Journal of Plant Pathology*, 107, 7–12.

Kumar, V., Mills, D.J., Anderson, J.D. and Mattoo, A.K. (2004) An alternative agriculture system is defined by a distinct expression profile of select gene transcripts and proteins. *PNAS*, 101, 10535–10540.

Lampkin, N. (1990) *Organic farming*, Farming Press Books, Ipswich.

Letourneau, D.K. (1988) *Soil management for pest control: a critical appraisal of the concepts, in Global perspectives on agroecology and sustainable agricultural systems*. Proceedings of the Sixth International Science Conference of IFOAM, Santa Cruz, CA, pp. 581–587.

Letourneau, D.K. and Goldstein, B.P. (2001) Pest damage and arthropod community structure in organic vs. conventional tomato production in California. *Journal of Applied Ecology*, 38, 557–450.

Letourneau, D.K., Drinkwater, L.E. and Shennon, C. (1996) Effects of soil management on crop nitrogen and insect damage in organic versus conventional tomato fields. *Agriculture Ecosystems and Environment*, 57, 174–187.

Lockeretz, W., Shearer, G. and Kohl, D.H. (1981) Organic farming in the corn belt. *Science*, 211, 540–547.

Luna, J.M. (1988) *Influence of soil fertility practices on agricultural pests, in Global perspectives on agroecology and sustainable agricultural systems*. Proceedings of the Sixth International Science Conference of IFOAM, Santa Cruz, CA, pp. 589–600.

Mäder P., Fliessbach A., Dubois D., Gunst L. Fried P. and Niggli U. (2002) Soil fertility and biodiversity in organic farming. *Science*, 296, 1694–1697.

Magdoff, F. (2007) Ecological agriculture: principles, practices, and constraints. *Renewable Agriculture and Food Systems*, 22, 109–117.

Magdoff, F. and van Es, H. (2000) *Building soils for better crops*, SARE, Washington DC.

Mattson, W.J., Jr. (1980) Herbivory in relation to plant nitrogen content. *Annual Review of Ecology and Systematics*, 11, 119–161.

McGuiness, H. (1993) *Living soils: sustainable alternatives to chemical fertilizers for developing countries*, Consumers Policy Institute, New York.

Meyer, G.A. (2000) Interactive effects of soil fertility and herbivory on *Brassica nigra*. *Oikos*, 22, 433–441.

Morales, H., Perfecto, I. and Ferguson, B. (2001) Traditional fertilization and its effect on corn insect populations in the Guatemalan highlands. *Agriculture Ecosystems and Environment*, 84, 145–155.

Moreby, S.J., Aebischer, N.J., Southway, S.E. and Sotherton, N.W. (1994) A comparison of flora and arthropod fauna of organically and conventionally grown winter wheat in southern England. *Annals of Applied Biology*, 12, 13–27.

Painter, R.H. (1951) *Insect resistance in crop plants*, University of Kansas Press, Lawrence.

Patriquin, D.G., Baines, D., Lewis, J. and MacDougall, A. (1988) Aphid infestation of fababeans on an organic farm in relation to weeds, intercrops and added nitrogen. *Agriculture, Ecosystems and Environment*, 20, 279–288.

Patriquin, D.G., Baines, D. and Abboud, A. (1995) Diseases, pests and soil fertility, in *Soil management in sustainable agriculture* (eds H.F. Cook and H.C. Lee), Wye College Press, Wye, pp. 161–174.

Pfiffner, L. and Niggli, U. (1996) Effects of biodynamic, organic and conventional farming on ground beetles (Coleoptera: Carabidae) and other epigaeic arthropods in winter wheat. *Biological Agriculture and Horticulture*, 12, 353–364.

Phelan, P.L., Mason, J.F. and Stinner, B.R. (1995) Soil fertility management and host preference by European corn borer, *Ostrinia nubilalis*, on *Zea mays*: a comparison of organic and conventional chemical farming. *Agriculture Ecosystems and Environment*, 56, 1–8.

Pimentel, D. and Warneke, A. (1989) Ecological effects of manure, sewage sludge and and other organic wastes on arthropod populations. *Agricultural Zoology Reviews*, 3, 1–30.

Pineda, A., Zheng, S.-J., van Loon, J.J.A., Pieterse, C.M. and Dicke, M. (2010) Helping plants to deal with insects: the role of beneficial soil-borne microbes. *Trends in Plant Science*, 15, 507–514.

Ponti, L., Altieri, M.A. and Gutierrez, A.P. (2007) Effects of crop diversification levels and fertilization regimes on abundance of *Brevicoryne brassicae* (L.) and its parasitization by *Diaeretiella rapae* (McIntosh) in broccoli. *Agricultural and Forest Entomology*, 9, 209–214.

Purvis, G. and Curry, J.P. (1984) The influence of weeds and farmyard manure on the activity of Carabidae and other ground-dwelling arthropods in a sugar beet crop. *Journal of Applied Ecology*, 21, 271–283.

Schuphan, W. (1974) Nutritional value of crops as influenced by organic and inorganic fertilizer treatments: results of 12 years' experiments with vegetables (1960–1972). *Qualitas Plantarum – Plant Foods for Human Nutrition*, 23, 333–358.

Scriber, J.M. (1984) Nitrogen nutrition of plants and insect invasion, in *Nitrogen in crop production* (ed. R.D. Hauck), American Society of Agronomy, Madison.

Settle, W.H., Ariawan, H., Astuti, E.T. *et al.* (1996) Managing tropical rice pests through conservation of generalist natural enemies and alternative prey. *Ecology*, 77, 1975–1988.

Slansky, F. (1990) Insect nutritional ecology as a basis for studying host plant resistance. *Florida Entomologist*, 73, 354–378.

Slansky, F. and Rodriguez, J.G. (1987) *Nutritional ecology of insects, mites, spiders and related invertebrates*, John Wiley & Sons, Inc., New York.

Staley, J.T., Stewart-Jones, A., Pope, T.W. *et al.* (2010) Varying responses of insect herbivores to altered plant chemistry under organic and conventional treatments. *Proceedings of the Royal Society B*, 277, 779–786.

Van Dam, N.M. and Heil, M. (2011). Multitrophic interactions below and above ground: en route to the next level. *Journal of Ecology*, 99, 77–88.

Van der Putten, W.H., Vet, L.E.M., Harvey, J.A. and Wackers, F.L. (2001) Linking above- and below-ground multitrophic interactions of plants, herbivores, pathogens, and their antagonists. *Trends in Ecology and Evolution*, 16, 547–554.

Van der Putten, W., Bardgett, R., de Ruiter, P. *et al.* (2009) Empirical and theoretical challenges in aboveground–belowground ecology. *Oecologia*, 161, 1–14.

Van Emden, H.F. (1966) Studies on the relations of insect and host plant. III. A comparison of the reproduction of *Brevicoryne brassicae* and *Myzus persicae* (Hemiptera: Aphididae) on brussels sprout plants supplied with different rates of nitrogen and potassium. *Entomologia Experimentalis et Applicata*, 9, 444–460.

Vestergard, M., Bjornlund, L. and Christensen, S. (2004) Aphid effects on rhizosphere microorganisms and microfauna depend more on barley growth phase than on soil fertilization. *Oecologia*, 141, 84–93.

Wardle, D.A., Bardgett, R.D., Klironomos, J.N., Setälä? H., van der Putten, W.H. and Wall, D.H. (2004) Ecological linkages between aboveground and belowground biota. *Science*, 304, 1629–1633.

Waring, G.L. and Cobb, N.S. (1991) The impact of plant stress on herbivore population dynamics, in *Insect–plant interactions*, Vol. 4 (ed. E. Bernays), CRC Press, Boca Raton, pp. 167–226.

Wermelinger, B. (1989) Respiration of the two spotted spider mite as affected by leaf nitrogen. *Journal of Applied Entomology*, 108, 208–212.

Wurst, S. (2010) Effects of earthworms on above- and belowground herbivores. *Applied Soil Ecology*, 45, 123–130.

Yardlm, E.N. and Edwards, C.A. (2003) Effects of organic and synthetic fertilizer sources on pest and predatory insects associated with tomatoes. *Phytoparasitica*, 31, 324–329.

Zehnder, G., Gurr, G.M., Kühne, S., Wade, M.R., Wratten, S.D. and Wyss, E. (2007) Arthropod pest management in organic crops. *Annual Review of Entomology*, 52, 57–80.

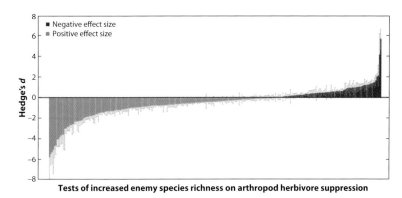

Plate 1.1 Results of a meta-analysis of studies of the response of herbivores to diversity of natural enemies (from Letourneau *et al.*, Effects of natural enemy biodiversity on the suppression of arthropod herbivores in terrestrial ecosystems. *Annual Review of Ecology, Evolution and Systematics*, 40, 573–592, (2009) (with permission)).

Tests of increased enemy species richness on arthropod herbivore suppression

Plate 1.2 Examples of agricultural features that offer scope to enhance natural enemy biodiversity of farmland: grassy, raised earth 'beetle bank', UK (top left); game bird habitat in Environmental Stewardship Scheme area, UK (top centre); diverse plants in a pollinator enhancement strip, UK (top right), native cover crop established to promote decomposition of oil palm prunings, Papua New Guinea (bottom left), young tree strip to shelter crops and livestock, Australia (bottom centre), wild flower mix in ecological compensation area bottom right, Switzerland (photos by G.M. Gurr).

Plate 2.1 A diamondback moth caterpillar hangs in midair from a silken thread. Dropping on a thread allows caterpillars to escape from predators, but comes at the cost of lost feeding opportunities for the caterpillar. Thus, predators that scare caterpillars off of the plant can reduce herbivory even when the herbivore is not killed (photo by Shawn Steffan).

Biodiversity and Insect Pests: Key Issues for Sustainable Management, First Edition. Edited by Geoff M. Gurr, Steve D. Wratten, William E. Snyder, Donna M.Y. Read.
© 2012 John Wiley & Sons, Ltd. Published 2012 by John Wiley & Sons, Ltd.

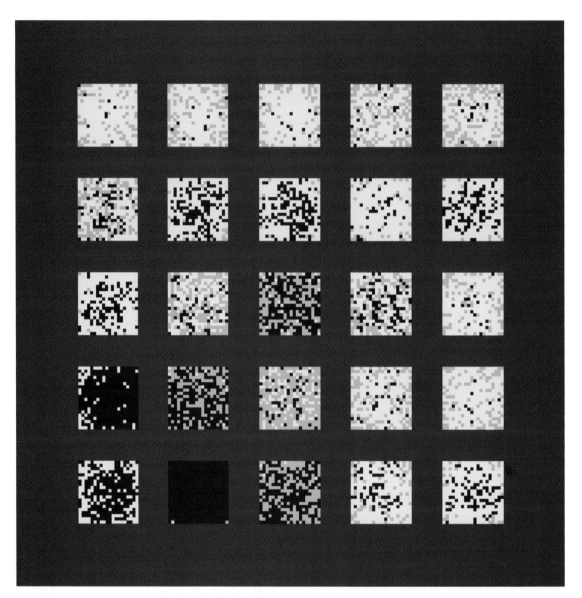

Plate 8.1 Snapshot of a virtual landscape of the scenario with low amount of habitat (habitat amount 2,500 cells, number of patches 25, patch distance 10 cells) during a simulation run. White: cells with only host population, dark pink: cells with host and parasitoid population, brown: matrix cells, green: empty habitat cells. Adapted from Visser *et al.*, 2009, Conservation biocontrol in fragmented landscapes: persistence and paratisation in a host–parasitoid model. *Open Ecology Journal*, 2, 52–61.

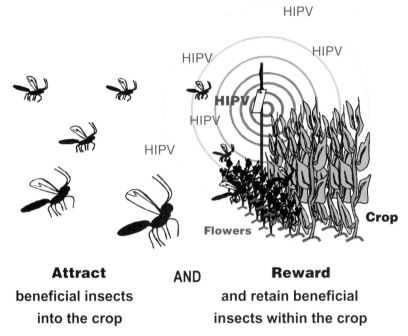

Attract
beneficial insects
into the crop

AND

Reward
and retain beneficial
insects within the crop

Plate 11.1 'Attract and reward' is a concept based on attracting natural enemies into a crop using an HIPV or blend of HIPVs and rewarding them (hopefully enticing them to stay in the crop) by providing nectar-rich plants like buckwheat as a ground cover.

Plate 12.1 Flowers on bunds beside rice fields in Cai Lay district, Tien Giang, Vietnam, as a nectar resource for parasitoids of rice pests.

Plate 12.2 Posters used in Ecological Engineering for rice pest management media campaign in Tien Giang, Vietnam.

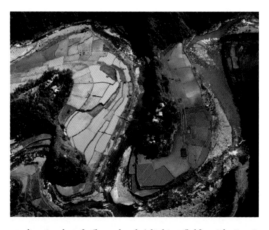

Plate 13.1 A typical rice landscape showing bunds (levee banks) linking fields with riparian vegetation and semi-natural vegetation sources of natural enemies, Ifugao Province, Philippines (from Hettel, 2009, Bird's-eye views of an enduring rice culture. *Rice Today*, 7, 4–19, reproduced with permission of IRRI).

Plate 14.1 Examples of 'Green Plant Protection'. a) yellow sticky traps in a rice paddy field; b) frequency trembler grid lamp beside a rice paddy field; c) sex pheromone trap for rice leaffolder, *Cnaphalocrocis medinalis* Guenee; d) sex pheromone trap for striped rice borer, *Chilo suppressalis* (photos: a, b) Zhongxian Lu; c, d) Jiangxing Wu).

Plate 14.2 Sticky traps employed in tea plantations: a) different coloured sticky traps in tea plantations, b) green sticky cards widely used in tea plantations, c) the pests attracted by green sticky traps, and d) the pests attracted by yellow sticky traps (photos: Baoyu Han).

Plate 14.3 Examples of pest management in vegetable fields: a) combination of sex pheromone and yellow sticky cards applied in the vegetable fields in Ningbo, Zhejiang Province of China, b) frequency trembler grid lamp which attracted vegetable pests in Xiaoshan, Zhejiang Province of China, c) sex pheromone traps for insect pests in the vegetable fields in Ningbo, Zhejiang Province of China (photos: a, c) Jiangxing Wu; b) Guorong Wang).

Plate 15.1 A: Results of a brown planthopper (*Nilaparvata lugens*) choice test on rice varieties. Four varieties showed resistance to the planthopper. All other varieties were highly susceptible, including several with the *Bph1* and *Bph2* genes against which planthoppers have already adapted. B: Susceptible hybrid rice (left) damaged by brown planthoppers at Santa Cruz, Philippines. The adjacent variety, IR74, (right) is relatively resistant and received considerably less damage. Farmers had heavily applied insecticides to both varieties. (Photos – Carmen Bernal IRRI)

Plate 17.1 Resource plants in New Zealand vineyards (clockwise from top left). The endemic New Zealand plants *Anaphalioides bellidioides* (G.Forst.) Glenny: *Acaena inermis* Hook f.; Stephen Wratten with author Tompkins in the vineyard; the non-native *Fagopyrun esculentum* Moench, grown between grapevines for CBC enhancement.

Plate 17.2 Native tallgrass prairie in Ingham County, Michigan established with the assistance of the US Department of Agriculture's Conservation Reserve Program. Overall planting contains over 80 native species of grasses, sedges and forbs. The yellow flowering plant is the native gray coneflower, *Ratibida pinnata* (Vent.) Barnhart, while the white plant in the foreground is the widespread invader wild carrot, *Daucus carota* L.

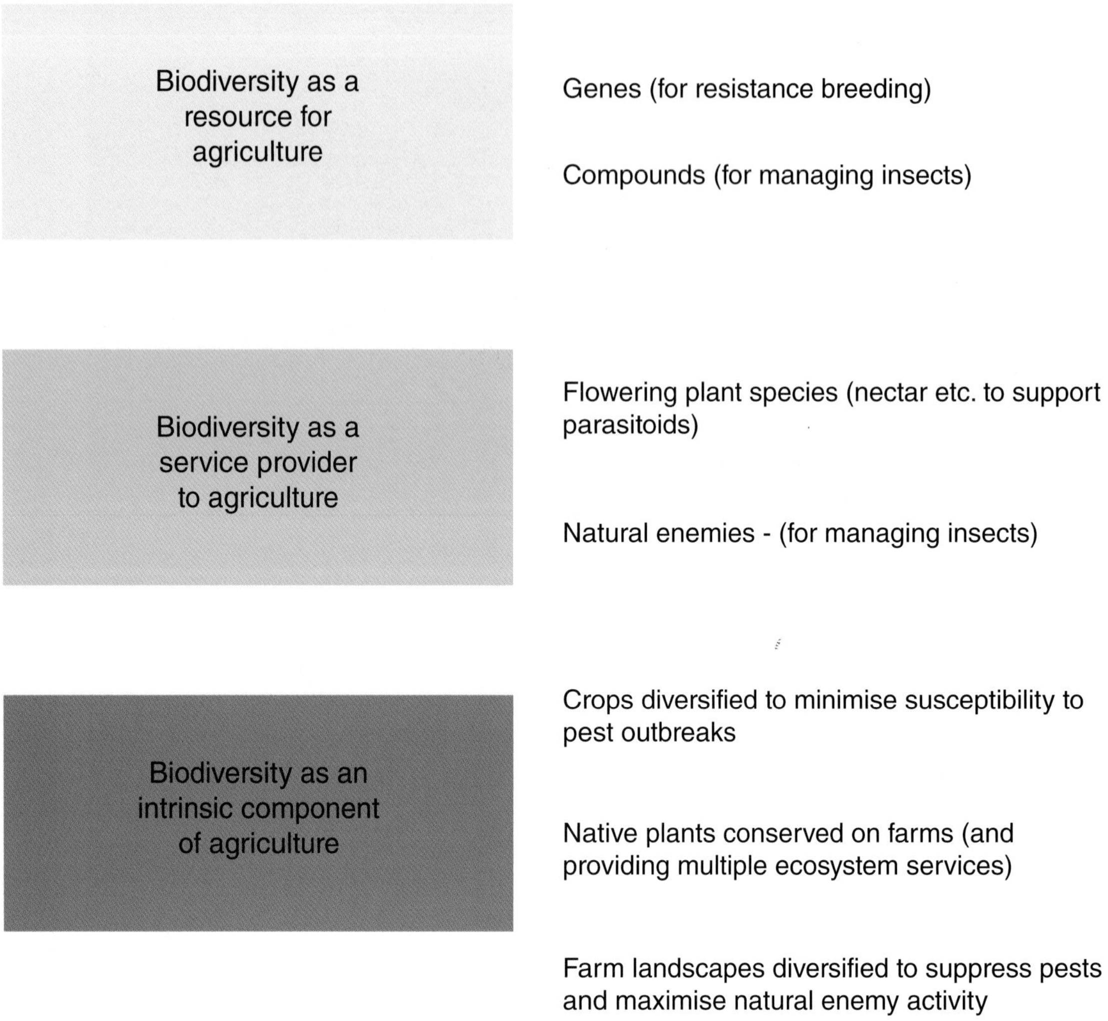

Biodiversity as a resource for agriculture	Genes (for resistance breeding)
	Compounds (for managing insects)

Biodiversity as a service provider to agriculture	Flowering plant species (nectar etc. to support parasitoids)
	Natural enemies - (for managing insects)

Biodiversity as an intrinsic component of agriculture	Crops diversified to minimise susceptibility to pest outbreaks
	Native plants conserved on farms (and providing multiple ecosystem services)
	Farm landscapes diversified to suppress pests and maximise natural enemy activity

Plate 20.1 Shades of green in concepts of biodiversity's role in agriculture (see text for full explanation).

Chapter 6

PLANT BIODIVERSITY AS A RESOURCE FOR NATURAL PRODUCTS FOR INSECT PEST MANAGEMENT

Opender Koul

Biodiversity and Insect Pests: Key Issues for Sustainable Management, First Edition. Edited by Geoff M. Gurr, Steve D. Wratten, William E. Snyder, Donna M.Y. Read.
© 2012 John Wiley & Sons, Ltd. Published 2012 by John Wiley & Sons, Ltd.

INTRODUCTION

The use of pesticides and synthetic fertilisers has increased dramatically over the past 60 years (McKinney *et al.*, 2007). In view of the adverse effects of synthetic organic pesticides, such as some organochlorins, organophosphates and carbamates, on non-target organisms and the environment (Wheeler, 2002), efforts to develop safer and more selective pesticides have increased in the last three decades. Pesticides termed 'botanical', 'low-risk' or 'biorational' include phytochemicals (see Box 6.1) that effectively control insect pests, and often have low toxicity to non-target organisms (such as humans, animals and natural enemies) and reduced environmental impact (Koul, 2005; Koul *et al.*, 2008; Koul and Walia, 2009). However, due to their specificity, the market potential of these chemicals is limited and the industry is increasingly reluctant to invest in the development and registration of new chemicals, which requires huge investment in terms of resources and time because of progressively more stringent registration regulations in most countries. Furthermore, many products available in the market are not of uniform quality and give inconsistent results (Isman, 2006). Recently identified products have exhibited broader-spectrum activity against insects without harming beneficial species, and suitable standards for quality and performance have been laid out for many compounds obtained from natural resources (such as essential oil compounds, Koul *et al.*, 2008; Isman *et al.*, 2010). Therefore, such

---

Box 6.1 Some terms used in relation to botanical insecticides

Biopesticides: Certain types of pesticides derived from such natural materials as animals, plants, bacteria, and certain minerals.
Biorational pesticides: Chemicals with greater selectivity and considerably lower risks to humans, wildlife and the environment compared to conventional synthetic insecticides.
Extractives: Substances present in vegetable and animal tissues that can be separated by successive treatment with solvents and recovered by evaporation of the solution.
Phytochemicals: These chemicals occur naturally in plants.

---

natural products are becoming more reliable and look set to play a vital role in management of crop pests in the future (Dhaliwal and Koul, 2010). In this context, there is a strong need to promote eco-friendly technologies in agriculture and natural products have to play a prominent role in sustainable crop production (Koul *et al.*, 2009). During the last two decades, phytochemicals from plant bio-resources have been heralded as desirable alternatives to synthetic chemical insecticides for pest management beause they reputedly pose little threat to the environment or to human health (Isman *et al.*, 2010).

BIODIVERSITY OF PLANTS: A RESOURCE OF INSECT CONTROL COMPOUNDS

Over 200,000 metabolites are currently known, but even this large number is estimated to account for just 10% of the possible number of these compounds in nature (Croteau *et al.*, 2000; Dixon and Strack, 2003). In most cases their structures, functions and uses have not been sufficiently evaluated. In general, these compounds do not take part in basic metabolism; instead they mediate plant–plant and plant–herbivore interactions (e.g. multitrophic and interguild interactions, which are frequently mediated by the plants' chemical defences against herbivores (Kessler, 2006)). Thus, several important types of insecticide are derived from or are analogues of plant products, but these are the 'tip of the iceberg' (Box 6.2). Discovery and use of phytochemicals is a highly active area of science but one that is often shrouded in commercial confidentiality. This chapter provides a synthesis of the available information to illustrate the potential for this approach to insect pest management, with an emphasis on the value of biodiversity as a bank of potentially useful bioactive phytochemicals.

PHYTOCHEMICALS FOR INSECT CONTROL

Commercial use of phytochemical biopesticides began in the nineteenth century with the introduction of nicotine from *Nicotiana tabacum* (L.), rotenone from *Lonchocarpus* sp., derris dust from *Derris elliptica* (Wallich) Benth and pyrethrum from *Tanacetum cinerariifolium* (Trevir) (previously *Chrysanthemum cinerari-*

Box 6.2 Current trends in the use of phytochemicals

Products based on plant products registered in the USA are pyrethrum, neem, rotenone, several essential oils, sabadilla, ryania and nicotine. Several azadirachtin-based (neem) insecticides are sold in the United States, and a number of plant essential oils are exempt from registration altogether.

Canada allows the use of pyrethrum, rotenone, nicotine, a few essential oils and neem (although they are yet to achieve full registration in Canada).

Mexico allows the use of most products sold in the United States, although there is no specific exemption for plant oils.

The European Union permits the use of pyrethrum, neem, rotenone and nicotine, along with components of etheric oils of plant origin; however, variations do exist among the European countries. For instance, neem has not yet been registered in the UK.

New Zealand has registrations for pyrethrum, rotenone and neem, whereas Australia has yet to approve neem in spite of almost two decades of research and development in that country.

In Asia, India appears to embrace botanicals more than any other country in the region, permitting all of the materials (except sabadilla). Neem alone has more than 100 products in India with both provisional and full registration status. Neem has yet to be approved for use in the Philippines, where pyrethrum is the only approved botanical insecticide. In China, rotenone, matrine, nicotine, toosendanin, veratridine and azadirachtin from neem are registered products.

Data on regulated insecticides are not readily available for most African countries. Among botanicals, only pyrethrum is approved for use in South and East Africa. As in Latin America, numerous crude plant extracts and oils are likely to be in local use in the poorer countries.

Table 6.1 Important plant families that have a number of species evaluated for anti-insect properties.

Plant family	Number of plant species
Annonaceae	12
Apiaceae	23
Apocyanaceae	39
Asteraceae	147
Bignoniaceae	13
Cryptogams	58
Cupressaceae	22
Euphorbiaceae	63
Fabaceae	157
Labiatae	52
Lamiaceae	24
Leguminosae	60
Meliaceae	>500
Moraceae	26
Myrtaceae	72
Pinaceae	52
Piperaceae	14
Poaceae	27
Ranunculaceae	55
Rosaceae	34
Rubiaceae	38
Rutaceae	42
Solanaceae	52
Verbenaceae	60

Modified from Koul 2003.

(Table 6.1); some of the most promising recent phytochemicals are discussed below.

Isobutylamides

The Piperaceae family is considered to be among the most archaic of pan-tropical flowering plants. The genus *Piper* belonging to this family contains approximately 1,000 species of herbs, shrubs, small trees and hanging vines. Several *Piper* species from India, Southeast Asia and Africa are of economic importance since they are used as spices and traditional medicines (Simpson and Ogorzaly, 1995). Many plant families have a global distribution, but few have the rich ethnobotanical and ethnopharmaceutical history of Piperaceae (Scott *et al.*, 2008).

The chemistry of members of the family Piperaceae is of great interest owing to the variety of biological properties displayed. A survey of structural diversity and bioactivity reveals that groups of species specialise

ifolium). The use of these compounds, their efficacy and commercial potential has been comprehensively discussed (Koul and Walia, 2009). This successful use of traditional botanicals has aroused further interest in exploring plant biodiversity for new bioactive phytochemicals and extractives as a possible source of pest control agents. In fact, several species of a wide diversity of plants are known to have anti-insect properties in the production of amides, phenylpropanoids, lignans and neolignans, benzoic acids and chromenes, alkaloids, polyketides, and a plethora of compounds of mixed biosynthetic origin. Since members of the Piperaceae represent such rich sources of bioactive compounds, a detailed knowledge of the dynamics of secondary compound biosynthesis during various stages of plant maturity would be of value. To this end, selected *Piper* species have been cultivated both in vitro (suspension cultures and differentiated plantlets) and in the field (seedlings and adult plants) in order to investigate changes in chemical composition during the developmental process (Kato and Furlan, 2007). A large number of unsaturated isobutylamides have been attributed to the defence of this family, specifically in *Piper* species. The compounds have been isolated from the fruits, stem and leaves of *Piper nigrum* (L.),

Piper acutisleginum (de Candolle), *Piper khasiana* (de Candolle), *Piper. longum* (L.), *Piper pedicellosum* (Wallich) and *Piper thomsonii* (de Candolle) (Parmar and Walia, 2001) (Figure 6.1). *Piper retrofractum* (Vahl) from Thailand, *Piper guineense* (Schumacher and Thon) from West Africa and *Piper tuberculatum* (Jacq.) from Central America reflect their global biodiversity and possess diverse active compounds (Isman, 2001). Some of the active compounds include piperine, piperlonguminine, pipericide, dihydropipericide, retrofractamide A and pellitorine; these occur mostly in the fruits of these plants (Figure 6.1). The wide variety of secondary plant compounds found in *Piper* were suggested as potential leads for novel insecticides (Miyakado *et al.*, 1989). Many varieties are used in traditional control of insects that are vectors of disease (Okorie and Ogunro, 1992) and many are toxic to number of insect

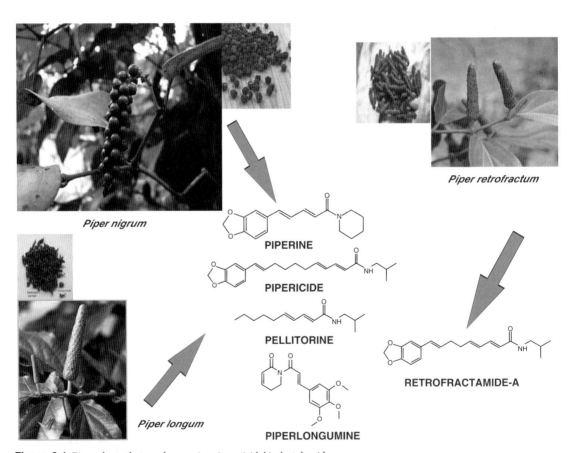

Figure 6.1 Piper plants that produce various insecticidal isobutylamides.

Table 6.2 Comparative toxicity (LC_{50} in ppm after 48 h) of isobutylamides against mosquito larvae.

Compound	*Culex pipiens pallans* (Coquillett)	*Aedes aegypti* (L.)	*Aedes togoi* (Theobald)
Pipericide	0.004	0.1	0.26
Retrofractamide A	0.028	0.039	0.01
Guineensine	0.17	0.89	0.75
Pellitorine	0.86	0.92	0.71
Piperine	3.21	5.1	4.6

Compiled from Park *et al*. 2002.

species (Scott *et al*., 2008). The behaviour modification (antifeedant and repellent effects) effects of *Piper*-based extracts have been determined in greenhouse trials; pepper seed extracts deterred lily leaf beetles, *Lilioceris lilii* (Scopoli) and striped cucumber beetles, *Acalymma vittatum* (Fabricius) from damaging leaves of lily and cucumber plants, respectively, at concentrations in the 0.1–0.5% range (Scott *et al*., 2004). The repellent activity was observed to benefit the plant for up to four days post-spraying. However, the residual repellent effect of *P. nigrum* was much less under full sunlight, and herbivore damage resumed shortly after application (Scott *et al*., 2003).

The efficacy of these compounds has been variable, too, and their activity appears to be species-specific (Scott *et al*., 2005). For instance, isobutylamides differ in their toxicity to mosquito larvae of various species (Table 6.2; Park *et al*., 2002). The comparative toxicities also vary. For example, acute toxicities to the velvet bean caterpillar, *Anticarsia gemmatalis* (Hübner), of these compounds show LD_{50} and LD_{90} values to be quite high at 31.3 and 104.5 mg/insect, respectively for pellitorine and 122.3 and 381.0 mg/insect for 4,5-dihydro piperlonguminine (Navickiene *et al*., 2007).

All of the unsaturated isobutylamides are neurotoxins that impair or block voltage-dependent sodium channels on nerve axons. Being neurotoxic, these amides with methylenedioxyphenyl group (MDP) (e.g. pipericide, Figure 6.1) were more toxic but did not have the knockdown toxicity of the piperamides without the MDP group (e.g, pellitorine, Figure 6.1). Actually the piperamides found in species of *Piper* are bifunctional due to the presence of two active functional groups, the isobutyl amide moiety and a MDP ring. These functional groups are responsible for the dual mode-of-action of these compounds (i.e. neurotoxic and inhibition of cytochrome P450 enzymes, respectively). These characteristics are useful to plants of the *Piper* genus as a defence strategy against herbivores (Scott

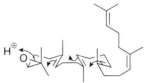

Figure 6.2 Squalene epoxide precursor for biosynthesis of quassinoids and limonoids.

et al., 2003). Fortunately, the risk to human health is likely to be much reduced because the active components have had a safe history as food additives and spices (Scott *et al*., 2003).

LIMONOIDS AND QUASSINOIDS

The plant families Rutaceae, Meliaceae, Cneoraceae, Simaroubaceae and to some extent Burseraceae are rich in metabolically altered triterpenes, the limonoids (tetranortriterpenoids) and the quassinoids (decanortriterpenoids), which are derived from the triterpenoid precursor euphol. Both groups of compounds are derived from condensation of a chair–chair–chair–boat configured squalene epoxide precursor (Figure 6.2). Most of the intermediates and enzymes in these pathways remain unstudied. There are at least 300 known members of this group of compounds. They are stereochemically homogeneous.

Quassinoids occur only in the family Simaroubaceae. This family. the quassia family of flowering plants, in the order Sapindales, comprises 25 genera of pantropical trees, including *Ailanthus* (Figure 6.3) or the tree of heaven. Members of the family have leaves that alternate along the stem and are composed of a number of leaflets arranged along an axis. Most species have small flowers, bitter bark and fleshy fruits that are sometimes winged. More than 120 quassinoids have

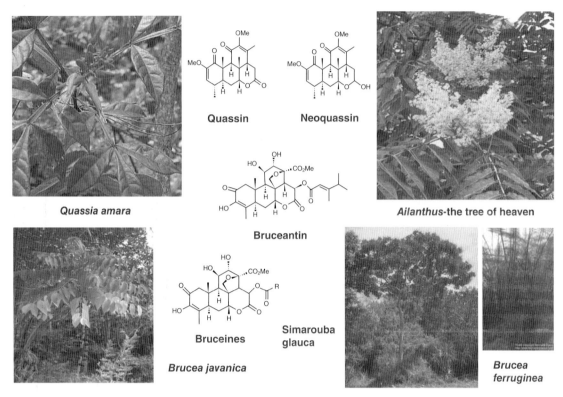

Figure 6.3 Some plants of family Simaroubaceae that produce quassionoids. The figure shows the structures of some anti-insect quassinoids.

been described. The biosynthetic precursors of this series are similar to those of limonoids. D7-euphol and/or D7-tirucallol appear to be involved. Quassinoids, which are more like limonoids rather than degraded triterpenes, also possess anti-insect properties (Dev and Koul, 1997; Koul, 2005). Quassin and neoquassin isolated from *Quassia amara* (L.) and *Picrasma quassinoides* (Bennet) (Figure 6.3) were the first quassinoids shown to be toxic to a number of insects in the 1970s (Dev and Koul, 1997). Compounds such as bruceantin, bruceine-A, bruceine-B and bruceine-C from *Brucea antidysenterica* (Mill) are antifeedant compounds for tobacco budworms, Mexican bean beetles and southern armyworms (Koul, 2005).

Among limonoids, the best-known anti-insect compounds are azadirachtins which occur in the seeds of the neem tree, *Azadirachta indica* (A. Juss) (Figure 6.4). Azadirachtin A (Figure 6.4) is active against a broad spectrum of insects and is a known potential insecti-

cidal, antifeedant and insect growth regulatory compound from neem (Koul, 1996; 2008; Schmutterer, 2002; Koul and Wahab, 2004; Isman, 2006). In fact, neem is the most commercially exploited plant for insect pest management (Schmutterer, 2002). Azadirachtin, and neem extracts and preparations containing it, are effective against a broad spectrum of pest species, with more than 400 species of insects reported as susceptible (Koul and Wahab, 2004). Among the more susceptible pests are foliar-feeding Lepidoptera and Coleoptera, with some filter-feeding larval Diptera also susceptible. Phloem-feeders (e.g. aphids) and other sucking insects vary widely in susceptibility, a consequence of both innate susceptibility and the availability of azadirachtin in phloem sap or cell cytoplasm for ingestion (Lowery and Isman, 1993; 1996). Although there are exceptions, neem is rarely effective against subterranean pests or those that bore into plant tissues (Schmutterer, 2002). Interspecific differences in sus-

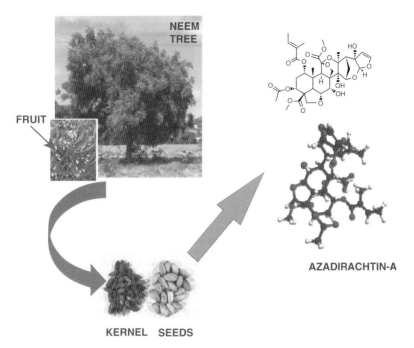

Figure 6.4 Indian neem tree, *Azadirachta indica* and its seeds and kernel. 'Azadirachin A' is the major insecticidal active ingredient. (Model structure source http://en.wikipedia.org/wiki/File:Azadirachtin_model.png)

ceptibility to the behavioural effects of neem (feeding, oviposition deterrence) are far greater than differences in susceptibility to the physiological actions (moult disruption) (Mordue, 2004; Koul, 2008). In addition to azadirachtins, there are several non-azadirachtin types of compounds in neem, which are also active against lepidopterans but moderate as compared to the azadirachtin A (Koul *et al.*, 2004a).

Other limonoids from the same plant, or in rutales in general, have many activities against insect pests (Champagne *et al.*, 1992). A similar series of compounds is found in a related plant, *Melia azedarach* (L.). The fruits of this species are quite toxic to livestock as well. These compounds, such as toosandanin and meliatoxins (Figure 6.5), have been reported to have potential for pest control (Macleod *et al.*, 1990; Koul *et al.*, 2002).

Another group is the citrus limonoids, which are a group of highly oxygenated tetranortriterpenoids found in the Rutaceae and Meliaceae plant families. In a number of citrus species (Figure 6.6), the bitterness causative factors are limonoids, and deacetylnomilinic acid is described as the most likely initial precursor of

all the known citrus (Rutaceae) limonoids (which itself may be biosynthesised from acetate, mevalonate and/or furanesyl pyrophosphate in the phloem region of stems) (Champagne *et al.*, 1992). With radioactive tracer work it has been shown that deacetylnomilinic acid converts into nomilin. Both deacetylnomilinic acid and nomilin are synthesised in the phloem region of stems and then are translocated to other plant tissues such as leaves, fruit, tissues and seeds (Roy and Saraf, 2006). Seed and fruit tissues are capable of biosynthesising other limonoids starting from nomilin independently, by at least four different pathways (Figure 6.7). Limonin and other citrus limonoids act as insect repellents, feeding deterrents, growth disrupters and reproduction inhibitors against several insect pest species across a wide range of agricultural crops (Alford and Murray, 2000). A few other citrus limonoids, including nomilin, nomilinic acid, ichangin and obacunoic acid are also bitter. Among these, limonin and nomilin (Figure 6.7) are known to deter feeding in lepidopterans and coleopterans with variable efficacies (Champagne *et al.*, 1992). It appears that the furan and epoxide groups play a major role in the activity of

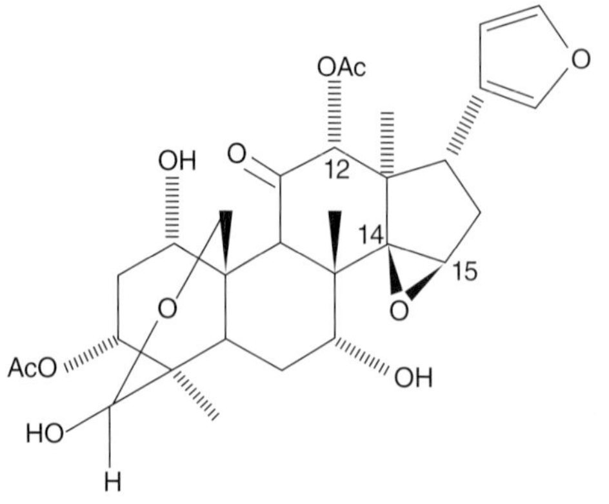

Toosandanin

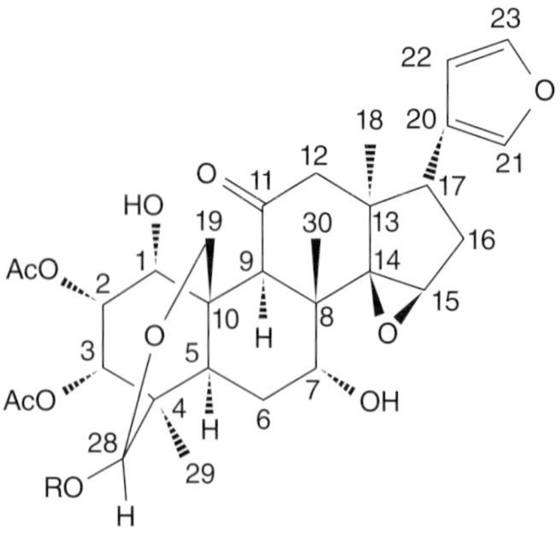

Meliatoxins

1 R = CH$_3$CH$_2$CH(CH$_3$)CO

2 R = (CH$_3$)$_2$CHCO

Figure 6.5 Anti-insect toosandanin and meliatoxins.

Figure 6.6 Some plants from the citrus family, rich in citrus limonoids.

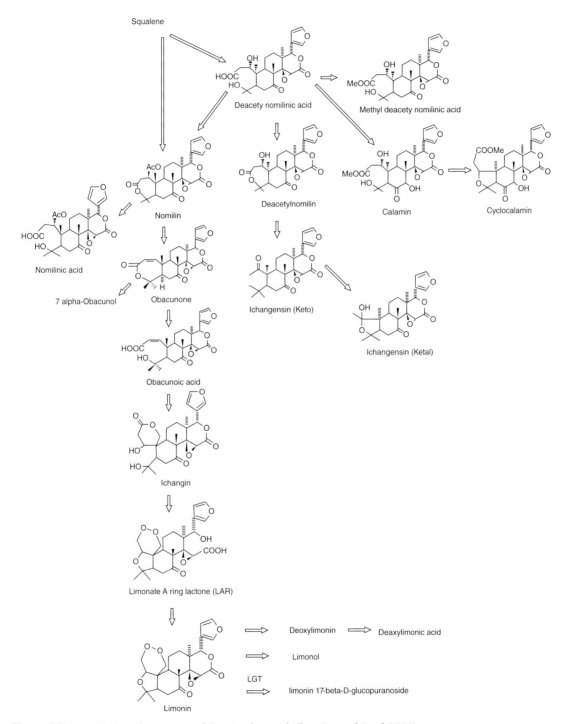

Figure 6.7 Biosynthetic pathway proposed for citrus limonoids (from Roy and Saraf, 2006).

these compounds. A possible role of C-7 is implied by the modest activity of the 7-hydroxylated de-epoxy system (Bentley *et al.*, 1988). For instance, highly reduced activity of deoxyepilimonol against limonin demonstrates the above conclusion. In certain cases, the cyclohexenone A ring and the α-hydroxy enone group in the B ring appear to be important for antifeed-ant activity (Champagne *et al.*, 1992; Koul, 2005). In addition, the absence of 14–45 epoxide may not drasti-cally reduce antifeedant activity (Govindachari *et al.*, 1995). Some structural-activity relationships have also been drawn by preparing semi-synthetic deriva-tives of citrus limonoids, suggesting the potential of functional groups for their activity (Ruberto *et al.*, 2002).

Naphthoquinones

Naphthoquinones are relatively widely occurring natural substances and products of secondary metabo-lism of some actinomycetes, fungi, lichens and higher plants. The importance of these substances is due to their broad biological activity. In most cases they act as phytoalexins. Naphthoquinones interact with mito-chondria, microsomes and cytoplasmic proteins, in the form of radicals they are bound to, and damage DNA and RNA (Babula *et al.*, 2009). Thus naphthoquinones are highly cytotoxic substances; their antimicrobial, antifungal, antiviral and antiparasitic effects have been observed. In traditional medicines, particularly in some parts of Asia (China) and South America, naphthoquinones-containing plants are widely used primarily in the treatment of various tumoral and parasitic diseases (Babula *et al.*, 2006). In higher plants naphthoquinones occur in families like Avicenniaceae, Bignoniaceae, Boraginaceae, Droseraceae, Ebenaceae, Juglandaceae, Nepenthaceae and Plumbagnaceae (Babula *et al.*, 2009). They commonly occur in the reduced and glycosidic forms. In some species, naph-thoquinones are present as monomers, as well as dimers or trimers. They are biosynthesised via a variety of pathways including acetate and malonate pathway (plumbagin), shikimate/succinyl CoA combined path-way (lawsone) and shikimate/mevalonate pathway (alkannin) (Babula *et al.*, 2009). Interest in these com-pounds is due to their broad range of biological activi-ties including antibacterial, fungicidal, antiparasitic and insecticidal properties. In addition, they have inhibitory effects on insect larval development (Babula

et al., 2009). Naphthoquinones, especially juglone, have been widely studied for their allelopathic activity (Willis, 2000). For example, the family Bignoniaceae includes several species whose wood is known to be resistant to termite attack (Castillo and Rossini, 2010). Naphthoquinone extracts isolated from *Catalpa bignon-ioides* (Walter) have shown activity against the termite *Reticulitermes flavipes* (Kollar) (Isoptera: Rhinotermiti-dae), specifically due to its inclusion of the compounds catalponol and catalponone (Becker *et al.*, 1972) (Figure 6.8). Some naphthoquinones are also active against Diptera species. For example, lapachol isolated from *Cybistax antisyphilitica* (Martius), and jacaranone (Figure 6.8) from *Jacaranda* wood extracts, are larvi-cidal against mosquitoes and houseflies (Kaushik and Saini, 2008). Two active principles from the Chilean plant *Calceolaria andina* (Bentham) (Scrophulariaceae), related to the familiar garden 'slipper' plant, have been identified as hydroxynapthoquinone and its acetate, designated as BTG 505 and BTG 504 (Figure 6.8). These compounds are effective against a range of com-mercially important pests, including the tobacco whitefly, *Bemisia tabaci* (Gennadius), aphids, and the two-spotted spider mite, *Tetranychus urticae* (Koch) (Khambay *et al.*, 1999). Their primary mode of action in insects is inhibition of complex III of the mitochon-drial respiratory chain (Khambay and Jewess, 2000).

Rocaglamides

The genus *Aglaia* of the family Meliaceae, consisting of some 130 species widely distributed in the Indo-Malaysian region (Nugroho *et al.*, 1999) has attracted considerable attention in recent years as a possible source of unique natural products. These trees occur in the tropical and subtropical forests of Southeast Asia, Northern Australia and the Pacific. Insecticidal activity has generally been attributed to rocaglamides. An outstanding property of these compounds is that they are effective against a range of resistant insect strains, including the notorious B-biotype of the tobacco whitefly, *B. tabaci*, which is devastating crops worldwide (Proksch *et al.*, 2001). Phytochemical inves-tigations of *Aglaia* have revealed the presence of a variety of compounds, including rocaglamides (Ishi-bashi *et al.*, 1993; Proksch *et al.*, 2001), aglains (Bacher *et al.*, 1999), bisamides (Brader *et al.*, 1998), triterpe-nes (Weber *et al.*, 2000) and lignans (Wang *et al.*, 2004), with interesting biological activities. More than

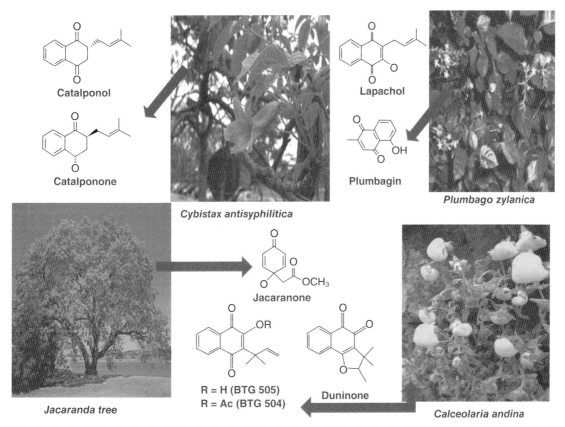

Figure 6.8 Naphthoquinone bearing plants. The figure shows structures of some potential natural insecticidal naphthoquinones.

50 naturally occurring rocaglamide derivatives have been isolated to date and rocaglamide was the first effective anti-insect compound identified (Figure 6.9) (Proksch *et al.*, 2001). There are many of them isolated from *Aglaia odorata* (Lour) and *Alaia elaeagnoidea* (Juss) (Figure 6.9). Rocaglamide derivatives are unusual aromatic compounds, featuring a cyclopentatetrahydrobenzofuran skeleton, and are confined to members of *Aglaia*. Recently, several novel rocaglamide derivatives isolated from different *Aglaia* species have been shown to have strong insecticidal activity (in some cases even comparable to azadirachtin), mostly against neonate larvae of *Spodoptera littoralis* (Boisduval), *Ostrinia* species and the gram pod borer, *Helicoverpa armigera* (Hübner) (Brader *et al.*, 1998; Gussregan *et al.*, 1999; Nugroho *et al.*, 1997a; 1997b; 1999; Koul *et al.*, 2004b). Their insecticidal mode-of-action, as well

as the potential anti-cancer activity of rocaglamides, results from inhibition of protein synthesis; this explains the long time-to-death in treated insects (Satasook *et al.*, 1993). The insecticidal activity of rocaglamides can be attributed to the presence of the furan ring system, since the closely related aglains, possessing a pyran ring, are devoid of insecticidal activity (Nugroho *et al.*, 1999). The nature of the substituents at C1, C2, C3 and C8 has also been suggested to be responsible for the bioactivity of respective derivatives (Nugroho *et al.*, 1997a; 1999; Schneider *et al.*, 2000). Acylation of the OH group (with formic or acetic acid) at C1 caused a reduction of insecticidal activity in neonate larvae of *S. littoralis* compared with other rocaglamide derivatives with a hydroxyl substituent isolated from the twigs of *Aglaia duperreana* (Pierre) (Nugroho *et al.*, 1997a). The strong bioactivity of

Aglaia elaeagnoidea

Aglaia odorata

Figure 6.9 *Aglaia* species that produce rocaglamide and aglaroxins.

rocaglamides against a number of insect pests suggests that they could be used as potential natural insecticides for plant protection if data at field level could be generated on a larger scale. Among the various compounds isolated from *A. odorata*, *Aglaia elliptica* (Blume) and *A. duppereana* (Meliaceae), rocaglamide is the most effective (EC$_{50}$= 0.8 ppm). It is slightly more potent than azadirachtin (EC$_{50}$=1.0 ppm) against some insect species (Janprasert *et al.*, 1993). As growth inhibitors, rocaglamide and methyl rocaglate are similar in their activity (EC$_{50}$=0.9 ppm) and quite comparable to azadirachtin (0.26 ppm) (Isman *et al.*, 1990), as are the

aglaroxins (Figure 6.9) isolated from other *Aglaia* species (Koul *et al.*, 2005a; 2005b).

Sugar esters

Plant glucose and sucrose esters occur naturally in glandular trichomes (Figure 6.10) of leaves of wild tobacco *Nicotiana gossei* (Domin), *Lycopersicon typicum* (Humb. and Bonpl.) and other solanaceous plants (King *et al.*, 1993; Neal *et al.*, 1994). These esters are composed of lower fatty acids (C2 to C10), and have been

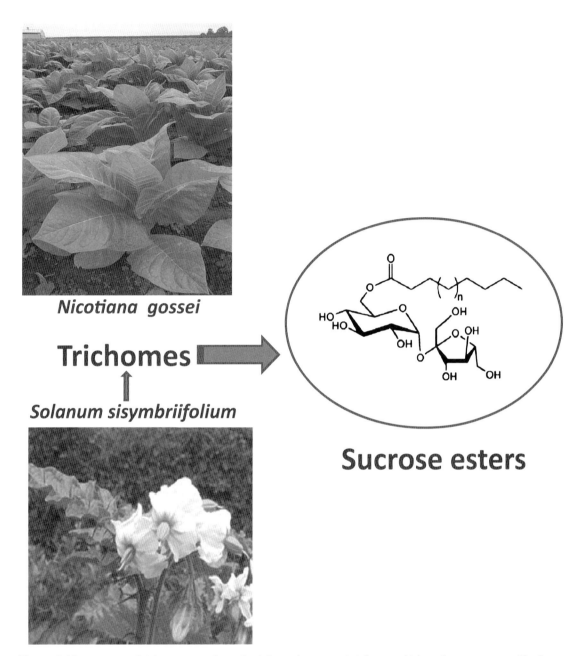

Figure 6.10 *Nicotiana* and *Solanum* species have glandular trichomes on their leaves, which produce sugar esters. The figure shows an insecticidal sucrose ester structure that can have different degrees of substitution.

found to be very effective against soft-bodied insects (Buta *et al.*, 1993; Puterka and Severson, 1995).

Phytochemical investigations of *Nicotiana* sp. have resulted in the isolation of a variety of glucose esters (Matsuzaki *et al.*, 1989a; 1989b) and acyl sugars (Neal *et al.*, 1990) that deter insects. Screening methods for sucrose esters from plant extracts are well established (Simonovska *et al.*, 2006). A series of sucrose esters (Figure 6.10) have been reported in the cuticular waxes of tobacco leaves (Severson *et al.*, 1994). For example, three sucrose esters were isolated from the surface lipids of leaves of *Nicotiana cavicola* (Burbidge) (Ohya *et al.*, 1994). Common features found in all three sucrose esters were the presence of one acetyl residue at the fructose ring, and free hydroxyl groups at positions 2 and 3 of the glucose ring. The presence of sucrose esters in wild tomato and wild potato species has also been related to aphid resistance (Goffreda *et al.*, 1989). Glucose and sucrose esters reportedly disrupt the integrity of cellular membranes and uncouple oxidative phosphorylation, similar to the action of insecticidal soaps (Parmar and Walia, 2001). According to Puterka and Severson (1995), sugar esters disrupt the structure of the insect cuticle. It has been stated that leaf surface moisture and ambient relative humidity affected the efficacy of *N. gossei* sugar esters. For example, the application of hygroscopic materials such as humectants at the site of application improves the toxicity of natural sugar esters from *N. gossei* and other *Nicotiana* species, as well as certain synthetic sugars against tobacco aphids (Xia *et al.*, 1997a; 1997b). Functionally, this product appears to differ little from insecticidal soaps based on fatty acid salts developed in the 1980s, particularly potassium oleate. Although useful in home and garden products and in greenhouse production, the utility of glucose and sucrose esters for agriculture remains to be seen, as no substantial activity has been recorded against lepidopterans (except for some growth inhibitory effects) (Koul *et al.*, unpublished data).

Phytochemicals from essential oils

Plant essential oils are produced commercially from several botanical sources, many of which are members of the mint (Lamiaceae), carrot (Apiaceae), myrtle (Myrtaceae) and citrus (Rutaceae) families. The oils are generally composed of complex mixtures of monoterpenes, biogenetically related phenols, and sesquiterpenes (Koul *et al.*, 2008). The composition of these oils can vary dramatically, even within species. Factors affecting the composition include the part of the plant from which the oil is extracted (i.e. leaf tissue, fruits, stem, etc.), the phenological state of the plant, the season, the climate, the soil type and other factors (Soković *et al.*, 2009). Examples include 1,8-cineole, the major constituent of oils from rosemary and eucalyptus; eugenol from clove oil; thymol from garden thyme; menthol from various species of mint; asarones from calamus; and carvacrol and linalool from many plant species (Koul *et al.*, 2008). A number of source plants have traditionally been used for protection of stored commodities, especially in the Mediterranean region and in southern Asia, but interest in the oils was renewed with emerging demonstration of their fumigant and contact insecticidal activities to a wide range of pests in the 1990s (Isman, 2000; Koul *et al.*, 2008). Rapid action against some pests is indicative of a neurotoxic mode of action, and there is evidence that some act through interference with the neuromodulator octopamine (Kostyukovsky *et al.*, 2002); other oils apparently interfere with GABA-gated chloride channels (Priestley *et al.*, 2003).

Essential oils are predominantly composed of terpenes (hydrocarbons) such as myrecene, pinene, terpinene, limonene, *p*-cymene, α- and β- phellandrene, etc.; and terpenoids (oxygen-containing hydrocarbons) such as acyclic monoterpene alcohols (geraniol, linalool), monocyclic alcohols (menthol, 4-carvomenthenol, terpineol, carveol, borneol), aliphatic aldehydes (citral, citronellal, perillaldehyde), aromatic phenols (carvacrol, thymol, safrol, eugenol), bicyclic alcohol (verbenol), monocyclic ketones (menthone, pulegone, carvone), bicyclic monoterpenic ketones (thujone, verbenone, fenchone), acids (citronellic acid, cinnamic acid) and esters (linalyl acetate). Some essential oils may also contain oxides (1,8-cineole), sulphur-containing constituents, and methyl anthranilate. Coumarins, zingiberene, curcumene, farnesol, sesquiphellandrene, termerone, nerolidol, etc. are examples of sesquiterpenes (C15) isolated from essential oils. Mono- and sesquiterpenoidal essential oil constituents are formed by the condensation of isopentenyl pyrophosphate units. Diterpenes usually do not occur in essential oils but are sometimes encountered as by-products. Many of these compounds from essential oils have been reported to possess potent biological activity and are responsible for bitter taste and toxicity (Koul *et al.*, 2008). Some of the compounds that have been made commercially available during the past decade are cinnamon oil and cinnamaldehyde (30% EC from Mycotech), although discontinued now; eugenol and 2-phenethylpropionate (from EcoSMART, Tennessee, USA); rosemary oil and 1,8-cineole (from Brandt, IL, USA); menthol as Apilife VAR™ (from Chemicals LAIF, Italy) and even the combinations of thymol, menthol, cineole and camphor are known from EcoSMART and Brandt.

Many monoterpenoids, phenylpropanoids and alcohols from essential oils are responsible for ovicidal activity or they alter the oviposition potential of insects (Koschier and Sedy, 2001; Chiasson *et al.*, 2004; Waliwitiya *et al.*, 2009). The essential oil of *Chloroxylon swietenia* (de Candolle) and its constituents, geijerene and pregeijerene, have been shown to deter oviposition by *Spodoptera litura* (Fabricius) in laboratory experiments (Kiran *et al.*, 2006). In addition, eight compounds from essential oils were shown to deter oviposition of *S. litura* and *Chilo partellus* (Swinhoe) in greenhouse conditions (Singh *et al.*, 2010; 2011). This

suggests that inhibition of behavioural responses from insects in response to natural phytochemicals needs to be exploited with a broader perspective than just toxic potential. This is supported by recent studies on plant latex, which inhibits oviposition in the cabbage looper moth, *Trichoplusia ni* (Hübner) (Shikano and Isman, 2009). Therefore, it is not only plant biodiversity that needs to be looked into as a bioresource, but also the diversity of modes-of-action that such plant products can provide.

Plant-derived cyclotides

To date nearly 200 cyclotides have been isolated from 30 plant species of the families Violaceae, Rubiaceae and Cucurbitaceae (Wang *et al.*, 2008). Common genera with these cyclotides are *Gloeospermum*, *Hybanthus*, *Melycitus* and *Viola* (Figure 6.11). Generally, these cyclotides are a family of small macrocyclic proteins of 29 to 31 amino acids mostly found in Violaceae and

Figure 6.11 Some plants from the families Violaceae and Rubiaceae. The principal structure of a cyclotide. (Image taken from http://en.wikipedia.org/wiki/File:Cyclotide_structure.jpg) derived from these plants is shown.)

Rubiaceae species and contain a cystine knot: two disulfide bonds together with their connecting peptide backbone forming an embedded ring which is penetrated by a third disulfide bond (Figure 6.11). The cyclotides are divided into two major subfamilies, Möbius and bracelet, depending on the presence or absence, respectively, of a cis-Pro peptide bond in loop 5 (Figure 6.11) (Craik *et al.*, 1999; 2010). The bracelet proteins have a higher net charge and are more cytotoxic than the Möbius ones. The cyclotides have been attributed a wide range of biological activities, which in combination with their chemical stability and structural plasticity have made them attractive tools for pharmaceutical applications (Gunasekera *et al.*, 2008; Thongyoo *et al.*, 2009; Craik, 2010). Insecticidal potential of these proteins has also been determined (Jennings *et al.*, 2001; 2005) and a cyclotide gene has even been transferred to crop plants in an attempt to improve the natural defence of the crop against pests (Gillon *et al.*, 2008). In particular, cyclotide varv A, E and kalata B1 are abundant in the *Viola* species. A cDNA clone that encodes the cyclotide kalata B1 has been isolated including three other clones for related cyclotides from the African plant *Oldenlandia affinis* (de Candolle) and shown to have potent inhibitory effect on the growth and development of larvae of the lepidopteran species, *Helicoverpa punctigera* (Hübner) (Jennings *et al.*, 2001).

From the large-scale mapping of cyclotides, it is evident that cyclotides are expressed in plants as a cocktail of up to 25 different cyclotides per species. The cyclotides in the cocktail have individually high activity against certain targets, less against others, but collectively excellent potency against multiple targets (Burman, 2010).

FUTURE OUTLOOK

The practice of using plant biodiversity as a bioresource allows us to develop and exploit naturally occurring plant defence mechanisms, thereby reducing the use of conventional pesticides. Biodiversity-rich countries like Brazil, Columbia, China and India should quickly survey their traditionally used flora to document pesticidal plants. Appropriate protection of species and ecological communities needs to be put in place. Such efforts will help protect the biodiversity resource from threats. A sound knowledge of the biodiversity resource is key to reduce bio-piracy (the com-

mercial development of naturally occurring biological materials, such as plant substances or genetic cell lines, by a technologically advanced country or organisation without fair compensation to the peoples or nations in whose territory the materials were originally discovered) and establish each country's sovereign right to any botanical pesticides developed from such plants. The Nagoya protocol (UNEP, 2011) emphasises the fair and equitable sharing of the benefits arising from the utilisation of genetic resources. This could be achieved by appropriate access to genetic resources and by appropriate transfer of relevant technologies, taking into account all rights over those resources and to technologies, and by appropriate funding, thereby contributing to the conservation of biological diversity and the sustainable use of its components.

Plant-based natural products have an important role in the future of integrated pest management either as new products directly; new-chemical frameworks for production; or for identifying new modes of action. These new products could potentially provide environmentally safe agents, although plants also produce compounds like alkaloids, cyanides and cardiac glycosides which are toxic to mammals. Nevertheless, plants possess a huge diversity of phytochemicals that have evolved partly as defence molecules against attacking organisms and this does provide a tremendous range of compounds that can provide promising new crop protection products. This is the current need as insects are becoming resistant to existing products at a greater rate than new insecticides can be developed. However, most of these new strategies need to be designed with four basic facts in mind.

First, natural plant sources need to be organised and those with potential should be grown with an industrial approach in order to obtain the raw material with greater ease and at lower cost. For example, neem seeds are an important source for the production of azadirachtin-based biopesticides but neem plantations are restricted to village homesteads and some farmlands. Therefore, industrialisation of neem plantation is required at global level. For instance, scores of leading enterprises are engaged in producing neem products developed in Yunnan province of China. There are over 400,000 plantations of neem in the province, developed with the support of government institutions and enterprises as well as local villagers. That makes Yunnan the biggest artificial area of neem planting globally and the raw material centre of neem products in China. This type of approach will need to be adopted globally for any plant that may have the potential to be developed as a biopesticide.

Second, to enhance the biodiversity of a given plant like neem would be to develop neem clones with required characteristics like faster maturation, high yield of seed with high oil content and yield of azadirachtin. This will provide the growers with a 'menu' from which they can choose desired characteristics according to their needs.

Third, quality control needs to be streamlined. Quality of the botanical material is essential in producing safe and premium-quality final products. It is very important to identify the proper genus and species and establish a traceable system to check and control each step from the harvesting of the raw material to the processing of the botanical product. Procured botanical materials need to be stored in temperature- and humidity-controlled conditions under good manufacturing practices, quarantined, validated and quantified for the active constituent for potency using advanced analytical equipment in order to obtain a standardised product. Following all the quality control measures would lead to a standard product for universal use. Standard strategies need to be developed for the delivery of all products. All plant-based products require a specific delivery system to be developed in relation to the mode-of-action of the active ingredient.

Fourth, regulatory constraints that have hindered biopesticides use need to be modified. Regulatory requirements are essential for the commercialisation of a product and thus applicable to all phytochemical-based products. The registration of new materials is a tough task because most of the regulatory parameters are based specifically on synthetic chemicals. Some organisations ask for toxicological data for every characterised active component, which will cost millions of dollars. However, some solace comes from the actions taken in Canada by the Pest Management Regulatory Agency, which approved an experimental use permit allowing the aerial application of neem for control of forest-defoliating sawflies based on HPLC analysis of the neem concentrate in which the major ten limonoids, accounting for 90% of the UV-visible material, were identified and quantified (Isman, 1997). In the United States, the regulatory changes have led to streamlining regulatory processes to favour products that are 'generally regarded as safe' (GRAS) and allow botanicals as a category different from conventional pesticides (EPA, 2006). In India, to achieve this goal,

provisional registrations have been given to manufacturers and the products are being sold in the market. However, it is imperative for producers to fulfil the requirements within the period stipulated by the regulatory authorities. Western countries should adopt this policy, if biological pesticides are to make any impact in the near future in the conventional insecticide market. Neem has already provided a modern paradigm for the development of biopesticides and others have to follow the direction.

All of the above-listed areas need substantial effort, if plant-based products are to be available in the future from plant biodiversity as well as being made into products that are successful and competitive in the marketplace.

ACKNOWLEDGEMENTS

Images of plants used in the compiled figures are from picture libraries at Shutterstock.com, Botanikfoto.com, flickr.com/photos/, fobhm.org/noframes/m_plant2.htm, wikipedia.org or as otherwise mentioned in captions.

REFERENCES

Alford, A.R. and Murray, K.D. (2000) Prospects for citrus limonoids in insect pest management, in *Citrus limonoids: functional chemicals in agriculture and food* (eds M.A. Berhow, S. Hasegawa and G.D. Manners), ACS Symposium Series 758, American Chemical Society, Washington, DC, pp. 201–211.

Babula, A., Mikelova, R., Adam, V., Kizek, R., Havel, L. and Sladky, Z. (2006) Naphthoquinones-biosynthesis, occurrence and metabolism in plants. *Ceska a Slovenska Farmacie*, 55, 151–159.

Babula, A., Adam, V., Havel, L. and Kizek, R. (2009) Noteworthy secondary metabolites naphthoquinones – their occurrence, pharmacological properties and analysis. *Current Pharmaceutical Analysis*, 5, 47–68.

Bacher, M., Hofer, O., Brader, G., Vajrodaya, S. and Greger, H. (1999) Thapsakins: possible biogenetic intermediates towards insecticidal cyclopenta[*b*]benzofurans from *Aglaia edulis*. *Phytochemistry*, 52, 253–263.

Becker, G., Lenz, M. and Dietz, S. (1972) Unterschiede im Verhalten und der Giftempfindlichkeit verschiedener Termiten-Arten gegenuber einigen Kernholzstoffen. *Zeitschrift fur Angewandte Entomologie*, 71, 201–214.

Bentley, M.D., Rajab, M.S., Alford, A.R., Mendel, M.J. and Hassanali, A. (1988) Structure-activity studies of modified citrus limonoids as antifeedants for Colorado potato beetle larvae, *Leptinotarsa decemlineata*. *Entomologia Experimentalis et. Applicata*, 49,189–193.

Brader, G., Vajrodaya, S., Greger, H., Bacher, M., Kalchhauser, H. and Hofer, O. (1998) Bisamides, lignans, triterpenes, and insecticidal cyclopenta[*b*]benzofurans from *Aglaia* species. *Journal of Natural Products*, 61, 1482–1490.

Burman, R. (2010) *Distribution and chemical diversity of cyclotides from Violaceae. Impact of structure on cytotoxic activity and membrane interactions.* Acta Universitatis Upsaliensis. Digital comprehensive summaries of Uppsala dissertations from the faculty of pharmacy, Uppsala University.

Buta, G.J., Lusby, W.R., Neal, J.W. Jr., Waters, R.M. and Pittarelli, G.W. (1993) Sucrose esters from *Nicotiana gossei* active against the greenhouse whitefly *Trialeuroides vaporaorium*. *Phytochemistry*, 32, 859–864.

Castillo, L. and Rossini, C. (2010) Bignoniaceae metabolites as semiochemicals. *Molecules*, 15, 7090–7105.

Champagne, D.E., Koul, O., Isman, M.B., Scudder, G.G.E. and Towers, G.H.N. (1992) Biological activities of limonoids from the Rutales. *Phytochemistry*, 31, 377–394.

Chiasson, H.N., Bostanian, J. and Vicent, C. (2004) Acaricidal properties of a *Chenopodium* based botanical. *Journal of Economic Entomology*, 97, 1373–1377.

Craik, D.J. (2010) Discovery and applications of the plant cyclotides. *Toxicon*, 56, 1092–1102.

Craik, D.J., Daly, N.L., Bond, T. and Waine, C. (1999) Plant cyclotides: a unique family of cyclic and knotted proteins that defines the cyclic cystine knot structural motif. *Journal of Molecular Biology*, 294, 1327–1336.

Craik, D.J., Mylne, J.S. and Daly, N.L. (2010) Cyclotides: macrocyclic peptides with applications in drug design and agriculture. *Cell and Molecular Life Sciences*, 67, 9–16.

Croteau, R., Kutchan, T.M. and Lewis, N.G. (2000) Natural products (secondary metabolites), in *Biochemistry and molecular biology of plants* (eds B. Buchanan, W. Gruissem and R. Jones), American Society of Plant Physiologists, Rockville, pp. 1250–1318.

Dev, S. and Koul, O. (1997) *Insecticides of plant origin*, Harwood Academic Publishers Gmbh, Amsterdam, The Netherlands.

Dhaliwal, G.S. and Koul, O. (2010) *Quest for pest management: from green revolution to gene revolution*, Kalyani Publishers, New Delhi.

Dixon, R.A. and Strack, D. (2003) Phytochemistry meets genome analysis and beyond. *Phytochemistry*, 62, 815–816.

EPA (2006) *Regulating biopesticides*. United States Environmental Protection Agency. http://www.epa.gov/oppbppd1/biopesticides/

Gillon A.D., Saska, I., Jennings, C.V., Guarino, R.F., Craik, D.J. and Anderson, M.A. (2008) Biosynthesis of circular proteins in plants. *Plant Journal*, 53, 505–515.

Goffreda, J.C., Mutschler, M.A., Ave, D.A., Tengey, W.M. and Steffens, J.C. (1989) Aphid deterrence by glucose esters in glandular trichome exudates of the wild tomato,

Lycopersicon pennellii. *Journal of Chemical Ecology*, 15, 2135–2147.

Govindachari, T.R., Narasimhan, N.S., Suresh, G., Partho, P.D., Gopalakrishnan, G. and Krishna Kumari, G.N. (1995) Structure-related insect antifeedant and growth regulating activities of some limonoids. *Journal of Chemical Ecology*, 21, 1585–1600.

Gunasekera, S., Foley, F.M., Clark, R.J. *et al.* (2008) Engineering stabilized vascular endothelial growth factor-A antagonists: synthesis, structural characterization, and bioactivity of grafted analogues of cyclotides. *Journal of Medicinal Chemistry*, 51, 7697–7704.

Gussregan, B., Puhr, M., Nugroho, B.W., Wray, V., Witte, I. and Proksch, P. (1999) New insecticidal rocaglamide derivative fromflower of *Aglaia odorata*. *Zeitschrift fur Naturforschung*, 52, 339–343.

Ishibashi, F., Satasook, C., Isman, M.B. and Towers, G.H.N. (1993) Insecticidal 1*H* cyclopenta tetrahydro[*b*]-benzofurans from *Aglaia odorata*. *Phytochemistry*, 32, 307–310.

Isman, M.B. (1997) Neem and other botanical insecticides: barriers for commercialisation. *Phytoparasitica*, 25, 339–344.

Isman, M.B. (2000) Plant essential oils for pest and disease management. *Crop Protection*, 19, 603–608.

Isman, M.B. (2001) Biopesticides based on phytochemicals, in *Phytochemical biopesticides* (eds O. Koul and G.S. Dhaliwal), Harwood Academic Publishers, Amsterdam, pp. 1–12.

Isman, M.B. (2006) Botanical insecticides, deterrents, and repellents in modern agriculture and an increasingly regulated world. *Annual Review of Entomology*, 51, 45–66.

Isman, M.B., Koul, O., Luczynski, A. and Kaminski, J. (1990) Insecticidal and antifeedant bioactivities of neem oils and their relationship to azadirachtin content. *Journal of Agricultural and Food Chemistry*, 38,1406–1411.

Isman, M.B., Miresmailli, S. and Machial, C. (2010) Commercial opportunities for pesticides based on plant essential oils in agriculture, industry and consumer products. *Phytochemistry Reviews*, DOI: 10.1007/s11101-010-9170-4.

Janprasert, J., Satasook, C., Sukumalanand, P. *et al.* (1993) Rocaglamide, a natural benzofuran insecticide from *Aglaia odorata*. *Phytochemistry*, 32, 67–69.

Jennings, C.V., West, J., Waine, C., Craik, D. and Anderson, M. (2001) Biosynthesis and insecticidal properties of plant cyclotides: the cyclic knotted proteins from *Oldenlandia affinis*. *Proceedings National Academy of Sciences USA*, 98, 10614–10619.

Jennings, C.V., Rosengren, K.J., Daly, N.L. *et al.* (2005) Isolation, solution structure, and insecticidal activity of kalata B2, a circular protein with a twist: do Möbius strips exist in nature? *Biochemistry*, 44, 851–860.

Kato, M.J. and Furlan, M. (2007) Chemistry and the evolution of Piperaceae. *Pure and Applied Chemistry*, 79, 529–538.

Kaushik, R. and Saini, P. (2008) Larvicidal activity of leaf extract of *Millingtonia hortensis* (Family: Bignoniaceae)

against *Anopheles stephensi*, *Culex quinquefasciatus* and *Aedes aegypti*. *Journal of Vector Borne Diseases*, 45, 66–69.

Kessler, A. (2006) Plant-insect interactions in the era of consolidation in biological sciences, in *Chemical ecology: from gene to ecosystem* (eds M. Dicke and W Takken), Springer, Netherlands, pp. 19–37.

Khambay, B.P.S. and Jewess, P. (2000) *The potential of natural naphthoquinones as the basis for a new class of pest control agents – an overview of research at IACR-Rothamsted*. *Crop Protection*, 19, 597–601.

Khambay, B.P.S., Batty, D., Cahill, M.R. *et al.* (1999) Isolation, characterization, and biological activity of naphthoquinones from *Calceolaria andina* L. *Journal of Agricultural and Food Chemistry*, 47, 770–775.

King, R.R., Calhoun, L.A., Singh, R.P. and Boucher, A. (1993) Characterization of 2,3,4,3'-tetra-O-acylated sucrose esters associated with the glandular trichomes of *Lycopersicon typicum*. *Journal of Agricultural and Food Chemistry*, 41, 469–473.

Kiran, S.R., Reddy, A.S., Devi, P.S. and Reddy, K.J. (2006) Insecticidal, antifeedant and oviposition deterrent effects of the essential oil and individual compounds from leaves of *Chloroxylon swietenia*. *Pest Management Science*, 62, 1116–1121.

Koschier, E.H. and Sedy, K.A. (2001) Effects of plant volatiles on the feeding and oviposition of *Thrips tabaci*, in *Thrips and Tospo-viruses* (eds R. Marullo and L. Mound), CSIRO, Canberra, Australia, pp. 185–187.

Kostyukovsky, M., Rafaeli, A., Gileadi, C., Demchenko, N. and Shaaya, E. (2002) Activation of octopaminergic receptors by essential oil constituents isolated from aromatic plants: possible mode of action against insect pests. *Pest Management Science*, 58, 1101–1106.

Koul, O. (1996) Mode of azadirachtin action, in *Neem* (eds M.S. Randhawa and B.S. Parmar), New Age International Publishers Ltd, New Delhi, pp. 160–170.

Koul, O. (2003) Utilization of plant products in pest management: a global perspective, in *Frontier Areas of Entomological Research* (eds B. Subrahmanyam, V.V. Ramamurthy and V.S. Singh), IARI, New Delhi, pp. 331–342.

Koul, O. (2005) *Insect antifeedants*. CRC Press, Boca Raton.

Koul, O. (2008) Phytochemicals and insect control: an antifeedants approach. *Critical Reviews in Plant Sciences*, 27, 1–24.

Koul, O. and Wahab, S. (2004) *Neem: today and in the new millennium*, Kluwer Academic Publishers, Dordrecht.

Koul, O. and Walia, S. (2009) Comparing impacts of plant extracts and pure allelochemicals and implications for pest control. *CAB Reviews, Perspectives in Agriculture, Veterinary Science, Nutrition and Natural Resources*, 4, 1–30.

Koul, O., Multani, J.S., Singh, G. and Wahab, S. (2002) Bioefficacy of toosendanin from *Melia dubia* (syn. *M. azedarach*) against gram pod borer, *Helicoverpa armigera* (Hubner). *Current Science*, 83, 1387–1391.

Koul, O., Multani, J.S., Goomber, S., Daniewski, W.M. and Berlozecki, S. (2004a) Activity of some non-azadirachtin limonoids from *Azadirachta indica* against lepidopteran larvae. *Australian Journal of Entomology*, 43,189–195.

Koul, O., Kaur, H., Goomber, S. and Wahab, S. (2004b) Bioefficacy and mode of action of rocaglamide from *Aglaia elaeagnoidea* (syn. *A. roxburghiana*) against gram pod borer, *Helicoverpa armigera* (Hubner). *Journal of Applied Entomology*, 128, 177–181.

Koul, O., Singh, G., Singh, R. and Multani, J.S. (2005a) Bioefficacy and mode-of-action of aglaroxin A from *Aglaia elaeagnoidea* (syn. *A. roxburghiana*) against *Helicoverpa armigera* and *Spodoptera litura*. *Entomologia Experimentalis et Applicata*, 114, 197–204.

Koul, O., Singh, G., Singh, R. and Singh, J. (2005b) Bioefficacy and mode-of-action of aglaroxin B and aglaroxin C from *Aglaiaelaeagnoidea* (syn. *A. roxburghiana*) against *Helicoverparmigera* and *Spodoptera litura*. *Biopesticides International*, 1, 54–64.

Koul, O., Walia, S. and Dhaliwal, G.S. (2008) Essential oils as green pesticides: potential and constraints. *Biopesticides International*, 4, 63–84.

Koul O., Dhaliwal G.S. and Kaul V.K. (2009) *Sustainable crop protection: biopesticide strategies*, Kalyani Publishers, New Delhi.

Lowery, D.T. and Isman, M.B. (1993) Laboratory and field evaluation of neem for the control of aphids (Homoptera: Aphididae). *Journal of Economic Entomology*, 86, 864–870.

Lowery, D.T. and Isman, M.B. (1996) Inhibition of aphids (Homoptera: Aphididae) reproduction by neem seed oil and azadirachtin. *Journal of Economic Entomology*, 89, 602–807.

Macleod, J.K., Moeller, P.D.R., Molinski, T.F. and Koul, O. (1990) Antifeedant activity and ^{13}C NMR spectral assignments of the meliatoxins. *Journal of Chemical Ecology*, 16, 2511–2518.

Matsuzaki, T., Shinozaki, Y., Suhara, S., Ninomiya, M., Shigematsu, H. and Koiwai, A. (1989a) Isolation of glycolipids from the surface lipids of *Nicotiana bigelovii* and their distribution in *Nicotiana* species. *Agricultural and Biological Chemistry*, 53, 3079–3082.

Matsuzaki, T., Shinozaki, Y., Suhara, S., Shigematsu, H. and Koiwai, A. (1989b) Isolation and characterization of tetra- and triacylglucose from surface lipids of *Nicotiana miersii*. *Agricultural and Biological Chemistry*, 53, 3343–3345.

McKinney, M.L., Schoch, R.M. and Yonavjak, L. (2007) *Environmental science: systems and solutions*, Jones and Bartlett Publishers Canada, Mississauga.

Miyakado, M., Nakayama, I. and Ohno, N. (1989) Insecticidal unsaturated isobutylamides. From natural products to agrochemical leads, in *Insecticides of plant origin* (eds J.T. Arnason, B.J.R. Philogene and P. Morand), American Chemical Society Symposium Series 387, Washington, DC, pp.173–187.

Mordue, A.J. (2004) Present concepts of the mode of action of azadirachtin from neem, in *Neem: today and in the new millennium* (eds O. Koul and S. Wahab), Kluwer Academic Publishers, Amsterdam, pp. 229–242.

Navickiene, H.M.D., Miranda, J.E., Bortoli, S.A., Kato, M.J., Bolzani, V.S. and Furlan, M. (2007) Toxicity of extracts and isobutyl amides from *Piper tuberculatum*: potent compounds with potential for the control of the velvetbean caterpillar, *Anticarsia gemmatalis. Pest Management Science*, 63, 399–403.

Neal, J.J., Tingey, W.M. and Steffens, J.C. (1990) Sucrose esters of carboxylic acids in glandular trichomes of *Solanum berthaultii* deter settling and probing by green peach aphid. *Journal of Chemical Ecology*, 16, 487–497.

Neal Jr., J.W., Buta, J.G., Pittarelli, G.W., Lusby, W.R. and Bentz, J.A. (1994) Novel sucrose esters from *Nicotiana gossei*: effective biorationals against selected horticultural insect pests. *Journal of Economic Entomology*, 87, 1600–1607.

Nugroho, B.W., Edrada, R.A., Güssregen, B., Wray, V., Witte, L. and Proksch, P. (1997a) Insecticidal rocaglamide derivatives from *Aglaia duppereana. Phytochemistry*, 44, 1455–1461.

Nugroho, B.W., Güssregen, B., Wray, V., Witte, L., Bringmann, G. and Proksch, P. (1997b) Insecticidal rocaglamide derivatives from *Aglaia elliptica and A. harmsiana. Phytochemistry*, 45, 1579–1585.

Nugroho, B.W., Edrada, R.A., Wray, V. *et al.* (1999) An insecticidal rocaglamide derivatives and related compounds from *Aglaia odorata* (Meliaceae). *Phytochemistry*, 51, 367–376.

Ohya, I., Shinozaki, Y., Tobita, T., Takahashi, H., Matsuzaki, T. and Koiwai, A. (1994) Sucrose esters from the surface lipids of *Nicotiana cavicola. Phytochemistry*, 37, 143–145.

Okorie, T.G. and Ogunro, O.F. (1992) Effects of extracts and suspensions of the black pepper *Piper guineense* on the immature stages of *Aedes agypti* (Linn) (Diptera: Culicidae) and associated aquatic organisms. *Discovery and Innovations*, 4, 59–63.

Park, I., Lee, S., Shin, S., Park, J. and Ahn, Y. (2002) Larvicidal activity of isobutylamides identified in *Piper nigrum* fruits against three mosquito species. *Journal of Agricultural and Food Chemistry*, 50, 1866–1870.

Parmar, B.S. and Walia, S. (2001) Prospects and problems of phytochemical biopesticides, in *Phytochemical Biopesticides* (eds O. Koul and G.S. Dhaliwal), Hardwood Academic Publishers, Amsterdam, pp. 133–210.

Priestley, C.M., Williamson, E.M., Wafford, K.A. and Sattelle, D.B. (2003) Thymol, a constituent of thyme essential oil, is a positive allosteric modulator of human GABA receptors and a homo-oligomeric GABA receptor from *Drosophila melanogaster. British Journal of Pharmacology*, 140, 1363–1372.

Proksch P., Edrada R., Ebel R., Bohnenstengel, I.F. and Nugroho, W.B. (2001) Chemistry and biological activity of rocaglamide derivatives and related compounds in *Aglaia* species (Meliaceae). *Current Organic Chemistry*, 5, 923–938.

Puterka, G.J. and Severson, R.F. (1995) Activity of sugar esters isolated from leaf trichomes of *Nicotiana gossei* to pear psylla (Homoptera: Psyllidae). *Journal of Economic Entomology*, 88, 615–619.

Roy, A. and Saraf, S. (2006) Limonoids: Overview of significant bioactive triterpenes distributed in plants kingdom. *Biological and Pharmaceutical Bulletin*, 29, 191–201.

Ruberto, G., Renda, A., Tringali,C., Napoli, E.M. and Simmonds, M.S.J. (2002) Citrus limonoids and their semi-synthetic derivatives as antifeedant agents against *Spodoptera frugiperda* larvae. A structure–activity relationship study. *Journal of Agricultural and Food Chemistry*, 50, 6766–6774.

Satasook, G., Isman, M.B. and Wiriyachita, P. (1993) Activity of rocaglamide, an insecticidal natural product, against the variegated cut worm, *Peridroma saucia* (Lepidoptera: Noctuidae). *Pesticide Science*, 36, 53–58.

Schmutterer, H. (2002) *The Neem Tree*, Neem Foundation, Mumbai.

Schneider, C., Bohnenstengel, F.I., Nugroho, B.W. *et al.* (2000) Insecticidal rocaglamide derivatives from *Aglaia spectabilis* (Meliaceae). *Phytochemistry*, 54, 731–736.

Scott, I.M., Jensen, H., Scott, J.G., Isman, M.B., Arnason, J.T. and Philogene, B.J.R. (2003) Botanical insecticides for controlling agricultural pests: piperamides and the Colorado potato beetle, *Leptinotarsa decemlineata* Say (Coleoptera: Chrysomelidae). *Archives of Insect Biochemistry and Physiology*, 54, 212–225.

Scott, I.M., Jensen, H., Nicol R. *et al.* (2004) Efficacy of *Piper* (Piperaceae) extracts for control of common home and garden insect pests. *Journal of Economic Entomology*, 97, 1390–1403.

Scott, I.M., Gagnon, N., Lesage, L., Philogene, B.J.R. and Arnason, J.T. (2005) Efficacy of botanical insecticides from *Piper* species (Piperaceae) extracts for control of European chafer (Coleoptera: Scarabaeidae). *Journal of Economic Entomology*, 98, 845–855.

Scott, I.M., Jensen, H.R., Philogene, B.J.R. and Arnason, J.T. (2008) A review of *Piper* spp. (Piperaceae) phytochemistry, insecticidal activity and mode of action. *Phytochemistry Reviews*, 7, 65–75.

Severson, R.F., Chortyk, O.T., Stephenson, M.G. *et al.* (1994) Characterisation of natural pesticides from *Nicotiana gossei*, in *Bioregulators for crop protection and pest control* (ed. P.A. Hedin), ACS Symposium Series 557, American Chemical Society, Washington, DC, pp. 109–121.

Shikano, I. and Isman, M.B. (2009) A sensitive period of larval gustatory learning influences subsequent oviposition choice by the cabbage looper moth. *Animal Behaviour*, 77, 247–251.

Simonovska, B., Srbinoska, M. and Vovk, I. (2006) Analysis of sucrose esters – insecticides from the surface of tobacco plant leaves. *Journal of Chromatography*, 1127, 273–277.

Simpson, B.B. and Ogorzaly, M.O. (1995) *Economic botany: plants in our world*, McGraw-Hill Inc., New York.

Singh, R., Koul, O. and Rup, P.J. (2010) Effect of some essential oil compounds on the oviposition and feeding behavior of the Asian armyworm, *Spodoptera litura* F. (Lepidoptera: Noctuidae). *Biopesticides International*, 6, 52–66.

Singh, R., Koul, O., Rup, P.J. and Jindal, J. (2011) Oviposition and feeding behavior of maize borer, *Chilo partellus*, in response to eight essential oil allelochemicals. *Entomologia Experimentalis et Applicata*, 138, 55–64.

Soković, M.D., Vukojevic, J., Marin, P.D., Brkic, D.D., Vajs, V. and van Griensven, L.J.L.D. (2009) Chemical composition of essential oils of Thymus and Mentha species and their antifungal activity. *Molecules*, 14, 238–249.

Thongyoo, P., Bonomelli, C., Leatherbarrow, R.J. and Tate, E.W. (2009) Potent inhibitors of β-tryptase and human leukocyte elastase based on the MCoTI-II scaffold. *Journal of Medicinal Chemistry*, 52, 6197–6200.

UNEP (2011) *Nagoya protocol*. United Nations Environment Programme (UNEP), Secretariat of the convention on biological diversity, Quebec, Canada.

Waliwitiya, R., Kennedy, C.J. and Lowenberger, C.A. (2009) Larvicidal and oviposition altering activity of monoterpenoids, *trans*-anithole and rosemary oil to the yellow fever mosquito *Aedes aegypti* (Diptera: Culicidae). *Pest Management Science*, 65, 241–248.

Wang, B.G., Ebel, R., Wang, C.Y., Edrada, R.A., Wray, V. and Proksch, P. (2004) Aglacins I-K, three highly methoxylated lignans from *Aglaia cordata*. *Journal of Natural Products*, 67, 682–684.

Wang, C.K., Kaas, Q., Chiche, L. and Craik, D.J. (2008) CyBase: a database of cyclic protein sequences and structures, with applications in protein discovery and engineering. *Nucleic Acids Research*, 36(Database issue):D206–210.

Weber, S., Puripattanavong, J., Brecht, V. and Frahm, A.W. (2000) Phytochemical investigation of *Aglaia rubiginosa*. *Journal of Natural Products*, 63, 636–642.

Wheeler, W.B. (2002) *Pesticides in agriculture and the environment*, CRC Press, Boca Raton.

Willis, R.J. (2000). *Juglans* spp., Juglone and allelopathy. *Allelopathy Journal*, 17, 1–55.

Xia, Y., Johnson, A.W. and Chortyk, O.T. (1997a) Effect of leaf surface moisture and relative humidity on the efficacy of sugar esters from *Nicotiana gossei* against the tobacco aphid (Homoptera: Aphididae). *Journal of Economic Entomology*, 90, 1010–1014.

Xia, Y., Johnson, A.W. and Chortyk, O.T. (1997b) Enhanced toxicity of sugar esters to the tobacco aphid using humectants. *Journal of Economic Entomology*, 90, 1015–1021.

Chapter 7

THE ECOLOGY AND UTILITY OF LOCAL AND LANDSCAPE SCALE EFFECTS IN PEST MANAGEMENT

Sagrario Gámez-Virués, Mattias Jonsson and Barbara Ekbom

Biodiversity and Insect Pests: Key Issues for Sustainable Management, First Edition. Edited by Geoff M. Gurr, Steve D. Wratten, William E. Snyder, Donna M.Y. Read.
© 2012 John Wiley & Sons, Ltd. Published 2012 by John Wiley & Sons, Ltd.

INTRODUCTION

Pest management is often studied at the field scale even though pest managers know that both pests and their natural enemies must recolonise a crop after disturbances such as tillage, harvest, and insecticide use. It is evident that these processes occur at a scale larger than that of the field. In this chapter we discuss ecological theory in relation to pest management at different spatial scales, examine empirical studies about the relationship between landscape components and abundance/diversity of pests and natural enemies, and give two examples of habitat manipulation to improve biological control at local and landscape scales.

An appreciation of the fact that the amount and arrangement of land use elements in the landscape is important for pest attack and management is not new. Almost 100 years ago Stephen A. Forbes in his annual address to the Entomological Society of America called for a greater application of ecological principles in the application of entomology to pest management, stressing in particular 'the ecological structure of the region and the probability of changes in such structure under gradually intensified human use' (Forbes, 1915).

Despite this early recognition of the significance of landscape structure for pest management, very little research along these lines took place for many years. It was known that pest insects could migrate long distances and that dispersal from overwintering sites was an important part of pest phenology, but these facts were seldom considered within the context of pest control. A great deal of attention was paid to interactions between pests and natural enemies, but early work showed little regard for spatial patterns. Also, the introduction of widespread chemical control of insects after the Second World War deferred the need for other methods.

Within a relatively short time, problems with insecticides became evident. Pests developed resistance to insecticides, which necessitated the development of new products. Broad-spectrum toxicity caused environmental and health problems as well as pest resurgence after killing natural enemies (Carson, 1962; Way, 1966). The concept of 'integrated control' (Stern *et al.*, 1959) was introduced in order to find a framework to harmonise management practices with biological and chemical control.

The successes of classical biological control taught us that when a pest escaped the natural control agents in its home environment, order could be restored by introducing appropriate natural enemies into the new environment (DeBach, 1964). This was perceived as restoring 'balance' to the system and brought to mind theories that diversity or complexity may avert insect outbreaks and create stability (Odum, 1963). Diversity could be achieved within the crop habitat by, for example, mixed plant species stands (Pimentel, 1961) or by diversification of habitats in proximity to crop fields (van Emden, 1965).

An early example of the importance of these non-crop habitats in pest management was the observation that the grape leafhopper was effectively controlled by a parasitoid (*Anagrus epos* (Girault)) in vineyards close to blackberry thickets (Doutt and Nakata, 1965). The parasitoid can reproduce on another, non-pest, leafhopper that occurs on the wild blackberries and then moves into vineyards to parasitise the economically important grape leafhopper. In addition the thickets provide an overwintering site for the parasitoid. Further details of this pioneering work are provided in chapter 1 of this volume. This example follows theory in that the closeness of the source (blackberries) increases the probability of the parasitoid arriving and establishing in the crop (vineyards). However, it does not answer the specific question of how close or how large the blackberry thickets need to be to achieve control of the grape leafhopper. Studies designed on the basis of theory may, however, provide a good estimate of habitat distance and size. In this chapter we will illustrate how using ecological theory will improve and enrich studies aimed at improving pest management.

LANDSCAPE ECOLOGY AND PEST MANAGEMENT

Although the application of ecological principles for pest control is an old idea, research concerning landscape aspects in agriculture did not gather speed until ecological theory development provided a framework for study. Price and Waldbauer (1975) pointed out the applicability of two ecological theories for the agricultural ecosystem; one theory concerning space (the theory of island biogeography (MacArthur and Wilson, 1967)) and the other concerning time (community succession (Odum, 1969)). A crop field can be likened to an island, where size and distance to a source of insects will determine colonisation patterns after crop establishment. Annual and perennial crops will show different succession patterns and influence arthropod community composition. Levins and Wilson (1980) summarised the scope for intervention with reference to three levels: 1) region or landscape, where land use patterns may determine colonisation; 2) local level, where provision of alternative habitats for natural enemies may augment their numbers; and 3) within-field manipulations, where choice of crop(s) and other agronomic practices may influence insect behaviour and community interactions.

No theory is forever, but one of the most important qualities of a good theory is that it stimulates a flood of empirical work (Laurance, 2008). Studies in both agricultural and natural environments have led to diversification, development, and enrichment of theory. These studies have also contributed to the development of tools to quantify and analyse the landscapes in which organisms live (Scherber *et al.*, chapter 8 of this volume). Theory that considers spatiotemporal scales and technological advances has contributed to the emergence of landscape ecology (Box 7.1) as an important research discipline (Turner *et al.*, 2001). How the size, shape and arrangement of different land use types (landscape components) influence ecological processes is a central question for landscape ecology and because the roots of this discipline lie in landscape planning and design it is natural to consider changes in these

components to achieve management goals. This development brings us closer to 1) understanding the pest–natural enemy interactions by studying them using a landscape perspective and 2) suggesting reliable management options to promote pest control.

One might think that armed with a theoretical framework, an ecological approach would quickly provide management models to solve pest problems in agriculture. The reality is, as expressed by Lawton (1999), that particular pest 'management questions will most reliably be solved by site- and location-specific studies'. But these studies must, of course, be guided by context provided by theory.

LANDSCAPE COMPLEXITY AND BIOLOGICAL CONTROL

During the past 20 years the number of studies considering the landscape context of biological pest control has increased dramatically. The most common approach of such studies has been to relate natural enemy and pest abundances, or in some cases natural enemy attack rates, in the crop to the composition of different habitat types in the landscape surrounding the crop (Figure 7.1). These studies have often used the proportion of non-crop vegetation in a landscape sector as a proxy for landscape complexity, because this variable is often strongly correlated with habitat type diversity and other measures of landscape heterogeneity (Box 7.2).

Effects on natural enemy abundance, condition and fecundity

Theories predict that biological control should be less effective in landscapes simplified by agriculture because natural enemies found at higher trophic levels should be more sensitive to disturbance (Pimm and Lawton, 1977) and habitat loss (Holt *et al.*, 1999). In a review, Bianchi *et al.* (2006) found some empirical support for this hypothesis; of the studies reviewed 74% reported higher abundance of natural enemies and 45% found lower pest populations in complex landscapes. Landscape composition and structure may also influence the condition and fecundity of natural enemies; Bommarco (1998) and Östman *et al.* (2001a) found that body condition and fecundity of some carabid beetles were higher in heterogeneous landscapes. In contrast,

---

Box 7.1 Landscape ecology

Turner *et al.* (2001) highlight 'the importance of spatial configuration for ecological processes' in landscape ecology. Broad spatial extents and the influence of human activities on landscape patterns are important components of landscape ecology. The study of landscape structure, function and change is broadly interdisciplinary. Some fundamental concepts within landscape ecology used in studies on pest management are defined below:

Landscape: A spatially heterogeneous area made up of different habitats or cover types. Landscapes may vary in size.

Composition: The proportion of different habitats or cover types found in the landscape.

Configuration: The specific spatial arrangement of different cover types on the landscape

Connectivity: Spatial continuity of a cover type across a landscape

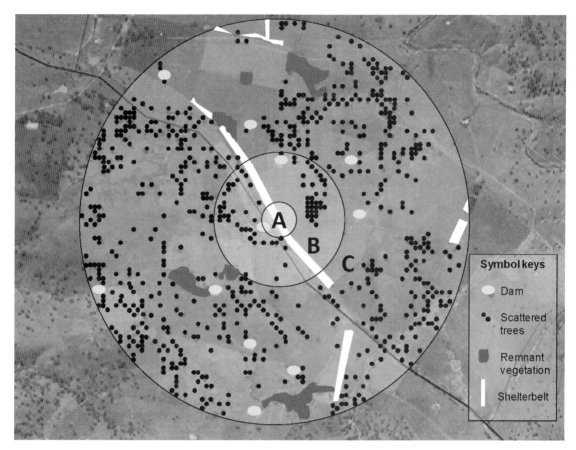

Figure 7.1 Composition of different habitat types at **A** = 150 m, **B** = 600 m and **C** = 1,800 m radii, in a grazing land.

Box 7.2 Landscape complexity

Landscape complexity is a commonly used term in landscape ecology. The term is related to the diversity of landscape elements, their spatial organisation and shape. A number of landscape metrics describe different dimensions of landscape complexity. Below are some of the more common terms listed (adapted from Ode and Miller, 2011):

Shannon diversity index: describes the habitat-type diversity by taking into account the number of land-cover classes and their proportional distribution. Another similar type of measure is the Simpson diversity index.

Shannon evenness index: describes the evenness of area among land-cover classes where a high value describes an even distribution of area among classes.

Mean shape index: describes the relationship between the perimeter and area of patches as a mean of all patches found. A higher value means a more complex shape (more perimeter per area).

Edge density: provides a measurement of the length of edge segments per hectare and is dependent on both patchiness and patch shape.

Contagion: describes the level of cell-like adjacencies (i.e. cells of a patch-type adjacent to cells of the same patch type). It is affected by both the dispersion and interspersion of land-cover classes.

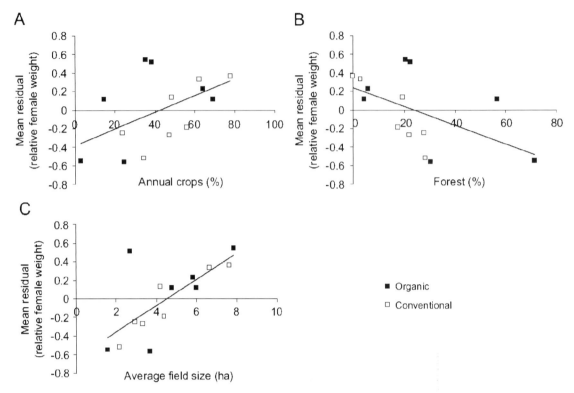

Figure 7.2 The importance of landscape features for the condition of wolf spider females (*Pardosa* spp.) on organic and conventional farms. Body condition is measured using the residuals from the regression of body mass (weight) against body size (width of cephalothorax). In this case wolf spider condition was lower in simple landscapes (dominated by large fields of annual crops) when compared to complex landscapes (after Öberg, 2009).

individuals of *Pardosa* sp. (a genus of wolf spiders, Lycosidae) were found to have superior body condition in simple landscapes dominated by agriculture (Öberg, 2009; Figure 7.2).

Effects on natural enemy diversity

Several studies have also found that complex landscapes have a higher diversity of natural enemies compared to simplified landscapes (Clough *et al.*, 2005; Purtauf *et al.*, 2005; Schmidt *et al.*, 2005; Öberg *et al.*, 2007). In fact, there is some evidence that a high landscape complexity may be more important for maintaining a high natural enemy diversity than a high natural enemy abundance (Tscharntke *et al.*, 2007). Schmidt *et al.* (2005), for example, showed that conversion to

organic farming at the local field level had a positive impact on the abundance of some common spiders, whereas spider diversity was driven by landscape complexity. However, not all studies have found higher natural enemy diversity in complex landscapes; Vollhardt *et al.* (2008) found no difference in parasitoid diversity between complex and simple landscapes and Weibull *et al.* (2003) found a higher diversity of carabid beetles in simple landscapes. The effect of landscape complexity may depend on prey and habitat specificity, with more specialised natural enemies more often being negatively affected by loss of complexity, whereas more generalist and highly vagile natural enemies in some cases may benefit from landscape simplification (Rand and Tscharntke, 2007; Haenke *et al.*, 2009). Maintaining a high landscape complexity may therefore be particularly important in situations where effective biological control depends on a high diversity of comparatively specialised natural enemies (Snyder et al., 2006; Tscharntke et al., 2007).

Landscape complexity may also have more long-term positive effects on biological control that are difficult to quantify. Beta diversity (i.e. the differences in species composition among fields; see chapter 1 of this volume) may be higher in complex landscapes (Tscharntke et al., 2007). High beta diversity may be particularly important as insurance against environmental change because different natural enemy species are likely to be effective during different environmental conditions (insurance hypothesis (Yachi and Loreau, 1999)). However, empirical evidence for this hypothesis is, so far, limited (Tscharntke et al., 2007).

Effects of landscape on biocontrol services

Even though the above studies suggest that complex landscapes are likely to have more effective biological control because the diversity, abundance and condition of natural enemies tend to be higher/better, studies that have explicitly quantified landscape effects on biological control are still rare. Östman et al. (2001b) quantified the effect of ground-living predators on the bird cherry-oat aphid in spring-sown cereals by estimating predation rates on establishing aphids using sentinel aphids and by comparing aphid growth rates inside and outside barriers that excluded predators. The results showed that biological control by ground-living predators was higher in complex landscapes during the aphid establishment phase in spring, but lower in complex landscapes during the aphid population growth phase during summer (Figure 7.3). More recently, Gardiner et al. (2009a) studied aphid growth rates inside and outside cages that excluded all types of natural enemies in soybean fields located in different landscapes. Biological control of soybean aphids was more effective in landscapes with a high diversity of habitat types and a low cover of corn fields. Based on these results, Landis et al. (2008) estimated how large the economic consequences of the calculated differences in biological control efficacy would be between different landscapes and extrapolated this to estimate the economic consequences of increased corn acreage for biofuel production across four states in the USA. It was predicted that this change in landscape composition would lead to a 24% reduction in biological control services in soybean production, corresponding to a

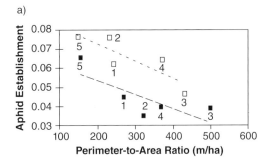

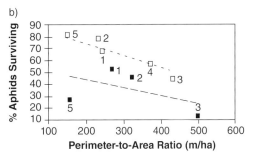

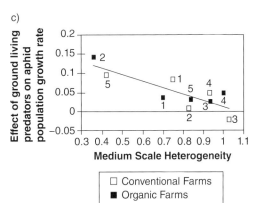

Figure 7.3 The effect of landscape and farming practice on the impact of ground-living predators on *Rhopalosiphum padi*, an aphid pest in cereals. At the beginning of the season when predators could eat aphids arriving to the field, the number of winged aphids establishing in the field a) and the proportion of aphids surviving predation b) decreased with increasing landscape complexity. Later in the season the effect of predators on aphid population growth c) was highest in simple landscapes. The numbers (1–5) next to the symbols denote farm pairs of one organic and one conventional farm that are less than 10 km apart and have similar size and landscape structure (after Östman et al., 2001b).

cost of $58 million a year due to reduced yield and increased pesticide use. This example illustrates the magnitude of the economic consequences of altered landscape composition for farmers depending on biological pest control. Overall, however, there is a lack of studies investigating how landscape effects on natural enemies translate into changes in biological control.

MECHANISMS BEHIND LANDSCAPE EFFECTS

Complex landscapes with a large amount of non-crop vegetation often have higher abundance and diversity of natural enemies, and relatively often also lower pest densities. But what are the underlying mechanisms for such patterns? The most commonly proposed mechanism is that the availability of various resources of importance for natural enemies varies among landscapes (Landis *et al.*, 2000; Bianchi *et al.*, 2006; Tscharntke *et al.*, 2007). For example, many natural enemies hibernate or aestivate in non-crop habitats and the availability of suitable sites to survive the winter or the dry season in the landscape may therefore have an impact on the number of natural enemies colonising crop fields. The abundance of ground-living predators such as spiders and carabid beetles has been found to be positively correlated with the availability of semi-natural grasslands (Purtauf *et al.*, 2005) and perennial crops (Öberg *et al.*, 2007) in the landscape, where many of these species overwinter. This pattern has been restricted to early spring (Schmidt and Tscharntke, 2005) or to spring-breeding species (Purtauf *et al.*, 2005), further indicating that the availability of overwintering sites is an underlying explanation. Another type of resource that may help explain landscape effects on natural enemies is availability of alternative food. For generalist natural enemies this includes alternative prey and for omnivorous natural enemies – such as parasitoids, lady beetles and hoverflies – nectar and pollen also constitute important food (Landis *et al.*, 2000). Bianchi *et al.* (2008) found that parasitism rates by *Diadegma* spp. on diamondback moth *Plutella xylostella* were higher in landscapes with an abundance of forest margins. Such margins often contain many flowering plants that provide nectar as a food source for adult parasitoids, and this is one likely explanation for the positive correlation between parasitism rates and this landscape variable (Bianchi *et al.*, 2008).

An alternative explanation for less effective biological control in highly simplified landscapes, that has received less attention, is that natural enemies are negatively influenced by the increasing scale and intensity of disturbance processes, such as ploughing, harvesting and insecticide application in such landscapes (Croft, 1990; Kruess and Tscharntke, 2002). A recent Europe-wide study showed that local pesticide application has persistent negative effects on biodiversity and natural enemy attack rates (Geiger *et al.*, 2010), whereas Gabriel *et al.* (2010) found that landscapes with a high proportion of organic farms had higher abundances of carabid beetles, suggesting that disturbances associated with land-use intensity (which is often higher in conventional farming) can be an important driver of landscape effects on natural enemy abundances and biological control.

Complex, mosaic landscapes may also improve biological control because such landscapes are highly connected and have many crop-noncrop borders that facilitate spillover of natural enemies among habitat types. Perovic *et al.* (2010) found that the inclusion of a measure of connectivity among habitats improved the explanatory power of an analysis of landscape effects on the abundance of a predatory beetle *Dicranolaius bellulus* (Guérin-Méneville) in Australian cotton crops. Other studies have found natural enemy attack rates to be positively related to average perimeter-to-area ratio of annual crop fields in the landscape (Östman *et al.*, 2001b; Figure 7.3) and the abundance of spiders to be positively related to the total length of boundaries between different habitat types (Öberg *et al.*, 2008).

Although pest abundance is often lower in complex landscapes, probably due to more effective biological control, in some cases pests also benefit from non-crop vegetation and landscape diversity (Bianchi *et al.*, 2006). Thies *et al.* (2005) found that the abundance of cereal aphids in wheat fields in spring was higher in landscapes with a high proportion of non-crop habitats. The most common aphid in this study was *Sitobion avenae* (Fabricius), and this species can overwinter on perennial grasses. Spring migration from non-crop habitats containing such vegetation may therefore explain the observed pattern. Zaller *et al.* (2008) found that pollen beetles were more abundant in oilseed rape fields located in landscapes with a high proportion of forest. This may be due to the presence of overwintering sites and alternative host plants in forest habitats.

SPATIAL SCALES OF LANDSCAPE EFFECTS

The spatial scale at which natural enemies and pests are affected by landscape composition depends on the dispersal ability of the species. Unfortunately it is very difficult to track marked individuals over large distances, so direct evidence of the distances that different species move is sparse. Marked natural enemies have been recaptured at distances of around 100 metres away from the refuges where they were marked (Corbett and Rosenheim, 1996; Schellhorn *et al.*, 2008) but many natural enemies are able to move much longer distances. Combining tracking of individuals with simulations, Baars (1979) and Firle *et al.* (1998) estimated that different carabid beetles would move over scales of between 7 and 49 ha per season. An approach that has repeatedly been used to compare the importance of different spatial scales is to study how the abundance or attack rate of natural enemies is related to landscape composition within buffers of different size around the study fields (Figure 7.1). With this approach the importance of different spatial scales is determined by comparing statistical models with different variables using R^2-values or information-theoretic metrics (e.g. AIC). This approach has shown that parasitism rates on pests are often most strongly correlated to landscape composition within a 1 km diameter around the crop (Thies *et al.*, 2003; 2005; Bianchi *et al.*, 2008), suggesting that dispersal over such distances is common for many parasitoids. For predators similar types of studies have been conducted for a range of groups. Schmidt *et al.* (2008) showed that many spiders that disperse by ballooning (being transported by wind on silken threads), for example species in the family Linyphiidae (sheet-web spiders), respond most strongly to a relatively large landscape scale, up to a radius of at least 3 km. In contrast, many spiders that disperse by walking on the ground respond more strongly to a smaller scale of a few hundred metres (Schmidt *et al.*, 2008). Gardiner *et al.* (2009b) showed that many lady beetles respond most strongly to landscape composition within a radius of approximately 2 km. Lacewings have been found to respond most strongly to a landscape scale at a radius of approximately 2.4 km (Elliott *et al.*, 1998).

Holt (1996) predicted that species operating at higher trophic levels (such as natural enemies) should be influenced by larger spatial scales than species operating at lower trophic levels (such as pest herbiv-

ores). This theory has, however, not found much support from empirical studies of pests and natural enemies in agroecosystems. Thies *et al.* (2003) found that parasitism of pollen beetles and pollen beetle herbivory was correlated with landscape complexity at the same spatial scale, whereas Thies *et al.* (2005) found that cereal aphids were affected by landscape composition at a larger scale than aphid parasitism rates. Just as for natural enemies, the scale at which pests are affected by landscape composition varies strongly among species. For example, Zaller *et al.* (2008) found that pod midges and stem weevils were influenced by landscape composition at a radius of 250–500 m whereas pollen beetles were most strongly influenced by landscape composition at 1–2 km scales. Aphids are often very strong dispersers (Taylor, 1977; Riley *et al.*, 1995) and they may be affected by landscape composition at even larger scales (Thies *et al.*, 2005).

HABITAT MANIPULATIONS TO IMPROVE BIOLOGICAL CONTROL: ATTEMPTS TO PLACE THE RIGHT DIVERSITY AT THE RIGHT SPATIAL SCALE

Habitat manipulation techniques have been seen as local management measures that effectively alleviate some of the effects of simplified landscapes and improve biological control. However, in order for this to work managers have to consider that interactions between individual species and different habitat types depend on the requirements and specialisation of each species (Krauss *et al.*, 2003). A complex mosaic of isolated habitats for one species may be a simple landscape for another. The complexity of the surrounding landscape at one spatial scale may disappear at either finer or broader scales of resolution (Chust *et al.*, 2004). The way that species experience the landscape depends on their body size, dispersal ability, functional group and trophic level (Kareiva and Wennergren, 1995). Below we present examples of how a greater level of complexity has been added to simplified landscapes in temperate ecosystems in order to improve pest management (for a contrasting example concerning habitat manipulation in tropical and subtropical rice see chapter 13 of this volume). We focus particularly on the establishment of non-crop habitats at local – flower strips – and landscape – woody non-crop habitats – scales and also discuss how the effect of local habitat manipulation may interact with the composition of the surrounding landscape.

Local manipulations: establishment of flower strips

Traditionally, the selection of plant species for the establishment of flower strips has focused on the pest reduction service they can provide (Fiedler *et al.*, 2008). Thus, the establishment of flower strips has typically been accomplished by selecting and planting exotic species along crop edges or between crop rows to supply alternative resources, such as nectar, pollen and shelter, for omnivorous arthropods (Landis *et al.*, 2000). For example, in vineyards in all major wine regions in New Zealand, the establishment of flower strips of buckwheat, *Fagopyrum esculentum* (Moench), an exotic species, has been adopted as a method to control the light-brown apple moth *Epiphyas postvittana* (Walker), which is an invasive species (for reviews on 'habitat manipulation to mitigate the impacts of invasive arthropod pests' see Jonsson *et al.*, 2010 and Steingröver *et al.*, 2010). Effective biological control in this case has been achieved primarily because parasitoids increase their fecundity and longevity by feeding on buckwheat nectar, and they are thereby able to parasitise more light-brown apple moth on adjacent vines (Scarratt, 2005).

Although buckwheat, and other exotic species such as *Phacelia tanacetifolia* (Bentham), *Lobularia maritima* (L.) and *Coriandrum sativum* (L.), have proven to be highly attractive to natural enemies, native plants may best enhance biodiversity while simultaneously providing other ecosystem services (Fiedler *et al.*, 2008; Landis *et al.*, chapter 17 of this volume). For a review of the role of native plants in maximising arthropod-mediated ecosystem services in agricultural landscapes see Isaacs *et al.* (2009). Additionally, chapter 17 of this volume looks broadly at the use of native plant species in insect pest management.

In order to achieve better pest management, flower strips should be deployed with consideration given to the distance that target natural enemies travel after nectar feeding. In the United Kingdom and New Zealand, for example, three species of hoverflies were found to move up to 200 m from their floral resource when there were no barriers such as field boundaries (between the flower strips and the traps used to catch the flies) (Wratten *et al.*, 2003). In this scenario, we could assume that 400 m is the maximum distance at which flower strips need to be deployed in the landscape to work as a source of hoverflies, so that they can move 200 m from each flower strip into the crop. However, before making such specific recommendations, other key factors such as crop permeability and level of pest reduction in crops under such management should also be investigated. Lavandero *et al.* (2005) observed that although *Diadegma semiclausum* (Hellen) adults were capable of moving over distances of 80 m, they were more effective as biological control agents at 60 m: the spatial scale at which floral resources were available. Thus, the distances that some parasitoids travel and the spatial scale at which they parasitise hosts are not always identical, which is particularly important for the deployment of floral resources at the right spatial scale. Additionally, as highlighted by Tillman *et al.* in chapter 19 of this volume, the need to deploy flower strips may also depend on the local conditions – some organic growers in California have benefited from the activity of hoverflies without establishing flower strips or insectary plants in their lettuce fields because of the presence of natural flowering vegetation associated with these farms.

Landscape manipulations: establishment of woody non-crop habitats

One habitat management measure that can be deployed at the landscape scale is the establishment of woody non-crop habitats within agricultural systems (agroforestry, Figure 7.4). This method has been acknowledged as an effective strategy to enhance conservation of biodiversity (Stamps and Linit, 1998; Griffiths *et al.*, 2008) and to alleviate the effects of habitat fragmentation. Woody non-crop habitats, such as shelterbelts and hedgerows, are usually linear plantations of perennial vegetation that delimit arable land, conserve soil moisture, help reduce wind erosion and provide protection for crops, pastures and livestock (Brouwer, 1998). However, several of these benefits are not evident until a few years after tree planting, which could be seen as a disadvantage by some landholders. The selection of tree species for the establishment of woody non-crop habitats requires consideration of the tree species' susceptibility to insect attack and availability of local sources of infestation. Tree species may also be consida)

b)

Figure 7.4 The establishment of woody non-crop habitats (agroforestry) can diversify agricultural landscapes: a) woody non-crop habitats; i.e. shelterbelts in grazed pastures in temperate Australia (photo: S. Gámez-Virués), b) landscape dominated by maize monoculture (left side of the road), and diversified by agroforestry (right side of the road) in the Trans-Nzoia district, Western Kenya (photo: M. Jonsson).

Several arthropod taxa that utilise woody non-crop habitats provide ecological services such as crop pollination (e.g. Steffan-Dewenter *et al.*, 2001) and pest suppression in crop- and non-crop habitats (e.g. Tuovinen 1994). Shelter, nectar, pollen, aphid honey-dew, non-pest prey and hosts are resources for natural enemies provided by woody non-crop habitats. They are complementary to the availability of crop pests and can therefore increase the spillover of natural enemies into crop lands (Rand *et al.*, 2006). For example, parasitic wasps require host insects for their larvae, but floral-feeding as adults increases their longevity and potential fecundity (e.g. Baggen and Gurr, 1998). In such cases, availability of floral resources in woody non-crop habitats can increase pest suppression (Box 7.3). The provision of complementary resources should take into account the behaviour of target species. For example, the attractiveness of food plants should be screened for both natural enemies and herbivores in order to choose those plants which encourage a better performance of natural enemies, but not of herbivores (Baggen and Gurr, 1998).

A study in temperate Australia that examined shelterbelt characteristics as factors regulating populations of natural enemies between shelterbelts and adjacent pastures found that abundance, activity and spillover of beneficial arthropods depended greatly on the quality (e.g. plant diversity and habitat structure), quantity (shelterbelt size) and proximity of woody non-crop habitats deployed up to 1.8 km radius in the surrounding landscape (the maximum spatial scale investigated) (Gámez-Virués, 2009). In that study, abundance of natural enemies correlated positively with high levels of plant diversity, which was represented – at its maximum value – by the relative abundance of 23 species of native woody plants within 3,000 m^2; whereas habitat structure – measured as the amount of lower vegetation strata, logs, litter and rocks available – correlated negatively with herbivory of experimental saplings. These findings suggest that planting shelterbelts with high plant diversity and habitat structure may enhance the impact of natural enemies.

The spatial arrangement of land uses, or habitat types, in agricultural landscapes is likely to have different effects on the dispersal of different species of natural enemies. To assess the composition of habitat types that allows a species to disperse most efficiently throughout the landscape at different spatial scales (in

ered with the multiple benefits they provide. In tropical agroforestry, for example in cacao and coffee plantations, the selection of tree species may also depend on the shade they provide – the planted shade trees can range from a monoculture to a polyculture of shade tree species that are deployed amongst crop plants (see Perfecto and Armbrecht, 2003; and Rice and Greenberg, 2000). Additionally, native trees may be used to increase the availability of niches for native fauna (including natural enemies), thus contributing to the conservation of biodiversity (for more details of the utility of native plants in habitat manipulation see chapter 17 of this volume).

Box 7.3 Flower strips in woody non-crop habitats

The establishment of *Lobularia maritima* ground-cover around *Eucalyptus blakelyi* Maiden trees in a shelterbelt increased the levels of parasitism of *Ardozyga stratifera* (Meyrick) (Lepidoptera: Gelechiidae) larvae, that were naturally infesting the eucalypts, indicating that *L. maritima* had a role in either attracting or retaining parasitoids (Gámez-Virués *et al.*, 2009).

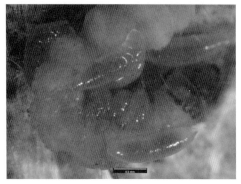

connectivity of wooded land uses within a 3 km radius from the crops.

Recommendations for a better spatial arrangement of woody non-crop habitats, for the benefit of biological control in farmlands, may rely on taxonomic identification of the key organisms which affect pest control, because the species comprising each functional guild differ in the impact they have on biological control (Tscharntke *et al.*, 2005). Thus, specific management of particular arthropod species requires more knowledge about their function and habitat requirements within agricultural systems, because certain species may exert strong dominance over ecological processes (Ricklefs *et al.*, 1984). Moreover, given that complex interactions (e.g. bottom-up and top-down feedback effects) are operating simultaneously, straightforward responses of manipulating diversity at a particular trophic level might not necessarily be expected (Meyer *et al.*, 2010). The proportion of the landscape that should be used to protect, or to establish, non-crop habitats at local and landscape scales has to be determined in order to sustain the functionally dominant natural enemies (Holland and Fahrig, 2000; Tscharntke *et al.*, 2007). Once a spatial arrangement of land uses has been identified as positively correlated with density and activity of natural enemies, this could be incorporated in agricultural policies such as agri-environment schemes (AES) to promote conservation of biodiversity and pest control, mainly in simple landscapes (see Tscharntke *et al.*, 2005).

Interactions between local habitat management and landscape composition

Landscape composition may influence the effectiveness of local habitat management measures designed to improve biological control. Tscharntke *et al.* (2005) hypothesised that the effectiveness of local AES (e.g. conversion to organic farming, encouragement of set-aside fields, and creation of crop field boundaries) would have a hump-shaped relationship with landscape complexity (Figure 7.5). In complex landscapes, with a high proportion of non-crop habitats, such schemes would have low efficacy, because diversity is already high everywhere. In relatively simple landscapes, with a lower proportion of non-crop habitats, the effect of AES would be greater, because such landscapes hold an intermediate pool of species that can respond to local management. Finally in cleared land-

other words, to assess functional landscape connectivity), the structural connectivity, the species-specific dispersal activity and the effects of the matrix on that dispersal should be taken into account (Tischendorf and Fahrig, 2000). Using this type of modelling, Perovic *et al.* (2010) found that, within cotton fields in Australia, the in-crop density of the predator *D. bellulus* was positively and significantly correlated with the

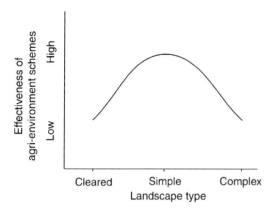

Figure 7.5 Effectiveness of agri-environmental schemes, measured as biodiversity enhancement, in relation to landscape type. Landscape type is classified as cleared (<1% non-crop habitat), simple (1–20% non-crop habitat), and complex (>20% non-crop habitat) (after Tscharntke *et al.*, 2005).

scapes with less than 1% non-crop vegetation remaining, AES would, again, have only a small effect, because such landscapes have only a very small species pool of natural enemies that can respond to the AES (Tscharntke *et al.*, 2005). Several empirical studies have compared the effects of local management measures and landscape-level complexity on natural enemy abundances and attack rates. Thies and Tscharntke (1999) found higher parasitism rates on pollen beetles close to the field edge compared to the centre, in structurally simple landscapes dominated by agriculture, but no such effect was found in complex landscapes. Haenke *et al.* (2009) found a higher abundance of hoverflies in flower strips located in simple landscapes, but no such effect in complex landscapes. Schmidt *et al.* (2005) found higher abundance of spiders in organic versus conventional fields in simplified landscapes only. These studies support the prediction by Tscharntke *et al.* (2005) that local management measures should have a greater effect in simple landscapes compared to complex landscapes. However, it appears that empirical support for a low effect of such measures in cleared landscapes is largely missing. It should also be noted that other studies have found no interactive effect between local management and landscape context (Östman *et al.*, 2001b; Purtauf *et al.*, 2005; Roschewitz *et al.*, 2005; Clough *et al.*, 2007).

CONCLUSION

Landscape-scale effects on the abundance, diversity and efficacy of natural enemies of pests in the agroecosystem are numerous, as shown by many empirical studies. These investigations provide not only tests of theory, but also information useful for designing management methods that include changing the structure of the landscape. In this chapter we have shown that complex landscapes often have higher abundance and diversity of natural enemies, and more effective biological control, than landscapes simplified by intensive agriculture. Different species respond to the landscape at different spatial scales, but it appears that the landscape composition within a radius of between a few hundred metres to a few kilometres is appropriate for the majority of natural enemies. We have also shown that different habitat management techniques – such as provision of flower strips and shelterbelts – can be effective for the conservation of biological control. However, we still know relatively little about the mechanisms explaining the relationship between landscape structure and the pest control patterns often observed, and we still have a limited understanding of the optimal spatial and temporal arrangement of habitat management measures required to improve biological control. To improve our understanding of the drivers of landscape effects on biological control, future studies will need to be designed so that the effect of different landscape properties can be disentangled from each other. In comparative mensurative experiments, this can be done by deliberately selecting landscapes where key landscape features are uncorrelated with each other (see Gabriel *et al.*, 2010). In addition, statistical methods such as structural equation modelling (SEM) can be used to test how effects of landscape metrics are mediated by proximate factors (e.g. insecticide application and floral resource availability). Ideally, though, manipulative experiments are needed where different variables are modified at the landscape scale, but such experiments are obviously difficult to manage within normal research budgets. For the future, we also need to begin to link the effects of agronomic practices, such as crop rotation and insecticide use, to our knowledge of landscape spatial context (Rusch *et al.*, 2010). The addition of non-crop habitat can facilitate biological pest control, but we also need to consider the possibilities that arise when crop rotation, crop diversity, tillage regimes and other aspects of cropping systems are included in landscape studies.

ACKNOWLEDGEMENTS

Financial support for M.J. and B.E. was provided by SAPES (Multifunctional Agriculture: Harnessing Biodiversity for Sustaining Agricultural Production and Ecosystem Services), from FORMAS (the Swedish Research Council for Environment, Agricultural Sciences and Spatial Planning) and from UD-40: an initiative from the Ministry of Foreign Affairs of Sweden on Food Security at the Swedish University of Agricultural Sciences.

REFERENCES

Baars, M.A. (1979) Patterns of movement of radioactive carabid beetles. *Oecologia*, 44, 125–140.

Baggen, L.R. and Gurr, G.M. (1998) The influence of food on *Copidosoma koehleri* (Hymenoptera: Encyrtidae), and the use of flowering plants as a habitat management tool to enhance biological control of potato moth, *Phthorimaea operculella* (Lepidoptera: Gelechiidae). *Biological Control*, 11, 9–17.

Bianchi, F.J.J.A., Booij, C.J.H. and Tscharntke, T. (2006) Sustainable pest regulation in agricultural landscapes: a review on landscape composition, biodiversity and natural pest control. *Proceedings of the Royal Society of London B*, 273, 1715–1727.

Bianchi, F.J.J.A., Goedhart, P.W. and Baveco, J.M. (2008) Enhanced pest control in cabbage crops near forest in The Netherlands. *Landscape Ecology*, 23, 595–602.

Bommarco, R. (1998). *Reproduction* and energy reserves of a predatory carabid beetle relative to agroecosystem complexity. *Ecological Applications*, 8, 846–853.

Brouwer, D.W. (1998) Plan for trees: a guide to farm revegetation on the Coast and Tablelands. New South Wales Agriculture Tocal, Paterson.

Carson, R. (1962) *Silent spring*, Houghton Mifflin, Boston.

Chust, G., Pretus, J.L., Ducrot, D. and Ventura, D. (2004) Scale dependency of insect assemblages in response to landscape pattern. *Landscape Ecology*, 19, 41–57.

Clough, Y., Kruess, A., Kleijn, D. and Tscharntke, T. (2005) Spider diversity in cereal fields: comparing factors at local, landscape and regional scales. *Journal of Biogeography*, 32, 2007–2014.

Clough, Y., Kruess, A. and Tscharntke, T. (2007) Organic versus conventional arable farming systems: Functional grouping helps understand staphylinid response. *Agriculture, Ecosystems and Environment*, 118, 285–290.

Corbett, A. and Rosenheim, J.A. (1996) Impact of a natural enemy overwintering refuge and its interaction with the surrounding landscape. *Ecological Entomology*, 21, 155–164.

Croft, B.A. (1990) *Arthropod biological control agents and pesticides*, John Wiley & Sons, Inc., New York.

DeBach, P. (1964) *Biological control of insect pests and weeds*, Chapman & Hall, London.

Doutt, R.L. and Nakata, J. (1965) Parasites for control of the grape leaf hopper. *California Agriculture*, 19, 3.

Elliott, N.C., Kieckhefer, R.W., Lee, J.-H. and French, B.W. (1998) Influence of within-field and landscape factors on aphid predator populations in wheat. *Landscape Ecology*, 14, 239–252.

Fiedler, A.K., Landis, D.A. and Wratten, S.D. (2008) Maximizing ecosystem services from conservation biological control: the role of habitat management. *Biological Control*, 45, 254–271.

Firle, S., Bommarco, R., Ekbom, B. and Natiello, M. (1998) The influence of movement and resting behavior on the range of three carabid beetles. *Ecology*, 79, 2113–2122.

Forbes, S.A. (1915) The ecological foundations of applied entomology. *Annals of the Entomological Society of America*, 8, 1–19.

Gabriel, D., Sait, S.M., Hodgson, J.A., Schmutz, U., Kunin, W.E. and Benton, T.G. (2010) Scale matters: the impact of organic farming on biodiversity at different spatial scales. *Ecology Letters*, 13, 858–869.

Gámez-Virués, S. (2009) *Conservation of biological control agents within shelterbelts and adjacent pastures*. Doctor of Philosophy thesis, University of Sydney.

Gámez-Virués, S., Gurr, G.M., Raman, A., La Salle, J. and Nicol, H.I. (2009) Effects of flowering groundcover vegetation on diversity and activity of wasps in a farm shelterbelt in temperate Australia. *BioControl*, 54, 211–218.

Gardiner, M.M., Landis, D.A., Gratton, C. *et al.* (2009a) Landscape diversity enhances the biological control of an introduced crop pest in the north-central US. *Ecological Applications*, 19, 143–154.

Gardiner, M.M., Landis, D.A., Gratton, C. *et al.* (2009b) Landscape composition influences patterns of native and exotic lady beetle abundance. *Diversity and Distributions*, 15, 554–564.

Geiger, F., Bengtsson, J. and Berendse, F. *et al.* (2010) Persistent negative effects of pesticides on biodiversity and biological control potential on European farmland. *Basic and Applied Ecology*, 11, 97–105.

Griffiths, G.J.K., Holland, J.M., Bailey, A. and Thomas, M.B. (2008) Efficacy and economics of shelter habitats for conservation biological control. *Biological Control*, 45, 200–209.

Haenke, S., Scheid, B., Schaefer, M., Tscharntke, T. and Thies, C. (2009) Increasing syrphid fly divserity and density in sown flower strips within simple vs. complex landscapes. *Journal of Applied Ecology*, 46, 1106–1114.

Holland, J. and Fahrig, L. (2000) Effect of woody borders on insect density and diversity in crop fields: a landscape-scale analysis. *Agriculture, Ecosystems and Environment*, 78, 115–122.

Holt, R.D. (1996) Food webs in space: an island biogeographic perspective, in *Food webs – Integration of patterns and dynamics* (eds G.A. Polis and K.O. Winemiller), Chapman & Hall, New York, pp. 313–323.

Holt, R.D., Lawton, J.H., Polis, G.A. and Martinez, N.D. (1999) Trophic rank and the species-area relationship. *Ecology*, 80, 1495–1504.

Isaacs, R., Tuell, J., Fiedler, A., Gardiner, M. and Landis, D.A. (2009) Maximizing arthropod-mediated ecosystem services in agricultural landscapes: the role of native plants. *Frontiers in Ecology and the Environment*, 7, 196–203.

Jonsson, M., Wratten, S.D., Landis, D.A., Tompkins, J-L.M. and Cullen, R. (2010) Habitat manipulation to mitigate the impacts of invasive arthropod pests. *Biological Invasions*, 12, 2933–2945.

Kareiva, P. and Wennergren, U. (1995) Connecting landscape patterns to ecosystem and population processes. *Nature*, 373, 299–302.

Krauss, J., Steffan-Dewenter, I. and Tscharntke, T. (2003) How does the landscape context contribute to effects of habitat fragmentation on diversity and population density of butterflies? *Journal of Biogeography*, 30, 889–900.

Kruess, A. and Tscharntke, T. (2002) Grazing intensity and the diversity of grasshoppers, butterflies, and trap-nesting bees and wasps. *Conservation Biology*, 16, 1570–1580.

Landis, D.A., Wratten, S.T. and Gurr, G.M. (2000) Habitat management to conserve natural enemies of arthropod pest in agriculture. *Annual Review of Entomology*, 45, 175–201.

Landis, D.A., Gardiner, M.M., van der Werf, W. and Swinton, S.M. (2008) Increasing corn for biofuel production reduces biocontrol services in agricultural landscapes. *Proceedings of the National Academy of Science*, 105, 20552–20557.

Laurance, W.F. (2008) Theory meets reality: how habitat fragmentation research has transcended island biogeographic theory. *Biological Conservation*, 141, 1731–1744.

Lavandero, B., Wratten, S., Shishehbor, P. and Worner, S. (2005) Enhancing the effectiveness of the parasitoid *Diadegma semiclausum* (Helen): movement after use of nectar in the field. *Biological Control*, 34, 152–158.

Lawton, J.H. (1999) Are there general laws in Ecology? *Oikos*, 84, 177–192.

Levins, R. and Wilson, M. (1980) Ecological theory and pest management. *Annual Review of Entomology*, 25, 287–308.

MacArthur, R.H. and Wilson, E.O. (1967) *The theory of island biogeography*, Princeton University Press, Princeton, New Jersey.

Meyer, K.M., Jopp, F., Münkemüller, T., Reuter, H. and Schiffersh, K. (2010) Crossing scales in ecology. *Basic and Applied Ecology*, doi:10.1016/j.baae.2010.08.003.

Öberg, S. (2009) Influence of landscape structure and farming practice on body condition and fecundity of wolf spiders. *Basic and Applied Ecology*, 10, 614–621.

Öberg, S., Ekbom, B. and Bommarco, R. (2007) Influence of habitat type and surrounding landscape on spider diversity in Swedish agroecosystems. *Agriculture, Ecosystems and Environment*, 122, 211–219.

Öberg, S., Mayr, S. and Dauber, J. (2008) Landscape effects on recolonisation patterns of spiders in arable fields. *Agriculture, Ecosystems and Environment*, 123, 211–218.

Ode, Å. and Miller, D. (2011) Analysing the relationships between indicators of landscape complexity and preference. *Environment and Planning B: Planning and Design*, 38, 24–40.

Odum, E.P. (1963) *Ecology*, Holt, Rinehart & Winston, New York.

Odum, E.P. (1969) The strategy of ecosystem development. *Science*, 164, 262–270.

Östman, Ö., Ekbom, B., Bengtsson, J. and Weibull, A.C. (2001a) Landscape complexity and farming practice influence the condition of polyphagous carabid beetles. *Ecological Applications*, 11, 480–488.

Östman, Ö., Ekbom, B. and Bengtsson, J. (2001b) Landscape heterogeneity and farming practice influence biological control. *Basic and Applied Ecology*, 2, 365–371.

Perfecto, I. and Armbrecht, I. (2003) The coffee agroecosystems in the neotropics: combining ecological and economic goals, in *Tropical Agroecosystems* (ed. J.H. Vandermeer), CRC Press, Boca Raton, pp. 159–194.

Perovic, D., Gurr, G.M., Raman, A. and Nicol, H.I. (2010) Effect of landscape composition and arrangement on biological control agents in a simplified agricultural system: a cost–distance approach. *Biological Control*, 52, 263–270.

Pimentel, D. (1961) species diversity and insect population outbreaks. *Annals of the Entomological Society of America*, 54, 76–86.

Pimm, S.L. and Lawton, J.H. (1977) Number of trophic levels in ecological communities. *Nature*, 268, 329–331.

Price, P.W. and Waldbauer, G.P. (1975) Ecological aspects of pest management, in *Introduction to insect pest management* (eds R.L. Metcalf and W.H. Luckmann), Wiley Interscience, New York. pp. 37–73.

Purtauf, T., Roschewitz, I., Dauber, J., Thies, C., Tscharntke, T. and Wolters, V. (2005) Landscape context of organic and conventional farms: influences on carabid beetle diversity. *Agriculture, Ecosystems and Environment*, 108, 165–174.

Rand, T.A. and Tscharntke, T. (2007) Contrasting effects of natural habitat loss on generalist and specialist aphid natural enemies. *Oikos*, 116, 1353–1362.

Rand, T., Tylianakis, J.M. and Tscharntke, T. (2006) Spillover edge effects: the dispersal of agriculturally-subsidized insect natural enemies into adjacent natural habitats. *Ecology Letters*, 9, 603–614.

Rice, R.A. and Greenberg, R. (2000) Cacao cultivation and the conservation of biological diversity. *Ambio*, 29, 167–173.

Ricklefs, R.E., Naveh, Z. and Turner, R.E. (1984) Conservation of ecological processes. *The Environmentalist*, 4, 6–16.

Riley, J.R., Reynolds, D.R., Mukhopadhyay, S., Ghosh, M.R. and Sarkar, T.K. (1995) Long-distance migration of aphids and other small insects in northeast India. *European Journal of Entomology*, 92, 639–653.

Roschewitz, I., Hücker, M., Tscharntke, T. and Thies, C. (2005) The influence of landscape context and farming practices on parasitism of cereal aphids. *Agriculture, Ecosystems and Environment*, 108, 218–227.

Rusch, A., Valantin-Morison, M., Sarthou, J.P. and Roger-Estrade, J. (2010) Integrating crop and landscape management into new crop protection strategies to enhance biological control of oilseed rape insect pests, in *Biocontrol-Based Integrated Management of Oilseed Rape Pests* (ed. I. Williams), Springer, Dordrecht, pp. 415–448.

Scarratt, S.L. (2005) Enhancing the biological control of leaf-rollers (Lepidoptera: Tortricidae) using floral resource subsidies in an organic vineyard in Marlborough, New Zealand. Doctor of Philosophy thesis, Lincoln University, Lincoln.

Schellhorn, N.A., Bellati, J., Paull, C.A. and Maratos, L. (2008) Parasitoid and moth movement from refuge to crop. *Basic and Applied Ecology*, 9, 691–700.

Schmidt, M.H. and Tscharntke, T. (2005) Landscape context of sheetweb spider (Araneae: Linyphiidae) abundance in cereal fields. *Journal of Biogeography*, 32, 467–473.

Schmidt, M.H., Roschewitz, I., Thies, C. and Tscharntke, T. (2005) Differential effects of landscape and management on diversity and density of ground-dwelling farmland spiders. *Journal of Applied Ecology*, 42, 281–287.

Schmidt, M.H., Thies, C., Nentwig, W. and Tscharntke, T. (2008) Contrasting responses of arable spiders to the landscape matrix at different spatial scales. *Journal of Biogeography* 35, 157–166.

Snyder, W.E., Snyder, G.B., Finke, D.L. and Straub, C.S. (2006) Predator biodiversity strengthens herbivore suppression. *Ecology Letters*, 9, 789–796.

Stamps, W.T. and Linit, M.J. (1998) Plant diversity and arthropod communities: Implications for temperate agroforestry. *Agroforestry Systems*, 39, 73–89.

Steffan-Dewenter, I., Müzenberg, U. and Tscharntke, T. (2001) Pollination, seed set, and seed predation on a landscape scale. *Proceedings of the Royal Society B*, 268, 1685–1690.

Steingröver, E.G., Geertsema, W. and van Wingerden, W.K.R.E. (2010) Designing agricultural landscapes for natural pest control: a transdisciplinary approach in the Hoeksche Waard (The Netherlands). *Landscape Ecology*, 25, 825–838.

Stern, V.M., Smith, R.F., van den Bosch, R. and Hagen, K.S. (1959) The integrated control concept. *Hilgardia*, 29, 81–101.

Taylor, L.R. (1977) Migration and spatial dynamics of an aphid, *Myzus persicae*. *Journal of Animal Ecology*, 46, 411–423.

Thies, C. and Tscharntke, T. (1999) Landscape structure and biological control in agroecosystems. *Science*, 285, 893–895.

Thies, C., Steffan-Dewenter, I. and Tscharntke, T. (2003) Effects of landscape context on herbivory and parasitism at different spatial scales. *Oikos*, 101, 18–25.

Thies, C., Roshewitz, I. and Tscharntke, T. (2005) The landscape context of cereal aphid–parasitoid interactions. *Proceedings of the Royal Society B*, 272, 203–210.

Tischendorf, L. and Fahrig, L. (2000) On the usage and measurement of landscape connectivity. *Oikos*, 90, 7–19.

Tscharntke, T., Klein, A.M., Kruess, A., Steffan-Dewenter, I. and Thies, C. (2005) Landscape perspectives on agricultural intensification and biodiversity – ecosystem service management. *Ecology Letters*, 8, 857–874.

Tscharntke, T., Bommarco, R., Clough, Y. *et al.* (2007) Conservation biological control and enemy diversity on a landscape scale. *Biological Control*, 43, 294–309.

Turner, M.G., Gardner, R.H. and O'Niell, R.V. (2001) *Landscape ecology in theory and practice: pattern and process*, Springer, New York.

Tuovinen, T. (1994) Influence of surrounding trees and bushes on the phytoseiid mite fauna on apple orchard trees in Finland. *Agriculture, Ecosystems and Environment*, 50, 39–47.

van Emden, H.F. (1965) The role of uncultivated land in the biology of crop pests and beneficial insects. *Scientific Horticulture*, 17, 126–136.

Vollhardt, I.M.G., Tscharntke, T., Wäckers, F.L., Bianchi, F.J.J.A. and Thies, C. (2008) Diversity of cereal aphid parasitoids in simple and complex landscapes. *Agriculture, Ecosystem and Environment*, 126, 289–292.

Way, M.J. (1966) The natural environment and integrated methods of pest control. *Journal of Applied Ecology* (Supplement), 3, 29–32.

Weibull, A.-C., Östman, Ö. and Granqvist, Å. (2003) Species richness in agroecosystems: the effect of landscape, habitat and farm management. *Biodiversity and Conservation*, 12, 1335–1355.

Wratten, S.D., Bowie, M.H., Hickman, J.M., Evans, A.M., Sedcole, J.R. and Tylianakis, J.M. (2003) Field boundaries as barriers to movement of hover flies (Diptera: Syrphidae) in cultivated land. *Oecologia*, 134, 605–611.

Yachi, S. and Loreau, M. (1999) Biodiversity and ecosystem productivity in a fluctuating environment: the insurance hypothesis. *Proceedings of the National Academy of Natural Science USA*, 96, 1463–1468.

Zaller, J.G., Moser, D., Drapela, T., Schmöger, C. and Frank, T. (2008) Insect pests in winter oilseed rape affected by field and landscape characteristics. *Basic and Applied Ecology*, 9, 682–690.

Methods

SCALE EFFECTS IN BIODIVERSITY AND BIOLOGICAL CONTROL: METHODS AND STATISTICAL ANALYSIS

Christoph Scherber, Blas Lavandero, Katrin M. Meyer, David Perovic, Ute Visser, Kerstin Wiegand and Teja Tscharntke

Biodiversity and Insect Pests: Key Issues for Sustainable Management, First Edition. Edited by Geoff M. Gurr, Steve D. Wratten, William E. Snyder, Donna M.Y. Read.
© 2012 John Wiley & Sons, Ltd. Published 2012 by John Wiley & Sons, Ltd.

INTRODUCTION AND DEFINITIONS OF SCALE

The structure of agricultural landscapes influences organisms living in these landscapes, and in particular insect pests and their natural enemies (Gámez-Virués et al., chapter 7 of this volume). Interactions at a local scale (for example an individual field) are likely to be influenced by processes acting at larger scales (for example the surroundings of that field; Figure 8.1). This is often called scale dependence or context dependence (Pearson, 2002).

This chapter serves as an introduction to the design and analysis of studies on biological control at different spatial scales. Spatial scale can be described by two factors: grain and extent (Wiens, 1989; Fortin and Dale, 2005). Grain is the size of an individual sampling unit (for example a plot measuring $4\,m^2$); extent is the total size of the study area (for example a landscape measuring $100\,ha$). The grain size used for individual study units should be carefully chosen to match the spatial structure of the phenomenon being studied. For example, a grain size of $0.5\,cm$ could be necessary in a study of insects inhabiting wheat stems (where the spatial arrangement of damaged vs. intact wheat stems is of interest). In addition, the grain size can also be important when it comes to data analysis – that is, when data are aggregated for statistical analysis.

Hence, 'spatial scale' can refer to an individual study organism, an individual sampling unit, or an individual unit of statistical analysis (see also Dungan et al., 2002).

Knowing now what we mean by 'scale', we may now ask: how can scaling effects be included in studies on pest control? Before addressing scale effects out in the landscape, it is often useful to start with smaller-scale laboratory systems where it is easier to control for confounding variables. We therefore start this chapter with an introduction to the problem of 'upscaling'; that is, the extrapolation from smaller to larger scales. We then move on to the landscape scale, and provide an overview of field methods used to study the movement of organisms through the landscape. This is followed by sections on data analysis and modelling. The chapter concludes with some guidelines likely to be useful for practitioners who want to incorporate scale effects in their own biological control studies.

FROM THE LABORATORY TO THE FIELD: UPSCALING PROBLEMS

In traditional biological control studies, it is often necessary to start with a series of smaller-scale laboratory experiments before moving to the field scale. For example, we need to understand the host specificity of

Figure 8.1 Scale transitions and landscape complexity in agroecosystems. a) Wheat spikes are attacked by pest insects (e.g. aphids) interacting with biological control agents on a local scale; b) a complex agricultural landscape near Holzminden (Central Germany); c) a simple agricultural landscape in the cereal plain of Chizé (France) (all photos by C. Scherber).

biological control agents, or the food plant spectrum of individual insect herbivores, before we can begin to understand what is happening in the field. Often, the underlying interactions between the biological control agent and the pest organisms occur at the individual level at a scale of centimetres and smaller. To develop efficient biological control measures, we need to understand individual-level ecological processes such as herbivory, parasitism, colonisation and competition and then upscale this knowledge to the level of whole plants or whole stands. However, upscaling is not a straightforward task for ecological and methodological reasons. The ecological processes that drive small-scale and large-scale patterns are usually not the same and do not necessarily overlap (Hartley *et al.*, 2004; Teodoro *et al.*, 2009; see also chapter 7 of this volume). For instance, the foraging pattern of gall-forming insects differs across scales from the leaf over the branch to the tree level (Lill, 1998). Similarly, parasitism by different parasitoids of the forest pest *Malacosoma disstria* (Hübner) is affected both by spatial scale and by parasitoid body size (Roland and Taylor, 1997). Hence, ecological mechanisms between scales cannot always be easily compared. The main methodological challenge is to maintain the high resolution (fine grain) of small-scale laboratory studies when increasing the extent of a study to the field scale (e.g. Xia *et al.*, 2003). This is often not possible due to logistical constraints such as limited labour, facilities or computing power. The methodological alternative is to decrease the resolution of a study when moving from the laboratory to the field scale. Aggregation procedures can be used to achieve this decrease in resolution. However, nonlinearities and thresholds often complicate aggregation procedures, so that aggregation provides no simple upscaling solution, either.

These difficulties of scaling up from small to large scales are reflected in the scarcity of upscaling approaches and of studies that adopt or test these approaches by using scales as explanatory variables. The simplest approach is to take samples at different scales, ideally in a nested manner (a hierarchical sampling approach). Due to the logistical constraints mentioned above, the resolution of the samples will in most cases change across scales (for an exception see Roland and Taylor, 1997). If the relationship between the ecological variable of interest and the scales on the x-axis is linear, upscaling of the ecological process can be performed based on this relationship. Unfortunately, most studies adopting this approach have found scale-

dependence of the ecological process, preventing straightforward upscaling (e.g. parasitism (Lill, 1998; Matsumoto *et al.*, 2004), mite predation (Zhang and Anderson 1993; 1997), foraging in multitrophic systems (Heisswolf *et al.*, 2006) and pathogenic nematode attack (Efron *et al.*, 2001)). We are aware of one exception, where upscaling of parasitoid foraging from the local to the landscape scale yielded consistent results (Fraser *et al.*, 2008).

Three general approaches can be taken to scale up from small to large scales: sampling at different scales, interpolating between local estimates to cover larger scales, and extrapolating from local estimates to larger scales (Table 8.1). The first approach of taking (hierarchical) samples at different scales is often analysed with scale-area plots to determine the scale-dependence of ecological processes (Table 8.1). When sampling is not possible at multiple scales, local estimates have to be used to reach larger spatial or temporal scales, either by interpolating or extrapolating. In the second approach, the space or time between estimates is interpolated to cover larger areas or time frames. Methods of spatial interpolation include Voronoi polygons (i.e. interpolation using a network of nearest-neighbour points; Table 8.1) and thin plate spline interpolation (i.e. interpolation using a smoothing function; Table 8.1). These methods have also been applied to species distribution modelling (Jarvis and Collier, 2002). Species distribution modelling is also the major field of application of the third approach in which bioclimatic models extrapolate local estimates to larger scales using regression techniques (Table 8.1). Non-climatic factors such as biotic interactions, rapid evolutionary change and dispersal may also affect species distributions, but are often not included in bioclimatic models (Pearson and Dawson, 2003).

For a successful extrapolation across scales, critical scale transitions (He and Hubbell, 2003) and the extent and direction of change in the interactions between organisms at these transitions have to be identified. Critical scale transitions are characterised by abrupt changes in a landscape parameter (e.g. field perimeter) with changing spatial scale (for details, see He and Hubbell, 2003).

A useful starting point to study such scale transitions is the biological control of microbial leaf pathogens. Population sizes of microorganisms on the leaf surface vary unpredictably across scales and are highly aggregated at all scales from leaf segments to tree stands (Kinkel *et al.*, 1995; Kinkel, 1997; Hirano and

Table 8.1 Commonly used methods of upscaling from smaller to larger scales.

Purpose	Method	Selected references	Applications
Analysing the impact of scales on an ecological process	Scale-area plots	Kunin, 1998	Hierarchical sampling Assessment of scale-dependence of range sizes of plant species (Hartley et al., 2004)
Interpolating between local estimates to cover larger scales	Voronoi polygon method (= Dirichlet tessellation)	Dale, 1999	Prediction of species distributions Interpolation of local temperature estimates to the landscape level to predict phenological events in the life cycle of three pest species (Jarvis and Collier, 2002)
	Thin plate spline interpolation	Hutchinson, 1991	Thin plate spline interpolation performed better than the Voronoi polygon method (Jarvis and Collier, 2002)
Extrapolating local estimates of ecological and climatic limits of a species to landscape and global scales	Bioclimatic modelling	Pearson and Dawson, 2003	Inference of actual or potential species distributions via climate envelopes ference of the distribution of the biological control agent *Podisus maculiventris* (Say) (Legaspi and Legaspi, 2007), cautioning against basing field-level decisions on bioclimatic models due to the lack of sufficient data for their parameterisation and validation

Upper, 2000). Hence, there is no optimal sampling scale from which population sizes at other scales can simply be extrapolated (Kinkel et al., 1995). This is also reflected in the variable efficiencies of biological control measures observed at the seed and at the field scale (Kildea et al., 2008). Microbial systems can be a worthwhile starting point to test the performance of current and new upscaling approaches before transferring the results to insect biological control agents.

The lack of overarching upscaling approaches indicates that each scale probably requires its own approach, so that we should advance the coupling of existing approaches rather than aiming to develop a universal upscaling approach (Meyer et al., 2010). One example of a coupled approach is the pattern-oriented modelling strategy (Grimm et al., 2005) where small-scale mechanisms are derived from large-scale patterns. Pattern-oriented modelling can be used to distinguish between alternative hypotheses on the transition from one scale to the other and thus identify the most appropriate upscaling approach for a particular biological control study.

Overall, upscaling studies show that it can be difficult to compare results obtained in laboratory systems to the field or landscape scale. It is therefore inevitable to move one step further and try to follow organisms out in the agricultural landscape. The next section discusses how we can track the movement of insects through real landscapes – a prerequisite for many approaches that follow.

FIELD METHODS FOR UNDERSTANDING LANDSCAPE-SCALE PATTERNS

Moving from smaller laboratory systems to the field and landscape scale, researchers often have to become detectives – simply because there is so much space available for study organisms to hide and escape. This is not so much of a problem under small-scale laboratory conditions, but it is central to the success of large-scale field studies. Upscaling from the laboratory to the field thus requires a whole new set of approaches to track arthropods at the large scale. During the last few decades, a series of different marking and tracking techniques have been developed to study arthropod movement and dispersal (Table 8.2). These techniques can be used to identify the land

Table 8.2 An overview of marking and tracking techniques commonly employed in landscape-scale biological control studies.

Characteristics				Recent examples		Reviews
Technique	**Simplicity**	**Cost**	**Requires specialist equipment**	**Movement studies**	**Resource use (self-marking) studies**	
Dyes	simple	low	no	Bianchi et al., 2009	–	Schellhorn et al., 2004
Rare earths	moderate	relatively low	yes	Prasifka et al., 2004a	Lavandero et al., 2005; Scarratt et al., 2008	Southwest Entomologist Special Issue, 14, 1991
Sugar analysis	moderate	relatively low	yes	Desouhant et al., 2010	Winkler et al., 2009	Heimpel et al., 2004
Stable Isotopes	moderate	relatively low	yes	Prasifka and Heinz, 2004	Wanner et al., 2006	Hood-Nowotny and Knols, 2007; Prasifka and Heinz, 2004
Protein marking	increasingly simple	relatively low	yes	Jones et al., 2006	See Jones et al., 2006	Hagler and Jones, 2010; Horton et al., 2009

uses that (1) act as sources of movement into crops, for both pests and natural enemies, and (2) act as alternative resources and resource subsidies for natural enemies. In the following brief overview of marking and tracking techniques we outline how different techniques have been used to investigate the movement and spatial ecology of arthropods and suggest areas for future focus. Due to the limits on space, however, the following section is by no means an in-depth review of this subject (more detailed reviews are highlighted in Table 8.1).

Following animals from one point to another is the basic requirement of any marking and tracking technique. The fact that 'old fashioned' techniques such as fluorescent dyes are still being used (e.g. Schellhorn et al., 2004; Bianchi et al., 2009b) despite the high-tech revolution of recent decades illustrates the power of the basic guidelines (e.g. outlined by Hagler and Jackson, 2001) that a marking technique should be simple to apply, readily detectable, inexpensive, safe and not affect the biology or ecology of the target species. Fluorescent dyes score well in all of these categories (see Table 8.1). For example, despite the relatively low recapture rates compared with rare-earth labels (Hagler and Jackson, 2001), fluorescent dyes are cheaper to apply and there is no need for specialised laboratory equipment with trained technicians to process the samples. And while rare-earth labelling techniques (Box 8.1) may offer much greater capture rates, in mark-capture trials (e.g. see Prasifka et al., 2004a), this form of labelling requires intensive background sampling before the mark-capture is conducted (in order to first establish the naturally occurring variation, within the local population, of the elements to be used as a marker (e.g. rubidium). Similarly, the enormous potential for mass mark-capture offered by marking with cheap proteins for ELISA analysis (described by Hagler and Jones, 2010) may be overshadowed, for many researchers, by the need for specialised equipment for identification. Although fluorescent dyes may offer a good, cheap, all-purpose type of marking solution, they are perhaps best suited to mark-release-recapture type investigations (where a large number of collected or laboratory-reared individuals are marked and release, en masse, from a central point and subsequently recaptured). The emerging potential of marking with cheap proteins (e.g. milk and egg protein as described in Hagler and Jones, 2010) offers the opportunity to apply the marker to unprecedentedly large areas of vegetation in order

to mark wild populations of arthropods in mark-capture type investigations.

Traditional mark-capture techniques suffer from several disadvantages. In particular, mark-recapture techniques require equal catchability of marked individuals, and often high numbers of individuals need to be marked. In many cases a technique described as 'self-marking' may be preferable, where arthropods obtain the mark, for example through foraging, rather than being directly and intentionally marked by the observer. The extra ecological information from such studies can be useful in habitat management and conservation biological control. For example, HPLC nectar analysis (Wäckers, 2007), pollen marking (Silberbauer et al., 2004) and the use of stable carbon isotopes (e.g. Prasifka et al., 2004b, Prasifka and Heinz, 2004) can identify the resources, resource subsidies and alternative habitats utilised by pests and natural enemies. However, these approaches may not have the critical information about the origin of the 'mark' (unless there is a unique source of pollen, nectar or C_3 plants in the area). It is here that rare-earth labels are perhaps most useful (e.g. Lavandero et al., 2005; Scarratt et al., 2008), because plants can be intentionally marked via the vascular system, leaving no doubt about how and

---

Box 8.1 Using rare-earth labelling techniques to investigate spatial population dynamics of insects exploiting a patchy food resource (Dempster et al., 1995)

Movements between plant patches were studied with the use of chemical markers (Rb, Sr, Dy and Cs) which were applied as chloride salts to individual patches, and which were translocated to the flower heads and so to insects feeding on the seed, and to their parasitoids.

These analyses showed that individuals of all species moved considerable distances, with movements of up to 2 km commonly recorded. Estimates of rates of immigration to patches showed that movement plays an important role in the population dynamics of these insects. There was some evidence that immigration was density-dependent: it was highest when the resident populations (numbers per flower head) were low.

where the mark had been obtained (stable isotopes can also be employed in this fashion (e.g. Wanner *et al.*, 2006; see Table 8.1). Rare-earth elements, such as rubidium and strontium, have the advantage of moving through trophic levels (as do stable isotopes). They may, therefore, provide information on the foraging habits of captured insects (Prasifka *et al.*, 2004a). The identification of sugars in the gut contents of natural enemies can also help to inform on the use of resource subsidies or the foraging of pest-originated sugars such as melezitose included in lepidopteran frass and homopteran honeydew (Heimpel *et al.*, 2004).

Perhaps the greatest potential for marking and tracking techniques in biological control, especially with a focus on biodiversity, is the use of multiple markers to adapt techniques to more complex field situations, so that different markers may be used for source habitats or resource subsidies. For example, milk proteins may be used to mark one field, or one prey species, and egg proteins to mark another field or prey species. The recent advances in identifying common proteins with ELISA (Hagler and Jones, 2010) offer great potential in this endeavour.

There is also considerable scope for combining different disciplines, for example in 'landscape genetics'. In recent years the use of landscape genetics, which is the combination of high-resolution genetic markers with spatial data analysis, has been particularly relevant when assessing the influence of landscape characteristics on the genetic variability and the identification of barriers to gene flow (Storfer *et al.*, 2007). Examples of the assessment of suppressive landscapes using landscape genetics are still scarce, although molecular markers are available for many species (Behura, 2006), and area-wide pest management programmes provide valuable information about landscape attributes (Calkins and Faust, 2003; Carrière *et al.*, 2004; Beckler *et al.*, 2005; Park *et al.*, 2006). Correctly identifying sinks and sources of pests and natural enemies can inform refuge placement and determine whether a landscape is pest suppressive or not. As different parasitoid races can be specific to different host species (for parasitoids with a great host range), genetic and allozyme studies have shown that there is gene flow between refuge-alternative hosts and the target pest on the target crop (Blair *et al.*, 2005; Forbes *et al.*, 2009; Stireman *et al.*, 2006). Thus, the ability of a parasitoid to control different hosts on different host plants may not be constant, even among different genotypes of a single species (Henry *et al.*, 2010). In a recent study in

central Chile's main apple production area, the relationships between aphid (*Eriosoma lanigerum* (Hausmann)) and parasitoid (*Aphelinus mali* (Haldemann)) population genetics were studied. Samples were taken from commercial apple orchards and from a different *E. lanigerum* host (*Pyracantha coccinea* (Roem)) in a farm hedge dominated by the plant genus *Pyracantha*. Prior studies had shown geographic barriers interrupting gene flow of the aphid host between neighbouring populations independently from geographical distances (Lavandero *et al.*, 2009). There was evidence of extensive gene flow between sites, but no evidence of reproductive barriers for the parasitoid, suggesting no host-plant related specialisation and therefore indicating that *Pyracantha* hedges are a source of parasitoids for the crop. Based on this knowledge, future integrated pest management programmes could rely on the use of refuges of alternative hosts to increase migration of parasitoids to areas where they are more rare, aiding the augmentation of the parasitoid population after disturbances.

Overall, the approaches highlighted in this section show a wide range of methods available to the researcher – from marking and tracking to landscape genetics. Another important area is experimental design and statistics, discussed below.

DESIGN AND STATISTICAL ANALYSIS OF LARGE-SCALE BIOLOGICAL CONTROL

Knowing how to mark and track insects in agricultural landscapes, we can now move on to think of how to apply this knowledge to conduct a biological control study on a landscape scale. First, we need to consider the spatial arrangement of study sites and treatments (experimental design). Second, we need to come up with sampling schemes that work for our study organisms (sampling design).

Experimental design

Of the wide variety of available experimental designs (e.g. Figure 1 in Hurlbert, 1984), the completely randomised design will probably be the least useful. It is almost certain that our study sites will need to be arranged in blocks in space and time. Blocks share similar abiotic conditions (e.g. soil parameters) and help reduce the unexplained variation in data. To reduce workload and costs, it is often advisable to apply split-plot designs in which smaller subplots are nested within larger plots. Experimental treatments (for example bagging, caging, pesticide application, etc.) are then applied at random at increasingly smaller spatial scales.

Sampling design

After deciding on the experimental design to be used in our biological control study, we need to define an appropriate sampling scheme to estimate organism abundance, species richness, predation rates and so forth. To decide on an appropriate sampling method, it is necessary to know our study organisms: How large are they, how mobile will they be, and how will they respond to landscape features (Wiens, 1989)? Secondly, we need to employ sampling, marking and tracking procedures that are as unbiased as possible (Hagler and Jackson, 2001). This requires setting up traps and other devices according to systematic or random schemes (Fortin and Dale, 2005; see Table 8.3). At this stage, we will also need to know which types of analyses we want to conduct with the data after they have been collected. For example, grid-based sampling will lead to different types of geostatistical procedures than random sampling (Fortin and Dale, 2005).

Combining observational and experimental approaches

In landscape-wide biological control studies, observational data ('mensurative experiments' *sensu* Hurlbert, 1984) should be combined with experiments to achieve what is called 'strong inference' (Platt ,1964). For example, if we study multitrophic interactions in oilseed rape, it is a good idea to establish experimental oilseed rape plots in addition to fields already existing in the landscape (Thies and Tscharntke, 1999). Additionally, experimental plant individuals ('phytometers') may be used to study local-scale phenomena (Gibson, 2002). Such approaches may help to standardise plant cultivars, soil conditions and other confounding variables. Experimental plots can then be used for specific treatments on a subplot scale (e.g. fertilisation, insecticide treatment, or caging experiments). In general, an 'ideal' landscape-scale study always involves experimentation ('manipulative experiments' *sensu* Hurlbert, 1984): experimental establishment of hedges (e.g. Girma *et al.*, 2000), experimental fragmentation of habitats (e.g. Lindenmayer *et al.*, 1999; Debinski and Holt, 2000), experimental application of herbicides, insecticides and biological control agents (e.g. Cochran and Cox, 1992). However, in many cases experimentation will be impossible for logistical reasons. Landscape-scale studies cover large areas, and individual fields often belong to landowners who manage their fields individually. Under these cir-

Table 8.3 Experimental or sampling designs employed in landscape-scale biological control studies.

	Experimental studies	**Observational studies**
Most frequent experimental or sampling designs applied	• Completely randomised designs • Randomised blocks designs • Paired designs	• Landscape gradients (e.g. gradients in landscape complexity) • Concentric circles designs (to study landscape context) • Grid sampling designs • Paired designs (e.g. paired comparisons between organic-conventional farms)
Main advantages	• Clear separation of response and explanatory variables • Classical hypothesis testing, strong inference	• Realism • Direct application to real-world scenarios possible
Main disadvantages	• Sometimes unrealistic • Small power if sample sizes is low • Upscaling problems	• Causes and effects may be difficult to separate • Unanticipated block-by-treatment interactions

cumstances, we can study gradients in landscape complexity, composition or configuration. Paired designs using 'pseudo-treatments' can also yield insights – for example if organic and conventional farming systems are studied (e.g. Kleijn *et al.*, 2006). Below, we list some of the most important features to consider for successful experimental design of biological control studies.

The importance of blocking

Blocks are still among the most useful features to control for variations in abiotic conditions in both experimental and observational studies on a landscape scale. For example, individual countries can form blocks in continent-wide studies (Billeter *et al.*, 2008; Dormann *et al.*, 2007). Likewise, pairs of farms can be considered as blocks (Kleijn *et al.*, 2006). Further, individual observers moving through the landscape can be 'applied' to different groups of study plots and 'observer effects' can then easily be incorporated into the block effect in statistical models.

Proper use of random effects

Every study site has its own characteristics, and it is never possible be sure which of these characteristics will be important for a given study. In the statistical design and analysis of landscape-wide studies, it is therefore important to be very clear about which factors should be treated as 'random' (McCulloch and Searle, 2001; Bolker *et al.*, 2009). For example, in a study with 30 study sites scattered through a larger landscape different experimenters would selected different sites. Hence, the population of possible sites may be almost infinitely large. The sites chosen by any one researcher just happened to be that particular 30. Hence, the sites are actually random effects, and this should be clear from the beginning of the study (Zuur *et al.*, 2009). Such random effects should always have at least two levels, and ideally as many as possible (Giovagnoli and Sebastiani, 1989; McCulloch and Searle, 2001).

How to incorporate the landscape context

Observations at a single site may be influenced by the surrounding landscape; these indirect influences are commonly termed the 'landscape context' (Pearson, 2002). The traditional approach has been to use individual sampling points, scattered through landscapes differing in landscape complexity. These points were then surrounded by concentric circles in which landscape parameters were assessed (Figure 8.2a, c). However, this means that landscape effects can only be guessed from correlations between what is observed at an individual plot, and some features of the landscape surrounding that point. It is more desirable to collect replicated samples in space as well, for example using replicated grids of sampling points at every study site (e.g. Billeter *et al.*, 2008; Dormann *et al.*, 2007; see Figure 8.2b, d). Note, however, that the grid cell size needs to match the cell size of the expected spatial pattern (Fortin and Dale, 2005). Alternatively, stratified random sampling may be employed; that is, each habitat forms a 'stratum' and is sampled separately. The sample size will then be a function of habitat area and costs of sampling (for details see Krebs, 1999).

Know your response and explanatory variables

It is always a good idea to set up an artificial dataset before the beginning of a study. You can then already try out different statistical models and do power analyses to estimate the sample sizes needed (e.g. Crawley, 2002). In biological control studies, we will often encounter count data (numbers of insects) or proportion data (proportion parasitised hosts). These data types usually require special types of statistical models such as generalised linear (mixed) models (McCulloch and Searle, 2001).

How to do the statistical analysis of landscape-scale biological control studies

After successful data collection, we usually want to draw inferences from these data using statistical techniques. In the past, many datasets have been analysed using standard regression techniques, although datasets actually had a clearly spatial nature (Dormann, 2007). The most important steps in the analysis of datasets on landscape-scale biological control are the following:

(1) Decide on how to deal with count and proportion data. Usually, you may wish to analyse them using

Sampling intensity

Landscape complexity

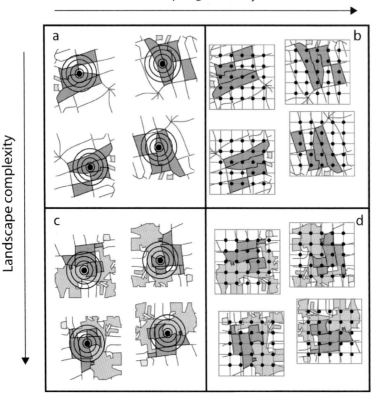

Figure 8.2 Sampling designs in biological control studies on a landscape scale. Sampling sites are indicated by filled black dots within landscapes; a) and c), low sampling intensity (N = 4 data points in 4 landscapes), landscape structure around each sampling site is measured in concentric circles with increasing radii. b) and d) high sampling intensity (N = 25 data points in 4 landscapes); landscape structure and spatial information about sampling locations are measured simultaneously. Landscape complexity increases from a) to c) and from b) to d). Figure created by C. Scherber.

generalised linear (mixed) models, but current software packages often lack methods to incorporate spatial and/or temporal autocorrelation into these models (for an overview see Bolker *et al.*, 2009). The best solution often is to transform the response variable, or to use variance functions to account for nonconstant variance.

(2) Decide on what to do with space and time. If you are interested in spatial trends, decide if you want to interpolate between sampling locations (kriging), or if you simply want to account for spatial autocorrelation (correlation structures in the residuals); a good introductory reference is Fortin and Dale (2005). If you are interested in temporal trends, ensure that observations are regularly spaced in time and that there is sufficient temporal replication (Zuur *et al.*, 2009). Treat temporal pseudoreplication, using time series analysis or by incorporating time, as a random slope. Avoid incorporating time as a pseudo-'subplot' because this may violate the sphericity assumption (sphericity is a

measure of variance homogeneity in repeated measures analyses; for details see von Ende, 2001).

(3) Plot the data, together with the model predictions, instead of plotting linear regressions provided by graphics software. Remember that model predictions from generalised linear models look nonlinear on the untransformed scale.

MODELLING SCALE EFFECTS IN BIOLOGICAL CONTROL

Even the most sophisticated statistical analysis often opens up new questions. For example, we may find that landscape context influences the distribution of a specialist parasitoid, but we may be unclear about the mechanisms. Modelling can be a useful tool to understand the spatiotemporal dynamics of pests and their biological control agents in the field. Modelling is also needed as a final step in designing pest-suppressive landscapes. In order to be able to give management recommendations towards promotion of biodiversity and biological control via design of pest-suppressive landscapes, a good understanding of the ecological processes acting at different scales is important (e.g. Levin, 2000; Turner 2005). Key questions are: Which species are promoted/threatened in a given landscape structure and what are the species and landscape characteristics making these species abundant/prone to extinction in such a landscape? How can a landscape be altered to promote beneficial species and suppress pest species?

The fundamental idea of ecological modelling is to reconstruct the basic features of ecological systems in simulation models. In other words, these models are a representation of all essential factors of the real system that are relevant with respect to the scientific question being addressed (Wissel, 1989). In case of rule-based simulation models, these essential factors and their interactions are being described using 'if-then rules' (Starfield *et al.*, 1994). For example, one rule in the model might be: *if* a parasitoid finds a host individual at a specific location, *then* the parasitoid lays an egg into the larva and at this location no host but a new parasitoid will develop. Experts who know from field experience which factors shape the system are a great help in modelling development.

Typically, several model variants are developed that can be used to test specific hypotheses on the functioning of the system. Factors can be added or removed, parameter values increased or decreased, and thereby our understanding of the system can be greatly improved. Models can also be used to help the planning of new field experiments. Using virtual experiments, different landscapes can be created and the (insect) species placed into these landscapes so that their populations develop according to the model rules. In such experiments, long time series can be investigated, which would not be possible in the field.

There are two main classes of models that are most frequently used to model large-scale spatiotemporal dynamics of organisms: individual-based models (IBM) and grid-based models. In IBM, each individual is tracked explicitly, along with its properties (e.g. size, sex, developmental stage). Population processes emerge from the combined behaviour of many individuals (e.g. Bianchi *et al.*, 2009a).

In grid-based models (e.g. Bianchi and van der Werf, 2003), space is represented as a grid of cells. This means each of these cells represents a small subunit of

space in a certain position and contains specific information, for example about its suitability for the regarded species (e.g. 'habitat') or the presence of the organisms to be studied (e.g. 'occupied by host population') (see also the grid-based sampling approach shown in Figure 8.2b, d). Within a cell, non-spatial processes such as reproduction can take place. Cells are interlinked via dispersal and this way the reproduction and spread of a local insect population can be depicted. Inspecting the landscape-level patterns emerging from such a model can help to scale up local insect dynamics to the landscape.

Visser *et al.* (2009) developed a grid-based host–parasitoid model based on the ecology of the rape pollen beetle *Meligethes aeneus* (Fabricius) and its specific parasitoids in semi-natural habitats. In fragmented landscapes, parasitoids have been found to go extinct before their hosts do, which suggests that species at different trophic levels experience a landscape differently (Kruess and Tscharntke, 1994; Tscharntke *et al.*, 2002). Parasitoids are often antagonists of important pest insects and therefore a good understanding of host–parasitoid systems in agricultural landscapes is of great interest to biological control.

One grid cell in the model represents a 100 m × 100 m area of an agricultural landscape which can be either suitable 'habitat' for the host (e.g. set-asides) or unsuitable 'matrix' (e.g. other crops, but not rape). Each cell can contain a subpopulation of host and parasitoid and is the site of the local processes reproduction, parasitism, and mortality. Local subpopulations are linked by dispersing host and parasitoid individuals. For model details see Visser *et al.* (2009).

Habitat fragmentation has been studied by varying the number, size of, and mutual distance between habitat patches in the virtual landscapes of the host–parasitoid model (Visser *et al.*, 2009). A habitat patch is defined as a continuous area of adjacent habitat cells. Across all scenarios, host–parasitoid dynamics in a given cell are oscillating in time (Figure 8.3). Generally, these local oscillations of host and parasitoid densities lead to a wave-like or chaotic spatial pattern (Plate 8.1) with increasing local host populations at the wave front, followed by increasing parasitoid populations (see also Hirzel *et al.*, 2007). These waves of hosts and parasitoids move across the landscape with time. As the parasitoid populations cause the local extinction of the host, they leave a zone of empty cells behind. Analyses across fragmentation scenarios show the following trends: (1) Parasitism rates decrease with

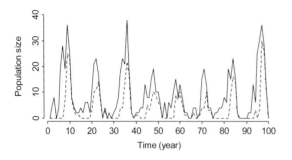

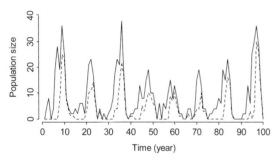

Figure 8.3 Population density of adult hosts (black line) and parasitoid larvae (dotted line) oscillating with time in one exemplary cell; simulation run with landscape parameters as visualised in Plate 8.1. Adapted from Visser *et al.*, 2009.

Box 8.2   Persistence of parasitoid populations and parasitism rate

Two measures are widely used to assess the performance of biological control: *persistence* (a measure of the parasitoid's reliability), and *parasitism rate*. The first measure is commonly used in theoretical studies and the latter in field studies.

Persistence of parasitoid populations and parasitism rate are often applied in theoretical and field studies, respectively. Each measure reveals important properties of biological control, namely reliability and effectiveness, respectively.

Visser *et al.* (2009) found that the amount of habitat in a landscape modulates the effect of fragmentation on parasitoid persistence. Parasitism rate, on the other hand, decreased with fragmentation regardless of the habitat amount in a landscape. Consequently, the effect of fragmentation and isolation on the performance of biological control as an ecosystem service hinges on whether the focus is on persistence or parasitism.

the number of patches and decrease with patch distance, and (2) host outbreak duration increases with the number of patches, and (3) parasitoid persistence is additionally modulated by habitat amount: if habitat is abundant, persistence decreases with the number of patches and with patch distance, if habitat is scarce persistence is highest at intermediate levels of fragmentation (Box 8.2; Visser *et al.*, 2009).

In summary, the amount of habitat in a landscape modulates the effect of fragmentation on parasitoid persistence. Parasitisation rates, on the other hand, decrease with fragmentation regardless of the habitat amount in a landscape. Consequently, the effect of fragmentation and isolation on the performance of biological control as an ecosystem service hinges on whether the focus is on persistence or parasitism rates.

Although the dispersal of both hosts and parasitoids is hindered by increasing fragmentation and isolation, this effect is much stronger for the parasitoid. This is because the parasitoid depends on a more ephemeral resource (host) than the host (habitat). With increasing fragmentation, the disadvantage of the parasitoid

increasingly leads to the decoupling of the host population from the control of the parasitoid, which results in prolonged host outbreak duration and decreased average parasitism. Thus, the modelling study by Visser *et al.* (2009) confirms the findings of several field studies that increasing fragmentation and isolation can decrease parasitism (Kruess and Tscharntke, 1994), increase prey outbreak duration (Kareiva, 1987) and reduce prey tracking at a certain scale (With *et al.*, 2002). It also reveals that the basic mechanism underlying their observations may be neither the difference in dispersal abilities of host and parasitoid (which were kept identical in the model) nor the predator searching behaviour interacting with landscape features (which was not incorporated in the model), but the decoupling of the population dynamics of pest and antagonist due to habitat structure.

The example of the host–parasitoid model illustrates that modelling can improve our understanding of complex systems beyond the possibilities of field studies. The model shows that landscape effects on biological control agents can be found without any significant differences in local dispersal abilities and even without any specific active response of the organisms to the landscape features. This was greatly facilitated by the fact that, within a model, properties such as dispersal ability and degree of interaction with landscape features can be changed while keeping all other properties constant.

CONCLUSIONS

Data collection, sampling design, tracking and marking techniques, statistics and modelling of data on a landscape scale can be challenging for the individual researcher. In this chapter, we have tried to cover the areas that we believe are most relevant for landscape-scale studies. As everywhere in science, innovation is often based on methodological or technological advancements. For example, landscape genetics would be unthinkable without the rapid developments in molecular biology. Likewise, analyses of landscape structure are greatly aided by advances in multiband satellite imagery and image processing and classification software. Finally, new types of sampling design, such as grid-based landscape-wide sampling, may provide new insights and opportunities for modelling. All in all, we think that there are several key steps that can be followed to make the most of an individual study:

(1) Start off with a small-scale study (for example with your favourite biological control agent and insect pest), and try to predict what might happen on larger spatial scales.

(2) Choose from selected marking and tracking techniques, and do preliminary studies in your type of landscape. Find out which spatial and temporal scales you can reasonably cover.

(3) Know your study organisms, their biology, life cycle and dispersal behaviour.

(4) Invest time in finding an appropriate sampling or experimental design. If your design is solid, your study will also be (provided you know your organisms). If you have too low replication, or block-by-treatment interactions, you can often not remedy this at the statistics stage.

(5) Use established, robust and well-documented statistical procedures for data analysis. This does not mean using 'canned' solutions, but it is important not to place too much faith in approaches that are still under development (such as generalised linear mixed models). Always graph your data before you start any analyses.

(6) Use the advantages of modelling and simulation techniques to derive predictions that extend across the scales of your study.

REFERENCES

Beckler, A.A., French, B.W. and Chandler, L.D. (2005) Using GIS in area-wide pest management: a case study in South Dakota. *Transactions in GIS*, 9, 109–127.

Behura, S.K. (2006) Molecular marker systems in insects: current trends and future avenues. *Molecular Ecology*, 15, 3087–3113.

Bianchi, F. and van der Werf, W. (2003) The effect of the area and configuration of hibernation sites on the control of aphids by *Coccinella septempunctata* (Coleoptera: Coccinellidae) in agricultural landscapes: a simulation study. *Environmental Entomology*, 32, 1290–1304.

Bianchi, F., Schellhorn, N.A. and van der Werf, W. (2009a) Foraging behaviour of predators in heterogeneous landscapes: the role of perceptual ability and diet breadth. *Oikos*, 118, 1363–1372.

Bianchi, F., Schellhorn, N.A. and van der Werf, W. (2009b) Predicting the time to colonization of the parasitoid *Diadegma semiclausum*: The importance of the shape of spatial dispersal kernels for biological control. *Biological Control*, 50, 267–274.

Billeter, R., Lllra, J., Bailey, D. *et al.* (2008) Indicators for biodiversity in agricultural landscapes: a pan-European study. *Journal of Applied Ecology*, 45, 141–150.

Blair, C.P., Abrahamson, W.G., Jackman, J.A. and Tyrrell, L. (2005) Cryptic speciation and host-race formation in a purportedly generalist tumbling flower beetle. *Evolution*, 59, 304–316.

Bolker, B.M., Brooks, M.E., Clark, C.J. *et al.* (2009) Generalized linear mixed models: a practical guide for ecology and evolution. *Trends in Ecology & Evolution*, 24, 127–135.

Calkins, C.O. and Faust, R.J. (2003) Overview of area-wide programs and the program for suppression of codling moth in the Western USA directed by the United States Department of Agriculture – Agricultural Research Service. *Pest Management Science*, 59, 601–604.

Carrière, Y., Dutilleul, P., Ellers-Kirk, C. *et al.* (2004) Sources, sinks, and the zone of influence of refuges for managing insect resistance to *Bt* crops. *Ecological Applications*, 14, 1615–1623.

Cochran, W.G. and Cox, G.M. (1992) *Experimental designs*. John Wiley & Sons, Ltd, Chichester.

Crawley, M.J. (2002) Power calculations, in *Statistical computing*, John Wiley & Sons, Ltd, Chichester, pp. 131–137.

Dale, M.R.T. (1999) *Spatial pattern analysis in plant ecology*, Cambridge University Press, Cambridge.

Debinski, D.M. and Holt, R.D. (2000) A survey and overview of habitat fragmentation experiments. *Conservation Biology*, 14, 342–355.

Dempster, J.P., Atkinson, D.A. and French, M.C. (1995) The spatial population dynamics of insects exploiting a patchy food resource. 2. Movements between patches. *Oecologia*, 104, 354–362.

Desouhant, E., Lucchetta, A.P., Giron, D. and Bernstein, C. (2010) Feeding activity pattern in a parasitic was when foraging in the field. *Ecological Research*, 25, 419–428.

Dormann, C.F. (2007) Effects of incorporating spatial autocorrelation into the analysis of species distribution data. *Global Ecology and Biogeography*, 16, 129–138.

Dormann, C.F., Schweiger, O., Augenstein, I. *et al.* (2007) Effects of landscape structure and land-use intensity on similarity of plant and animal communities. *Global Ecology and Biogeography*, 16, 774–787.

Dungan, J.L., Perry, J. N., Dale, M.R.T. *et al.* (2002) A balanced view of scale in spatial statistical analysis. *Ecography*, 25, 626–640.

Efron, D., Nestel, D. and Glazer, I. (2001) Spatial analysis of entomopathogenic nematodes and insect hosts in a citrus grove in a semi-arid region in Israel. *Environmental Entomology*, 30, 254–261.

Forbes, A.A., Powell, T.H.Q., Stelinski, L.L., Smith, J.J. and Feder, J.L. (2009) Sequential sympatric speciation across trophic levels. *Science*, 323, 776–779.

Fortin, M.J. and Dale, M. (2005) *Spatial analysis: a guide for ecologists*, Cambridge University Press, Cambridge.

Fraser, S.E.M., Dytham, C. and Mayhew, P.J. (2008) Patterns in the abundance and distribution of ichneumonid parasitoids within and across habitat patches. *Ecological Entomology*, 33, 473–483.

Gibson, D.J. (2002) *Methods in comparative plant population ecology*, Oxford University Press, Oxford.

Giovagnoli, A. and Sebastiani, P. (1989) Experimental designs for mean and variance estimation in variance components models. *Computational Statistics and Data Analysis*, 8, 21–28.

Girma, H., Rao, M. and Sithanantham, S. (2000) Insect pests and beneficial arthropods population under different hedgerow intercropping systems in semiarid Kenya. *Agroforestry Systems*, 50, 279–292.

Grimm, V., Revilla, E., Berger, U. *et al.* (2005) Pattern-oriented modelling of agent-based complex systems: lessons from ecology. *Science*, 310, 987–991.

Hagler, J. and Jackson, C. (2001) Methods for making insects: current techniques and future prospects. *Annual Review of Entomology*, 46, 511–543.

Hagler, J.R. and Jones, V.P. (2010) A protein-based approach to mark arthropods for mark-capture type research. *Entomologia Experimentalis et Applicata*, 135, 177–192.

Hartley, S., Kunin, W.E., Lennon, J.J. and Pocock, M.J.O. (2004) Coherence and discontinuity in the scaling of specie's distribution patterns. *Proceedings of the Royal Society of London B*, 271, 81–88.

He, F. and Hubbell, S.P. (2003) Percolation theory for the distribution and abundance of species. *Physical Review Letters*, 91, 198103 (epub, 4 pages).

Heimpel, G.E., Lee, J., Wu, Z., Weiser, L., Wäckers, F. and Jervis, M. (2004) Gut sugar analysis in field-caught parasitoids: adapting methods originally developed for biting flies. *International Journal of Pest Management*, 50, 193–198.

Heisswolf, A., Poethke, H.J. and Obermaier, E. (2006) Multitrophic influences on egg distribution in a specialized leaf beetle at multiple scales. *Basic and Applied Ecology*, 7, 565–576.

Henry, L.M., May, N., Acheampong, S., Gillespie, D.R. and Roitberg, B.D. (2010) Host-adapted parasitoids in biological control: Does source matter? *Ecological Applications*, 20, 242–250.

Hirano, S.S. and Upper, C.D. (2000) Bacteria in the leaf ecosystem with emphasis on *Pseudomonas syringae* – a pathogen, ice nucleus, and epiphyte. *Microbiology and Molecular Biology Reviews*, 64, 624–653.

Hirzel, A.H., Nisbet, R.M. and Murdoch, W.W. (2007) Host–parasitoid spatial dynamics in heterogeneous landscapes. *Oikos*, 116, 2082–2096.

Hood-Nowotny, R. and Knols, B.G.J. (2007) Stable isotope methods in biological and ecological studies of arthropods. *Entomologia Experimentalis et Applicata*, 124, 3–16.

Horton, D.R., Jones, V.P.and Unruh, T.R. (2009) Use of a new immunomarking method to assess movement by generalist predators between a cover crop and tree canopy in a pear orchard. *American Entomologist*, 55, 49–56.

Hurlbert, S.H. (1984) Pseudoreplication and the design of ecological field experiments. *Ecological Monographs*, 54, 187–211.

Hutchinson, M.F. (1991) The application of thin plate smoothing splines to continent-wide data assimilation, in *Data assimilation systems* (ed. J.D. Jasper), BMRC Research Report No. 27., Bureau of Meteorology, Melbourne, pp. 104–113.

Jarvis, C.H. and Collier, R.H. (2002) Evaluating an interpolation approach for modelling spatial variability in pest development. *Bulletin of Entomological Research*, 92, 219–231.

Jones, V.P., Hagler, J.R., Brunner, J.F., Baker, C.C. and Wilburn, T.D. (2006) An inexpensive immunomarking technique for studying movement patterns of naturally occurring insect populations. *Environmental Entomology*, 35, 827–836.

Kareiva, P. (1987) Habitat fragmentation and the stability of predator–prey interaction. *Nature*, 326, 388–390.

Kildea, S., Ransbotyn, V., Khan, M.R., Fagan, B., Leonard, G., Mullins, E. and Doohan, F.M. (2008) *Bacillus megaterium* shows potential for the biocontrol of *Septoria tritici* blotch of wheat. *Biological Control*, 47, 37–45.

Kinkel, L.L. (1997) Microbial population dynamics on leaves. *Annual Review of Phytopathology*, 35, 327–47.

Kinkel, L.L., Wilson, M. and Lindow, S.E. (1995) Effect of sampling scale on the assessment of epiphytic bacterial populations. *Microbial Ecology*, 29, 283–297.

Kleijn, D., Tscharntke, T., Steffan-Dewenter, I. *et al.* (2006) Mixed biodiversity benefits of agri-environment schemes in five European countries. *Ecology Letters*, 9, 243–254.

Krebs, C.J. (1999) *Ecological Methodology*, 2nd edn, Addison Wesley Longman, Inc., Menlo Park, California.

Kruess, A. and Tscharntke, T. (1994) Habitat fragmentation, species loss, and biological control. *Science*, 264, 1581–1584.

Lavandero, B., Wratten, S.D., Shishehbor, P. and Worner, S. (2005) Enhancing the effectiveness of Diadegma semiclausum (Helen): quantifying movement after use of nectar in the field. *Biological Control*, 34, 152–158.

Lavandero, B., Miranda, M., Ramirez, C.C. and Fuentes-Contreras, E. (2009) Landscape composition modulates population genetic structure of *Eriosoma lanigerum* (Hausmann) on *Malus domestica* Borkh in Central Chile. *Bulletin of Entomological Research*, 99, 97–105.

Legaspi, J.C. and Legaspi, B.C. (2007) Bioclimatic model of the spined soldier bug (Heteroptera: Pentatomidae) using CLIMEX: testing model predictions at two spatial scales. *Journal of Entomological Sciences*, 42, 533–547.

Levin, S.A. (2000) Multiple scales and the maintenance of biodiversity. *Ecosystems*, 3, 498–506.

Lill, J.T. (1998) Density-dependent parasitism of the Hackberry Nipplegall Maker (Homoptera: Psyllidae): a multiscale analysis. *Environmental Entomology*, 27, 657–661.

Lindenmayer, D.B., Cunningham, R.B. and Pope, M.L. (1999) A large-scale 'experiment' to examine the effects of landscape context and habitat fragmentation on mammals. *Biological Conservation*, 88, 387–403.

Matsumoto, T., Itioka, T. and Nishida, T. (2004) Is spatial density-dependent parasitism necessary for successful biological control? Testing a stable host–parasitoid system. *Entomologia Experimentalis et Applicata*, 110, 191–200.

McCulloch, C.E. and Searle, S.R. (2001) *Generalized, linear, and mixed models*, John Wiley & Sons, Inc., New York.

Meyer, K.M., Schiffers, K. and Muenkemueller, T. *et al.* (2010) Predicting population and community dynamics – the type of aggregation matters. *Basic and Applied Ecology*, 11, 563–571.

Park, Y.L., Perring, T.M., Farrar, C.A. and Gispert, C. (2006) Spatial and temporal distributions of two sympatric *Homalodisca* spp. (Hemiptera : Cicadellidae): implications for area-wide pest management. *Agriculture Ecosystems and Environment*, 113, 168–174.

Pearson, S.M. (2002) Landscape context, in *Learning landscape ecology* (eds S.E. Gergel and M.G. Turner), Springer, New York, pp. 199–207.

Pearson, R.G. and Dawson, T.P. (2003) Predicting the impacts of climate change on the distribution of species: are bioclimatic envelope models useful? *Global Ecology and Biogeography*, 12, 361–371.

Platt, J.R. (1964) Strong Inference. *Science*, 146, 347–35.

Prasifka, J.R. and Heinz, K.M. (2004) The use of C_3 and C_4 plants to study natural enemy movement and ecology, and its application to pest management. *International Journal of Pest Management*, 50, 177–181.

Prasifka, J.R., Heinz, K.M. and Sansone, C.G. (2004a) Timing, magnitude, rates, and putative causes of predator movement between cotton and grain sorghum fields. *Environmental Entomology*, 33, 282–290.

Prasifka, J.R., Heinz, K.M. and Winemiller, K.O. (2004b) Crop colonisation, feeding, and reproduction by the predatory beetle, *Hippodamia convergens*, as indicated by stable carbon isotope analysis. *Ecological Entomology*, 29, 226–233.

Roland, J. and Taylor, P.D. (1997) Insect parasitoid species respond to forest structure at different spatial scales. *Nature*, 386, 710–713.

Scarratt, S.L., Wratten, S.D. and Shishehbor, P. (2008) Measuring parasitoid movement from floral resources in a vineyard. *Biological Control*, 46, 107–113.

Schellhorn, N.A., Siekmann, G., Paull, C., Furness, G. and Baker, G. (2004) The use of dyes to mark populations of beneficial insects in the field. *International Journal of Pest Management*, 50, 153–159.

Silberbauer, L., Yee, M., Del Socorro, A., Wratten, S., Gregg, P. and Bowie, M (2004) Pollen grains as markers to track the movements of generalist predatory insects in agrocosystems. *International Journal of Pest Management*, 50, 165–171.

Starfield, A.M., Smith, K.A. and Bleloch, A.L. (1994) *How to model it – Problem solving for the computer age*, Burgess International Group Inc., Edina.

Stireman, J.O., Nason, J.D., Heard, S.B. and Seehawer, J.M. (2006) Cascading host-associated genetic differentiation in parasitoids of phytophagous insects. *Proceedings of the Royal Society B*, 273, 523–530.

Storfer, A., Murphy, M.A., Evans, J.S. *et al.* (2007) Putting the 'landscape' in landscape genetics. *Heredity*, 98, 128–142.

Teodoro, A.V., Tscharntke, T. and Klein, A.M. (2009) From the laboratory to the field: contrasting effects of multi-trophic interactions and agroforestry management on coffee pest densities. *Entomologia Experimentalis et Applicata*, 131, 121–129.

Thies, C. and Tscharntke, T. (1999) Landscape structure and biological control in agroecosystems. *Science*, 285, 893–895.

Tscharntke, T., Steffan-Dewenter, I., Kruess, A. and Thies, C. (2002) Contribution of small habitat fragments to conservation of insect communities of grassland-cropland landscapes. *Ecological Applications*, 12, 354–363.

Turner, M.G. (2005) Landscape ecology: What is the state of the science? *Annual Review of Ecology, Evolution, and Systematics*, 36, 319–344.

Visser, U., Wiegand, K., Grimm, V. and Johst, K. (2009) Conservation biocontrol in fragmented landscapes: persistence and paratisation in a host–parasitoid model. *Open Ecology Journal*, 2, 52–61.

von Ende, C.N. (2001) Repeated-measures analysis: growth and other time-dependent measures, in *The design and analysis of ecological experiments* (eds S. Scheiner and I. Gurevitch), Oxford University Press, New York. pp. 134–157.

Wäckers, F. (2007) Using HPLC sugar analysis to study nectar and honeydew feeding in the field. *Journal of Insect Science*, 7, 23–23.

Wanner, H., Gu, H.N., Gunther, D., Hein, S. and Dorn, S. (2006) Tracing spatial distribution of parasitism in fields with flowering plant strips using stable isotope marking. *Biological Control*, 39, 240–247.

Wiens, J.A. (1989) Spatial scaling in ecology. *Functional Ecology*, 3, 385–398.

Winkler, K., Wäckers, F.L. and Kaufman, L. (2009) Nectar exploitation by herbivores and their parasitoids is a function of flower species and relative humidity. *Biological Control*, 50, 299–306.

Wissel, C. (1989) *Theoretische ökologie – eine einführung* (Theoretical ecology – an introduction), Springer, Berlin.

With, K.A., Pavuk, D.M., Worchuck, J.L., Oates, R.K. and Fisher, J.L. (2002) Threshold effects of landscape structure on biological control in agroecosystems. *Ecological Applications*, 12, 52–65.

Xia, J.Y., Rabbinge, R. and van der Werf, W. (2003) Multistage functional responses in a ladybeetle–aphid system: scaling up from the laboratory to the field. *Environmental Entomology*, 32, 151–162.

Zhang, Z.-Q. and Anderson, J.P. (1993) Spatial scale of aggregation in three acarine predator species with different degrees of polyphagy. *Oecologia*, 96, 24–31.

Zhang, Z.-Q. and Anderson, J.P. (1997) Patterns, mechanisms and spatial scale of aggregation in generalist and specialist predatory mites (Acari: Phytoseiidae). *Experimental and Applied Acarology*, 21, 393–404.

Zuur, A.F., Ieno, E.N., Walker, N.J., Saveliev, A.A. and Smith, G.M. (2009) *Mixed effects models and extensions in ecology with R*, Springer, New York.

Chapter 9

PICK AND MIX: SELECTING FLOWERING PLANTS TO MEET THE REQUIREMENTS OF TARGET BIOLOGICAL CONTROL INSECTS

Felix L. Wäckers and Paul C.J. van Rijn

Biodiversity and Insect Pests: Key Issues for Sustainable Management, First Edition. Edited by Geoff M. Gurr, Steve D. Wratten, William E. Snyder, Donna M.Y. Read.

INTRODUCTION

Many pollinators and entomophagous arthropods rely on floral food (pollen and nectar) at some point during their life cycle (Baggen *et al.*, 1999; Pontin *et al.*, 2006; Wäckers, 2005). The lack of floral resources in modern intensified agricultural systems has long been suspected to be an important bottleneck for natural pest control and pollination (Illingworth, 1921; van Emden, 1962; Hagen, 1986; Biesmeijer *et al.*, 2006). In conservation biological control, diversification of the agroecosystem with flowering vegetation is seen as an important tool to support the broad range of predators and parasitoids that require nectar and pollen sources to survive and reproduce. However, direct and quantitative evidence for the impact of such landscape management approaches has been scarce (Heimpel and Jervis, 2005). New biochemical techniques allow us to analyse the gut content of field collected insects (Heimpel *et al.*, 2004). These methods not only quantify the nutritional state of even the smallest individual predator or parasitoid in the field, but also provide information on their feeding history and food source use (Wäckers and Steppuhn, 2003). This methodology has generated the first proof that insects in agricultural landscapes lacking floral resources can be severely food-deprived (Olsen and Wäckers, 2007; Winkler *et al.*, 2009a). It has also proven to be an effective tool to quantify the impact of landscape management strategies on the nutritional state of predators and parasitoids in the field. This provides sound data for the optimisation of conservation biological control programmes (Box 9.1). This shows that enhancing beneficial arthropods through diversification of agroecosystems is not a function of increased botanical diversity per se, but depends on the selection of the 'right' flowering plants. We know from pollination ecology that plant–pollinator interactions can be often highly specific and that plants have evolved many mechanisms through which they can exclude visitors other than the intended specialised pollinators (Waser *et al.*, 1996). Specificity of flower exploitation can be based on the apparency and accessibility of flowers, as well as the nectar/pollen composition (Wäckers, 2005). It is thus not surprising that groups of entomophagous arthropods may differ in the range of nectar and pollen sources they can exploit. The impact of floral resources on biological control can be optimised by selecting those flowers whose availability, appearance, accessibility and chemical com-

> ### Box 9.1 The 'right kind' of diversity
>
> Olsen and Wäckers (2007) demonstrated that *Meteorus autographae* parasitoids collected from cotton fields bordered by pure stands of cahaba white vetch (*Vicia sativa* × *Vicia cordata* L.), a species selected on the basis of its suitability in providing nectar, had threefold higher energy reserves as compared to unfed individuals. In sharp contrast, *M. autographae* collected from cotton fields bordered by botanically diverse bird conservation margins showed no elevation in energy levels and were actually starving. This shows that enhancing beneficial arthropods through diversification of agroecosystems is not a function of increased botanical diversity per se, but depends on the selection of the 'right' flowering plants.

position matches the behaviour, morphology and physiology of target organisms (Wäckers, 2005; Fiedler *et al.*, 2008).

While recent studies have increased insight into the suitability of flowering plants for entomophagous arthropods, seed mixes for conservation biological control programmes have long been selected more or less arbitrarily (Gurr *et al.*, 2005). These shotgun approaches have been 'hit and miss' in terms of their effectiveness in supporting beneficial arthropods (Andow, 1991). An uninformed choice of non-crop vegetation not only means missing out on potential benefits but may also actually generate negative effects. Figure 9.1 illustrates the complexities inherent in the notion that floral resources might benefit biological control; achieving this outcome depends on avoiding a series of potential negative effects. It is, therefore, no surprise that arbitrarily composed floral vegetation can increase pest populations (Baggen and Gurr, 1998; Wäckers *et al.*, 2007; Winkler *et al.*, 2010) and populations of higher trophic level organisms (Araj *et al.*, 2008), while these structures can also serve as a sink, attracting beneficial species away from the crop (Dunning *et al.*, 1992).

The selection of floral vegetation to maximise ecosystem services, such as biological control and pollination, requires an understanding of the biology and

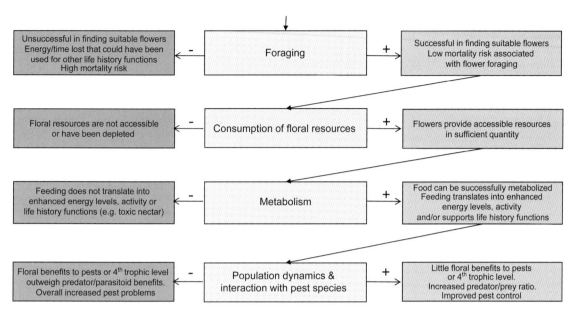

Figure 9.1 The interactions with and at floral resources can have positive as well as negative effects on predators and parasitoids at various levels. At each level examples of positive and negative impacts are presented.

ecology of the beneficial species delivering these services. Based on existing insights in insect–plant interactions, specific floral seed mixtures can be developed which target specific pollinators and/or flower-visiting biocontrol agents (Wäckers *et al.*, 1996; Pontin *et al.*, 2006; Van Rijn and Wäckers, 2010). This habitat management strategy has been demonstrated to be effective in enhancing the nutritional state of targeted beneficial insect groups (Olsen and Wäckers, 2007; Lee *et al.*, 2006), as well as their local abundance (Baggen *et al.*, 1999; Vattala *et al.*, 2006; Pontin *et al.*, 2006; Carvell *et al.*, 2007).

The first part of this chapter gives an overview of the various methods that have been used to study the exploitation of specific flowering plants by parasitoids and predators, while discussing the respective advantages and drawbacks of the various methods. In the second part, the specific floral requirements of different groups of biological control agents are discussed. The focus will be on parasitoids, syrphids, lacewings, predatory bugs and ladybeetles, representing important taxa of biological control agents. The level of specificity that emerges from this comparison underscores the importance of selecting the right floral resources to optimise the biological control services of agroecosystems.

METHODS OF STUDYING FLOWER EXPLOITATION

Previously, Jervis *et al.* (2004) reviewed approaches used to study the effect of habitat manipulation on parasitoids and biological control. Here this review is extended to include methods used to study various predators, with the focus on methods to select flowering species for use in such habitat manipulation programmes. Various methods have been employed to assess the suitability of individual flowering plant species as insect food sources (Table 9.1). These can be divided into five main categories:
• Recordings of flower choice
• Morphometric studies (i.e. measurements of flower and arthropod mouthpart morphology to assess whether the mouthparts of the arthropods would theoretically allow access to nectaries and anthers)
• Establishing consumption, either by assessing weight increase and/or fitness benefits in laboratory studies or through (palynological or biochemical) analyses of field collected individuals
• Study of nutritional suitability through lifetable/performance experiments
• Impact of mono-flower plots on insect populations in the field

Table 9.1 Compilation of studies showing the range of methods employed to assess suitability of flowering plant species as insect food sources.

Type of study	Methodology	Field/Lab	Establishes			References
			Attractiveness	**Nectar/pollen accessibility/consumption**	**Nutritional suitability**	
Recording of flower choice	Olfactometer experiments	Lab	Possible given adequate choice experiments (only olfactory response)	–	–	Wäckers and Swaans, 1993; Wäckers, 2004
	Flower visits in flight cage	Lab	Possible given adequate choice experiments (olfactory as well as visual response)	Only if observations of nectar or pollen feeding are included	–	Patt et al., 1997; Begum et al., 2004; Van Rijn and Wäckers, 2010
	Direct observations	Field	Often implied but not established	Only if observations of nectar or pollen feeding are included	–	Gilbert, 1981; Weiss and Stettmer,1991; Ambrosino et al., 2006
	Sampling flowers (for small arthropods)	Field/Lab	Often implied but not established	Only if observations of nectar or pollen feeding are included	– Can be implied when juveniles are found	Silveira et al., 2005; Fiedler and Landis, 2007a; Bosco and Tavella, 2008; Atakan and Tunc, 2010
Morphometric studies	Measurements of flower morphology and insect mouthparts	Lab	–	theoretical	–	Gilbert, 1985; Winkler et al., 2009b; van Rijn and Wäckers, 2010
Establishing consumption	Weighing by microbalance	Lab/semi-field	–	+ (quantitative) (non-destructive)	If combined with survival experiment	Wäckers et al., 1996; Wäckers, 2004; Winkler et al., 2009b
Analysis of field-collected individuals	Sugar analysis	Field/Lab	– May be implied but not established	+ (quantitative and to some extent qualitative on basis of sugar profiles)	–	Wäckers and Steppuhn, 2003; Heimpel et al., 2004; Lee et al., 2006
	Pollen analysis	Field/Lab	– May be implied but not established	+ (quantitative and qualitative)	–	Jones and Rowe, 1999; Golding and Edmunds, 2003; Silberbauer et al., 2003; Villenave et al., 2006; Davidson and Evans, 2010

		Setting				References
	Plant secondary metabolites	Field/Lab	– May be implied but not established	+ (quantitative and qualitative)	–	Ferreres et al., 1996
	DNA analysis	Field/Lab	– May be implied but not established	+ (quantitative and qualitative)	–	Weber and Lundgren, 2011; Wilson et al., 2010
Marking of nectar or pollen	Stable isotope marking	Field/Lab	– May be implied but not established	+ (quantitative and qualitative)	–	Gu et al., 2001; Patt et al., 2003; Scarratt et al., 2008; Wanner et al., 2006
	Protein marking	Field/Lab	– May be implied but not established	+ (quantitative and qualitative)	–	DeGrandi-Hoffman and Hagler, 2000
Life history/ Performance studies	Development of juveniles provided with floral resources	Lab	–	+	+	Vacante et al., 1997; Patt et al., 2003; Lundgren and Wiedenmann, 2004
	Survival (longevity) of individuals provided with floral resources	Lab/semi-field	–	+ (if survival is prolonged compared to control)	+ (see text for pitfalls)	Wäckers, 2004; Baggen and Gurr, 1998; Begum et al., 2006; Irvin and Hoddle, 2007; Winkler et al., 2009b; Van Rijn and Wäckers, 2010
	Fecundity	Lab/semi-field	–	+	+	Van Rijn et al., 2006; Venzon et al., 2006; Winkler et al., 2006; Bertolaccini et al., 2008
	Flight capacity following flower/ food exposure	Lab	–	+	+ (convertibility of sugar/ nectar to flight energy)	Hausmann et al., 2005; Wanner et al., 2006
Impact of mono-flower plots on insect field populations	Comparison of population responses to mono-flower plots	Field	– May be implied but not established	– May be implied but not established		Pemberton and Lee, 1996; Chaney, 1998

These methods differ in terms of required input (labour and equipment) as well as the informational output they provide (Table 9.1). Below the different methodologies are described in more detail and their advantages and limitations are discussed.

Recordings of flower choice

Flower choice can be addressed in a number of laboratory, semi-field or field experiments. Laboratory experiments include olfactometer studies which establish responses to flower odour, as well as flight chamber and cage experiments. Cages can also be used in semi-field experiments, whereas field experiments typically record and count arthropods on flowers in their (semi-) natural environment.

Olfactometer studies assess responses to flower odours. This is achieved by giving an organism a choice between two or multiple air plumes, which should be of equal size and have equal laminar flow and humidity. Humidifying the air improves odour detection. Based on the number of air plumes, we distinguish for instance between two-arm (Y-tube), four-arm, or eight-arm olfactometers. At least one arm should feature a control (clean air). Air pumps, in combination with flow-meters, should be used to attune airflow. The insect should be introduced downwind from the place where the odour plumes join. At this point it should be able to perceive the odour from both sources without too much mixing, which can be checked with smoke. Above-average choice for an odour alternative indicates olfactory attraction. Above-average choice for the control can indicate odour repellency. Most olfactometers require that the insect tested moves upwind in response to an attractive odour. Sometimes this requires specific adjustments. For *Orius* flower bugs for instance, the olfactometer has to be positioned vertically (Venzon *et al.*, 1999). While arthropods commonly exhibit upwind orientation in response to an attractive odour source, it has to be considered that organisms may also orient to odour gradients or show an arrestment in response to an odour. Most olfactometers are not designed to assess the latter two responses.

A second type of laboratory set-up is the use of flight chambers or flight cages where the insects can respond to chemical, as well as visual, flower stimuli. Visual stimuli are likely to be affected by the light conditions chosen. Artificial light, plastic, plexiglass and glass modify the light spectrum and intensity, which may affect how flowers are perceived by the insect. To establish flower preference the use of an appropriate choice set-up is required, where factors other than the flower species tested are excluded or standardised as much as possible. For instance, it would be advisable to standardise the biomass of the flowers used in choice trials to ensure that increased responses are not simply a reflection of an increased chance of randomly landing on a larger flower. Another aspect to consider is that the behavioural response of an insect is the result of the interplay between its innate preference and previous experience (associative learning). Feeding on a particular flower will not only affect the response to this particular flower species, but also to other flowers. Interference from prior feeding experience can be excluded when unfed insects from laboratory rearing are used in the tests. If individuals need to be fed, it is important to ensure that their nutritional state is standardised, given that this factor is crucial to the expression of behavioural responses to flower stimuli (Wäckers, 1994).

Results of flower choice experiments are often interpreted in terms of flowers being 'attractive' (i.e. being able to draw in insects from a distance). While this attractiveness is usually presented as an absolute quality, bioassays frequently assess relative responses. When giving an insect the choice between flowers, the one receiving most visits may be attractive, or the alternatives may be repellent. Actual attraction can be established in choice experiments in which flowers are offered against an appropriate control, such as humidified clean air.

Field experiments on larger mobile species typically record the number of visits to particular flowering species within a predefined time frame (Kevan, 1973; Jervis *et al.*, 1993; Winkler *et al.*, 2005). Such recordings will be more difficult for smaller, less mobile species such as flower bugs and small parasitoids as they tend to remain on a given flower/plant for much longer periods. For these insects, as well as for non-flying arthropods such as mites, flower choice can be studied by assessing their distribution among flower species. The number of insects present per flower can be determined by vacuuming the flowers (Fiedler and Landis, 2007a; 2007b), by tapping flowers onto a white surface and subsequently collecting the insects with an aspirator (Bosco and Tavella, 2008), or by collecting the flowers in bags and extracting the arthropods in the lab (van Rijn, unpublished results). Generally the numbers are expressed per flower or per unit of soil surface area (m²).

Interpreting these field recordings is often difficult, as floral visits depend largely on the relative abundance of the flower species tested, the background vegetation, and the presence of arthropod competitors. Moreover, in the field it usually remains unknown whether an observed response to a flower represents an innate attraction or is the result of associative learning during previous successful or unsuccessful feeding events. Rewarding feeding experiences can change innate repellency or neutral responses to attraction, whereas lack of reinforcement may extinguish a response to stimuli that are innately attractive. Furthermore, aggregation at particular plant species is not necessarily proof of attraction, as it may also be a result of random flower visits followed by arrestment after successful feeding. Spatial, rather than olfactory or visual learning may enhance this aggregation, as it enables insects to return to rewarding food sites (Wäckers et al., 2007). These mechanisms can underlie variations in responses to particular flowers between different studies and at different sites.

A final point to be considered in any study on nectar or pollen foraging is the fact that insects may be visiting flowers for resources other than nectar or pollen. These may include shelter, mates, prey or heat, as flowers may reflect/concentrate infrared radiation from the sun or even produce heat themselves through chemical reactions. Actual observations of nectar or pollen feeding can help eliminate these confounding factors.

Morphometric studies

Flower morphology has been recognised as an important factor limiting floral nectar use (Faegri and van der Pijl, 1979; Jervis et al., 1993). Floral nectaries and the pollen-bearing anthers are often hidden within complex floral structures. Accessing these floral resources requires that the (mouthpart) morphology and the behaviour of the flower visitor match the floral architecture (Patt et al., 1997; Jervis, 1998). Most hymenopteran parasitoids (Jervis et al., 1993; Gilbert and Jervis, 1998) and many predators have short mouthparts that largely restrict their feeding to exposed nectar and pollen. For these species head width is the limiting factor hampering nectar exploitation from deep, narrow flower corollas. Small parasitoids and predators may not be restricted by flower morphology, but complex floral structures might nevertheless prevent them from finding nectar and pollen (Patt et al., 1997).

In morphometric studies the length of the mouth-parts and head width of the insect are measured, as well as the depth and width of the corolla (Vattala et al., 2006; Winkler et al., 2009b; van Rijn and Wäckers, 2010). The corolla depth is measured from the most proximal point of the corolla to the location of the nectaries or anthers. By combining these measurements it is possible to identify whether nectar or pollen from a particular flower could potentially be reached by a particular insect.

While this method is attractive due to the fact that it requires little more than a microscope, its drawback lies in the fact that it determines potential accessibility rather than establishing actual access. A number of issues are not accounted for. These include the fact that nectar or pollen foragers might be thwarted by complex flower morphology (Patt et al., 1997), nectar viscosity (Winkler et al., 2009b) or competition by other flower visitors (see Box 9.2). Finally, consumption may be inhibited by particular flower odours (Wäckers, 2004) or through nectar/pollen chemistry (Feinsinger and Swarm, 1978). Behavioural recording of insects foraging on real or artificial flowers can be used to test the predictions on nectar and pollen accessibility (Patt et al., 1997).

On the other hand, flower visitors may be able to exploit floral resources that appear inaccessible on the basis of morphological measurements. This applies when nectar or pollen has been spilled from the nectar/stamen either by previous visitors or due to physical disturbance (wind) or simply by pollen dropping during flower maturation. Also, in a number of flowers the corolla opening is flexible. In the Fabaceae, for example, the wing and keel petals can be pushed apart to create a widened access to the floral resources. Here the strength and weight of the flower visitor as much as its size can determine foraging success. Some insects are also known to circumvent the floral structures and use their biting mouthparts to gnaw their way to the hidden resources (Inouye, 1983; Idris and Grafius, 1997).

Establishing consumption

To test whether floral resources are actually consumed a number of methods are available (Table 9.1). The approaches include gravimetrical methods, as well as methods identifying food source markers in the arthropods. Depending on the method chosen quantitative and/or qualitative information on food uptake is obtained.

Box 9.2 Competition for floral resources

Under field conditions a broad range of insects compete for the often limited nectar and pollen resources. Insects can compete either through physical interactions (interference competition) (Kikuchi, 1963; Morse, 1981; Beattie, 1985; Blüthgen et al., 2000) or through depletion of the resource (exploitative competition) (Comba et al., 1999; Hansen et al., 2002).

Interference competition occurs when flower visitors monopolise floral resources through aggressive behaviour towards other flower visitors, or simply because they are bigger and avoided by the other species. Kikuchi (1963) established dominance rankings between flower visitors, showing that bumblebees chase away syrphids, which in turn outcompete butterflies. Even among syrphid species interference competition may occur. Ambrosino et al. (2006) observed that where the Eristalis spp. were numerous they displaced the smaller zoophagous species from their preferred Coriandrum flowers.

Exploitative competition is based on depletion of the nectar or pollen resources by the earlier flower

visitor (first come, first served). In this way even bigger or more aggressive insects can suffer from the actions of smaller species when foraging on the same flowers (Reader et al., 2005).

Competition for nectar is likely to be most fierce at accessible nectaries, as these can be exploited by a large pool of nectar feeders. Chances for direct interference competition to occur are further increased by the fact that exposed nectar tends to be more viscous, increasing intake time.

Resource partitioning is a way to avoid or alleviate competition. Deep and narrow flowers cater for a more restricted subset of nectar feeders and may thus provide a nectar refuge for specialised pollinators (Comba et al., 1999). An example of temporal avoidance of competition is provided by Morse (1981), who showed that the hoverfly Melanostoma mellinum (L.) performs its highest foraging activity early in the morning when bumblebee activity is still very low.

Microbalances can be used on individual arthropods to establish weight increase following flower visits (Wäckers et al., 1996; Winkler et al., 2009b). Such a weight increase can indicate successful access to and consumption of floral nectar or pollen. In those species that do not feed on pollen (e.g. most parasitoids), any weight increase can be attributed to nectar feeding. Before being tested, individuals should be kept with water only to ensure that they are food deprived and motivated to search for food. Any external contamination with pollen should be excluded by assessing the insect under a loupe or microscope for external pollen deposition before the second weight assessment. The established weight increase provides a quantitative yet conservative measure of food consumption, as control individuals exposed to non-flowering plants typically lose weight during the exposure. It is important to ensure that test plants are not contaminated with honeydew or plant sap (exuding from damage), as this could invalidate conclusions. To ensure nectar availability flowers can be caged or covered with nylon bags for a few hours before releasing the test arthropod.

Another way of proving consumption is based on the detection of food-specific markers in the arthropod.

The advantage of these methods is that they can be used to study food consumption under various field conditions. Various markers can be used, including the food itself (specifically pollen grains), naturally occurring food compounds, such as specific carbohydrates, secondary metabolites, or DNA. The use of carbohydrate analyses to identify and quantify feeding on nectar and other sugar sources has been covered in previous reviews (Heimpel et al., 2004; Steppuhn and Wäckers, 2004; Wäckers et al., 2006a). Markers may also be added to the food plant in the form of stable isotopes or protein (Table 9.1). While these methods can be powerful tools to establish whether a particular nectar or pollen source is consumed, not all generate quantitative data. Quantitative data on pollen feeding can be derived from identifying the pollen grains found in the gut (e.g. Villenave et al., 2006) or in the frass (Davidson and Evans, 2010). The latter, non-destructive method may also be interesting for educational projects (Golding and Edmunds, 2003).

When using these methods under non-standardised field conditions, the possibility of contamination has to be considered. Contamination can occur when omnivorous predators consume prey that has been feeding on pollen or nectar. Arthropods may also be exposed to markers through feeding on plant-derived material other than floral nectar and pollen, such as plant sap, fruits or honeydew (Wäckers, 2005). Various types of pollen grains might be stuck in nectar or honeydew and be consumed inadvertently.

Life history/performance studies

The best proof that floral food resources benefit a biological control agent is by showing an effect on its lifetable parameters (development, longevity, fecundity) or its (flight) activity. These parameters are typically studied in laboratory bioassays in which caged insects are provided with a particular flower species, or with nectar/pollen. Many landscape management projects employing floral resources base their selection of flowering plants on such laboratory studies.

Longevity (or survival) is the parameter most commonly studied as the experiments require generally little more than the insect and the floral resource. However, for long-lived arthropods, assessing longevity can require long-term commitment. Longevity is strongly affected by temperature. Within the range of temperatures acceptable for a given organism, a temperature increase/decrease of $10\,°C$ will roughly increase/decrease longevity by a factor 2. This can be used to extend survival in longevity studies for short-lived organisms, or reduce longevity for those that are long-lived.

Impact of nectar or pollen on fecundity can be seen through an increase in daily fecundity, oviposition time, and/or lifetime fecundity (Winkler et al., 2006). Studying fecundity in zoophagous insects often requires a supply of suitable prey or hosts. With a few exceptions, parasitoids only oviposit in or on suitable hosts, but also aphidophagous hoverflies and lady beetles require (cues from) their aphid prey (especially honeydew) to trigger oviposition (Scholz and Poehling, 2000). For flowerbugs, who insert their eggs in green plant tissue, a suitable oviposition substrate is essential.

To assess if juvenile development is affected by pollen or nectar, a supply of prey is required in the positive control (Patt et al., 2003). In some cases the commercially produced eggs of Ephestia can be a suitable substitute.

Another way to show that floral resources are utilised by the insect is by assessing its flight capacity following flower exposure. Flight capacity can be assessed in simple flight bioassays measuring the time it takes for flight ability to be restored in food deprived individuals (Hausmann et al., 2005) or on flight mills assessing flight duration (Wanner et al., 2006).

When using actual flowering plants, enhanced performance indicates that a food source is both accessible and nutritionally suitable. On the other hand, in cases of poor performance it usually remains unclear whether this is due to limited food intake or poor nutritional value. When the interest lies primarily with the nutritional suitability of floral resources, insects can be tested with collected nectar or pollen. When collected nectar and pollen are not used directly, they should be stored frozen. Also, the way the pollen is treated affects its quality as a food source (van Rijn and Tanigoshi, 1999). When testing nectar, it is advisable to offer this at high relative humidity (RH), to avoid limited consumption due to high viscosity. When testing pollen, high humidity should be avoided as this can lead to deterioration of pollen quality (van Rijn and Tanigoshi, 1999), while free water should always be available to compensate for the low water content of pollen (Michaud and Grant, 2005). Bee-collected pollen is nutritionally different from fresh pollen, due to nectar and enzymes being added by the bees (Human and Nicolson, 2006). For arthropods that feed on the content of individual pollen grains feeding success can be limited not only by nutritional content but also by pollen wall morphology (Ouyang et al., 1992). Conclusions on nutritional suitability may also be confounded when a food source fails to stimulate feeding or is actually repellent. One way to control for this is to measure both food intake and survival. Wäckers (2001) determined the quantity of sugar consumption during a single feeding bout and subsequently assessed the effect of this known quantity on parasitoid survival. Nutritionally suitable sugars such as sucrose, fructose and melibiose showed a positive correlation between amount of sugar consumed and longevity with R^2 values between 0.7 and 0.81, whereas unsuitable sugars such as rhamnose, lactose, galactose and raffinose showed no such correlation.

Lifetable or performance studies should ideally include two control treatments: a water-only treatment, as a negative control, and a known suitable food (e.g. sucrose for nectar studies) as a positive control. When a treatment results in poorer performance relative to the negative control, this could indicate that the food source is toxic (Wäckers, 2001). Some foraging studies also show that certain flowers may have a negative impact on both insect energy levels and survival (Winkler *et al.*, 2009b). This can occur when flowers stimulate nectar foraging without providing accessible nectar or pollen (Wäckers, 2004).

Data from laboratory survival studies are unlikely to be fully representative of the effect of food under natural conditions. On the one hand, they may underestimate the impact of nectar feeding as most experiments are conducted with caged individuals that are restricted in their mobility and are presented with *ad libitum* oviposition sites and food. Under natural conditions, mobile arthropods may need to cover long distances searching for oviposition and foraging sites. As a result they are likely to use considerably more energy, which would increase the need for adult feeding and enhance its impact on longevity and fecundity (Steppuhn and Wäckers, 2004). On the other hand, laboratory studies may also overestimate fitness benefits of feeding. Arthropods in the field are subject to a range of biotic and abiotic mortality factors that may limit fitness irrespective of feeding.

Impact of mono-flower plots on insect field populations

A final method to assess the impact of particular flower species on predators/parasitoids are field studies comparing the impact of plots with individual flowering species on the populations of the biological control agents or the pests they control. The appeal of such field studies lies in the fact that they are closest to the actual objective of using flowering plants to boost biological pest control. However, performing independent replicates in this set-up requires a lot of space and labour, and this will often limit the number of flower species that can be compared.

Under field conditions, the availability of floral resources depends not only on floral traits but also on competition by other flower visitors such as bees. The level of this competition can vary significantly from site to site and could be an important factor in determining the impact of landscape management programmes (see Box 9.2).

Typically, field studies monitor insect populations in the flowering plots themselves or the adjacent crop, comparing this with plots lacking the floral vegetation. Flower suitability is usually deduced based on enhanced predator/parasitoid numbers or increased levels of predation/parasitism. However, these population responses may also occur when the flowering

plants simply act as a sink (Dunning *et al.*, 1992), drawing in predators or parasitoids from the field, without necessarily providing any fitness benefit. In this case, the impact of a flowering field margin might actually be counterproductive, as the concentrating of biological control agents would result in a depletion of predators/parasitoids in the crop field. To substantiate that the biological control agents actually obtain nutritional benefits from the flowering vegetation it is advisable to combine field tests with studies demonstrating nectar/pollen consumption or quantifying energy reserves (Olsen and Wäckers, 2007). Nutritional benefits are also expected to be reflected in enhanced longevity in the field. A number of methods have been used to establish a relative estimate of age or activity in field-collected arthropods (e.g. by assessing mandible wear for carabids and wing wear for flying insects) (Hayes and Wall, 1999; Lee and Heimpel, 2008).

Choosing and combining methods

The listed approaches represent a broad arsenal of methodologies that can be used to study the suitability of flowering plants for conservation biological control. The specific characteristics of the listed methods and their respective advantages and limitations highlight the importance of choosing the right experimental approach and conducting the tests under ecologically relevant experimental conditions.

As some of the methodologies are costly and labour-intensive, it could be sensible to examine whether methods requiring lower inputs can serve as an adequate proxy. Van Rijn and Wäckers (2010) assessed hoverfly survival on 30 different flowers and correlated this with flower depth. With very few exceptions, flower depth explained hoverfly survival with the cut-off point for successful flower feeding being at 2 mm corolla depth. Winkler *et al.* (2009b) compared three methods to study the exploitation of 19 flowering plants by two lepidopteran pests (*Pieris rapae* (L.) and *Plutella xylostella* (L.)), and their respective hymenopteran parasitoids, *Cotesia glomerata* (L.) and *Diadegma semiclausum* (Hellen). First, theoretical nectar accessibility was established on the basis of floral architecture and the mouthpart structure of the arthropods. Second, it was tested whether the arthropods could actually access the nectar by quantifying weight gain of individual insects when exposed to individual flowering species. Finally, the impact of various flowers on longevity of the herbivores and parasitoids was assessed. For the parasitoids tested the theoretical nectar accessibility was shown to be suitable as a first step in selecting plant species. Also there was a good fit between results from the short-term flower exposure bioassay and the results from the longevity study. This means that for these species, the theoretical nectar accessibility and/or the short-term bioassay would have been sufficient to assess flower suitability. For the herbivores, however, there were more disagreements between the methods. In this case, the gravimetric bioassay provided information that would have been missed on the basis of morphological measurements or longevity experiment alone. These results indicate that measurements of floral architecture and insect mouthpart structure can provide a good first assessment of potential nectar/pollen feeding. However, for an accurate understanding of the contribution of individual flowering plant species to the survival of arthropods in the field, these data need to be complemented with more detailed studies.

Since laboratory studies are at best an approximation of the complex interactions acting in the field, ultimately results from laboratory studies have to be confirmed in semi-field or field studies. For instance, the impact of floral nectar on longevity and fecundity of *D. semiclausum* has been found to be much more pronounced in the field than in the laboratory (Steppuhn and Wäckers, 2004; Winkler *et al.*, 2006). Other factors may also affect the exploitation of flowers under field conditions. For instance *P. rapae* was shown to be unable to feed on a number of exposed floral nectar sources in dry climatic conditions (45% RH), whereas the nectar from these plants was successfully exploited at 90% RH, presumably due to the resulting decrease in nectar viscosity (Winkler *et al.*, 2009b).

FLORAL FOOD REQUIREMENTS FOR DIFFERENT GROUPS OF BIOLOGICAL CONTROL AGENTS

A number of studies have shown that there can be clear differences in the exploitation of floral foods between arthropod pests and their natural enemies (Baggen *et al.*, 1999; Wäckers *et al.*, 2007; Winkler *et al.*, 2009b). These differences can be exploited to select flowering plants for conservation control programmes that provide nutritional resources for target beneficial insects, while minimising or excluding benefits to crop pests.

Given the reported specificity of flower exploitation patterns between pests and their natural enemies, one

can also expect differences in flower associations among the various groups of entomophagous arthropods. To explore these potential differences, the floral requirements of a range of nectar/pollen feeding biological control agents are reviewed below with a focus on parasitoids, syrphids, lacewings, coccinelids and predatory bugs. Table 9.2 shows results from survival studies with representatives of these groups. To allow for comparisons only those flowers are included which had been tested with at least two groups.

Hymenopteran parasitoids

While most parasitoid species are highly specialised in terms of their larval food requirements (host associations), the feeding requirements of the adult stages are less specific. The majority of hymenopteran parasitoids are dependent on sugar sources to cover their energy needs. For many parasitoid species the adult diet is restricted to nectar or other sugar-rich substrates. However, synovigenic parasitoid females may also feed on their host whose haemolymph provides protein and lipids for egg maturation. However, as haemolymph usually contains relatively low levels of carbohydrates (often trehalose and glycogen) carbohydrate sources are usually still required for energy. Moreover, carbohydrates found in haemolymph are typically poor in terms of supporting survival (Wäckers, 2001; Williams and Roane, 2007).

Parasitoids can be further divided according to the association between host and carbohydrate sources (Wäckers *et al.*, 2008). One group includes those parasitoid species whose hosts are closely linked to carbohydrate-rich food sources. This applies to species whose hosts excrete sugars (e.g. honeydew) or whose hosts occur on sugar-rich substrates like fruits or nectar-bearing plant structures. For these parasitoids the task of locating hosts and carbohydrates is linked. Parasitoids from this group may show few specific adaptations to the exploitation of additional carbohydrate sources and little or no task differentiation between food foraging and host search (Wäckers *et al.*, 2008). The second group comprises those parasitoids whose hosts are not reliably associated with a suitable carbohydrate source. These parasitoids have to alternate their search for hosts (reproduction) with bouts of food foraging, which requires a clear task differentiation. The latter group must decide whether to stay in a host patch, thereby optimising short-term reproductive success, or leave the host patch in search of food

Table 9.2 Effect of flowering plant species on adult flower choice (for hoverflies) and adult longevity for three types of natural enemies.

Family	Species	Floral Nectar depth	Choice	Longevity (AFLI)			References parasitoids (species)
			Hoverfly *E. balteatus*	Hoverfly *E. balteatus*	Lacewing *C. carnea*	Parasitoids	
Apiaceae	*Ammi majus* (L.)	0	+	+	+	–	Geneau et al., unpubl. (*Microplitis mediator* (Haliday))
Apiaceae	*Coriandrum sativum* (L.)	0	+	+		+/–	Vattala et al., 2006 (*Microctonus hyperodae* (Loan))
Apiaceae	*Daucus carota* (L.)	0	+/–	+	+	+	Winkler et al., 2009b (*Cotesia glomerata* (L.))
Apiaceae	*Foeniculum vulgare* (Miller)	0	+	+		+	Winkler et al., 2009b (*Cotesia glomerata*)
Apiaceae	*Heracleum spondylium* (L.)	0	+/–	+		+/–	Winkler et al., 2009b (*Cotesia glomerata*)
Apiaceae	*Pastinaca sativa* (L.)	0	+	+	++	+/–	Foster and Ruessink, 1984 (*Meteorus rubens* (Nees))
Polygonaceae	*Fagopyrum esculentum* (Moench)	0	+	+	+	+	Winkler et al., 2009b (*Cotesia glomerata*)
Boraginaceae	*Borago officinalis* (L.)	0	–	+	++	–	Nilsson et al., unpubl. (*Trybliographa rapae* (Westwood))
Ranunculaceae	*Ranunculus acris* (L.)	0	+	+		–	Kehrli and Bacher, 2008 (*Minotetrastichus frontalis* (Nees))
Caryophyllaceae	*Gypsophila elegans* (Bieb)	1	+/–	+	+		
Asteraceae	*Matricaria chamomilla* (L.)	1	+/–	+	+/–	–	Nilsson et al., unpubl. (*Trybliographa rapae*)
Asteraceae	*Achillea millefolium* (Yarrow)	1	+/–	+	+/–	–	Wäckers, 2004 (*Cotesia glomerata*)

Family	Species	Floral nectar depth	Flower choice (Episyrphus)	AFLI (Episyrphus)	AFLI (Chrysoperla)	Parasitoid	Reference
Asteraceae L	*Cichorium intybus* (L.)	1	−	−	+/−		
Asteraceae	*Chrysanthemum segetum* (L.)	2	+/−	+	+/−		
Asteraceae	*Anthemis tinctoria* (L.)	2	−	+/−	+/−		
Asteraceae	*Leucanthemum vulgare* (Lamarck)	2	+/−	+/−	+	−	Wäckers, 2004 (*Cotesia glomerata*)
Asteraceae	*Tanacetum vulgare* (L.)	2	−	−	+/−		
Asteraceae	*Calendula officinalis* (L.)	3	−	−		−	Rahat *et al.*, 2005 (*Trissolcus basalis* (Wollaston))
Asteraceae	*Centaurea cyanus* (L.) (+EFN)	3	+/−	+	+	+/−	Winkler *et al.*, 2009b (*Cotesia glomerata*)
Asteraceae	*Helianthus annuus* (L.) (+EFN)	3	+	+	+		
Asteraceae	*Cosmos bipinnatus* (Cosmos)	4	−	−	+/−	+	Rahat *et al.*, 2005 (*Trissolcus basalis*)
Malvaceae	*Malva sylvestris* (L.)	4	−	−			
Boraginaceae	*Phacelia tanacetifolia* (Bentham)	4	+/−	+/−	+/−	−	Irvin and Hoddle, 2007 (*Gonatocerus* spp.)
Fabaceae	*Medicago sativa* (L.)	4	−	−		−	Kehrli and Bacher, 2008 (*Minotetrastichus frontalis*)
Fabaceae	*Vicia sativa* (L.) (+EFN)	4	−	+		++	Geneau *et al.*, unpubl. (*Microplitis mediator*)
Fabaceae	*Lotus corniculatus* (L.)	4	−	−	−		

Plants are ranked according to corolla depth measured up to floral nectaries. Data for *Episyrphus balteatus* (DeGeer) and *Chrysoperla carnea* (Stephens) from van Rijn and Wäckers (2010) and van Rijn (unpublished results). Data for various hymenopteran parasitoids from literature indicated in last column. Floral nectar depth, 0: <0.2 mm, 1: 0.2–1.0 mm, 2: 1.0–2.0 mm, 3: 2.0–3.0 mm, 4: >3.0 mm. Flower choice, −: less than average, +/−: average, +: more than average. Adult longevity increase relative to control ('Adult food longevity index', AFLI), −: not significant, +/−: 2–4 fold, +: 4–8 fold, ++: more than 8 fold. L: Asteraceae without tubular florets. EFN: plant with extrafloral nectaries.

sources – a strategy that may optimise reproduction in the long term.

Carbohydrates can have a strong impact on several key lifetable parameters. Numerous studies have shown that sugar feeding is indispensable to parasitoid survival; a factor applying both to females and males (Azzouz *et al.*, 2004; Wyckhuys *et al.*, 2008). In addition, sugar feeding can also raise a female's propensity to search for herbivorous hosts and increase her daily fecundity. When combined, these factors can have a considerable cumulative effect on parasitisation rates and parasitoid reproductive success. Using semi-field Brassica patches Winkler *et al.* (2006) showed that the addition of flowering buckwheat increased the average reproductive lifespan of *D. semiclausum* from 1.2 days (control) to 28 days. In the absence of the floral resource the majority of wasps failed to attack any diamondback moth larvae but all parasitoids in the Brassica/buckwheat plots parasitised in excess of 300 larvae. These results demonstrate that access to carbohydrate-rich food may be indispensable to parasitoid fecundity. The impact of food sources on *D. semiclausum* fecundity was more pronounced in the semi-field experiments compared with laboratory studies, emphasising the importance of studying lifetable parameters under more natural conditions.

Parasitoids can fulfil their energy requirements by feeding on a broad range of accessible sugar sources such as floral nectar, extrafloral nectar or honeydew. While honeydew is often the most prevalent sugar source, especially in agricultural ecosystems, it is usually less suitable for supporting parasitoid survival when compared to nectar (Wäckers *et al.*, 2008). Parasitoids appear to be able to select sugar sources on the basis of their nutritional suitability (Wäckers, 2001; Vollhardt *et al.*, 2010).

As far as flower associations are concerned, most parasitoids are limited to the exploitation of exposed, easily accessible nectaries. This includes flower species such as buckwheat (*Fagopyron esculentum* (Moench)) and sweet alyssum (*Lobularia maritima* (L.)) as well as most species from the families Euphorbiaceae and Apiaceae. The flowers from the latter family are well known among taxonomists, who single them out to collect parasitic Hymenoptera. Tooker and Hanks (2000), based on records of flower visitations collected by Robertson (1928), showed that among 112 flowering plant species included in the records the top five in terms of parasitoid visits were all Apiaceae. Interestingly, Euphorbiaceae, while featuring similarly exposed

and accessible floral nectar, were visited far less frequently. This indicates that factors other than accessibility drive flower choice by hymenopteran parasitoids. Other plant species that are frequently visited by hymenopteran parasitoids are those species that produce extrafloral nectar (EFN) (Bugg *et al.*, 1989). EFN glands can usually be successfully exploited due to their exposed nature. EFN is available on a number of flowering herbs used in conservation biological control, such as cornflowers, sunflowers and some vetches (Table 9.2). In addition extrafloral nectaries are found on some crops, including some beans, peach, cherry, plum, cotton and zucchini.

Parasitoids are equipped with a number of mechanisms that enable them to forage effectively for floral nectar. They possess innate preferences for certain floral odours as well as for common flower colours (Wäckers, 1994). Following feeding, parasitoids ignore these floral cues and start responding to host-associated cues (Wäckers, 1994). Associative learning of food-associated stimuli enables parasitoids to concentrate foraging on those flowers where they previously experienced successful feeding (Takasu and Lewis, 1993; Wäckers *et al.*, 2006b).

Hoverflies (Syrphidae)

Hoverflies or flowerflies (Syrphidae) are a diverse group of insects regarding the feeding habits of the larvae. Some species feed on plants while many are saprophagous (i.e. feeding on decaying plant material) either on land or in fresh water. More than one-third of all species are zoophagous, mostly aphidophagous (feeding on aphids) and sometimes also feeding on other small insects (Reemer *et al.*, 2009). Some of these zoophagous species are relatively common in agricultural areas, and can under suitable conditions play an important role in suppressing aphid populations in agricultural crops. Despite the large differences in feeding habits of the larvae, all adult hoverflies feed on pollen and nectar from flowers (Figures 9.2 and 9.3). The protein-rich pollen is needed by both the males and the females for sexual maturation (Haslett, 1989; van Rijn *et al.*, 2006), whilst nectar is an important energy source. The typical hovering flight is an important part of male territorial courtship behaviour. It is therefore suggested that males demand more energy (nectar) but less pollen than females (Gilbert, 1981; Haslett, 1989; Hickman *et al.*, 1995). Further, between species, large

Figure 9.2 *Episyrphus balteatus* feeding on pollen of *Phacelia tanacetifolia* (photo by Paul van Rijn).

Figure 9.3 *Syrphus ribesii* feeding on nectar and/or pollen of *Chrysanthemum segetum* (photo by Paul van Rijn).

differences seem to occur in the relative need for nectar vs. pollen. Some *Melanostoma* and *Platycheirus* species are observed to contain mainly pollen from non-nectar flowers such as grasses and plantain (Ssymank and Gilbert, 1993). These small species are apparently able to obtain enough energy from pollen only. For larger hoverflies nectar makes up a larger part of the diet (Gilbert, 1985).

Although most studies consider hoverflies as one pollination guild (e.g. Comba *et al.*, 1999), large differences exist between species in the types of flowers used as a nectar source (Gilbert, 1981; 1985). Tongue length is one characteristic that can explain such differences, as it clearly limits the range of flowers from which nectar can be obtained (see Figures 9.2 and Figure 9.3). The data from Gilbert (1981; 1985) as well as from Branquart and Hemptinne (2000) indicate (within the nectar feeding species) a correlation between tongue length and the average depth of the flowers they visit. These data also show that most zoophagous hoverflies have much shorter mouthparts than common saprophagous species (e.g. *Eristalis* spp.). Conclusions on food plant ranges for hoverflies in general may therefore not always be applicable for the group of zoophagous hoverflies.

In a similar way to parasitoids, hoverflies show a propensity to visit umbelliferae (Apiaceae). In the analysis by Tooker *et al.* (2006), seven out of the ten plant species most visited by zoophagous hoverflies were from the family Apiaceae, even though they made up only a small proportion of all plant species sampled.

In several other American studies the visit frequencies were studied under more standardised conditions for a small range of flower species. In Hogg *et al.* (2011) hoverflies and other pollinators were observed on nine plant species that were sown in the experimental plots. In all periods where sweet alyssum was flowering this flower received most hoverfly visits (predominantly *Toxomerus marginatus* (Say)), with the exception that buckwheat was visited equally during the short period it was flowering. Alyssum and buckwheat have open flowers. However, the same applies for some of the alternative species (*Brassica* sp., *Diplotaxis muralis* (L.), *Borago officinale* (L.)) that attracted far fewer hoverflies. Some of the latter attracted more bees, which may have interfered with the (smaller) hoverflies (see Box 9.2). In the study by Colley and Luna (2000), featuring 11 plant species, alyssum, buckwheat and mustard were highly visited, but here the most visits were recorded on coriander (*Coriandrum sativum* (L.)). The latter result was confirmed by Ambrosino *et al.* (2006).

Until recently the range of suitable flowers for zoophagous hoverflies has been inferred from direct behavioural observations in the field only. In recent studies, using *Episyrphus balteatus* (DeGeer) as a model organism, results for flower choice studies in flight cages were compared with survival experiments in laboratory cages and morphometric measurements for about 30 plant species (van Rijn and Wäckers, 2010). Females of *E. balteatus* start reproducing within six days of emergence. Therefore, a food source was considered adequate when females survived for at least six days (van Rijn *et al.*, 2006). In control experiments with only water *E. balteatus* lives for a mere two days, emphasising the importance of floral resources for hoverflies. The study by van Rijn and Wäckers (2010) showed that all plants with exposed nectaries, such as buckwheat and Apiaceae, were suitable food sources. On the other hand, the composites (Asteraceae) tested, featuring tubular florets, showed large differences in suitability for *E. balteatus*. Out of 15 species, 6 were insufficient food sources; 3 species were marginal food sources (resulting in a mean survival just above 6 days) and 6 species allowed the females to survive for 10 days or more. The morphometric data from these plants showed that longevity is generally well correlated with the depth of the tubular florets (see Table 9.2 for a selection of plants). When the florets are 1.6 mm deep or less (e.g. *Matricaria chamomilla* (L.) and *Achillea millefolium* (yarrow)) the flowers were suitable *for E. balteatus*. When the florets were 2.1 mm deep or more (as in *Calendula officinalis* (L.)) the flowers were unsuitable as food sources. The only exception to this latter category was *Jacobaea vulgaris* (Gaertn) where the hoverflies survived despite the floret depth of 2.8 mm. The presence of extrafloral nectaries (EFN) allowed hoverflies to survive long enough, even when the floral nectar is beyond reach. This is the case in cornflowers (*Centaurea cyanus* (L.)), sunflowers (*Helianthus annuus* (L.)) and the common vetch (*Vicia sativa*). Since pollen in most species tested (except Fabaceae species) is well exposed, the results clearly indicate that it is their nectar accessibility that determines hoverfly survival. Floral choice in flight cages appears to be correlated with the longevity performance (excluding plants with EFN: $R^2=0.28$, n=25, p < 0.01). When floral choice is categorised in 'preferred', 'neutral', and 'avoided', 12 out of 14 species with accessible floral nectar are classified as neutral or 'preferred'. Of the plants with apparent inaccessible nectar only 2 out of 11 species were neutral or 'preferred' (see Table 9.2 and van Rijn and Wäckers, 2010).

These results indicate that flower choice mainly reflects nectar accessibility rather than pollen accessibility. This is in agreement with the field studies by Gilbert (1981) and Branquart and Hemptinne (2000) showing a correlation between (relative) tongue length of the hoverfly species and the average depth of the flowers visited. At the same time, it is in contrast with the observations by Gilbert (1981; 1985) that in urban gardens *E. balteatus* spends most of its time feeding on pollen: only 10% of the overall feeding time is spent consuming nectar. In fact, since these studies by Gilbert, *E. balteatus* and several other zoophagous species (including *Syrphus ribesii* (L.)) have been categorised as pollen feeders or pollen specialists (Gilbert and Owen, 1990; Branquart and Hemptinne, 2000; Reemer *et al.*, 2009). Accepting that the basic time allocation observations are correct, we have to assume that either pollen feeding requires a longer handling time as compared to the consumption of nectar required survival and flight (at least in the urban garden environment), or feeding on honeydew (which occur on vegetative plant parts not observed by the authors) is an important way of obtaining sugars. Hogervorst *et al.* (2007) concluded on the basis of sugar composition of field-collected adults that 40% of *E. balteatus* collected from one field had been feeding on honeydew, while no honeydew indicating sugars could be detected from those collected in a second field.

Lacewings (Chrysopidae)

Green lacewings (Chrysopidae) and brown lacewings (Hemerobiidae) are families within the order of Neuroptera. The larvae of these lacewings are predators of aphids, caterpillars and other soft-bodied insects. Some species are common within crop habitats, and are considered as important natural enemies of crop pests. Brown lacewing adults and some green lacewing adults feed on prey as well on floral food, while adults of other species (e.g. *Chrysoperla* and *Dichochrysa*) feed on pollen, nectar and honeydew only (Stelzl, 1991). Adult lacewings of *Chrysoperla* species (see Figure 9.4) require a sugar source (nectar or honeydew) and pollen for maximal survival and reproduction. With only a sugar source (sucrose or honey) no oviposition occurs. With only pollen some reproduction occurs, but at a much lower rate than when pollen and sugar are provided together. Moreover, survival is reduced in the absence of a sugar source (Venzon *et al.*, 2006; Li *et al.*, 2010). *Chrysoperla* adults house yeasts in their crop (diverticulum) that are assumed to provide the insects with essential amino acids that may not be present in their floral or honeydew diet (Hagen *et al.*, 1970; Gibson and Hunter, 2005).

Figure 9.4 *Chrysoperla carnea* (s.l.) feeding on pollen of *Chrysanthemum segetum* (photo by Paul van Rijn).

The range of plants that are used as pollen source can be assessed from crop content analyses of field collected lacewings. Females of *Chrysoperla carnea* (Stephens) collected in flowering maize fields contained around 5,000 pollen grains each (Li *et al.*, 2010). Adult lacewings (*C. externa*) collected in vegetable fields in Brazil contained much lower numbers (on average 252 grains/predator) which appeared to be mainly (greater than 99%) pollen from grasses (Medeiros *et al.*, 2010). Possibly these anemophilous (i.e. wind dispersed) pollen grains are collected from leaves rather than from the flowers. Villenave *et al.* (2005; 2006) found pollen from a large range of plant families within the crops of *Chrysoperla* species in western France from field crops and surrounding vegetation. Based on the assumed activity pattern of lacewings, scrubs and trees were sampled in daytime and herbs during twilight. They concluded that these lacewings were opportunistic foragers feeding on all flowers that are readily available, including shrubs from *Sambucus* and *Corylus* and Rosaceae, and herbs from the families Brassicaceae, Cayophylaceae and Asteraceae (Villenave *et al.*, 2005). Other plant families that are fed upon in proportion to their presence are Chenopodiaceae, grasses (Poaceae) and Liliaceae (Villenave *et al.*, 2006). However some families are absent or clearly underrepresented in the lacewing crop. In the case of Convolvulaceae, Cucurbitaceae, Geraniaceae and Malvaceae this could be explained by the closure of the flowers at twilight, when the lacewings are active. For other families (e.g. Rubiaceae, Resedaceae, Primulaceae, Onagraceae, Verbenaceae, Violaceae) no direct explanation is available without complementary experimental studies.

Various laboratory studies have shown the suitability of different pollen for survival and reproduction of *Chrysoperla* adults. Elbadry and Fleschner (1965) found reproduction was higher when *Chrysoperla* adults were kept on a diet of *Mesembryanthemum* pollen compared to pollen of *Capsicum* or *Cedrus*. Venzon *et al.* (2006) showed that pollen from leguminous cover crops (*Cajanus cajan* (L.) and *Crotalaria juncea* (L.)) allows for high reproduction rates, especially when complemented with honey, whereas pollen of the euphorbic castor bean *Ricinus communis* (L.) was less suitable.

Pollen analysis may reveal which plant species are exploited as source of pollen, but these do not necessarily constitute a suitable nectar source. Since the mouthparts of lacewings are rather small, only a limited number of plant species have nectar that is accessible to them. The sugar content of *C. carnea* adults collected in wheat fields by Hogervorst *et al.* (2007) indicated that 98% of the adults has been feeding on a sugar source, with a remarkably high average sugar level that was 45–90% of the maximum level measured in lacewings that had unlimited access to sucrose. The sugar spectrum, however, was not conclusive to establish whether these sugars originated from honeydew or nectar.

Survival experiments performed by van Rijn (unpublished results) to assess the suitability of various flowering plants as food sources for adult lacewings show the combined impact of pollen and nectar. Adult longevity was higher for all species with exposed nectaries compared to all other plant species (Table 9.2), which indicates the importance of nectar accessibility for flower suitability. However, on many plants with deeper corollas or inaccessible nectar, lacewings live on pollen long enough to start ovipositing, making the range of flowers suitable for lacewings broader as compared to those for hoverflies.

While larvae of *Chrysoperla* primarily feed on aphids and other prey, they will also consume (extrafloral) nectar and pollen when available. Nectar feeding is especially prevalent among neonate larvae and when prey is scarce (Limburg and Rosenheim, 2001), but occurs even when access to prey is unlimited (Hogervorst *et al.*, 2008).The nectar (of cotton) by itself does not allow for larval development but supports their survival and searching activity considerably (Limburg and Rosenheim, 2001). When the prey diet is relatively poor, the addition of pollen and nectar enhances larval growth, and beyond the second stage allows further development even in the absence of prey (Patt *et al.*, 2003).

Flowerbugs (Anthocoridae)

Flowerbugs or Anthocoridae have, like all Heteroptera, a pointed rostrum with which they feed on mites and small insects such as thrips, aphids, psyllids, small caterpillars and eggs. Various *Anthocoris* and *Orius* species are considered important predators of insect pests. In addition to prey, many anthocorids also feed on plant material, such as plant juices, pollen and nectar. By feeding on plant juices they increase their longevity when prey is scarce (Salas-Aguilar and Ehler, 1977; Kiman and Yeargan, 1985; Coll, 1996). Other predatory bugs (from the related family Miridae) are thought to feed on plant sap only. However, a recent study showed that survival of *Macrolophus pygmaeus* (Rambur) is prolonged on broad bean plants providing extrafloral nectar as compared to broad bean with EFN removed. It also demonstrated that a greater proportion of mirid females laid eggs when extrafloral nectar was available compared to those confined on nectariless plants (Portillo *et al.*, 2012).

Feeding on pollen occurs in many anthocorids, but its impact on life history varies. *Orius pallidicornis* (Carayon) seem to feed almost exclusively on pollen (Carayon and Steffan, 1959). *Orius insidiosus* (Say), an important predator of corn pests, is reported to complete development and to oviposit on a diet of corn pollen only (Kiman and Yeargan, 1985), but this result was not confirmed by other studies (Richards and Schmidt, 1996). *Orius sauteri* (Poppius) is able to develop and oviposit on a diet of pollen, but at much lower rates than with prey diets (Funao and Yoshiyasu, 1995; Yano, 1996). In other species (*Orius tristicolor* (White), *Orius laevigatus* (Fieber) and *Orius albidipennis* (Reuter)) pollen increases longevity but does not allow for full development or oviposition (Salas-Aguilar and Ehler, 1977; Cocuzza *et al.*, 1997; Vacante *et al.*, 1997). However, supplementing prey diet with pollen increases their development and oviposition. These results explain why various *Orius* spp. are generally more successful in pollen-bearing crops (e.g. strawberry, eggplant, sweet pepper), as opposed to crops without pollen (e.g. cucumber) (Van den Meiracker and Ramakers, 1991; Dissevelt *et al.*, 1995). It also explains why

these anthocorids often become more abundant in periods of increased pollen availability within the crop (Dicke and Jarvis, 1962; Isenhour and Yeargan, 1981; Coll and Bottrell, 1995).

In contrast with the previously discussed insect groups, the anthocorids are true omnivores that can feed both on prey and pollen in each life stage. Consequently, feeding on pollen may also directly affect their feeding on prey through satiation. Indeed, Skirvin *et al.* (2007) found that the presence of pollen led to a 40% reduction in thrips predation by *O. laevigatus*. Corey *et al.* (1998) concluded from electrophoresis of gut contents that in flowering corn, *O. insidiosus* fed mostly on corn pollen and much less on prey. In this way pollen can diminish the pest control capacity of the predator in the short term. On longer time scales, however, the enhanced production or attraction of predators is likely to tip the balance in the other direction (van Rijn *et al.*, 2002).

Flowers produced by plants other than crop plants can affect the anthocorid populations in and around crop fields as well. Letourneau and Altieri (1983) found that intercropping squash with corn and cowpea increased the number of Orius, possibly due to corn pollen. Frescata and Mexia (1995) observed more *O. leavigatus* and less thrips in a strawberry patch where a composite weed (*Chamaemelum mixtum* (L.)) was flowering compared to all other patches. Atakan (2010) found that numbers of *Orius niger* (Wolff) were significantly greater in faba bean plots with weedy margins than in weed-free plots.

The suitability of plants species for supporting anthocorids is little studied. Due to their small size direct observations on flower choice is not possible, but by tapping or collecting flowers from vegetation in and around crop fields and sorting out its content the distribution over various plant species can be assessed. Silveira *et al.* (2003) collected anthocorids on crops and weeds in south-east Brazil. The main species, *O. insidiosus*, was found on plants belonging to the families Poaceae, Fabaceae, Asteraceae and Amaranthaceae. Since they also found an association between Orius and thrips, its main prey, this species distribution of Orius may both be a direct and an indirect effect (through prey distribution) of plant features (Silveira *et al.*, 2005). Fiedler and Landis (2007a; 2007b) vacuumed flowers in plots of 43 native perennial plant species and found that *O. insidiosus* made up 30% of all natural enemies on these flowers, with chalcidoid parasitoids as the second most important group (25%).

The total number of natural enemies (all groups taken together) was positively correlated with the flowering period and with the flower surface. Within each seasonal period the highest numbers of natural enemies were found on species with well exposed nectaries belonging to the families of Apiaceae (3/3), Rosaceae (3/4), and Ranunculaceae (1/1), as well as species with less exposed nectaries including Asteraceae (5/13), Apocynaceae (1/1), Lamiaceae (1/2), Scrophulariaceae (1/2) and Onagraceae (1/1). No patterns are provided for the individual groups of natural enemies. Bosco and Tavella (2008) sampled the wild vegetation surrounding pepper greenhouses in northwestern Italy for *O. niger, Orius majusculus* (Reuter) and *Orius minutus* (L.). They showed that *Orius* exhibits species-specific plant associations, with *O. niger* exhibiting a marked preference for Fabaceae. Van Rijn (unpublished results) regularly sampled flowers from flower strips along onion fields in the Netherlands. He found *Orius* spp. (mainly *O. minutus*) almost exclusively on Asteraceae plants (see Figure 9.5), with the highest numbers on *H. annuus* and *Cosmos bipinnatus* (Cavanilles). These results suggest that these bugs prefer flowers with deeper (and wider) corollas, possibly because these can serve as a hiding place as well.

Figure 9.5 A ladybird *Harmonia axyridis* and three *Orius* feeding on *Chrysanthemum segetum* (photo by Paul van Rijn).

Lady beetles (Coccinellidae)

Lady(bird) beetles are probably the most emblematic biological control agents. In some agroecosystems their presence may contribute considerably to natural pest control. Although they seem to lack the ability to regulate aphid populations in summer (Dixon, 2000) early-season pest suppression may benefit from conservation measures that promote lady beetle subsistence on prey and non-prey food in spring (Obrycki *et al.*, 2009).

Both larvae and adults have chewing mouthparts and can utilise similar food sources. Most entomophagous species of Coccinellidae feed on aphids or scale insects. Although floral food sources are not essential for these predators, they may be dependent on them when prey is temporarily scarce (Lundgren, 2009). Feeding on nectar, especially from extrafloral nectaries, is well known among coccinellids (Pemberton and Vandenberg, 1993; see also Figure 9.6). In absence of prey these sugar sources can strongly support survival (Putman, 1955) as well as flight capacity (Nedved *et al.*, 2001). Pollen provides the beetles with enough

Figure 9.6 *Coccinella septempunctata* feeding on extrafloral nectaries of cornflower (photo by Paul van Rijn).

proteins and other nutrients for some to complete their development on a diet of pollen only. The development rate and the final weight, however, are always lower than on a suitable prey diet (Lundgren, 2009). Pollen allows adults to survive longer, but generally does not support reproduction. An exception is *Coleomegilla maculata* (DeGeer), which can complete its life cycle on a diet of pollen only (Lundgren and Wiedenmann, 2004). In the presence of prey, both nectar and pollen generally enhance development and reproduction (Lundgren, 2009). The fact that combining floral food and prey can benefit biological control is shown by Harmon *et al.* (2000), who observed that alfalfa plots with flowering *Taraxacum* (dandelion) plants had higher numbers of *C. maculata* and lower number of aphids than plots without these flowers.

Intensive inspection of more than 60 plant species in weed strips within a wheat field in Switzerland revealed 20 species that were frequently visited by local coccinelids (Schmid, 1992). For 11 plant species this could be attributed to the presence of abundant aphid populations. On other plant species floral resources seems to be the main food source for the lady beetles. The relative importance of the two food sources appears to be different for the various species of coccinelids. *Coccinella septempunctata* (L.) mainly occurs on plants with high number of aphids, such as *Symphytum officinale* (L.), *Silene alba* (Miller) and *Urtica dioica* (L.), on which also the juveniles are found. *Adonia variegate* (Goeze) on the other hand is mainly abundant on plants with floral resources, such as *Tripleurospermum inodora* (L.), *Myosotis arvensis* (L.), *Leucanthemum vulgare* (Lamarck), *Daucus carota* (L.), and *Verbascum thapsus* (Mullein). The same is true for *Propylea quatuordecimpunctata* (L.) showing high numbers on *Lamium purpureum* (L.) and *Plantago major* (L.). Juveniles are only observed when aphids are present as well, as on *T. inodora*. The large variation in flower types and nectar accessibility among these plants suggests that the beetles feed on pollen more than on floral nectar. Extrafloral nectar was present only on *C. cyanus* (see Figure 9.6), which showed intermediate numbers of three coccinelid species.

Few other studies have considered the association between plant species and cocinellids (Honěk, 1985; Burgio *et al.*, 2004), and even fewer have separated the impact of flowers from that of prey on these plants. In a mono-flower plot set-up comparing different Apiacaea, Lixa *et al.* (2010) found higher densities of various coccinelid species on *Anethum graveolens* (L.) (dill) compared to *C. sativum* (coriander) and *Foeniculum vulgare* (Miller) (fennel). In olfactometer tests *Harmonia axyridis* (Pallas) prefers odours from sunflower and dill over eight other plant species (Adedipe and Park, 2010). Visual preference tests were in concurrence with these results showing that this species

prefers the common flower colour yellow over all other colours. In an experimental set-up Bertolaccini *et al.* (2008) found that the egg production of *Hippodamia variegata* (Goeze) on plants with aphids increased when flowers of *Brassica* and *Sonchus* were present but not when flowers of *Daucus* were present. There is an obvious need for more of these experimental studies to establish which plant species and plant features can support lady beetle survival and reproduction in the absence of prey.

CONCLUSION

The detailed information on flower suitability generated by various laboratory and field studies has been used to underpin the selection of seed mix prescriptions for programmes aiming at arthropod conservation and/or enhancement of ecosystem services, such as pollination and biological control. These seed mixes typically target one particular group of arthropods (Wäckers, 2004; Carvell *et al.*, 2007; Coll, 2009). It is often implicitly assumed that flower-rich vegetation that has been selected to cater for one particular target group will generate benefits to nectar and pollen feeders across the board. This ignores the fact that flower associations can often be highly specific, and that plants have evolved many mechanisms through which they can exclude visitors other than the intended specialised pollinators (Faegri and van der Pijl, 1979; Kevan and Baker, 1998; Waser and Ollerton, 2006). Comparing pollinator and biological control targeted seed mixes, Campbell, Biesmeijer and Wäckers (unpublished) showed recently that flower visitation differed significantly between pollinators and biological control agents. Bumblebees almost completely refrained from visiting biological control seed mixes, whilst parasitoids were all but absent from flowers in the pollinator plots. These results highlight that insects providing ecosystem services differ distinctly with regard to their flower associations and that flower mixes targeting particular insect groups are not necessarily effective in supporting other beneficial arthropods.

The studies described in this chapter demonstrate that there can also be considerable variation within one functional group (i.e. among various categories of entomophagous arthropods). The data presented in Table 9.2 show the survival of hoverflies, lacewings and parasitoids on a range of flowering plants ranked by the depth of floral nectaries. The overview shows that parasitoids are more constrained by the accessibility of floral nectaries, exhibiting enhanced survival only on those flowers where the nectaries are fully exposed. The examples where parasitoids show enhanced longevity on flowers with deeper corollas can be explained by the fact that *V. sativa* and *C. cyanus* feature EFN. The fact that the parasitoid *Trissolcus basalis* (Wollaston) is able to exploit nectar from *C. bipinnatus* can be explained by the small size of the (egg) parasitoid, allowing it to enter the relatively deep flower.

Zoophagous hoverflies can access flowers up to a nectar depth of 2 mm, but appear to be constrained by flowers with deeper lying nectaries. Exceptions to this are *Phacelia tanacetifolia* (Bentham) and plants which feature EFNs (Table 9.2).

Chrysoperla lacewings can probably use nectar from a smaller range of flowers than indicated in Table 9.2. The study of lacewing nectar exploitation can be skewed by the fact that *Chrysoperla* can also survive for some time by feeding on the often more accessible pollen (Venzon *et al.*, 2006). The lower nectar dependency of lacewings relative to hoverflies may be explained by the much reduced flight activity of the former.

Anthocorids (especially *Orius* spp.) are thought to be even less dependent on nectar feeding (but see examples described above). Moreover, their much smaller size and prolonged mouthpart structure relative to hoverflies and lacewings results in very different flower type relationships. Their specific morphology allows them to access or enter flowers with deep corollas as well as flowers of the flag type (Fabaceae) that are inaccessible to other beneficial insect groups. They even seem to prefer these flowers, possibly as these provide shelter, or prey, such as thrips, commonly associated with these flowers (Silveira *et al.*, 2005).

The specificity that emerges from these studies underscores the importance of selecting the right flowers when composing seed mixtures to optimise the biological control services of agroecosystems. Earlier approaches using flowering plants to support biological pest control could not draw upon the specific information that is currently available. As these studies make up a substantial part of the examples included in meta-analyses of the impact of habitat diversification on pest control, such analyses probably still underestimate the potential of using informed selection of flowering vegetation in conservation biological control.

When selecting flowering plant mixes that effectively support biological control in a particular crop, a sound approach would be to:
- identify the main pests in that particular crop as well as their key natural enemies,
- identify floral resource requirement of these target organisms, both in terms of which floral resource they need and when they need it,
- identify plant species that are effective in providing these resources to the natural enemies at the right time, while preferably excluding nutritional benefits to the pests.

Additional criteria should be considered when selecting flowering plants for targeted seed mixes (see the Ecostac website for more information: http://www.ecostac.co.uk/seed_selection.php). These criteria could be soil and climate requirements of the flowering plants; plant phenology (e.g. height); whether plants are annual/biannual/perennial; native/naturalised/non-native; seed availability/quality/price. Plants with negative traits such as weed potential, or those serving as potential alternative host to crop diseases or pests should be excluded. The hosting of non-pest herbivores, on the other hand, is often advantageous as it allows populations of natural enemies to develop in absence of the crop pest (banker plants). Competitive interactions between flowering plants and crop plants can occur when flowering vegetation is used in close proximity to the crop, such as in cases of flowering undergrowth in orchards, mixed cropping, or companion planting. These competitive interactions can be averted through the selection of specific compatible combinations.

The combining of flowering plants in flower mixes requires some other points to be considered. For instance, the compatibility of the separate plants and their competitive strength in the mixture can be issues. Important is the issue of whether individual plant species are complementary in their traits, for instance in terms of resources provided, accessibility, or flowering time. While it is usually of little benefit to have too much redundancy between floral offerings, some level of redundancy can be useful as an insurance policy in case some plants do not germinate or establish.

Overall, the informed choice of flowering plants and the spatial setting of floral vegetation relative to crops and other landscape elements gives us a powerful tool to shape the composition of the agricultural arthropod fauna and to maximise the ecosystem services they provide. It is all in the mix!

REFERENCES

Adedipe, F. and Park, Y.L. (2010) Visual and olfactory preference of Harmonia axyridis (Coleoptera: Coccinellidae) adults to various companion plants. *Journal of Asia-Pacific Entomology*, 13, 319–323.

Ambrosino, M.D., Luna, J.M., Jepson, P.C. and Wratten, S.D. (2006) Relative frequencies of visits to selected insectary plants by predatory hoverflies (Diptera : Syrphidae), other beneficial insects, and herbivores. *Environmental Entomology*, 35, 394–400.

Andow, D.A. (1991) Vegetational diversity and arthropod population response. *Annual Review of Entomology*, 36, 561–586.

Araj, S.E., Wratten, S., Lister, A. and Buckle, H. (2008) Floral diversity, parasitoids and hyperparasitoids – A laboratory approach. *Basic and Applied Ecology*, 9, 588–597.

Atakan, E. (2010) Influence of weedy field margins on abundance patterns of the predatory bugs *Orius* spp. and their prey, the western flower thrips (*Frankliniella occidentalis*), on faba bean. *Phytoparasitica*, 38, 313–325.

Atakan, E. and Tunc, I. (2010) Seasonal abundance of hemipteran predators in relation to western flower thrips *Frankliniella occidentalis* (Thysanoptera: Thripidae) on weeds in the eastern Mediterranean region of Turkey. *Biocontrol Science and Technology*, 20, 821–839.

Azzouz, H., Giordanengo, P., Wäckers, F.L. and Kaiser, L. (2004) Effects of sugar availability and concentration on behavior and longevity of the aphid parasitoid *Aphidius ervi* (Haliday) (Hymenoptera: Braconidae). *Biological Control*, 31, 445–452.

Baggen, L.R. and Gurr, G.M. (1998) The influence of food on *Copidosoma koehleri* (Hymenoptera: Encyrtidae), and the use of flowering plants as a habitat management tool to enhance biological control of potato moth, *Phthorimaea operculella* (Lepidoptera: Gelechiidae). *Biological Control*, 11, 9–17.

Baggen, L.R., Gurr, G.M. and Meats, A. (1999) Flowers in tri-trophic systems: mechanisms allowing selective exploitation by insect natural enemies for conservation biological control. *Entomologia Experimentalis et Applicata*, 91, 155–61.

Beattie, A.J. (1985) *The evolutionary ecology of ant-plant mutualisms*, Cambridge University Press, Cambridge.

Begum, M., Gurr, G.M., Wratten, S.D. and Nicol, H.I. (2004) Flower color affects tri-trophic-level biocontrol interactions. *Biological Control*, 30, 584–590.

Begum, M., Gurr, G.M., Wratten, S.D., Hedberg, P.R. and Nicol, H.I. (2006) Using selective food plants to maximize biological control of vineyard pests. *Journal of Applied Ecology*, 43, 547–554.

Bertolaccini, I., Núñez-Pérez, E. and Tizado, E.J. (2008) Effect of wild flowers on oviposition of Hippodamia variegata (Coleoptera: Coccinellidae) in the laboratory. *Journal of Economic Entomology*, 101, 1792–1797.

Biesmeijer, J.C., Roberts, S.P.M., Reemer, M. *et al.* (2006) Parallel declines in pollinators and insect-pollinated plants in Britain and the Netherlands. *Science*, 313, 351–354.

Blüthgen, N., Verhaagh, M., Goitia, W., Jaffé, K., Morawetz, W. and Barthlott, W. (2000) How plants shape the ant community in the Amazonian rainforest canopy: the key role of extrafloral nectaries and homopteran honeydew. *Oecologia*, 125, 229–240.

Bosco, L. and Tavella, L. (2008) Collection of Orius species in horticultural areas of northwestern Italy. *Bulletin of Insectology*, 61, 209–210.

Branquart, E. and Hemptinne, J.L. (2000) Selectivity in the exploitation of floral resources by hoverflies (Diptera: Syrphinae). *Ecography*, 23, 732–742.

Bugg, R.L., Ellis, R.T. and Carlson, R.W. (1989) Ichneumonidae (Hymenoptera) using extrafloral nectar of faba bean (Vicia Faba L., Fabaceae) in Massachusetts. *Biological Agriculture and Horticulture*, 6, 107–114.

Burgio, G., Ferrari, R., Pozzati, M. and Boriani, L. (2004) The role of ecological compensation areas on predator populations: an analysis on biodiversity and phenology of Coccinellidae (Coleoptera) on non-crop plants within hedgerows in Northern Italy. *Bulletin of Insectology*, 57, 1–10.

Carayon, J. and Steffan, J. (1959) Observations sur le regime alimentaire des Orius et particulierment d' Orius pallidicornis (Peuter) (Hereroptera: Anthocoridae). *Cahiers des Naturalistes*, 15, 53–63.

Carvell, C., Meek, W.R., Pywell, R.F., Goulson, D. and Nowakowski, M. (2007) Comparing the efficacy of agri-environment schemes to enhance bumble bee abundance and diversity on arable field margins. *Journal of Applied Ecology*, 44, 29–40.

Chaney, W.E. (1998) Biological control of aphids in lettuce using in-field insectaries, in *Enhancing Biological Control: Habitat management to promote natural enemies of agricultural pests* (eds C.H. Pickett and R.L. Bugg), University of California Press, Berkeley, pp.73–83.

Cocuzza, G.E., DeClercq, P., VandeVeire, M., DeCock, A., Degheele, D. and Vacante, V. (1997) Reproduction of Orius laevigatus and Orius albidipennis on pollen and Ephestia kuehniella eggs. *Entomologia Experimentalis et Applicata*, 82, 101–104.

Coll, M. (1996) Feeding and oviposition on plants by an omnivorous insect predator. *Oecologia*, 105, 234–220.

Coll, M. (2009) Conservation biological control and the management of biological control services: are they the same? *Phytoparasitica*, 37, 205–208.

Coll, M. and Bottrell, D.G. (1995) Predator–prey association in mono and dicultures: effect of maize and bean vegetation. *Agriculture, Ecosystems and Environment*, 54, 115–125.

Colley, M.R. and Luna, J.M. (2000) Relative attractiveness of potential beneficial insectary plants to aphidophagous hoverflies (Diptera: Syrphidae). *Environmental Entomology*, 29, 1054–1059.

Comba, L., Corbet, S.A., Hunt, L. and Warren, B. (1999) Flowers, nectar and insect visits: evaluating British plant species for pollinator-friendly gardens. *Annals of Botany*, 83, 369–383.

Corey, D., Kambhampati, S. and Wilde, G. (1998) Electrophoretic analysis of *Orius insidiosus* (Hemiptera: Anthocoridae) feeding habits in field corn. *Journal of the Kansas Entomological Society*, 71, 11–17.

Davidson, L.N. and Evans, E.W. (2010) Frass analysis of diets of aphidophagous lady beetles (Coleoptera: Coccinellidae) in Utah alfalfa fields. *Environmental Entomology*, 39, 576–582.

DeGrandi-Hoffman, G. and Hagler, J. (2000) The flow of incoming nectar in a honey bee (*Apis mellifera* L.) colony as revealed by a protein marker. *Insectes Sociaux*, 47, 302–306.

Dicke, F.F. and Jarvis, J.L. (1962) The habits and seasonal abundances of Orius insidiosus (Say) (Hemiptera-Heteroptera: Anthocoridae) on corn. *Journal of the Kansas Entomological Society*, 35, 339–344.

Disselvelt, M., Altena, K. and Ravensberg, W J. (1995) Comparison of different Orius species for control of Frankliniella occidentalis in glasshouse vegetable crops in the Netherlands. *Mededelingen Faculteit Landbouwwetenschappen, Universiteit Gent*, 60/3a, 839–845.

Dixon, A.F.G. (2000) *Insect predator–prey dynamics – ladybird beetles and biological control*, Cambridge University Press, Cambridge.

Dunning, J.B, Danielson, B.J. and Pulliam, H.R. (1992) Ecological processes that affect populations in complex landscapes. *Oikos*, 65, 169–175.

Elbadry, E.A. and Fleschner, C.A. (1965) The feeding habits of adults of *Chrysopa california* Coquillett. (Neuroptera: Chrysopidae). *Bulletin de la Societe Entomologique d'Egypte*, 49, 359–366.

Faegri, K. and van der Pijl, L. (1979) *The Principles of Pollination Ecology*, Pergamon Press, Oxford.

Feinsinger, P. and Swarm, L.A. (1978) How common are ant-repellent nectars? *Biotropica*, 10, 238–239.

Ferreres, F., Andrade, P., Gil, M.I. and Tomas-Barberan, F.A. (1996) Floral nectar phenolics as biochemical markers for the botanical origin of heather honey. *Zeitschrift fur Lebensmitteluntersuchung und Forschung A*, 202, 40–44.

Fiedler, A.K. and Landis, D.A. (2007a) Attractiveness of Michigan native plants to arthropod natural enemies and herbivores. *Environmental Entomology*, 36, 751–765.

Fiedler, A.K. and Landis, D.A. (2007b) Plant characteristics associated with natural enemy abundance at Michigan native plants. *Environmental Entomology*, 36, 878–886.

Fiedler, A.K., Landis, D.A. and Wratten, S.D. (2008) Maximizing ecosystem services from conservation biological control: the role of habitat management. *Biological Control*, 45, 254–271.

Foster, M.A. and Ruesink, W.G. (1984) Influence of flowering weeds associated with reduced tillage in corn on a black cutworm (Lepidoptera: Noctuidae) parasitoid, *Meteorus rubens* (Nees von Esenbeck). *Environmental Entomology*, 13, 664–668.

Frescata, C. and Mexia, A. (1995) Biological control of western flower thrips with Orius laevigatus (Heteroptera: Anthocoridae) in organic strawberries in Portugal, in *Thrips biology and management* (eds B.L. Parker, M. Skinner and T. Lewis), NATO ASI Series A, Life Sciences Vol. 276, Plenum Press, New York, pp. 249–250.

Funao, T. and Yoshiyasu, Y. (1995) Development and fecundity of *Orius sauteri* (Poppius) (Hemiptera: Anthocoridae) reared on *Aphis gossypii* Glover and corn pollen. *Japanese Journal of Entomology and Zoology*, 39, 84–85.

Gibson, C.M. and Hunter, M.S. (2005) Reconsideration of the role of yeasts associated with Chrysoperla green lacewings. *Biological Control*, 32, 57–64.

Gilbert, F.S. (1981) Foraging ecology of hoverflies: morphology of the mouthparts in relation to feeding on nectar and pollen in some common urban species. *Ecological Entomology*, 6, 245–262.

Gilbert, F.S. (1985) Ecomorphological Relationships in Hoverflies (Diptera, Syrphidae). *Proceedings of the Royal Society B*, 224, 91–105.

Gilbert, F.S. and Jervis, M.A. (1998) Functional, evolutionary and ecological aspects of feeding-related mouthpart specializations in parasitoid flies. *Biological Journal of the Linnean Society*, 63, 495–535.

Gilbert, F.S. and Owen, O. (1990) Size, shape, competition, and community structure in hoverflies (Diptera: Syrphidae). *Journal of Animal Ecology*, 59, 21–39.

Golding, Y. and Edmunds, M. (2003) A novel method to investigate the pollen diets of hoverflies. *Journal of Biological Education*, 37, 182–185.

Gu, H., Wäckers, F.L., Steindl, P., Günther, D. and Dorn, S. (2001) Different approaches to labelling parasitoids using strontium. *Entomologia Experimentalis et Applicata*, 99, 173–181.

Gurr, G.M., Wratten, S.D., Tylianakis, J., Kean, J. and Keller, M. (2005) Providing plant foods for natural enemies in farming systems: balancing practicalities and theory, in *Plant-Provided Food for Carnivorous Insects A Protective Mutualism and its Applications* (eds F.L. Wäckers and P.C.J. van Rijn), Cambridge University Press, Cambridge, pp. 326–347.

Hagen, K.S. (1986) Ecosystem analysis: Plant cultivars, entomophagous species and food supplements, in *Interactions of Plant Resistance and Parasitoids and Predators of Insects* (eds D.J. Boethel and R.D. Eikenbary), John Wiley & Sons, Inc., New York, pp. 153–197.

Hagen, K.S., Tassan, R.L. and Sawall Jr., E.F. (1970) Some ecophysiological relationships between certain Chrysopa, honeydews and yeasts. *Bollettino Laboratorio Entomologia Agraria Filippo Silvestri*, 28, 113–134.

Hansen, D.M., Olesen, J.M. and Jones, C.G. (2002) Trees, birds and bees in Mauritius: exploitative competition between introduced honey bees and endemic nectarivorous birds? *Journal of Biogeography*, 29, 721–734.

Harmon, J.P., Ives, A.R., Losey, J.E., Olson, A.C. and Rauwald, K.S. (2000) *Colemegilla maculata* (Coleoptera: Coccinellidae) predation on pea aphids promoted by proximity to dandelions. *Oecologia*, 125, 543–548.

Haslett, J.R. (1989) Adult feeding by holometabolous insects – pollen and nectar as complementary nutrient sources for *Rhingia campestris* (Diptera, Syrphidae). *Oecologia*, 81, 361–363.

Hausmann, C., Wäckers, F.L. and Dorn, S. (2005) Sugar convertibility in the parasitoid *Cotesia glomerata* (Hymenoptera: Braconidae). *Archives of Insect Biochemistry and Physiology*, 60, 223–229.

Hayes, E.J. and Wall, R. (1999) Age-grading adult insects: a review of techniques. *Physiological Entomology*, 24, 1–10.

Heimpel, G.E. and Jervis, M.A. (2005) Does floral nectar improve biological control by parasitoids? In *Plant-provided food for carnivorous insects: a protective mutualism and its applications* (eds F.L. Wäckers, P.C.J. van Rijn and J. Bruin), Cambridge University Press, Cambridge, pp. 267–304.

Heimpel, G.E., Lee, J.C., Wu, Z. *et al.* (2004) Gut Sugar Analysis in Field-Caught Parasitoids: Adapting Methods Used on Biting Flies. *International Journal of Pest Management*, 50, 193–198.

Hickman, J.M., Lovei, G.L. and Wratten, S.D. (1995) Pollen feeding by adults of the hoverfly *Melanostoma fasciatum* (Diptera: Syrphidae). *New Zealand Journal of Zoology*, 22, 387–392.

Hogervorst, P.A.M., Wäckers, F.L. and Romeis, J. (2007) Detecting nutritional state and food source use in field-collected insects that synthesize honeydew oligosaccharides. *Functional Ecology*, 21, 936–946.

Hogervorst, P.A.M., Wäckers, F.L., Carette, A.-C. and Romeis, J. (2008) The importance of honeydew as food for larvae of *Chrysoperla carnea* in the presence of aphids. *Journal of Applied Entomology*, 132, 18–25.

Hogg, B.N., Bugg, R.L. and Daane, K.M. (2011) Attractiveness of common insectary and harvestable floral resources to beneficial insects. *Biological Control*, 56, 76–84.

Honěk, A. (1985) Habitat preferences of aphidophagous coccinellids [Coleoptera]. *BioControl*, 30, 253–264.

Human, H. and Nicolson, S.W. (2006) Nutritional content of fresh, bee-collected and stored pollen of Aloe greatheadii var. davyana (Asphodelaceae). *Phytochemistry*, 67, 1486–1492.

Idris, A.B. and Grafius, E. (1997) Nectar-collecting behavior of *Diadegma insulare* (Hymenoptera: Ichneumonidae), a parasitoid of diamondback moth (Lepidoptera: Plutellidae). *Environmental Entomology*, 26, 114–120.

Illingworth, J.F. (1921) Natural enemies of sugar-cane beetles in Queensland, Queensland Bureau of Sugar Experiment Stations, Division of Entomology, *Bulletin* 13, 1–47.

Inouye, D.W. (1983) The ecology of nectar robbing, in *The Biology of Nectaries* (eds B. Bentley and T. Elias), Columbia University Press, New York, pp. 153–173.

Irvin, N.A. and Hoddle, M.S. (2007) Evaluation of floral resources for enhancement of fitness of *Gonatocerus ashmeadi*, an egg parasitoid of the glassy-winged sharpshooter, *Homalodisca vitripennis*. *Biological Control*, 40, 80–88.

Isenhour, D.J. and Yeargan, K.V. (1981) Predation by Orius insidiosus (Hemiptera: Anthocoridae) on the soybean thrips Sericothrips variabilis (Thysanoptera: Thripidae): effect of prey stage and density. *Environmental Entomology*, 10, 496–500.

Jervis, M. (1998) Functional and evolutionary aspects of mouthpart structure in parasitoid wasps. *Biological Journal of the Linnean Society*, 63, 461–493.

Jervis, M.A., Kidd, N.A.C., Fitton, M.G., Huddleston, T. and Dawah, H.A. (1993) Flower-visiting by hymenopteran parasitoids. *Journal of Natural History*, 27, 67–105.

Jervis, M.A., Lee, J.C. and Heimpel, G.E. (2004) Use of behavioural and life-history studies to understand the effects of habitat manipulation, in *Ecological engineering for pest management advances in habitat manipulation for arthropods* (eds G.M. Gurr, S.D. Wratten and M.A. Altieri), CSIRO Publishing, Collingwood, pp. 65–101.

Jones, T.P. and Rowe, N.P. (1999) *Fossil Plants and Spores*, The Geological Society of London, London.

Kehrli, P. and Bacher, S. (2008) Differential effects of flower-feeding in an insect host-parasitoid system. *Basic and Applied Ecology*, 9, 709–717.

Kevan, G. (1973) Parasitoid wasps as flower visitors in the Canadian high artic. *Anz Schädlingskd Pflanz Umweltschutz*, 46, 3–7.

Kevan, P.G. and Baker, H.G. (1998) Insects on flowers, in *Ecological Entomology* (eds C.B. Huffaker and A.P. Gutierrez), John Wiley & Sons, Inc., New York, pp. 553–83.

Kikuchi, T. (1963) Studies on the coaction among insects visiting flowers. III. Dominance relationship among flower-visiting flies, bees and butterflies. *Scientific Reports of the Tohoku University*, 29, 1–8.

Kiman, Z.B. and Yeargan, K.V. (1985) Development and reproduction of the predator Orius insidiosus (Hemiptera: Anthocoridae) reared on diets of selected plant material and arthropod prey. *Annals of the Entomological Society of America*, 78, 464–467.

Lee, J.C. and Heimpel, G.E. (2008) Floral resources impact longevity and oviposition rate of a parasitoid in the field. *Journal of Animal Ecology*, 77, 565–572.

Lee, J.C., Andow, D.A. and Heimpel, G.E. (2006) Influence of floral resources on sugar feeding and nutrient dynamics of a parasitoid in the field. *Ecological Entomology*, 31, 470–480.

Letourneau, D.K. and Altieri, M.A. (1983) Abundance patterns of a predator, *Orius tricolor* (Hemiptera: Anthocoridae), and its prey, *Frankliniella occidentalis* (Thysanoptera:

Thripidae): habitat attraction in polycultures versus mono-cultures. *Environmental Entomology*, 12, 1464–1469.

Li, Y., Michael, M. and Romeis, J. (2010) Use of maize pollen by adult Chrysoperla carnea (Neuroptera: Chrysopidae) and fate of Cry proteins in *Bt*-transgenic varieties. *Journal of Insect Physiology*, 56, 157–164.

Limburg, D.D. and Rosenheim, J.A. (2001) Extrafloral nectar consumption and its influence on survival and development of an omnivorous predator, larval Chrysoperla plorabunda (Neuroptera: Chrysopidae). *Environmental Entomology*, 30, 595–604.

Lixa, A.T., Campos, J.M., Resende, A.L., Silva, J.C., Almeida, M.M. and Aguiar-Menezes, E.L. (2010) Diversity of Coccinellidae (Coleoptera) using aromatic plants (Apiaceae) as survival and reproduction sites in agroecological system. *Neotropical Entomology*, 39, 354–359.

Lundgren, J.G. (2009) Nutritional aspects of non-prey foods in the life histories of predaceous Coccinellidae. *Biological Control*, 51, 294–305.

Lundgren, J.G. and Wiedenmann, R.N. (2004) Nutritional suitability of corn pollen for the predator *Coleomegilla maculata* (Coleoptera: Coccinellidae). *Journal of Insect Physiology*, 50, 567–575.

Medeiros, M.A., Ribeiro, P.A., Morais, H.C., Castelo, B.M., Sujii, E.R. and Salgado-Labouriau, M.L. (2010) Identification of plant families associated with the predators *Chrysoperla externa* (Hagen) (Neuroptera: Chrysopidae) and *Hippodamia convergens* Guerin-Meneville (Coleoptera: Coccinelidae) using pollen grain as a natural marker. *Brazilian Journal of Biology*, 70, 293–300.

Michaud, J.P. and Grant, A.K. (2005) Suitability of pollen sources for the development and reproduction of *Coleomegilla maculata* (Coleoptera: Coccinellidae) under simulated drought conditions. *Biological Control*, 32, 363–370.

Morse, D.H. (1981) Interactions among syrphid flies and bumblebees on flowers. *Ecology*, 62, 81–88.

Nedved, O., Ceryngier, P., Hodkova, M. and Hodek, I. (2001) Flight potential and oxygen uptake during early dormancy in *Coccinella septempunctata*. *Entomologia Experimentalis et Applicata*, 99, 371–380.

Obrycki, J.J., Harwood, J.D., Kring, T.J. and O'Neil, R.J. (2009) Aphidophagy by Coccinellidae: Application of biological control in agroecosystems. *Biological Control*, 51, 244–254.

Olsen, D. and Wäckers, F.L. (2007) Management of field margins to maximize multiple ecological services. *Journal of Applied Ecology*, 44, 13–21.

Ouyang, Y., Grafton-Cardwell, E.E. and Bugg, R.L. (1992) Effects of various pollens on development, survivorship, and reproduction of *Euseius tularensis* (Acari: Phytoseiidae) *Environmental Entomology*, 21, 1371–1376.

Patt, J.M., Hamilton, G.C. and Lashomb, J.H. (1997) Foraging success of parasitoid wasps on flowers: interplay of insect morphology, floral architecture and searching behavior. *Entomologia Experimentalis et Applicata*, 83, 21–30.

Patt, J.M., Wainright, S.C., Hamilton, G.C. *et al.* (2003) Assimilation of carbon and nitrogen from pollen and nectar by a predaceous larva and its effects on growth and development. *Ecological Entomology*, 28, 717–728.

Pemberton, R.W. and Lee, J.H. (1996) The influence of extra-floral nectaries on parasitism on an insect herbivore. *American Journal of Botany*, 83, 1187–1194.

Pemberton, R.W. and Vandenberg, N.J. (1993) Extrafloral nectar feeding by ladybird beetles (Coleoptera: Coccinellidae). *Proceedings of the Entomological Society of Washington*, 95, 139–151.

Pontin, D.R., Wade, M.R., Kehrli, P. and Wratten, S.D. (2006) Attractiveness of single and multiple species flower patches to beneficial insects in agroecosystems. *Annals of Applied Biology*, 148, 39–47.

Portillo, N., Alomar, O. and Wäckers, F.L. (2012) Nectarivory by the plant-tissue feeding predator Macrolophus pygmaeus Rambur (Heteroptera: Miridae): nutritional redundancy or nutritional benefit? *Journal of Insect Physiology* (in press).

Putman, W.L. (1955) Bionomics of *Stethorus punctillum* Weise (Coleoptera: Coccinellidae) in Ontario. *Canadian Entomologist*, 87, 9–33.

Rahat, S., Gurr, G.M., Wratten, S.D., Mo, J.H. and Neeson, R. (2005) Effect of plant nectars on adult longevity of the stinkbug parasitoid, *Trissolcus basalis*. *International Journal of Pest Management*, 51, 321–324.

Reader, T., MacLeod, I., Elliott, P.T., Robinson, O.J. and Manica, A. (2005) Inter-order interactions between flower-visiting insects: Foraging bees avoid flowers previously visited by hoverflies. *Journal of Insect Behavior*, 18, 51–57.

Reemer, M., Renema, W., van Steenis, W. *et al.* (2009) *De Nederlandse zweefvliegen (Diptera: Syrphidae) – Nederlandse Fauna 8*. Leiden. Nationaal Natuurhistorisch Museum Naturalis, KNNV Uitgeverij, European Invertebrate Survey Nederland.

Richards, P.C. and Schmidt, J.M. (1996) The effects of selected dietary supplements on survival and reproduction of *Orius insidiosus* (Say) (Hemiptera: Anthocoridae). *Canadian Entomologist*, 128, 171–176.

Robertson, C. (1928) *Flowers and insects: lists of visitors of four hundred and fifty three flowers*, The Science Press, Lancaster.

Salas-Aguilar, J. and Ehler, L.E. (1977) Feeding habits of Orius tristicolor. *Annals of the Entomological Society of America*, 70, 60–62.

Scarratt, S.L., Wratten, S.D. and Shishehbor, P. (2008) Measuring parasitoid movement from floral resources in a vineyard. *Biological Control*, 46, 107–113.

Schmid, A. (1992) Untersuchungen zur Attraktivität von Ackerwildkräutern für aphidophage Marienkäfer (Coleoptera, Coccinellidae). *Agrarökologie*, 5. Paul Haupt, Bern.

Scholz, D. and Poehling, H.M. (2000) Oviposition site selection of *Episyrphus balteatus*. *Entomologia Experimentalis et Applicata*, 94, 149–158.

Silberbauer, L., Yee, M., Del Socorro, A., Wratten, S., Gregg, P. and Bowie, M. (2004) Pollen grains as markers to track the movements of generalist predatory insects in agroecosystems. *International Journal of Pest Management*, 50, 165–171.

Silveira, L.C.P., Bueno, V.H.P., Pierre, L.S.R. and Mendes, S.M. (2003) Crops and weeds as host plants *Orius* species (Heteroptera: Anthocoridae). *Bragantia Campinas*, 62, 261–265.

Silveira, L.C.P., Bueno, V.H.P., Louzada, J.N.C. and Carvalho L.M. (2005) Species of *Orius* (Hemiptera, Anthocoridae) and thrips (Thysanoptera): interaction in the same habitat? *Revista Árvore Viçosa-MG*, 29, 767–773.

Skirvin, D.J., Kravar-Garde, L., Reynolds, K., Jones, J., Mead, A. and Fenlon, J. (2007) Supplemental food affects thrips predation and movement of *Orius laevigatus* (Hemiptera: Anthocoridae) and *Neoseiulus cucumeris* (Acari: Phytoseiidae). *Bulletin of Entomological Research*, 97, 309–315.

Ssymank, A. and Gilbert, F. (1993) Anemophilous pollen in the diet of Syrphid flies with special reference to the leaf feeding strategy occurring in Xylotini (Diptera, Syrphidae). *Deutsche Entomologiste Zeitschrift*, 40, 245–258.

Stelzl, M. (1991) Investigations on food of Neuroptera-adults (Neuropteroidea, Insecta) in Central-Europe – with a short discussion of their role as natural enemies of insect pests. *Journal of Applied Entomology*, 111, 469–477.

Steppuhn, A. and Wäckers, F.L. (2004) HPLC sugar analysis reveals the nutritional state and the feeding history of parasitoids. *Functional Ecology*, 18, 812–819.

Takasu, K. and Lewis, W.J. (1993) Host- and food-foraging of the parasitoid *Microplitis croceipes*: learning and physiological state effects. *Biological Control*, 3, 70–74.

Tooker, J.F. and Hanks, L.M. (2000) Flowering plant hosts of adult hymenopteran parasitoids of central Illinois. *Annals of the Entomological society of America*, 93, 580–588.

Tooker, J.F., Hauser, M. and Hanks, L.M. (2006) Floral host plants of Syrphidae and Tachinidae (Diptera) of central Illinois. *Annals of the Entomological Society of America*, 99, 96–112.

Vacante, V., Cocuzza, G.E., De Clercq, P., van de Veire, M. and Tirry, L. (1997) Development and survival of *Orius albidipennis* and *O. laevigatus* (Het.: Anthocoridae) on various diets. *Entomophaga*, 42, 493–498.

van den Meiracker, R.A.F. and Ramakers, P.M.J. (1991) Biological control of the western flower thrips, *Frankliniella occidentalis*, in sweet pepper with the anthocorid predator *Orius insidiosus*. *Mededelingen van de Faculteit Landbouwwetenschappen, Rijksuniversiteit Gent*, 56, 241–249.

van Emden, H.F. (1962) Observations on the effect of flowers on the activity of parasitic Hymenoptera. *Entomologist's Monthly Magazine*, 98, 265–270.

van Rijn, P.C.J. and Tanigoshi, L.K. (1999) Pollen as food for the predatory mites *Iphiseius degenerans* and *Neoseiulus cucumeris* (Acari: Phytoseiidae): dietary range and life history. *Experimental and Applied Acarology*, 23, 785–802.

van Rijn, P.C.J. and Wäckers, F.L. (2010) The suitability of field margin flowers as food source for zoophagous hoverflies. *IOBC/WPRS Bulletin*, 56, 125–128.

van Rijn, P.C.J., van Houten, Y.M. and Sabelis, M.W. (2002) How plants benefit from providing food to predators even when it is also edible to herbivores. *Ecology*, 83, 2664–2679.

van Rijn, P.C.J., Kooijman, J. and Wäckers, F.L. (2006) The impact of floral resources on syrphid performance and cabbage aphid biological control. *IOBC/WPRS Bulletin*, 29, 149–152.

Vattala, H.D., Wratten, S.D., Phillips, C.B. and Wäckers, F.L. (2006) The influence of flower morphology and nectar quality on the longevity of a parasitoid biological control agent. *Biological Control*, 39, 179–185.

Venzon, M., Janssen, A. and Sabelis, M.W. (1999) Attraction of a generalist predator towards herbivore-infested plants. *Entomologia Experimentalis et Applicata*, 93, 305–314.

Venzon, M., Rosado, M.C., Euzebio, D.E., Souza, B. and Schoereder, J.H. (2006) Suitability of leguminous cover crop pollens as food source for the green lacewing *Chrysoperla externa* (Hagen) (Neuroptera: Chrysopidae). *Neotropical Entomology*, 35, 371–376.

Villenave, J., Thierry, D., Al Mamun, A., Lode, T. and Rat-Morris, E. (2005) The pollens consumed by common green lacewings *Chrysoperla* spp. (Neuroptera: Chrysopidae) in cabbage crop environment in western France. *European Journal of Entomology*, 102, 547–552.

Villenave, J., Deutsch, B., Lode, T. and Rat-Morris, E. (2006) Pollen preference of the *Chrysoperla* species (Neuroptera: Chrysopidae) occurring in the crop environment in western France. *European Journal of Entomology*, 103, 771–777.

Vollhardt, I.M.G., Bianchi, F.J.J.A., Wäckers, F.L., Thies, C. and Tscharntke, T. (2010) Nectar versus honeydew feeding by aphid parasitoids: does it pay to have a discriminating palate? *Entomologia Experimentalis et Applicata*, 137, 1–10.

Wäckers, F.L. (1994) The effect of food deprivation on the innate visual and olfactory preferences in *Cotesia rubecula*. *Journal of Insect Physiology*, 40, 641–649.

Wäckers, F.L. (2001) A comparison of nectar- and honeydew sugars with respect to their utilization by the hymenopteran parasitoid *Cotesia glomerata*. *Journal of Insect Physiology*, 47, 1077–1084.

Wäckers, F.L. (2004) Assessing the suitability of flowering herbs as parasitoid food sources: flower attractiveness and nectar accessibility. *Biological Control*, 29, 307–314.

Wäckers, F.L. (2005) Suitability of (extra-)floral nectar, pollen, and honeydew as insect food sources, in *Plant-provided food for carnivorous insects: a protective mutualism and its applications* (eds F.L. Wäckers, P.C.J. van Rijn and J. Bruin), Cambridge University Press, Cambridge, pp. 17–74.

Wäckers, F.L. and Steppuhn, A. (2003) Characterizing nutritional state and food source use of parasitoids collected in fields with high and low nectar availability. *IOBC/WPRS Bulletin*, 26, 203–208.

Wäckers, F.L. and Swaans, C.P.M. (1993) Finding floral nectar and honeydew in *Cotesia rubecula*: random or directed? *Proceedings of Experimental and Applied Entomology*, 4, 67–72.

Wäckers, F.L., Björnsen, A. and Dorn, S. (1996) A comparison of flowering herbs with respect to their nectar accessibility for the parasitoid *Pimpla turionellae*. *Proceedings of Experimental and Applied Entomology*, 7, 177–182.

Wäckers, F.L., Lee, J.C., Heimpel, G.E., Winkler, K. and Wagenaar, R. (2006a) Hymenopteran parasitoids synthesize "honeydew-specific" oligosaccharides. *Functional Ecology*, 20, 790–798.

Wäckers, F.L., Bonifay, C., Vet, L.M. and Lewis, W.J. (2006b) Gustatory response and appetitive learning in *Microplitis croceipes* in relation to sugar type and concentration. *Animal Biology*, 56, 193–203.

Wäckers, F.L., Romeis, J. and van Rijn, P.C.J. (2007) Nectar and pollen feeding by insect herbivores and implications for multitrophic interactions. *Annual Review of Entomology*, 52, 301–323.

Wäckers, F.L., van Rijn, P.C.J. and Heimpel, G.E. (2008) Exploiting honeydew as a food source. Making the best of a bad meal? *Biological Control*, 45, 176–185.

Wanner, H., Gu, H., Gunther, D., Hein, S. and Dorn, S. (2006) Tracing spatial distribution of parasitism in fields with flowering plant strips using stable isotope marking. *Biological Control*, 39, 240–247.

Waser, N.M. and Ollerton, J. (eds) (2006) . *Plant–pollinator interactions: from specialization to generalization*, The University of Chicago Press, Chicago.

Waser, N.M., Chittka, L., Price, M.V., Williams, N.M. and Ollerton, J. (1996) Generalization in pollination systems, and why it matters. *Ecology*, 77, 1043–1060.

Weber, D.C. and Lundgren, J.G. (2011) Effect of prior diet on consumption and digestion of prey and non-prey food by adults of the generalist predator *Coleomegilla maculate*. *Entomologia Experimentalis et Applicata*, DOI: 10.1111/j.1570-7458.2011.01141.x.

Weiss, E. and Stettmer, C. (1991) Unkräuter in der Agrarlandscahft locken blütenbesuchende Nutzinsekten an. *Agrarökologie*, 1, 1–104.

Williams, L. and Roane, T.M. (2007) Nutritional ecology of a parasitic wasp: food source affects gustatory response, metabolic utilization and survivorship. *Journal of Insect Physiology*, 53, 1262–1275.

Wilson, E.E., Sidhu, C.S., Levan, K.E. and Holway, D.A. (2010) Pollen foraging behaviour of solitary Hawaiian bees revealed through molecular pollen analysis. *Molecular Ecology*, 19, 4823–4829.

Winkler, K., Wäckers, F.L., Buitriago, L. and van Lenteren, J.C. (2005) Herbivores and their parasitoids show differences in abundance on eight different nectar producing plants. *Proceedings of Experimental Entomology*, 16, 36–42.

Winkler, K., Wäckers, F.L., Bukovinszkine-Kiss, G. and van Lenteren, J.C. (2006) Nectar resources are vital for *Diadegma semiclausum* fecundity under field conditions. *Basic and Applied Ecology*, 7, 133–140.

Winkler, K., Wäckers, F.L. and Pinto, D. (2009a) Nectar-providing plants enhance the energetic state of herbivores as well as their parasitoids under field conditions. *Ecological Entomology*, 34, 221–227.

Winkler, K., Wäckers, F.L., Kaufman, L.V., Larraz, V.G. and van Lenteren, J.C. (2009b) Nectar exploitation by herbivores and their parasitoids is a function of flower species and relative humidity. *Biological Control*, 50, 299–306.

Winkler, K., Wäckers, F.L., Termorshuizen, A.J. and van Lenteren, J.C. (2010) Assessing potential risks and benefits of floral supplements in conservation biological control. *Biological Control*, 55, 719–727.

Wyckhuys, K.A.G., Strange-George, J.E., Kulhanek, C.A., Wäckers, F.L. and Heimpel, G.E. (2008) Sugar feeding by the aphid parasitoid *Binodoxys communis*: how does honeydew compare to other sugar sources? *Journal of Insect Physiology*, 54, 481–491.

Yano, E. (1996) Biology of *Orius sauteri* (Poppius) and its potential as a biocontrol agent for *Thrips palmi*. *IOBC/WPRS Bulletin*, 19, 203–206.

Chapter 10

THE MOLECULAR REVOLUTION: USING POLYMERASE CHAIN REACTION BASED METHODS TO EXPLORE THE ROLE OF PREDATORS IN TERRESTRIAL FOOD WEBS

William O.C. Symondson

Biodiversity and Insect Pests: Key Issues for Sustainable Management, First Edition. Edited by Geoff M. Gurr, Steve D. Wratten, William E. Snyder, Donna M.Y. Read.

INTRODUCTION

The complexity and diversity of invertebrate communities on temperate agricultural land, even in the middle of arable crops within conventionally managed fields, can be surprising. The soil itself is full of earthworms, slugs, beetles and other macroorganisms, teaming with microarthropods such as Collembola and mites (Seastedt, 1984; Brennan *et al.*, 2006), swarming with nematodes and replete with bacteria and other microorganisms (Curry, 1994). Above the soil, crop plants, and the soil surface from which they are growing, are patrolled by a diversity of predators, including carabid, cantharid, staphylinid and coccinellid beetles, plus a diverse range of spiders, predatory bugs, lacewings and predatory flies. These are all feeding on herbivores on the crop itself (including pests), on soil organisms and on each other (intraguild predation and cannibalism). In addition, many of the herbivores (and some of the predators) will be under attack from parasitoids. This process is not one-way, because the predators may also be catching and consuming the adult parasitoids as well as hosts that contain immature parasitoids (Brodeur and Rosenheim, 2000; Snyder and Ives, 2001; Traugott and Symondson, 2008; Chacon and Heimpel, 2010; Traugott *et al.*, 2011). This tangled web of interconnections forms a food web and the population densities that arise are the end product of all these complex trophic interactions. The web is sustained both by the crop itself and by the soil beneath, with all the associated fungi, bacteria and detritivores that influence nutrient release and cycling. The soil, the crop and the associated food webs are all affected in turn by the history of the field, by abiotic factors (climate, geology, etc.) and by farming practices (chemical or organic fertilisers, pesticides, agronomic factors). The complexity of the system may increase significantly where organic, low input, conservation or no-tillage systems are instituted, or where 'beetle banks' or flowering strips along field margins are added to attract predators. All of these can increase the diversity of plants, invertebrates and microhabitats (Bengtsson *et al.*, 2005; Hole *et al.*, 2005; Gibson *et al.*, 2007; Letourneau & Bothwell, 2008) (Figure 10.1). These webs are highly dynamic, changing over time and space. In different arable crops, in horticulture, in orchards and in forestry, indeed in any situations where pests are problematic, the food webs surrounding those pests will be different.

How, therefore, can we possibly hope to understand and quantify the interactions within such food webs, particularly those that affect pest numbers? How too can we investigate the effects that different levels of species diversity within such webs will affect outcomes? It is known that specialist natural enemies such as lacewings or aphidophagous coccinellid beetles eat aphids. These predators are unlikely to be affected much by a diversity of other non-aphid prey and will continue to eat the aphids on the crop (or associated weeds) regardless. Such coupled predator–prey relationships can to some degree be parametised and modelled. The problem comes with generalist predators.

Figure 10.1 Even in conventionally managed cereal fields a) the complexity of invertebrate food webs can be considerable. However, in farming systems where chemicals such as herbicides are avoided b), the diversity of plants and invertebrates, and the complexity of the food webs therein, can be much greater.

Groups such as spiders and, to an even greater extent, carabid beetles (many of which eat an even wider range of invertebrates, including molluscs and earthworms) will eat almost anything smaller than themselves that they are capable of capturing and killing. What factors affect their prey choice? How will greater prey diversity affect prey choice? Will greater non-pest prey biomass or diversity in the crop divert the predators from feeding on the pests (Halaj and Wise, 2002; Koss and Snyder, 2005; Rypstra and Marshall, 2005)? Or will a greater diversity and/or abundance of non-pest foods improve predator fitness (Toft and Wise, 1999; Oelbermann and Scheu, 2002; Mayntz *et al.*, 2005; Harwood *et al.*, 2009a) and hence help to sustain higher populations of predators that kill more pests? Both processes may occur simultaneously (Symondson *et al.*, 2006). These critical questions are discussed in detail in Symondson *et al.* (2002) and in several of the chapters in this volume. In particular, Welch *et al.* (chapter 3) focus on generalist predators whilst Snyder and Tylianakis (chapter 2) provide a broader ecological perspective.

At present we understand very little about what happens within food webs that results in favourable outcomes (fewer pests) or unfavourable outcomes (more pests). In practice predators do not feed at random on whatever is available to them. Instead, prey choice is affected by a complex of interacting factors including relative prey densities, encounter rates between predators and prey, prey escape strategies and defensive mechanisms, the nutritional value of the prey, spatial and temporal dynamics of predators and prey, scavenging, intraguild predation and many other factors. Where prey availability is low predators will generally be less selective, eating whatever they can find (Anderson, 1974; Symondson *et al.*, 2002). The feeding history of the predator may affect its prey choices as it attempts to balance its nutritional requirements (Greenstone, 1979; Mayntz *et al.*, 2005). Unless we can start to quantify precisely who is eating whom, and in what quantities, we are unlikely to go far beyond the connectance food webs stage (built upon laboratory studies of who is capable of eating whom, rather than upon quantitative field data on actual prey choices) (Sunderland *et al.*, 2005).

The problem with invertebrates is that they are difficult to observe. Most are very small, live within thick vegetation (including crops) or beneath the soil, or may be active mainly at night. Trying to observe predation events directly is difficult if not impossible in many instances. The act of trying to observe such interactions (by parting the vegetation or using lights at night) will very often disturb the animals involved. Animals high in the vegetation may be knocked to the ground, providing easy (but unnatural) pickings for ground predators. Long periods of observation would be needed to gather limited data of uncertain quality. An alternative is to collect predators from the field and dissect them to see what they have been eating (Ingerson-Mahar, 2002; Sunderland *et al.*, 2005). This can provide useful data on the diets of invertebrates that chew up and consume the whole of their prey, such as most carabid, coccinellid and staphylinid beetles (e.g. Forsythe, 1983). Such work, however, requires the rare ability to identify species from highly fragmented, semi-digested remains and takes a long period of tedious work to get results. Worse, many soft-bodied prey leave few if any hard remains possessing diagnostic features (earthworms, slugs, eggs, many kinds of insect larvae), making microscopic analyses necessarily biased. Worse still, the majority of invertebrate predators are fluid feeders (spiders, bugs, carabid larvae, etc.), making this approach impossible for them.

The alternative is to use more objective, biochemical means to identify prey remains in the guts of predators. The subject of this chapter, therefore, is how we can use 'molecular' techniques to shed light on the contents of the 'black box' from which invertebrate population densities, diversities and rates of growth emerge. It will concentrate upon the use of molecular diagnostics to analyse trophic links between generalist predators and their many different potential prey. The term 'molecular' can have a variety of meanings but here is defined in the same sense as in 'molecular biology', encompassing techniques that involve DNA. The term 'molecular ecology' on the other hand more loosely includes a range of mainly DNA-based and protein-based techniques that are, or have been, used to analyse phylogenetic relationships and trophic interactions. The advent of PCR (polymerase chain reaction) based methods for tracking predator–prey interactions has largely displaced all protein-based approaches (for very good reasons, described below). However, protein-based methods were once widely used and still appear occasionally in the literature. The most widely used of these were protein electrophoresis and assays based upon polyclonal antisera or monoclonal antibodies. The latter are still a good choice for screening large numbers of predators quickly and efficiently (Harwood, 2008; Griffiths *et al.*, 2008) and, unlike DNA-base techniques, can even be targeted at particular stages (e.g. eggs, larval instars, reviewed in Symondson,

2002). Reviews of these techniques, and their application, can be found in Murray *et al.* (1989), Solomon *et al.* (1996), Sunderland (1988), Greenstone (1996), Symondson and Hemingway (1997), Symondson (2002a; 2002b), Sheppard and Harwood (2005) and Sunderland *et al.* (2005). The rest of this chapter will, however, concentrate upon the rapidly developing field of PCR-based techniques and applications, with particular emphasis on the study of predation within biodiverse communities.

WHY USE MOLECULAR MARKERS?

The DNA revolution is penetrating multiple areas of ecology, from evolutionary process to phylogenetics, population genetics, conservation genetics and trophic relationships. Part of this revolution has been the development of molecular barcoding (Hebert *et al.*, 2003a; 2003b). Vast databases (Genbank, the Barcoding of Life Database (BOLD)) of 'barcoding' gene sequences are being constructed. The standard 'barcode genes' used for most invertebrates are found within a region of the mitochondrial cytochrome oxidase I gene (COI) (Folmer *et al.*, 1994; Hebert *et al.*, 2003b). These 'barcodes' are invaluable as species-specific genetic markers as they are (mostly) conserved at the species level. If the COI gene fails to distinguish between closely related species, or does not provide suitable primer sites, then sites on other mitochondrial genes, including cytochrome oxidase II, or the ribosomal 12S and 16S genes, have been found to be useful for invertebrates. These can be initially amplified with published general primers and then sequenced. Similarly, multiple-copy sites within the nuclear ribosomal gene clusters, particularly 18S and Internal Transcribed Spacer Regions (ITS1 and ITS2), have proven to be useful (Table 10.1). This means that new species-specific genetic markers can be easily designed. These markers, in the form of PCR primers, then have to go through a thorough testing and optimisation process (King *et al.*, 2008). They are much quicker to make, more sensitive and usually more specific than any of the older protein-based markers, and cheaper too. By contrast monoclonal antibodies, the most specific of the protein markers, took many months to create a single specificity in a dedicated tissue culture facility. Unless one could beg or buy some of these antibodies they were not available to anyone else. Primer sequences, however, are published and available for use by anyone. Their rapidity of production means that

we are no longer limited to looking simply at the range of predators attacking a target prey species, but can begin to develop a whole library of markers that can be used by anyone, anywhere, to study the prey ranges of predators. Table 10.1 lists most of the primers that have been developed to date for analysis of predation on invertebrate pests and other non-pest invertebrate prey in the same ecosystem, but this list is growing rapidly and by the time this chapter is published it will be out of date.

THE BASICS: DETECTING PREDATION ON A TARGET PREY SPECIES

Full details of the processes involved, the need for adequate controls and solutions to the problems that may be encountered can be found in King *et al.* (2008). Here the aim is not to go into the details of protocols, but rather to give an overview of the main techniques involved and how they can be used to address questions related to analysing pest control by generalist predators within diverse communities.

If existing primers are unavailable for the target species new ones must be designed. The aim may be to produce a marker that will amplify a single species, perhaps a pest. To maximise the chances that this happens DNA is extracted from the target species, the predators species that will be screened and any species in the same ecosystem that are closely related to the target species (Figure 10.2). The DNA is then amplified using general invertebrate primers, usually those targeting the COI barcoding region (Folmer *et al.*, 1994), but many other general primers are available targeting different regions of the COI or other genes (e.g. Simon *et al.*, 1994; 2006). It was shown at an early stage (Zaidi *et al.*, 1999) that 'multiple-copy' genes need to be targeted to improve sensitivity. Degraded, semi-digested DNA in the guts of predators can be detected for significant time periods post-ingestion by targeting mitochondrial DNA, with hundreds or thousands of copies per cell, or the multiple-copy nuclear ribosomal gene clusters. The sequences for the target prey, the predators and related non-target prey are then aligned and new species-specific primers designed targeting sites unique to the prey sequence (King *et al.*, 2008). Many studies have shown that shorter fragments survive digestion better and will be detected for longer post-ingestion periods, and therefore new primers are generally designed to amplify fragments around 100–300 bp in length (Table 10.1). This is a conservative

Table 10.1 Summary of published primers that are currently available for molecular analyses of predation. These include invertebrate pests plus non-pest prey in the same ecosystems, with details of target gene regions and amplicon sizes.

Target groups and species	Target region*	Amplicon size (bp)	Reference
Lepidoptera			
Helicoverpa armigera (Hübner)	SCAR	254, 600, 1,100	Agustí *et al.*, 1999
Ostrinia nubilalis (Hübner)	ITS-1	150, 156, 369, 492	Hoogendoorn and Heimpel, 2001
Scotorythra rara (Butler)	COI	141	Sheppard *et al.*, 2004
General *Eupithecia*	COI	151	
General Geometridae	COI	170	
Plutella xylostella (L.)	ITS-1	275	Ma *et al.*, 2005
Plutella xylostella (L.)	COI	293	Hosseini *et al.*, 2008
General Lepidoptera	COI	648	Clare *et al.*, 2009[1]
Hemiptera – 1. Aphids			
Schizaphis graminum (Rondani)	COII	111, 166, 386	Chen *et al.*, 2000
Diuraphis noxia (Mordvilko)	COII	100, 137, 348	*Primers also used in:*
Rhopalosiphum padi (L.)	COII	77, 148, 331	*Greenstone and Shufran, 2003*
Rhopalosiphum maidis (Fitch)	COII	198, 246, 339	*Harper* et al., *2005*
Sipha flava (Forbes)	COII	291, 326	*McMillan* et al., *2007*
Sitobion avenae (Fabricius)	COII	159, 231	*Kuusk* et al., *2008*
General aphids	COII	181	
Rhopalosiphum insertum (Walker)	ND1, 16S	283	Cuthbertson *et al.*, 2003
Megoura viciae (Buckton)	COI	148	Harper *et al.*, 2005
Myzus persicae (Sulzer)	COI	160	*Primers also used in:*
Aphis fabae (Scopoli)	COI	212	*Foltan* et al., *2005*
General aphids	COI	242	*Sheppard* et al., *2005*
			Harper et al., *2006*
			Schmidt et al., *2009*
Sitobion avenae (Fabricius)	COI	110	Sheppard *et al.*, 2005[2]
General aphids	COI	242	Harper *et al.*, 2006
General aphids	COI	101	Chapman *et al.*, 2010
Aphis glycines (Matsumura)	COI	255	Harwood *et al.*, 2007
			Primers also used in:
			Harwood et al., *2009b*
			Lundgren and Weber, 2010
Sitobion avenae (Fabricius)	COI	113	King *et al.*, 2010a
Sitobion avenae (Fabricius)	COI	85, 231, 317, 383	von Berg *et al.*, 2008a
			Primers also used in:
			von Berg et al., *2008b*
			Birkhofer et al., *2008*
			Traugott et al., *2008*
Aphis fabae (Scopoli)	COI	122,369	Traugott and Symondson, 2008
Myzus persicae (Sulzer)	Esterase	105	Schmidt *et al.*, 2009
Hemiptera – 2. Others			
Trialeurodes vaporariorum (Westwood)	SCAR	310	Agustí *et al.*, 2000
Cacopsylla pyricola (Foerster)	COI	188, 271	Agustí *et al.*, 2003b
Homalodisca coagulate (Say)	SCAR, COI,	166–302	de Leon *et al.*, 2006
Homalodisca liturata (Ball)	COII	166–295	
Stephanitis pyrioides (Scott)	COI	116	Rinehart and Boyde, 2006
Bemisia tabaci (Gennadius)	SCAR	240	Zhang *et al.*, 2007a
Bemisia tabaci (Gennadius)	SCAR	93	Zhang *et al.*, 2007b
Homalodisca vitripennis (Germar)	COI	197	Fournier *et al.*, 2008

Table 10.1 *(Continued)*

Target groups and species	Target region*	Amplicon size (bp)	Reference
Collembola			
Isotoma anglicana (Lubbock)	COI	276	Agustí *et al.*, 2003a
Lepidocyrtus cyaneus (Tullberg)	COI	216	
Entomobrya multifasicata (Tullberg)	COI	211	
General Collembola	18S	177, 272	Kuusk and Agustí, 2008, *Primers also used in:* Kuusk and Ekbom, 2010 King et al., 2010a Chapman et al., 2010
Coleoptera			
Sitona sp.	COI	151	Harper *et al.*, 2005
Melolontha melolontha (L.)	COI	175, 273, 387, 585	Juen & Traugott, 2005
Amphimallon solstitiale (L.)	COI	127, 463	Juen and Traugott, 2006
Phyllopertha horticola (L.)	COI	291	Juen and Traugott, 2007
Leptinotarsa decemlineata (Say)	COI	214	Greenstone *et al.*, 2007 *Primers also used in:* Weber and Lundgren, 2009 Lundgren and Weber, 2010 Greenstone et al., 2010 Szendrei et al., 2010 Greenstone et al., 2011
Leptinotarsa juncta (Germar)	COI	219	
Enaphalodes rufulus (Haldeman)	16S	342	Muilenburg *et al.*, 2008
Diabrotica virgifera (LeConte)	COI, tRNA-Leu	119	Lundgren *et al.*, 2009
Notiophilus biguttatus (Fabricius)	COI	274	King *et al.*, 2010a
Trechus quadristriatus (Schrank)	COI	101	
Hypothenemus hampei (Ferrari)	COI	145	Jaramillo *et al.*, 2010
Harmonia axyridis (Pallas)	COI	261	Harwood *et al.*, 2007 *Primers also used in:* Harwood et al., 2009b
Meligethes aeneus (Fabricius)	COI	65–578	Cassel-Lindhagen *et al.*, 2009
Diptera			
Culex quinquefasciatus (Say)	α esterase	146, 263	Zaidi *et al.*, 1999
Anopheles gambiae (Giles)	ITS	290	Morales *et al.*, 2003
Anopheles gambiae (Giles)	IGS	202	Schielke *et al.*, 2007
Anopheles gambiae (Giles)	IGS	390	Ohba *et al.*, 2010
Anopheles arabiensis (Patton)	IGS	315	
Anopheles merus (Dönitz)	IGS	466	
Aedes aegypti (L.)	ITS	550	Ohba *et al.*, 2011[3]
Aedes albopictus (Skuse)	ITS	950	
Ceratitis capitata (Wiedemann)	ITS1	130, 333	Monzo *et al.*, 2010
Sitodiplosis mosellana (Géhin)	COI	271	King *et al.*, 2010a
General Diptera (Brachycera and Cyclorrhapha)	18S	171–175	
Thysanoptera			
Neohydatothrips variabilis (Beach)	COI	160	Harwood *et al.*, 2007 *Primers also used in:* Harwood et al., 2009b
Hymenoptera			
Lysiphlebus testaceipes (Cresson)	COI	291	Traugott and Symondson, 2008
Myriapoda			
Cylindroiulus fulviceps (Latzel)	COI	104	Seeber *et al.*, 2010
Cylindroiulus meinerti (Verhoeff)		96	

(Continued)

Table 10.1 *(Continued)*

Target groups and species	Target region*	Amplicon size (bp)	Reference
Nematoda			
Phasmarhabditis hermaphrodita (Schneider)	COI	154	Read *et al.*, 2006
Heterorhabditis megidis (Poinar, Jackson & Klein)	COI	150	
Steinernema feltiae (Filipjev)	COI	203	
Mollusca			
General *Arion* spp. (species within genus separation by amplicon length)	12S	204–221	Harper *et al.*, 2005 *Primers also used in:* Dodd et al., *2003; 2005*
Arion hortensis (Férussac)	12S	130	Foltan et al., *2005*
Deroceras reticulatum (Müller)	12S	109, 226, 294	Bell et al., *2010*
Vallonia pulchella (Müller)	12S	117	King et al., *2010a*
Candidula intersecta (Poiret)	12S	137	
Annelida			
General earthworms	12S	225–236	Harper *et al.*, 2005 *Primers also used in:* Harper et al., *2006*
General earthworm	COI	523	Admassu *et al.*, 2006
Allolobophora chlorotica (Savigny) (five separate lineages)	COI	126–261	King *et al.*, 2010b *Primers also used in:* Bell et al., *2010*
Aporrectodea longa (Ude)	COI	213	
Aporrectodea caliginosa (Savigny)	16S	116	
Aporrectodea rosea (Savigny) (two separate lineages)	COI	167, 171	
Lumbricus castaneus (Savigny)	COI	189	
Lumbricus rubellus (Hoffmeister)	COII	164	
Lumbricus terrestris (L.)	COII	256	
General primers			
General invertebrates	12S	165–400	Sutherland, 2000
General invertebrates and vertebrates	16S	500–650	Kasper *et al.*, 2004
General arthropods	Cyt b	358	Pons, 2006
General invertebrates	COI	332	Harper *et al.*, 2006
General invertebrates	COI	157	Zeale *et al.*, 2010

*COI and COII (cytochrome oxidase I and II genes, mtDNA), 12S and 16S (ribosomal RNA genes, mtDNA), Cyt b (cytochrome b, mtDNA), ND1 (NADH dehydrogenase 1, mtDNA), 18S (ribosomal RNA gene, nuclear), ITS-1 (internal transcribed spacer 1 rDNA, nuclear), IGS (intergenic spacer rDNA, nuclear), SCAR (sequence characterised amplified region markers, mainly nuclear).
[1] Using primers developed by Hebert *et al.* (2004).
[2] Using primer developed by Read (2002).
[3] Using primers developed by Higa *et al.* (2010).

target range because, as can be seen from Table 10.1, primers amplifying longer amplicons can sometimes be effective.

Once the primers have been designed, however, they need to be tested to ensure that they do not amplify anything else within the community under study. Recent studies, for example, have screened 77–93 different prey species to ensure specificity (Juen and Traugott, 2007; Harwood *et al.*, 2007; King *et al.* 2010b). The more diverse the community, the more

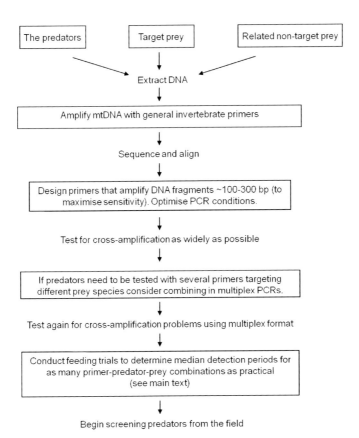

Figure 10.2 Steps required to design and test primers before starting to use them for field-screening of predators. Modified from King *et al.* (2008).

testing should be done. Even if a primer pair already exists, developed by another group, this testing may still need to be repeated for work at a different location where a different spectrum of prey species may be present.

The rates at which predators digest targeted prey DNA fragments need to be calibrated (Figure 10.2). At its simplest this is to define the period within which predation must have occurred. There are two principle variables: (a) different predators digest prey at different rates, and (b) different amplicons will survive digestion for different lengths of time. The first of these relates to the physiology of the predators. For example, sit-and-wait predators such as most spiders have lower metabolic rates (Anderson, 1970) than actively hunting carabid beetles and therefore DNA may be detected for a longer period in the former. The survival during digestion of prey amplicons may be affected by amplicon length, primer efficiency or other factors (King

et al., 2008). Ideally, each primer should be tested with each predator species consuming each prey species, to determine detection periods in feeding trials (large groups of predators are fed on the target prey and then killed in batches over successive time periods). A median detection period (often referred to as the detectability half-life, Chen *et al.*, 2000) is then calculated as the time at which half the predators are still testing positive for the prey. These median detection periods can be used as correction factors, allowing rates of predation by different predators on the same or different prey species (using different primer pairs) to be compared statistically (Greenstone *et al.*, 2007; 2010). The need to do such testing, which involves a lot of work and potentially hundreds of predators, is a major unresolved problem facing those studying more complex food webs. There is an urgent need for predictive modelling of the factors involved, incorporating additional variables such as temperature

(Hoogendoorn and Heimpel, 2001; von Berg et al., 2008a).

At their simplest pest-specific primers may be used to search for potentially useful biocontrol agents (e.g. Juen and Traugott, 2007; Kuusk et al., 2008, Lundgren et al., 2009; Szendrei et al., 2010). Increasingly, however, it is being realised that predation on any target species, pest or otherwise, is highly dependent on the density, diversity and biomass of other prey present in the same crop at the same time and available to the predators (Welch et al., chapter 3 of this volume). This has led to more complex field studies in which predators are tested with multiple primers for predation on a range of different prey species (e.g. Harper et al., 2005; Bell et al., 2010; King et al., 2010a).

MULTIPLEXING

Determining levels of predation in complex, multiprey systems requires analysis of predators using multiple primers. However, it would be impractical to test each predator with each primer pair separately, when hundreds of predators need to be screened. At its simplest, multiplexing has been used to co-amplify predator DNA at the same time as that of the prey in its gut (Juen and Traugott, 2006). Amplification of DNA from the predator provides an effective way of checking for PCR success (predator DNA should always amplify) and distinguishing between true negatives and failed PCRs (Zaidi et al., 1999). However, the main benefit of multiplexing comes from studies of the prey range of a predator. Harper et al. (2005) demonstrated that it was possible to amplify and detect 12 different prey in a single PCR and variants of this approach are being used increasingly by others to detect both predation and parasitism (Traugott et al., 2008; King et al., 2010a; 2010b). The primers used by Harper et al. (2005) and King et al. (2010a; 2010b) had fluorescent labels attached, having the dual advantages of: (a) increasing sensitivity compared with stained bands on gels and (b) allowing the amplicons for all the target species to be separated on a sequencer and presented in the form of an electropherogram (Figure 10.3). This technique allows a production-line approach, with prey species identities simply read and recorded. Although this approach requires some calibration, and primer interactions must be avoided (where forward primers for one species may amplify a spurious fragment in combination with the reverse primer for another species) (King et al., 2010b), this technique may become the standard when screening for multiple different prey with separate primers. Depending upon the protocols used, multiplexing can sometimes result in lower sensitivity than singleplexing and this needs to be checked (e.g. Traugott & Symondson, 2008).

ANALYSIS OF PREY CHOICE

PCR cannot, in most instances, tell us how many prey were consumed as the tests are usually qualitative (but see below). However, having screened predators for multiple different prey in their guts, including pest species, then the numbers of predators testing positive for each prey becomes a surrogate for rates of predation. If predation rates on a particular target prey within a diverse prey community are very high then, simply by chance, the proportion of individual predators that have eaten more than a single individual of that prey species is likely to be high. Conversely if predation rates are low, then, again by chance, a lower proportion of predators is likely to have eaten more than a single individual of the target species and recorded rates of positive tests becomes a reasonably accurate conservative estimation of predation rates. In other words at low predation rates, each positive can be assumed to have eaten a single individual for modelling purposes, with a reasonable degree of accuracy. In practice low rates are far more common, especially in biodiverse communities. However, this still does not tell us much about prey choice. The latter depends critically upon what is available in the field and, as we saw above, many other interacting factors. Intensive sampling of prey in the field using pitfall trapping, vacuum sampling (Figure 10.4), sweep netting, sticky traps, or chemical extraction from soil samples (whatever is appropriate to the system being studied) (Sunderland et al., 2005) can provide detailed information on prey densities. Prey choice can be analysed by comparing ratios of different prey in the field with the rates at which predators test positive for these species. There are several ways in which this might be done. Monte Carlo simulations provide one such approach, in which expected rates of predation, based upon random selection of prey by predators based upon relative densities in the field, can be compared with recorded rates of predation from PCR analysis. This approach was used by Agustí et al. (2003a) to study predation on different species of collembola by linyphiid spiders and by King et al. (2010b)

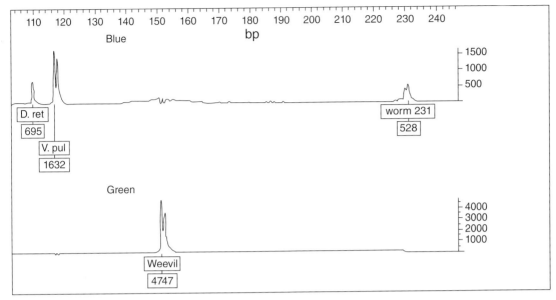

Figure 10.3 Electropherogram of the DNA fragments amplified from a single carabid beetle (*Pterostichus melanarius* (Fabricius)). The scale along the top is the fragment size (in base pairs), the scales on the right hand side are of fluorescent units. Primers with two fluorescent labels were used: the top trace shows peaks for the slug *Deroceras reticulatum* (Müller), the snail *Vellonia pulchella* (Müller) and earthworms, the lower trace shows a peak for weevil (*Sitona* bean weevils). Thus using multiplex PCR with fluorescent primers revealed four prey in the gut of a single beetle simultaneously.

Figure 10.4 Vacuum sampling, with a converted leaf blower, for invertebrates (predators and prey) along a field margin.

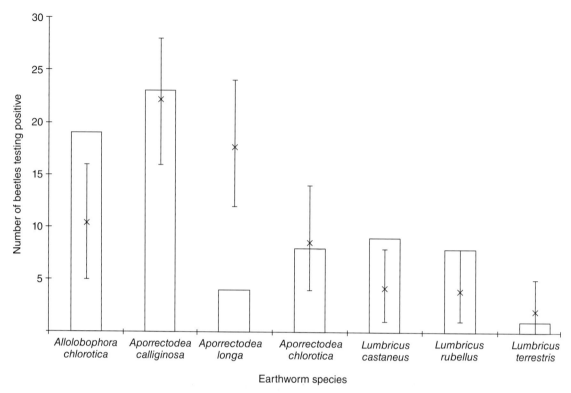

Figure 10.5 Monte Carlo simulations can be used to compare observed rates at which beetles test positive for each species (columns) with the simulated expected rates, which assume different prey species will be eaten in proportion to their occurrence in the field (×). Vertical bars are ±95% confidence intervals. In this dataset, *Allobophora chlorotica* (Savigny) was eaten in significantly greater numbers than expected (probably because it is a small species living near the surface, easily predated by the beetles), while *Aporrectodea longa* (Ude) was eaten in significantly lower than expected numbers (adults of this species and too large for the beetles and deep living). Taken from King *et al.* (2010b).

to study predation by carabid beetles on a community of earthworms (Figure 10.5). Most studies that seek to identify and compare generalist predator species as potential biocontrol agents do not take alternative prey densities (and diversities), or rates of predation on these prey, into account. Unless this is done the effectiveness of such predators at controlling pests within diverse communities cannot be properly assessed.

QUANTITATIVE ('REAL TIME') PCR

A question frequently asked is 'how many prey individuals are being taken by a particular predator?' As we saw above, there are ways of estimating this and a number of the models that have been developed to try to do so. These incorporate factors such as the proportion of a population of predators that test positive, the predator density, the daily feeding rate or mean meal size of the predator (measured in the laboratory), the rates at which DNA (or target proteins) become digested and the period within which prey remains are detectable in the gut of a predator (e.g. Sopp *et al.*, 1992; Mills, 1997). However, the prey mass found in the gut of a predator is not the same as the mass consumed (let alone the number of prey killed); rather it is what remains after an unknown period of digestion. However, this parameter alone can be useful. In earlier studies using antibodies the quantity of undigested target protein could be measured using enzyme-linked immunosorbent assays (ELISA), providing a useful *relative* (not absolute) measure of predation on a target species between field treatments or over time (Symondson *et al.*, 2000). Recent invertebrate predation studies have shown that something similar is possible using quantitative (or 'real time') PCR (qPCR), which can measure how much prey DNA is present in a sample (Zhang *et al.*, 2007b; Schmidt *et al.*, 2009; Lundgren *et al.*, 2009; Weber and Lundgren, 2009). Like PCR with fluorescent primers (Harper *et al.*, 2005), qPCR is also intrinsically more sensitive than standard PCR, avoids the time-consuming use of gels and offers considerable potential. Quantitative PCR depends critically upon copy number of the target gene sequence in prey cells, which varies considerably between candidate genes and species. Copy number can to be calibrated for different life stages (e.g. eggs, nymphs, adults) in order to obtain approximations of numbers of undigested prey in the gut of a predator (Zhang *et al.*, 2007). Calculated numbers of undigested prey in the gut are by no means the same things as number of prey ingested and significant calibration work is therefore needed before it can be applied for quantification purposes in practice.

CLONING AND SEQUENCING

An alternative approach to determining diet is to amplify DNA from the gut of a predator using general invertebrate, or group-specific, primers followed where necessary by cloning and sequencing of the amplified product. This approach has the advantage that no species-specific primers need to be developed, but has a high sequencing requirement which adds to costs. It has mainly been used for the analysis of the diets of vertebrates, where numbers of predators screened tend to be lower than in invertebrate predator studies. For example Sutherland (2000) used group-specific primers to measure feeding by birds (blue and great tits) on Lepidoptera through faecal analysis. Where a cloning step is necessary the numbers of clones provide a reasonable quantitative measure of the biomass of different prey in the sample. Clare *et al.* (2009) were able to avoid a cloning step by using general invertebrate primers to amplify DNA from prey fragments within the faeces of bats, followed by sequencing and identification on BOLD. Kasper *et al.* (2004) used a similar technique to compare the diets of alien and native wasps in Australia. Masticated balls of prey were removed from the jaws of wasps returning to the nest, from which DNA was extracted and sequenced. As the balls usually contained only one prey species, the DNA could be directly sequenced without cloning and identity sought by comparison with sequences on GenBank.

A significant problem with these approaches is that general invertebrate primers may amplify the predator too, so that the majority of clones will not be from the prey. Blocking primers have been used recently to effectively prevent this happening with marine invertebrate predators (krill – Vestheim and Jarman, 2008) and could in theory be widely used in any systems where predators and prey are not closely related (e.g. spiders feeding on insects).

SOURCES OF ERROR

Trophic relationships in highly biodiverse communities are complex and all molecular diagnostics can do is detect the presence of prey DNA within predators. It would be a mistake to assume that each positive resulting from a gut-content PCR represents at least one prey individual killed. Box 10.1 lists some of the main areas of concern arising from either ecological or technical sources. Reviews and discussion of these sources of

---

Box 10.1 Sources of error

Molecular analysis of the gut contents of predators must take into account wherever possible ecological sources of error (in addition to technical issues). These include:
- Secondary predation – false positive for direct predation (see main text)
- Scavenging – false positive for predation (see main text)
- Wasteful killing (e.g. by spider webs) – predators may be more effective than molecular analyses suggest
- Several predators feeding on the same prey individual – molecular analysis over-estimates rates of predation
- Prey is wounded but not killed – if prey tissue is ingested but prey does not die false positives for predation occur
- Prey is wounded and may die but not ingested – false negative for predation
- Predation on moribund prey – false positives for prey suppression. Killing diseased prey may actually benefit prey populations, including pests
- Pre-oral digestion (prey digested externally before being consumed) – false negatives for predation if DNA is denatured
- Predation occurs during sampling – this can happen in pitfall traps, sweep nets, pooters and the collection bags of vacuum samplers, generating ecologically false positives

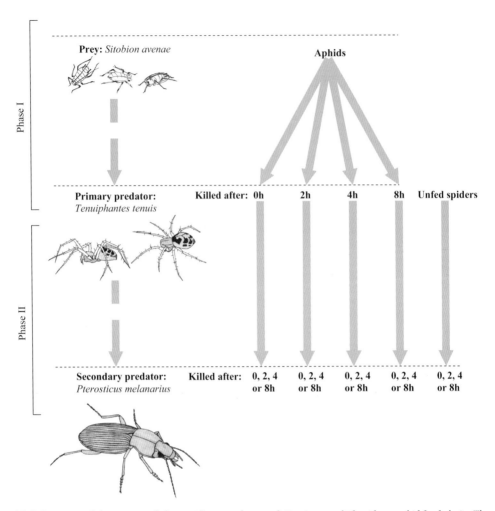

Figure 10.6 Summary of the steps needed to test for secondary predation in an aphid-spider-carabid food chain. There were 16 replicates for each treatment and time period. Only a single carabid tested positive for aphid (using primers that amplified a 110 bp fragment) when the spider had digested the aphid for 4 hours then the carabid digested the spider (which had eaten the aphid) for a further 4 hours. For secondary predation to be detected, the aphid DNA has to be passed along the food chain relatively rapidly. Taken from Sheppard *et al.* (2005).

error can be found in Sunderland (1996) and Sunderland *et al.* (2005). Of these, the most commonly mentioned causes of concern are secondary predation and scavenging.

Secondary predation is where a predator does not consume a target prey species directly, but rather feeds on another predator that has the DNA of the prey in its gut (Harwood *et al.* 2001; Sheppard *et al.*, 2005). As we have seen, different predators digest their prey at different rates and different amplicons survive diges-

tion for shorter or longer periods, depending upon their length and other factors. The study by Sheppard *et al.* (2005) showed, in a carabid-spider-aphid food chain, that detection of aphid DNA in a carabid that had eaten a spider that had consumed an aphid was possible. The structure of the experiment is shown in Figure 10.6. In this instance one of the predators (the spiders) had a low metabolic rate, and a separate analysis had shown that detection of the target aphid DNA in spiders had a long median detection period (60 hours).

However, the results showed that even under these circumstances secondary predation is likely to be a relatively rare source of error compared with scavenging.

Rates of scavenging will clearly depend upon availability of dead prey, something that is not easily measurable. Foltan *et al.* (2005) showed that dead slugs on the soil retained amplifiable DNA for more than a week and if eaten by a carabid during this period the DNA could be readily detected in the carabid's gut. Even dead aphids retained amplifiable DNA after more than five days on the soil. Similar results were obtained by Juen and Traugott (2005) for chafer (*Melolontha melolontha* (L.)) grubs. No way has yet been devised for distinguishing between feeding on live, dead or moribund prey. The source of nutrients and energy can be tracked, however, which for some studies may be relevant.

PYROSEQUENCING AND FUTURE DEVELOPMENTS

DNA sequencing technology has advanced rapidly in recent years. King *et al.* (2008) predicted in their review that pyrosequencing would one day become cheap enough to be applied to the analysis of diet. Within a year this approach was used by Soininen *et al.* (2009) and Valentini *et al.* (2009) to study feeding on plants by a range of vertebrate and invertebrate herbivores. The first use of pyrosequencing to study carnivory was by Deagle *et al.* (2009), who analysed the diets of seals and used sequence numbers to provide a quantitative measure of the biomass of each dietary component in faecal samples.

Pyrosequencing technology (Margulies *et al.*, 2005) allows mass sequencing of the hundreds of thousands of strands of DNA in a sample simultaneously (Box 10.2), providing data on the diversity and (more importantly) the abundance of each unique sequence in a sample. General and group-specific primers can be used to amplify DNA from gut samples and the mixture of amplicons put through the pyrosequencer. The sequences are then sorted into taxonomic units and identified by comparison with those on GenBank or BOLD using the basic local alignment search tool (BLAST) (Altschul *et al.*, 1990). No cloning step is needed to identify or quantify the separate sequences, making the whole process much more rapid.

This approach is ideal for any study of the effects of prey diversity on dietary choice; as long as the primers

---

Box 10.2 What is pyrosequencing?

In this 'massive parallel sequencing' approach, DNA is amplified from a predator gut sample using PCR. The resulting DNA fragments (amplicons) are bound to tiny beads (one fragment per bead), then each bead is incorporated within a droplet of emulsion containing PCR reagents. PCR takes place within each droplet generating millions of identical copies, attached to the bead. Each bead then goes into a picolitre-sized well on an optical fibre slide containing more than a million such wells. Each of the four bases (A, C, G and T) are sequentially added, one at a time, to a mix of chemicals containing the cloned strands of DNA that are to be sequenced. When a complementary base binds to the template, light is emitted and detected. If two or more bases (e.g. TT or CCC) are added at the same time, the light emitted is proportionately stronger. As the order in which the bases sequentially added is known, the sequence of the complementary strand can be recorded as it is constructed, known as 'sequencing by synthesis'. In this way hundreds of thousands of strands of DNA can be sequenced, in parallel, within a pyrosequencer. After bioinformatic processing and BLAST searching on GenBank, the ratios of DNA from different prey species in the gut samples can be quantified.

---

are general enough to amplify all possible prey, a quantitative measure will be obtained of the relative biomass of prey, pest and non-pest, which have been consumed and remain undigested. To date it has not been used to analyse the diets of invertebrate predators. However, it was used by Brown (2010) to study predation by legless lizards (slow worms, *Anguis fragilis* (L.)) on different species and lineages of earthworms in a range of habitats. The only technical obstacle to using this approach to study predation by invertebrates on other invertebrates is that general invertebrate primers would amplify the predator DNA too, unless blocking primers were designed that could prevent this (Vestheim and Jarman, 2008) or group-specific primers (that did not include the predator in the group) were applied, as in the study by Brown (2010). The potential of the pyrosequencing approach is enormous, but there are sources of error. For example the copy number of target genes, whether nuclear or mitochondrial, may differ between species and this would need to be calibrated (Deagle and Tollit, 2007). It would also be necessary to check that there was no PCR dominance by common prey in the gut samples, masking the presence of rarer species.

CONCLUSION

Manipulation experiments in the laboratory or in field cages under controlled conditions will continue to provide a wealth of information on ecological processes, including responses of predators to pests in the presence of intraguild predators and alternative prey. However, in the field, where things are more 'messy', outcomes, especially in terms of pest control by generalist predators, are harder to predict. Molecular diagnostics provide invaluable tools for detecting and quantifying trophic links within the tangled network of interactions found in diverse communities. They can help to inform us about the network structures that can lead to positive outcomes (pest suppression) within systems that are highly dynamic (both temporally and spatially). Newly developing technologies (especially next generation sequencing) are facilitating such work and making it open to all. In the longer term pyrosequencing may be rapidly overtaken by recent developments in nanopore sequencing (Clarke *et al.*, 2009). We need to grasp these opportunities to ask new questions. They will help us in our quest to develop more intelligent and sustainable agriculture systems that maximise the potential of natural systems to regulate pests.

REFERENCES

Admassu, B., Juen, A. and Traugott, M. (2006) Earthworm primers for DNA-based gut content analysis and their cross-reactivity in a multi-species system. *Soil Biology and Biochemistry*, 38, 1308–1315.

Agustí, N., De Vicente, M.C. and Gabarra, R. (1999) Development of sequence amplified characterized region (SCAR) markers of *Helicoverpa armigera*: a new polymerase chain reaction-based technique for predator gut analysis. *Molecular Ecology*, 8, 1467–1474.

Agustí, N., de Vicente, M.C. and Gabarra, R. (2000) Developing scar markers to study predation on *Trialeurodes vaporariorum*. *Insect Molecular Biology*, 9, 263–268.

Agustí, N., Shayler, S.P., Harwood, J.D., Vaughan, I.P., Sunderland, K.D. and Symondson, W.O.C. (2003a) Collembola as alternative prey sustaining spiders in arable ecosystems: prey detection within predators using molecular markers. *Molecular Ecology*, 12, 3467–3475.

Agustí, N., Unruh, T.R. and Welter, S.C. (2003b) Detecting *Cacopsylla pyricola* (Hemiptera: Psyllidae) in predator guts using COI mitochondrial markers. *Bulletin of Entomological Research*, 93, 179–185.

Altschul, S.F., Gish, W., Miller, W., Myers, E.W. and Lipman, D.J. (1990) Basic local alignment search tool. *Journal of Molecular Biology*, 215, 403–410.

Anderson, J.F. (1970) Metabolic rates of spiders. *Comparative Biochemistry and Biophysics*, 33, 51–72.

Anderson, J.F. (1974) Responses to starvation in spiders *Lycosa lenta* Hentz and *Filistata hibernalis* (Hentz). *Ecology*, 55, 576–585.

Bell, J.R., King, R.A., Bohan, D.A. and Symondson, W.O.C (2010) Spatial co-occurrence networks predict the feeding histories of polyphagous arthropod predators at field scales. *Ecography*, 33, 64–72.

Birkhofer, K., Gavish-Regev, E., Endkweber, K. *et al.* (2008) Cursorial spiders retard initial aphid population growth at low densities in winter wheat. *Bulletin of Entomological Research*, 98, 249–255.

Bengtsson, J., Ahnstrom, J. and Weibull, A.C. (2005) The effects of organic agriculture on biodiversity and abundance: a metaanalysis. *Journal of Applied Ecology*, 42, 261–269.

Brennan, A., Fortune, T. and Bolger, T. (2006) Collembola abundances and assemblage structures in conventionally tilled and conservation tillage arable systems. *Pedobiologia*, 50, 135–145.

Brodeur, J. and Rosenheim, J.A. (2000) Intraguild interactions in aphid parasitoids. *Entomologia Experimentalis et Applicata*, 97, 93–108.

Brown, D.S. (2010) *Molecular analysis of the diet of British reptiles*. Doctor of Philosophy thesis, Cardiff University.

Cassel-Lundhagen, A., Öberg, S., Högfeldt, C. and Ekbom, B. (2009) Species-specific primers for predation studies of the pollen beetle, *Meligethes aeneus* (Coleoptera, Nitidulidae). *Molecular Ecology Resources*, 9, 1132–1134.

Chacon, J.M. and Heimpel, G.E. (2010) Density-dependent intraguild predation of an aphid parasitoid. *Oecologia*, 164, 213–220.

Chapman, E.G., Romero, S.A. and Harwood, J.D. (2010) Maximising collection and minimising risk: does vacuum suction sampling increase the likelihood for misinterpretation of food web connections. *Molecular Ecology Resources*, 10, 1023–1033.

Chen, Y., Giles, K.L., Payton, M.E. and Greenstone, M.H. (2000) Identifying key cereal aphid predators by molecular gut analysis. *Molecular Ecology*, 9, 1887–1898.

Clare, E.L., Fraser, E.E., Braid, H.E., Fenton, M.B. and Hebert, P.D.N. (2009) Species on the menu of a generalist predator, the eastern red bat (*Lasiurus borealis*): using a molecular approach to detect arthropod prey. *Molecular Ecology*, 18, 2532–2542.

Clarke, J., Wu, H.C., Jayasinghe, L., Patel, A., Reid, S. and Bayley, H. (2009) Continuous base identification for single-molecule nanopore DNA sequencing. *Nature Nanotechnology*, 4, 265–270.

Curry, J. (1994) *Grassland invertebrates: ecology, influences on soil invertebrates and effects on plant growth*, Chapman & Hall, New York.

Cuthbertson, A.G.S., Fleming, C.C. and Murchie, A.K. (2003) Detection of *Rhopalosiphum insertum* (apple-grass aphid) predation by the predatory mite *Anystis baccarum* using molecular gut analysis. *Agricultural and Forest Entomology*, 5, 219–225.

Deagle, B.E. and Tollit, D.J. (2007) Quantitative analysis of prey DNA in pinniped faeces: potential to estimate diet composition. *Conservation Genetics*, 8, 743–747.

Deagle, B.E., Kirkwood, R. and Jarman, S.N. (2009) Analysis of Australian fur seal diet by pyrosequencing prey DNA in faeces. *Molecular Ecology*, 18, 2022–2038.

de León, J.H., Fournier, V., Hagler, J.R. and Daane, K.M. (2006) Development of molecular diagnostic markers for sharp-shooters *Homalodisca coagulata* and *Homalodisca liturata* for use in predator gut content examinations. *Entomologia Experimentalis et Applicata*, 119, 109–119.

Dodd, C.S., Bruford, M.W., Symondson, W.O.C. and Glen, D.M. (2003) Detection of slug DNA within carabid predators using prey-specific PCR primers, in *Slug and snail Pests: agricultural, veterinary and environmental perspectives* (ed. G.B.J. Dussard), British Crop Protection Council, Alton, pp. 13–20.

Dodd, C.S., Bruford, M.W., Glen, D.M. and Symondson, W.O.C. (2005) Molecular detection of slug DNA within carabid predators, in *Insect pathogens and entomoparasitic nematodes: slugs and snails* (ed. D.A. Bohan), *IOBC Bulletin*, 28 (6), 131–134.

Folmer, O., Black, M., Hoeh, W., Lutz, R. and Vrijenhoek, R. (1994) DNA primers for the amplification of mitochondrial cytochrome c oxidase subunit I from diverse metazoan invertebrates. *Molecular Marine Biology and Biotechnology*, 3, 294–299.

Foltan, P., Sheppard, S., Konvicka, M. and Symondson, W.O.C. (2005) The significance of facultative scavenging in generalist predator nutrition: detecting decayed prey in the guts of predators using PCR. *Molecular Ecology*, 14, 4147–4158.

Forsythe, T.G. (1983) Mouthparts and feeding of certain ground beetles (Coleoptera: Carabidae). *Zoological Journal of the Linnean Society*, 79, 319–376.

Fournier, V., Hagler, J.R., Daane, K.M., de Leon, J.H. and Groves, R. (2008) Identifying the predator complex of *Homalodisca vitripennis* (Hemiptera: Cicadellidae): a comparative study of the efficacy of an ELISA and PCR gut content assay. *Oecologia*, 157, 629–640.

Gibson, R.H., Pearce, S., Morris, R.J., Symondson, W.O.C. and Memmott, J. (2007) Plant diversity and land use under organic and conventional agriculture: a whole-farm approach. *Journal of Applied Ecology*, 44, 792–803.

Greenstone, M.H. (1979) Spider feeding behaviour optimises dietary essential amino acid composition. *Nature*, 282, 501–503.

Greenstone, M.H. (1996) Serological analysis of arthropod predation: past, present and future, in *The ecology of agricultural pests: biochemical approaches* (eds W.O.C. Symondson and J.E. Liddell), Chapman & Hall, London, pp. 265–300.

Greenstone, M.H. and Shufran, K.A. (2003) Spider predation: species-specific identification of gut contents by polymerase chain reaction. *Journal of Arachnology*, 31, 131–134.

Greenstone, M.H., Rowley, D.L., Weber, D.C., Payton, M.E. and Hawthorne, D.J. (2007) Feeding mode and prey detectability half-lives in molecular gut-content analysis: an example with two predators of the Colorado potato beetle. *Bulletin of Entomological Research*, 97, 201–209.

Greenstone, M.H., Szendrei, Z., Payton, M.E., Rowley, D.L., Coudron, T.C. and Weber, D.C. (2010) Choosing natural enemies for conservation biological control: use of the prey detectability half-life to rank key predators of Colorado potato beetle. *Entomologia Experimentalis et Applicata*, 136, 97–107.

Greenstone, M.H., Weber, D.C., Coudron, T.C. and Payton, M.E. (2011) Unnecessary roughness? Testing the hypothesis that predators destined for molecular gut-content analysis must be hand-collected to avoid cross-contamination. *Molecular Ecology Resources*, 11, 286–293.

Griffiths G.J.K., Alexander C., Perry J.N., Holland J.M., Symondson W.O.C., Kennedy P. and Winder L. (2008) Monoclonal antibodies reveal changes in predator efficiency with prey spatial pattern. *Molecular Ecology*, 17, 1828–1839.

Halaj, J. and Wise, D.H. (2002) Impact of a detrital subsidy on trophic cascades in a terrestrial grazing food web. *Ecology*, 83, 3141–3151.

Harper, G.L., King, R.A., Dodd, C.S. *et al.* (2005) Rapid screening of invertebrate predators for multiple prey DNA targets. *Molecular Ecology*, 14, 819–827.

Harper, G.L., Sheppard, S.K., Harwood, J.D. *et al.* (2006) Evaluation of temperature gradient gel electrophoresis for the analysis of prey DNA within the guts of invertebrate predators. *Bulletin of Entomological Research*, 96, 295–304.

Harwood, J.D. (2008) Are sweep net sampling and pitfall trapping compatible with molecular analysis of predation? *Environmental Entomology*, 37, 990–995.

Harwood, J.D., Phillips, S.W., Sunderland, K.D. and Symondson, W.O.C. (2001) Secondary predation: quantification of food chain errors in an aphid-spider-carabid system using monoclonal antibodies. *Molecular Ecology*, 10, 2049–2057.

Harwood, J.D., Desneux, N., Yoo, H.J.S. *et al.* (2007) Tracking the role of alternative prey in soybean aphid predation by *Orius insidiosus*: a molecular approach. *Molecular Ecology*, 16, 4390–4400.

Harwood, J.D., Phillips, S.W., Lello, J. *et al.* (2009a) Invertebrate biodiversity affects predator fitness and hence potential to control pests in crops. *Biological Control*, 51, 499–506.

Harwood, J.D., Yoo, H.J.S., Greenstone, M.H., Rowley, D.L. and O'Neil, R.J. (2009b) Differential impact of adults and nymphs of a generalist predator on an exotic invasive pest demonstrated by molecular gut-content analysis. *Biological Invasions*, 11, 895–903.

Hebert, P.D.N., Ratnasingham, S. and deWaard, J.R. (2003a) Barcoding animal life: cytochrome c oxidase subunit 1 divergences among closely related species. *Proceedings of the Royal Society B*, 270 (Suppl. 1), S96–S99.

Hebert, P.D.N., Cywinska, A., Ball, S.L. and deWaard, J.R. (2003b) Biological identifications through DNA barcodes. *Proceedings of the Royal Society B*, 270, 313–321.

Hebert, P.D.N., Penton, E.H., Burns, J.M., Janzen, D.H. and Hallwachs, W. (2004) Ten species in one: DNA barcoding reveals cryptic species in the neotropical skipper butterfly *Astraptes fulgerator*. *Proceedings of the National Academy of Sciences*, 101, 14812–14817.

Higa, Y., Toma, T., Tsuda, Y. and Miyagi, I. (2010) A multiplex PCR-based molecular identification of five morphologically related, medically important subgenus *Stegomyia* mosquitoes from the genus *Aedes* (Diptera: Culicidae) found in the Ryukyu Archipelago, Japan. *Japanese Journal of Infectious Diseases*, 63, 312–316.

Hole, D.G., Perkins, A.J., Wilson, J.D., Alexander, I.H., Grice, F. and Evans, A.D. (2005) Does organic farming benefit biodiversity? *Biological Conservation*, 122, 113–130.

Hoogendoorn, M. and Heimpel, G.E. (2001) PCR-based gut content analysis of insect predators: using ribosomal ITS-1 fragments from prey to estimate predation frequency. *Molecular Ecology*, 10, 2059–2067.

Hosseini, R., Schmidt, O. and Keller, M.A. (2008) Factors affecting detectability of prey DNA in the gut contents of invertebrate predators: a polymerase chain reaction-based method. *Entomologia Experimentalis et Applicata*, 126, 194–202.

Ingerson-Mahar, J. (2002) Relating diet and morphology in adult carabid beetles, in *The agroecology of carabid beetles* (ed. J. Holland), Intercept, Andover, pp. 111–136.

Jaramillo, J., Chapman, E.G., Vega, F.E. and Harwood, J.D. (2010) Molecular diagnosis of previously unreported predator-prey association in coffee: *Karnyothrips flavipes* Jones (Thysanoptera: Phlaeothripidae) predation on the coffee berry borer. *Naturwissenschaften*, 97, 291–298.

Juen, A. and Traugott, M. (2005) Detecting predation and scavenging by DNA gut-content analysis: a case study using a soil insect predator–prey system. *Oecologia*, 142, 344–352.

Juen, A. and Traugott, M. (2006) Amplification facilitators and multiplex PCR: tools to overcome PCR-inhibition in DNA-gut-content analysis of soil-living invertebrates. *Soil Biology and Biochemistry*, 38, 1872–1879.

Juen, A. and Traugott, M. (2007) Revealing species-specific trophic links in soil food webs: Molecular identification of scarab predators. *Molecular Ecology*, 16, 1545–1557.

Kasper, M.L., Reeson, A.F., Cooper, S.J.B, Perry, K.D. and Austin, A.D. (2004) Assessment of prey overlap between a native (*Polistes humilis*) and an introduced (*Vespula germanica*) social wasp using morphology and phylogenetic analyses of 16s rDNA. *Molecular Ecology*, 13, 2037–2048.

King, R.A., Read, D.S., Traugott, M. and Symondson, W.O.C. (2008) Molecular analysis of predation: a review of best practice for DNA-based approaches. *Molecular Ecology*, 17, 947–963.

King, R.A., Moreno-Ripoll, R., Agustí, N. *et al.* (2010a) Multiplex reactions for the molecular detection of predation on pest and non-pest invertebrates in agroecosystems. *Molecular Ecology Resources*, 11, 370–373.

King, R.A., Vaughan, I.P., Bell, J.R., Bohan, D.A. and Symondson, W.O.C. (2010b) Prey choice by carabid beetles feeding on an earthworm community analysed using species- and lineage-specific PCR primers. *Molecular Ecology*, 19, 1721–1732.

Koss, A.M. and Snyder, W.E. (2005) Alternative prey disrupt biocontrol by a guild of generalist predators. *Biological Control*, 32, 243–251.

Kuusk, A.K. and Agustí, N. (2008) Group-specific primers for DNA-based detection of springtails (Hexapoda: Collembola) within predator gut contents. *Molecular Ecology Resources*, 8, 678–681.

Kuusk, A.K. and Ekbom, B. (2010) Lycosid spiders and alternative food: feeding behaviour and implications for biological control. *Biological Control*, 55, 20–26.

Kuusk, A.K., Cassel-Lundhagen, A., Kvarnheden, A. and Ekbom B. (2008) Tracking aphid predation by lycosid spiders in spring-sown cereals using PCR-based gut-content analysis. *Basic and Applied Ecology*, 9, 718–725.

Letourneau D.K. and Bothwell S.G. (2008) Comparison of organic and conventional farms: challenging ecologists to make biodiversity functional. *Frontiers in Ecology and the Environment*, 6, 430–438.

Lundgren, J.G. and Weber, D.C. (2010) Changes in digestive rate of a predatory beetle over its larval stage: implications for dietary breadth. *Journal of Insect Physiology*, 56, 431–437.

Lundgren, J.G., Ellsbury, M.E. and Prischmann, D.A. (2009) Analysis of the predator community of a subterranean herbivorous insect based on polymerase chain reaction. *Ecological Applications*, 19, 2157–2166.

Ma, J., Li, D., Keller, M., Schmidt, O. and Feng, X. (2005) A DNA marker to identify predation of *Plutella xylostella* (Lep., Plutellidae) by *Nabis kinbergii* (Hem., Nabidae) and *Lycosa* sp (Aranaea, Lycosidae). *Journal of Applied Entomology*, 129, 330–335.

Margulies, M., Egholm, M., Altman, W.E. *et al.* (2005) Genome sequencing in microfabricated high-density picolitre reactors. *Nature*, 437, 376–380.

Mayntz, D., Raubenheimer, D., Salomon, M., Toft, S. and Simpson, S.J. (2005) Nutrient-specific foraging in invertebrate predators. *Science*, 307, 111–113.

McMillan, S., Kuusk, A-K., Cassel-Lundhagen, A. and Ekbom, B. (2007) The influence of time and temperature on molecular gut content analysis: *Adalia bipunctata* fed with *Rhopalosiphum padi*. *Insect Science*, 14, 353–358.

Mills, N. (1997) Techniques to evaluate the efficacy of natural enemies, in *Methods in ecological and agricultural entomology* (eds D.R. Dent and M.P. Walton), CAB International, Wallingford, pp. 271–291.

Monzo, C., Sabater-Munoz, B., Urbaneja, A. and Castanera, P. (2010) Tracking medfly predation by the wolf spider, *Pardosa cribata* Simon, in citrus orchards using PCR-based gut-content analysis. *Bulletin of Entomological Research*, 100, 145–152.

Morales, M.E., Wesson, D.M., Sutherland, I.W. *et al.* (2003) Determination of *Anopheles gambiae* larval DNA in the gut of insectivorous dragonfly (Libellulidae) nymphs by polymerase chain reaction. *Journal of the American Mosquito Control Association*, 19, 163–165.

Muilenburg, V.L., Goggin, F.L., Hebert, S.L., Jia, L. and Stephen, F.M. (2008) Ant predation on red oak borer confirmed by field observation and molecular gut-content analysis. *Agriculture and Forest Entomology*, 10, 205–213.

Murray, R.A., Solomon, M.G. and Fitzgerald, J.D. (1989) The use of electrophoresis for determining patterns of predation in arthropods, in *Electrophoretic studies on agricultural pests* (eds H.D. Loxdale and J. den Hollander), Clarendon Press, Oxford, pp. 467–483.

Oelbermann, K. and Scheu, S. (2002) Effects of prey type and mixed diets on survival, growth and development of a generalist predator, *Pardosa lugubris*. *Basic and Applied Ecology*, 3, 285–291.

Ohba, S., Kawada, H., Dida, G.O. *et al.* (2010) Predators of *Anopheles gambiae* sensu lato (Diptera: Culicidae) larvae in wetlands, Western Kenya: confirmation by polymerase chain reaction method. *Journal of Medical Entomology*, 47, 783–787.

Ohba, S., Huynh, T., Kawada, H. *et al.* (2011) Heteropteran insects as mosquito predators in water jars in southern Vietnam. *Journal of Vector Ecology*, 36, 170–174.

Pons, J. (2006) DNA-based identification of preys from non-destructive, total DNA extractions of predators using arthropod universal primers. *Molecular Ecology Notes*, 6, 623–626.

Read, D.S. (2002) Sequencing of aphid DNA and primer design for the detection of aphid remains in predators. Undergraduate dissertation, Cardiff University.

Read, D.S., Sheppard, S.K., Bruford, M.W., Glen, D.M. and Symondson, W.O.C. (2006) Molecular detection of predation by soil micro-arthropods on nematodes. *Molecular Ecology*, 15, 1963–1972.

Rinehart, T.A. and Boyd, D.W. (2006) Rapid, high-throughput detection of azalea lace bug (Hemiptera: Tingidae) predation by *Chrysoperla rufilabris* (Neuroptera: Chrysopidaae), using fluorescent-polymerase chain reaction primers. *Journal of Economic Entomology*, 99, 2136–2141.

Rypstra, A.L. and Marshall, S.D. (2005) Augmentation of soil detritus affects the spider community and herbivory in a soybean agroecosystem. *Entomologia Experimentalis et Applicata*, 116, 149–157.

Schielke, E., Costantini, C., Carchini, G., Sagnon, N., Powell, J. and Caccone, A. (2007) Development of a molecular assay to detect predation on *Anopheles gambiae* complex larval stages. *American Journal of Tropical Medicine and Hygiene*, 77, 464–466.

Schmidt, J.E.U., Almeida, J.R.M., Rosati, C. and Arpaia, S. (2009) Identification of trophic interactions between *Macrolophus caliginosus* (Heteroptera: Miridae) and *Myzus persicae* (Homoptera: Aphididae) using real time PCR. *BioControl*, 54, 383–391.

Seastedt, T.R. (1984) The role of microarthropods in decomposition and mineralisation processes. *Annual Review of Entomology*, 29, 25–46.

Seeber, J., Rief, A., Seeber, G.U.H., Meyer, E. and Traugott, M. (2010) Molecular identification of detritivorous soil invertebrates from their faecal pellets. *Soil Biology and Biochemistry*, 42, 1263–1267.

Sheppard, S.K. and Harwood, J.D. (2005) Advances in molecular ecology: tracking trophic links through predator-prey food-webs. *Functional Ecology*, 19, 751–762.

Sheppard, S.K., Henneman, M.L., Memmott, J. and Symondson, W.O.C. (2004) Infiltration by alien predators into invertebrate food webs in Hawaii: a molecular approach. *Molecular Ecology*, 13, 2077–2088.

Sheppard, S.K., Bell, J., Sunderland, K.D., Fenlon, J., Skirvin, D. and Symondson, W.O.C. (2005) Detection of secondary predation by PCR analyses of the gut contents of invertebrate generalist predators. *Molecular Ecology*, 14, 4461–4468.

Simon, C., Frati, F., Beckenbach, A., Crespi, B., Liu, H. and Flook, P. (1994) Evolution, weighting, and phylogenetic utility of mitochondrial gene-sequences and a compilation of conserved Polymerase Chain Reaction primers. *Annals of the Entomological Society of America*, 87, 651–701.

Simon, C., Buckley, T.R., Frati, F., Stewart, J.B. and Beckenbach, A.T. (2006) Incorporating molecular evolution into phylogenetic analysis, and a new compilation of conserved polymerase chain reaction primers for animal mitochondrial DNA. *Annual Review of Ecology, Evolution and Systematics*, 37, 545–579.

Snyder, W.E. and Ives, A.R. (2001) Generalist predators disrupt biological control by a specialist parasitoid. *Ecology*, 82, 705–716.

Soininen, E.M., Valentini, A., Coissac, E. *et al.* (2009) Analysing diet of small herbivores: The efficiency of DNA barcoding coupled with high-throughput pyrosequencing for deciphering the composition of complex plant mixtures. *Frontiers in Zoology*, 6, 16.

Solomon, M.G., Fitzgerald, J.D. and Murray, R.A. (1996) Electrophoretic approaches to predator-prey interactions, in *The Ecology of agricultural pests: biochemical approaches* (eds W.O.C. Symondson and J.E. Liddell), Chapman & Hall, London, pp. 457–468.

Sopp, P.I., Sunderland, K.D., Fenlon, J.S. and Wratten, S.D. (1992) An improved quantitative method for estimating invertebrate predation in the field using an enzyme-linked immunosorbent assay. *Journal of Applied Ecology*, 29, 295–302.

Sunderland, K.D. (1988) Quantitative methods of detecting invertebrate predation occurring in the field. *Annals of Applied Biology*, 112, 201–224.

Sunderland, K.D. (1996) Progress in quantifying predation using antibody techniques, in *The Ecology of agricultural pests: biochemical approaches* (eds W.O.C. Symondson and J.E. Liddell), Chapman & Hall, London, pp. 419–455.

Sunderland, K.D., Powell, W. and Symondson, W.O.C. (2005) Populations and communities, in *Insects as natural enemies: a practical perspective* (ed. M.A. Jervis), Springer, Berlin, pp. 299–434.

Sutherland, R.M. (2000) *Molecular analysis of avian diet.* Doctor of Philosophy thesis, University of Oxford.

Symondson, W.O.C. (2002a) Molecular identification of prey in predator diets. *Molecular Ecology*, 11, 627–641.

Symondson, W.O.C. (2002b). Diagnostic techniques for determining carabid diets, in *The agroecology of carabid beetles* (ed. J. Holland), Intercept, Andover, pp. 137–164.

Symondson, W.O.C. and Hemingway, J. (1997) Biochemical and molecular techniques, in *Methods in ecological and agricultural entomology* (eds D.R. Dent and M.P. Walton), CAB International, Oxford, pp. 293–350.

Symondson, W.O.C., Glen, D.M., Erickson, M.L., Liddell, J.E. and Langdon, C.J. (2000) Do earthworms help to sustain the slug predator *Pterostichus melanarius* (Coleoptera: Carabidae) within crops? Investigations using a monoclonal antibody-based detection system. *Molecular Ecology*, 9, 1279–1292.

Symondson, W.O.C., Sunderland, K.D. and Greenstone, M.H. (2002) Can generalist predators be effective biocontrol agents? *Annual Review of Entomology*, 47, 561–594.

Symondson, W.O.C., Cesarini, S., Dodd, P.W. *et al.* (2006) Biodiversity vs. biocontrol: positive and negative effects of alternative prey on control of slugs by carabid beetles. *Bulletin of Entomological Research*, 96, 637–645.

Szendrei, Z., Greenstone, M.H., Payton, M.E. and Weber, D.C. (2010) Molecular gut-content analysis of a predator assemblage reveals the effect of habitat manipulation on biological control in the field. *Basic and Applied Ecology*, 11, 153–161.

Toft, S. and Wise, D.H. (1999) Growth, development and survival of a generalist predator fed single- and mixed-species diets of different quality. *Oecologia*, 119, 191–197.

Traugott, M. and Symondson, W.O.C. (2008) Molecular analysis of predation on parasitized hosts. *Bulletin of Entomological Research*, 98, 223–231.

Traugott, M., Bell, J.R., Broad, G.R. *et al.* (2008) Endoparasitism in cereal aphids: molecular analysis of a whole parasitoid community. *Molecular Ecology*, 17, 3928–3938.

Traugott, M., Bell, J.R., Raso, L., Sint, D. and Symondson, W.O.C. (2011) Generalist predators disrupt parasitoid aphid control by direct and coincidental intraguild predation. *Bulletin of Entomological Research*, doi: 10.1017/S0007485311000551.

Valentini, A., Miquel, C., Nawaz, M.A. *et al.* (2009) New perspectives in diet analysis based on DNA barcoding and parallel pyrosequencing: the trnL approach. *Molecular Ecology Resources*, 9, 51–60.

Vestheim, H. and Jarman, S.N. (2008) Blocking primers to enhance PCR amplification of rare sequences in mixed samples – a case study on prey DNA in Antarctic krill stomachs. *Frontiers in Zoology*, 5, 12.

von Berg, K., Traugott, M., Symondson, W.O.C. and Scheu, S. (2008a) The effects of temperature on detection of prey DNA in two species of carabid beetle. *Bulletin of Entomological Research*, 98, 263–169.

von Berg K., Traugott M., Symondson W.O.C. and Scheu S. (2008b) Impact of abiotic factors on predator-prey interactions: DNA-based gut content analysis in a microcosm experiment. *Bulletin of Entomological Research*, 98, 257–261.

Weber, D.C. and Lundgren, J.G. (2009) Detection of predation using qPCR: effect of prey quantity, elapsed time, chaser diet, and sample preservation on detectable quantity of prey DNA. *Journal of Insect Science*, 9, Article 41.

Zaidi, R.H., Jaal, Z., Hawkes, N.J., Hemingway, J. and Symondson, W.O.C. (1999) Can multiple-copy sequences of prey DNA be detected amongst the gut contents of invertebrate predators? *Molecular Ecology*, 8, 2081–2087.

Zeale, M.R.K., Butlin, R.K., Barker, G.L.A. and Lees, D.C. (2010) Taxon-specific PCR for DNA barcoding arthropod prey in bat faeces. *Molecular Ecology Resources*, 11, 236–244.

Zhang, G.F., Lü, Z.C. and Wan, F.H. (2007a) Detection of *Bemisia tabaci* remains in predator guts using a sequence-characterised amplified region marker. *Entomologia Experimentalis et Applicata*, 123, 81–90.

Zhang, G.F., Lü, Z.C., Wan, F.H. and Lövei, G.L. (2007b) Real-time PCR quantification of *Bemisia tabaci* (Homoptera: Aleyrodidae) B-biotype remains in predator guts. *Molecular Ecology Notes*, 7, 947–954.

Chapter 11

EMPLOYING CHEMICAL ECOLOGY TO UNDERSTAND AND EXPLOIT BIODIVERSITY FOR PEST MANAGEMENT

David G. James, Sofia Orre-Gordon, Olivia L. Reynolds (née Kvedaras) and Marja Simpson

Biodiversity and Insect Pests: Key Issues for Sustainable Management, First Edition. Edited by Geoff M. Gurr, Steve D. Wratten, William E. Snyder, Donna M.Y. Read.

INTRODUCTION

Research on improving biological control in crop systems is often directed towards strengthening the natural enemy community both in terms of population density and species diversity (Cardinale *et al.*, 2003). In the past decade the exploitation of semiochemicals has emerged as a potentially powerful tool for increasing numbers of predators and parasitoids in crops and enhancing biological control. Research on the role of semiochemicals produced by plants and used for defence in mediating the behaviour of herbivores and their natural enemies has expanded greatly in recent years, especially applied studies in the context of pest management. Variation in induced plant defences resulting from attack mediated by different herbivores is in itself a driver of biodiversity in higher trophic levels (Poelman *et al.*, 2008). Prior to 2003, most research on plant semiochemicals and arthropod behaviour was conducted in the laboratory (Dicke *et al.*, 2003) but since then many field studies have been published on efforts to exploit this aspect of chemical ecology for managing pest arthropod populations in crops (Khan *et al.*, 2008). The focus of most of these studies has centred on the use of herbivore-induced plant volatiles (HIPVs) to attract and retain natural enemies of pests, thereby improving conservation biological control (CBC). This chapter deals with recent efforts to understand and exploit HIPVs for pest management. Push–pull strategies were reviewed by Cook *et al.* (2007) and are covered by Khan *et al.* (chapter 16 of this volume).

HERBIVORE-INDUCED PLANT VOLATILES (HIPVs)

Plants have evolved various direct and indirect defence mechanisms against attacking organisms (Pieterse and Dicke, 2007; Heil and Ton, 2008; Dicke, 2009) and release volatile compounds (semiochemicals) as part of constitutive and induced defence mechanisms from plant leaves, flowers and fruits into the atmosphere and from roots into the soil (Dudareva *et al.*, 2006). Semiochemicals emitted by plants in response to herbivorous damage are known as herbivore-induced plant volatiles (HIPVs) (Dicke and Sabelis, 1988). Emitted HIPVs may directly affect herbivores negatively due to their toxic, repelling and deterring properties and result in death or retarded development

---

Box 11.1 HIPVs and plant protection

Plants attacked by herbivores emit chemical distress signals known as herbivore-induced plant volatiles (HIPVs) (Dicke and Sabelis, 1988). Emitted HIPVs may directly impair herbivores, resulting in death or retarded development (Lou and Baldwin, 2003). However, the main function of HIPVs appears to be to recruit plant 'bodyguards'. Attacked plants emit a bouquet of HIPVs, 'words' of a complex language that call in predatory and parasitic arthropods to fight off the attackers (Dicke, 1999; Kessler and Baldwin, 2001; Turlings and Ton, 2006; Halitschke *et al.*, 2008; Mumm and Dicke, 2010). In addition, neighbouring plants 'eavesdrop' on the chemical conversation and mobilise their own defences against incoming herbivores. Benefits are evenly divided: arthropod bodyguards responding to plant distress signals benefit by obtaining food, while plants benefit from reduced herbivory.

(Lou and Baldwin, 2003). HIPVs are also the 'words' of a complex language used to 'warn' other plants of impending attack and to recruit predatory/parasitic arthropods for 'bodyguard' services (Dicke, 1999; Dicke *et al.*, 1999; Kessler and Baldwin, 2001; Lou and Cheng, 2003; Dudareva *et al.*, 2006; Turlings and Ton, 2006; Halitschke *et al.*, 2008; Mumm and Dicke, 2010). Such plant 'bodyguards' respond to the language of plants in distress, and benefit from the food/ host resources available (Box 11.1). The induction of HIPVs occurs not only in response to herbivore feeding on plant parts above ground but also following physical damage and the deposition of insect eggs, or from insect feeding on plant roots (Hilker and Meiners, 2006; Turlings and Ton, 2006). HIPV emission takes place at the site of damage but may also issue systemically from other uninfested plant parts (Turlings and Tumlinson, 1992). HIPVs that attract predators and parasitoids are volatile organic compounds including monoterpenes and sesquiterpenes, green leaf volatiles of the fatty acid/lipoxygenase pathway, and aromatic metabolites (e.g. indole and methyl salicylate) of the shikimate/tryptophan pathway (Pare and Tumlinson, 1996). The qualitative and quantitative characteristics of HIPVs can vary according to the herbivore involved, the plant species and even plant genotype (Turlings *et al.*, 1993; Takabayashi *et al.*, 1994; van den Boom *et al.*, 2004). HIPVs may function as direct attractants of natural enemies of pests and/or as plant-plant signals. Methyl jasmonate (MeJA), methyl salicylate (MeSA), ethylene (ET) and green leaf volatiles (GLVs) act as plant-plant signals (Farmer, 2001). These HIPVs activate jasmonic acid (JA), salicylic acid (SA) and ET dependent defence reactions in other parts of the plant or in neighbouring undamaged plants, alerting and priming them for impending attack by boosting production of aromatic and terpenoid volatiles that enhance induced defence (Engelberth *et al.*, 2004; Baldwin *et al.*, 2006; Turlings and Ton, 2006; Yan and Wang, 2006; Beckers and Conrath, 2007; Frost *et al.*, 2008; Heil and Ton, 2008). There is also evidence that MeSA and hexenyl acetate may function as plant signals (Shulaev *et al.*, 1997; Ozawa *et al.*, 2000; Engelberth *et al.*, 2004; Abdella, 2010). The use of HIPVs as signallers, elicitors or release primers of 'correct' and complete blends of natural enemy attracting emissions, is an attractive possibility for manipulating predator and parasitoid populations in pest management.

The first demonstration of the impact of HIPVs in the field came from research on psyllids in pear orchards (Drukker *et al.*, 1995), which showed that densities of predatory bugs (Anthocoridae) increased with the density of caged psyllids, increased herbivory and increased emission of HIPVs. Shimoda *et al.* (1997) recorded more predatory thrips on sticky cards near spider mite-infested bean plants than on traps near uninfested plants. Bernasconi *et al.* (2001) trapped more natural enemies near plants damaged and treated with caterpillar regurgitant, than near undamaged, untreated plants. Cabbage cultivars that emitted optimal HIPV blends for caterpillar parasitoids in laboratory tests were also confirmed as highly attractive to parasitoids in the field (Poelman *et al.*, 2009).

Using synthetic HIPV to attract natural enemies and improve biological control in crops

The first demonstrations of the potential of a *synthetic* HIPV as a *direct* field attractant for beneficial insects were provided by James (2003a; 2003b; 2003c) (Box 11.2). These studies showed attraction of a number of insect species and families to MeSA and (Z)-3-hexenyl acetate (HA) in a Washington (USA) hop yard. Insects

Box 11.2 The promise of synthetic HIPVs

A desire to increase the early-season community of natural enemies in Washington State hop yards led in 2002 to the first field demonstrations of the potential of synthetic HIPVs as attractants for beneficial insects (James, 2003a; 2003b; 2003c). The existence and function of HIPVs had earlier been described in a plethora of elegant laboratory studies conducted primarily in the Netherlands by Marcel Dicke, Maurice Sabelis and co-workers, who in 1990 suggested prospects for application in pest control (Dicke et al., 1990). The 2002–03 studies of James in Washington showed attraction of a number of beneficial insect species and families to methyl salicylate (MeSA) and (Z)-3-hexenyl acetate. In follow-up studies, the large population of predatory insects in MeSA-baited hops was associated with a dramatic reduction in spider mites and aphids, the major pests of hops (James and Price, 2004).

attracted to MeSA included the green lacewing, *Chrysopa nigricornis* (Burmeister) (Chrysopidae), the bigeyed bug, *Geocoris pallens* (Stål). (Geocoridae), the mite-eating lady beetle, *Stethorus punctum picipes* (Casey) (Coccinellidae) and hoverflies (Syrphidae) (Figure 11.1). Three beneficial species were attracted to HA: a predatory mirid, *Deraeocoris brevis* (Uhler), a minute

Figure 11.1 Hoverflies (Syrphidae) are one of the most readily attracted groups of beneficial insects to methyl salicylate.

Figure 11.2 Minute pirate bugs, *Orius tristicolor* (White), are reliably attracted to vineyards and hop yards in Washington State baited with methyl salicylate (Predalure™).

Figure 11.3 A commercially available controlled release dispenser (Predalure™) of methyl salicylate attached to a hop plant.

pirate bug, *Orius tristicolor* (White) (Figure 11.2) and *S. punctum picipes*. In Oregon strawberries, Lee (2010) showed significant attraction to synthetic MeSA by lacewings (Chrysopidae) and *O. tristicolor*. However, Jones *et al.* (2011) working in Washington State apple orchards found no attraction by lacewings to MeSA, although volatile release rates in this study were low. Other synthetic HIPV/trapping studies by James (2005) revealed at least 11 species and families of beneficial insects responded to one or more synthetic HIPV. Thirteen HIPVs attracted at least one or more species/family of beneficial insect. Jones *et al.* (2011) demonstrated substantial attraction by male lacewings (*C. nigricornis*) to the HIPV, squalene. Yu *et al.* (2008) conducted similar trapping studies with seven synthetic HIPVs in cotton in China and showed responses by various natural enemies to six of them. James and Price (2004) presented evidence for recruitment and retention of beneficial insects in grapes and hops using controlled-release dispensers of MeSA (Figure 11.3). In a replicated experiment conducted in a juice grape vineyard, sticky cards in blocks baited with MeSA captured significantly greater numbers of five species of predatory insects than cards in unbaited blocks. Canopy shake samples and sticky card monitoring conducted in a MeSA-baited, unsprayed hop yard indicated development and maintenance of a beneficial arthropod population that was nearly four times greater than that present in an unbaited reference

yard. The large population of predatory insects in MeSA-baited hops was associated with a dramatic reduction in spider mites and aphids, the major pests of hops. Mallinger *et al.* (2011) showed that abundance of soybean aphids (*Aphis glycines* (Matsumura)) was lower in MeSA-treated blocks than in untreated blocks, apparently due to greater numbers of lacewings and syrphid flies. Further data on the effects of MeSA dispensers in enhancing biological control in hops and grapes were provided in James *et al.*, (2005) and James (2006).

To date, field research in the western US has generally not demonstrated significant attraction of pest herbivores to MeSA and other synthetic HIPVs. An exception is the leaf-mining fly family Agromyzidae which was attracted to MeSA in a field experiment in Washington State (James, 2005). Molleman *et al.* (1997) reported attraction of pestiferous Lepidoptera species to MeSA in pear orchards in the Netherlands. In New Zealand deployment of synthetic MeSA in controlled release dispensers in a field of turnip *Brassica rapa* (L.) (Brassicaceae) attracted a herbivore, a parasitoid of a herbivore and a parasitoid of a predator (i.e. insects from the second, third and fourth trophic levels) (Orre *et al.*, 2010). MeSA increased the abundance of *Diadegma semiclausum* (Hellén) (Hymenoptera: Ichneumonidae), a parasitoid of the crucifer pest, diamondback moth (DBM) *Plutella xylostella* (L.) (Lepidoptera: Plutellidae). It also increased the abundance of a leaf mining fly, *Scaptomyza flava* (Fallén) (Diptera: Drosophilidae), a herbivore commonly found in Brassicaceae in New Zealand (Martin *et al.*, 2006). *Scaptomyza flava* causes sufficient damage in European vegetable crops for an integrated pest management (IPM) strategy to have been developed for its control in New Zealand (Cameron and Walker, 2000). MeSA also increased abundance of the lacewing parasitoid, *Anacharis zealandica* Ashmead (Hymenoptera: Figitidae) (Orre *et al.*, 2010). Lacewings are important predators of many small soft-bodied pests like mites and aphids and are an important component of CBC. In this system, benefits accruing from the attraction of a parasitoid of DBM to MeSA may be compromised by enhancement of populations of a potential pest and natural enemy of a natural enemy. This illustrates the need to determine the impacts of synthetic HIPVs on a regional and crop basis. Results from one area/crop cannot necessarily be extrapolated to another situation. The potential for HIPVs to be used for pest management in some crops may depend on the ability to manage such compromising factors.

USING SYNTHETIC HIPVs TO TRIGGER HIPV EMISSIONS FROM CROP PLANTS

The use of HIPVs as elicitors of 'correct' and complete blends of natural enemy attracting emissions has also been examined in field studies. Synthetic JA applied directly to tomato plants elicits production of HIPVs and increases parasitism of caterpillar pests (Thaler, 1999). Kessler and Baldwin (2001) showed that synthetic HIPVs incorporated in lanolin paste applied near moth eggs increased predation by a predatory bug. Airborne MeSA, MeJA and HA appeared to act as elicitors of HIPVs in a vineyard study (James and Grasswitz, 2005) (Figure 11.4). Numbers of parasitoids (*Metaphycus* sp. and *Anagrus* spp.) were higher in HIPV-baited blocks than in unbaited blocks, despite not being directly responsive to MeSA, MeJA and HA (James, 2005). James and Grasswitz (2005) concluded that the parasitoids responded to HIPVs produced by grape plants exposed to the synthetic HIPVs. Field experiments in Australia investigated the impact of spray applications of several synthetic HIPVs (with the canola oil-based adjuvant Synertrol®) in wine grapes, sweet corn and broccoli in attracting natural enemies (Simpson *et al.*, 2011a; Box 11.3). A number of these HIPVs resulted in increased numbers of natural enemies, primarily parasitic wasps, in the three crop systems. In wine grapes the abundance of wasps in the family Trichogrammatidae was increased near

Figure 11.4 Methyl salicylate dispensers (Predalure™) attached to posts in a vineyard.

Box 11.3 'Attract and reward' can boost the effect of HIPVs

Field experiments in Australia investigated the impact of spray applications of several synthetic HIPVs (with the adjuvant Synertrol®) in wine grapes, sweet corn and broccoli in attracting natural enemies (Simpson *et al.*, 2011a; 2011b). A number of these HIPVs resulted in increased numbers of natural enemies, primarily parasitic wasps, in the three crop systems. Subsequent experiments combined synthetic HIPVs with a floral plant reward (buckwheat, *Fagopyrum esculentum*) in an effort to increase attraction and retention of natural enemies. 'Attract and reward' sometimes resulted in greater densities of various parasitoids than in HIPV- or reward-only treatments.

---

benzaldehyde (Be) and methyl anthranilate (MeA)-treated plants at 0.5%, whilst wasps in the families Encyrtidae and Bethylidae responded to MeA at 1.0%. In sweet corn, encyrtids were more numerous near MeA-treated plants (0.5%) and in broccoli trichogrammids responded to Be (0.5%), z-3-hexen-1-ol (He) (0.5, 1.0%), MeJA (1.0% percent) and MeSA (0.5%). Simpson *et al.* (2011a) also showed that some HIPVs like Be and MeA have the potential to attract similar insects in multiple crop species. The abundance of herbivorous thrips was also increased near sweet corn and broccoli plants treated with MeSA, MeA, MeJA, Be, HA or He at 0.5 and 1.0%. However, thrips are not regarded

Figure 11.5 Monitoring insects attracted to a sticky card adjacent to an HIPV dispenser in a stand of buckwheat.

as pests of economic significance in Australia in these crop systems and they may provide alternative hosts or prey for attracted natural enemies. Subsequent experiments combined synthetic HIPVs with a floral plant reward (buckwheat, *Fagopyrum esculentum* (Moench)) in an effort to increase attraction and retention of natural enemies (Simpson *et al.* 2011b; Plate 11.1 and Figure 11.5). 'Attract and reward' resulted in greater densities of various parasitoids than in HIPV or reward-only treatments; however, this was statistically significant in only one instance. Scelionid egg parasitoids were more abundant near broccoli plants treated with MeSA and provided with a floral reward than MeSA without reward (Simpson *et al.*, 2011b). While a single application of an HIPV had a short-lived (approximately six days) impact in attracting natural enemies, the addition of flowering buckwheat helped prolong this impact (Simpson *et al.*, 2011a; 2011b). Although 'attract and reward' effects were only modestly synergistic the use of both techniques is warranted. Synthetic HIPVs could be applied early while nectar plants are being established and terminated when flowering occurs. Although some direct attraction of natural

enemies to HIPV-treated plants may have occurred in these experiments, it is likely that most of the observed attraction resulted from induced emission of HIPVs in the crop plants. Most of the HIPVs used by Simpson *et al.* (2011a; 2011b) are known to induce HIPV emission in plants (Yan and Wang, 2006; Tamogami *et al.*, 2008). The profile of volatiles emanating from MeSA/Synertrol®-treated broccoli plants differs significantly from the profile provided by untreated plants (Simpson, 2011c). Conversely, attract and reward experiments using MeSA and buckwheat conducted in a turnip crop in New Zealand did not show synergism between the two components (Orre, 2009); instead, different beneficial arthropods responded either to MeSA or floral rewards. The predatory hoverfly, *Melanostoma fasciatum* (Macquart), responded significantly to MeSA and rates of aphid parasitism were significantly higher in MeSA treatments. Attract and reward may still be of benefit in New Zealand turnip crops as the techniques complement each other by increasing the abundance of different species and guilds of natural enemies. The use of different HIPVs in different crops will result in different volatile emission profiles, attractive to different groups of predators and/or parasitoids. The available natural enemy fauna in different geographic or crop environments will also naturally impact outcomes. Thus, the potential benefits of 'attract and reward' will be strongly influenced by geography, crop type and the endemic arthropod fauna and need to be explored and assessed for specific crop/pest situations.

The use of botanical oil-based products like Synertrol™, in combination with synthetic HIPVs like MeSA, as a strategy for alerting plants to a herbivore threat and inducing natural defences may have great potential for improving CBC in crop pest management. In three field experiments conducted on hops and wine grapes, plants sprayed with botanical oil (canola, peppermint, rosemary), pesticides formulated with small concentrations of MeSA or HA, attracted significantly greater numbers of predatory and parasitic insect species, than unsprayed plants (James, unpublished). Hop plant cultivar strongly influenced the results obtained in this study, suggesting that botanical oil/HIPV induced plant volatile emissions varied qualitatively and/or quantitatively according to cultivar (Gouinguené *et al.*, 2001; Lou *et al.*, 2006). Charleston *et al.* (2006) showed application of a botanical extract made from the syringa tree (*Melia azedarach* (L.)) to cabbage plants increased volatile emission and attracted the parasitoid, *Cotesia plutellae* (Kurdjumov).

SILICON ENHANCES HIPV EMISSION AND NATURAL ENEMY ATTRACTION

The possibility that plant nutrients may play a role in modifying emissions of HIPVs has been confirmed in recent studies (Gouinguené and Turlings, 2002; Lenardis *et al.*, 2007; Szpeiner *et al.*, 2009). Although the function of silicon (Si) in plant nutrition is still not clear, its role as a nutrient that is beneficial for plants under a range of stressors, both abiotic (e.g. salinity and drought) and biotic (e.g. insects and pathogens) is indisputable (see reviews by Epstein, 1999; 2009; Reynolds *et al.*, 2009). Silicon's importance in plant defence was first reported in maize against the Hessian fly, *Mayetiola destructor* (Say) (Diptera: Cecidomyiidae) (McColloch and Salmon, 1923). For many years, the main established mechanism for Si increasing plant resistance against arthropod pests involved constitutive plant defence (Reynolds *et al.*, 2009). Sasamoto (1958) showed that larval rice stem borers, *Chilo suppressalis* (Walker) (Lepidoptera: Pyralidae) favoured untreated rice stalks compared with Si-treated rice stalks, and noted that host selection by an insect depended not only on its physical, but also its chemical properties. More recently, the role of Si in induced plant defence against arthropod pests and subsequent natural enemy attraction has become apparent (Reynolds *et al.*, 2009).

Silicon is known to act as a regulator of plant defence mechanisms and may interact with key components of plant stress signalling systems leading to induced resistance (Fauteux *et al.*, 2005). Silicon-accumulating plants, supplemented with silicon, translocate silicic acid and, when attacked, produce systemic stress signals such as SA and JA (Fauteux *et al.*, 2005) which are key to plant-induced defences (Gatehouse, 2002). Endogenous JA, a plant hormone synthesised when herbivores feed on a plant, induces putatively defensive phytochemicals and proteins such as proteinase inhibitors and oxidative enzymes (Thaler *et al.*, 2002). Studies by Gomes *et al.* (2005) first exhibited the importance of silicon for induced plant chemical defences against herbivorous insects. Application of calcium silicate to wheat plants that were infested with the aphid *Schizaphis graminum* (Rondani) (Hemiptera: Aphididae), elevated activity levels of three plant enzymes: peroxidase, polyphenoloxidase and phenylalanine ammonia-lyase, important in plant defence, and suppressed aphid reproduction. More recently, Kvedaras *et al.*, (2010) reported an exciting development

> ### Box 11.4 Silicon treatment boosts natural enemies on plants
>
> In Y-tube olfactometer studies, Kvedaras *et al.* (2010) showed that adult red and blue beetle, *Dicranolaius bellulus* (Guérin-Méneville) (Coleoptera: Melyridae) were significantly more attracted to Si-treated (Si+) plants upon which *Helicoverpa armigera* (Hübner) (Lepidoptera: Noctuidae) larvae had fed compared with Si-untreated (Si-), pest-infested plants. In the field, *H. armigera* egg baits stapled to plants, showed that greater predation occurred on Si+, pest-infested plants than on Si+, uninfested and Si- infested and uninfested plants. These results suggest that Si applied to plants subsequently infested by caterpillars increases the plants' attractiveness to natural enemies; an effect reflected in elevated biological control in the field.

whereby plants grown in a Si-rich environment were able to mount a greater HIPV-based defensive reaction when challenged by a herbivore, resulting in greater attraction of a predator in laboratory bioassays and elevated pest mortality in the field (Box 11.4). Kvedaras *et al.* (2010) proposed that ongoing studies may mark the opening of a new opportunity in biological control whereby inexpensive Si-containing materials could be used to augment HIPV production by various crops and support host plant resistance traits operating via the third trophic level.

PROSPECTS

The potential of chemical ecology via HIPVs to help us understand and exploit biodiversity for pest management during the coming decades is considerable. We have barely explored the 'tip of the iceberg', yet the information and results gleaned so far are promising and have begun stimulating research on HIPVs in crop ecosystems around the world. Like all good IPM tools, HIPVs should become a solid and synergistic component of the crop protection 'toolbox' (Gurr and Kvedaras, 2010). However, as Gurr and Kvedaras (2010) warn, a number of questions and issues need resolving before the full commercial potential of HIPVs for improving CBC in crop production can be realised.

Many of these questions and issues need resolving on a case-by-case basis. One of these questions is whether the attraction of natural enemies to a specific crop is mediated by direct response to synthetic HIPVs, or is a response to plant-produced HIPVs stimulated by exogenous HIPVs. This is likely to differ between different crop types and will determine appropriate strategies for HIPV deployment. Another issue that will vary in importance according to crop type is the metabolic cost incurred by plants in producing HIPVs. In some instances these costs may be so great that yield reductions and/or reduced crop quality may occur (Cipollini *et al.*, 2003). However, very few studies have addressed this issue. Kessler and Heil (2011) concluded that the balance of empirical evidence suggests that HIPV emission has a negligible impact on most plants in terms of metabolic cost.

Effective monitoring will be an important component of any IPM system using HIPVs to enhance natural enemy abundance and diversity, and will guide the timing of HIPV deployment and removal. Interestingly, the majority of natural enemies attracted to MeSA or MeSA-treated grapes and hops in Washington and Oregon are generalist species, not restricted to a single prey type (James, 2003a; 2003b; 2003c; 2005; James and Price, 2004; Lee, 2010). In biological control programmes based on specialist natural enemies, MeSA could play a valuable complementary role by enhancing the impact of generalist natural enemies. Exploiting both generalist and specialist natural enemies within crop ecosystems is generally regarded as an optimal approach to CBC (Welch *et al.*, chapter 3 of this volume). If MeSA or HIPVs generally have greater impact on generalist predators and parasitoids it would further highlight the need for more research on these natural enemies.

A key area for future research will be identifying linkages and synergism between HIPVs and components of crop landscaping. Ground covers and refugia designed to provide natural enemy habitat and resources are destined to become integral components of modern crop ecosystems. Integrating these habitat modifications with HIPVs for optimal overall benefits should provide major dividends in terms of enhancing and, perhaps more importantly, sustaining CBC. The initial studies on 'attract and reward' in New Zealand and Australia are very promising and suggest this strategy may become a major component of future CBC programmes. However, much location-specific research is needed to determine the optimal com-

position of plant species and degree of connectivity between crop and non-crop landscapes. A key consideration is whether sufficient communities of desired natural enemies exist in the landscapes near to crop ecosystems. HIPV-based attract and reward strategies should work optimally in fragmented landscapes of crop and natural areas with good connectivity. In contrast, crop monocultures extending over large areas with little natural habitat are not good candidates for exploiting natural enemy diversity. Further research is also needed on whether attract and reward strategies in crops can result in an overall increase in local populations of natural enemies, or simply divert populations to farms practising the technique at the expense of farms that are not. Another issue needing study is the sustainability of attract and reward systems. For example, do natural enemy population densities increase for the first few years then stabilise at a level that provides optimal CBC?

The future of chemical ecology as a tool to understand and exploit natural enemy abundance and biodiversity in pest management is bright. To date, very few known HIPVs have been investigated for their potential as natural enemy attractants and enhancers of CBC. It is possible that the HIPVs with most promise in pest management have yet to be evaluated. In addition, the HIPV profiles produced by most crop plants after attack by key pests remain unknown. Information on this profile is the first step needed for development of an HIPV-based strategy tailored to the crop. It is clear from the research conducted to date that HIPV-based strategies need to be tailored for specific crops and situations, reflecting plant chemistry and site-associated arthropod biodiversity.

REFERENCES

Abdella, R.M. (2010) Assessment of herbivore induced plant volatiles in juvenile Hops by methyl salicylate. Unpublished Master of Philosophy thesis, Department of Chemistry, Washington State University.

Baldwin, I.T., Halitschke, R., Paschold, A., von Dahl, C.C. and Preston, C.A. (2006) Volatile signaling in plant-plant interactions: 'talking trees' in the genomics era. *Science*, 311, 812–815.

Beckers, G.J.M. and Conrath, U. (2007) Priming for stress resistance: from the lab to the field. *Current Opinion in Plant Biology*, 10, 425–431.

Bernasconi, M.L., Turlings, T.C.J., Edwards, P.J. *et al.* (2001) Response of natural populations of predators and parasitoids to artificially induced volatile emissions in maize plants (*Zea mays* L.). *Agricultural and Forest Entomology*, 3, 201–209.

Cardinale, B.J., Harvey, C.T., Gross, K. and Ives, A.R. (2003) Biodiversity and biocontrol: emergent impacts of a multi-enemy assemblage on pest suppression and crop yield in an agroecosystem. *Ecology Letters*, 6, 857–865.

Cameron, P.J. and Walker, G.P. (2000) *Integrated Pest Management for Brassicas*, New Zealand Institute for Crop and Food Research Limited, Christchurch.

Charleston, D.S., Gols, R., Hordijk, K.A., Kfir, R., Vet, L.E.M. and Dicke, M. (2006) Impact of botanical pesticides derived from *Melia azedarach* and *Azadirachta indica* plants on the emission of volatiles that attract parasitoids of the diamondback moth to cabbage plants. *Journal of Chemical Ecology*, 32, 325–349.

Cipollini, D., Purrington, C.B. and Bergelson, J. (2003) Costs of induced responses in plants. *Basic and Applied Ecology*, 4, 79–85.

Cook, S.M., Khan, Z.R. and Pickett, J.A. (2007) The use of push–pull strategies in integrated pest management. *Annual Review of Entomology*, 52, 375–400.

Dicke, M. (1999) Evolution of induced indirect defense of plants, in *The ecology and evolution of inducible defenses* (eds R. Tollrian and C.D. Harvell), Princeton University Press, Princeton, pp. 62–88.

Dicke, M. (2009) Behavioural and community ecology of plants that cry for help. *Plant, Cell and Environment*, 32, 654–665.

Dicke, M. and Sabelis, M.W. (1988) How plants obtain predatory mites as bodyguards. *Netherlands Journal of Zoology*, 38, 148–165.

Dicke, M., Sabelis, M.W., Takabayashi, J., Bruin, J. and Posthumus, M.A. (1990) Plant strategies for manipulating predator–prey interactions through allelochemicals: prospects for application in pest control. *Journal of Chemical Ecology*, 16, 3091–3118.

Dicke, M., Gols, R., Ludeking, D. and Posthumus, M.A. (1999) Jasmonic acid and herbivory differentially induce carnivore-attracting plant volatiles in lima bean plants. *Journal of Chemical Ecology*, 25, 1907–1922.

Dicke, M., van Poecke, R.M.P. and de Boer, J.G. (2003) Inducible indirect defence of plants: from mechanisms to ecological functions. *Basic and Applied Ecology*, 4, 27–42.

Drukker, B., Scutareanu, P. and Sabelis, M.W. (1995) Do anthocorid predators respond to synomones from *Psylla*-infested pear trees under field conditions? *Entomologia Experimentalis et Applicata*, 77, 193–203.

Dudareva, N., Negre, F., Nagegowda, D.A. and Orlova, I. (2006) Plant volatiles: recent advances and future perspectives. *Critical Reviews in Plant Sciences*, 25, 417–440.

Engelberth, J., Alborn, H.T., Schmelz, E.A. and Tumlinson, J.H. (2004) Airborne signals prime plants against insect herbivore attack. *Proceedings of the National Academy of Sciences of the United States of America*, 101, 1781–1785.

Epstein, E. (1999) Silicon. *Annual Review of Plant Physiology and Plant Molecular Biology*, 50, 641–664.

Epstein, E. (2009) Silicon: its manifold roles in plants. *Annals of Applied Biology*, 155, 155–160.

Farmer, E.E. (2001) Surface-to-air signals. *Nature*, 411, 854–856.

Fauteux, F., Rémus-Borel, W., Menzies, J.G. and Bélanger, R.R. (2005) Silicon and plant disease resistance against pathogenic fungi. *FEMS Microbiology letters*, 249, 1–6.

Frost, C.J., Mescher, M.C., Carlson, J.E. and de Moraes, C.M. (2008) Plant defense priming against herbivores: getting ready for a different battle. *Plant Physiology*, 146, 818–824.

Gatehouse, J.A. (2002) Plant resistance towards insect herbivores: a dynamic interaction. *New Phytologist*, 156, 145–169.

Gomes, F. B., Moraes, J.C.d., Santos, C.D.d. and Goussain, M.M. (2005) Resistance induction in wheat plants by silicon and aphids. *Scientia Agricola*, 62, 547–551.

Gouinguené, S.P., Degen, T. and Turlings, T.C.J. (2001) Variability in herbivore-induced odour emissions among maize cultivars and their wild ancestors (teosinte). *Chemoecology*, 11, 9–16.

Gouinguené, S.P. and Turlings, T.C.J. (2002) The effects of abiotic factors on induced volatile emissions in corn plants. *Plant Physiology*, 129, 1296–1307.

Gurr, G.M. and Kvedaras, O.L. (2010) Synergizing biological control: scope for sterile insect technique, induced plant defences and cultural techniques to enhance natural enemy impact. *Biological Control*, 52, 198–207.

Halitschke, R., Stenberg, J.A., Kessler, D., Kessler, A. and Baldwin, I.T. (2008) Shared signals – 'alarm calls' from plants increase apparency to herbivores and their enemies in nature. *Ecology Letters*, 11, 24–34.

Heil, M. and Ton, J. (2008) Long-distance signalling in plant defence. *Trends in Plant Science*, 13, 264–272.

Hilker, M. and Meiners, T. (2006) Early herbivore alert: insect eggs induce plant defense. *Journal of Chemical Ecology*, 32, 1379–1397.

James, D.G. (2003a) Field evaluation of herbivore-induced plant volatiles as attractants for beneficial insects: Methyl salicylate and the green lacewing, *Chrysopa nigricornis*. *Journal of Chemical Ecology*, 29, 1601–1609.

James, D.G. (2003b) Synthetic herbivore-induced plant volatiles as attractants for beneficial insects. *Environmental Entomology*, 32, 977–982.

James, D.G. (2003c) *Synthetic herbivore-induced plant volatiles as field attractants for beneficial insects*, The British Crop Protection Council International Congress-Crop Science and Technology 2003, Glasgow, pp. 1217–1222.

James, D.G. (2005) Further evaluation of synthetic herbivore-induced plant volatiles as attractants for beneficial insects. *Journal of Chemical Ecology*, 31, 481–495.

James, D.G. (2006) Methyl salicylate is a field attractant for the goldeneyed lacewing, *Chrysopa oculata*. *Biocontrol Science and Technology*, 16, 107–110.

James, D.G. and Grasswitz, T.R. (2005) Field attraction of parasitic wasps, *Metaphycus* sp. and *Anagrus* spp. using synthetic herbivore-induced plant volatiles. *Biocontrol*, 50, 871–880.

James, D.G. and Price, T.S. (2004) Field-testing of methyl salicylate for recruitment and retention of beneficial insects in grapes and hops. *Journal of Chemical Ecology*, 30, 1613–1628.

James, D.G., Castle, S.C., Grasswitz, T. and Reyna, V. (2005) *Using synthetic herbivore-induced plant volatiles to enhance conservation biological control in hops and grapes*. Proceedings of the 2nd International Symposium on Biological Control of Arthropods, September 12–16, 2005, Davos, Switzerland, pp. 192–205.

Jones, V.P., Steffan, S., Wiman, N.G., Horton, D., Miliczky, E., Zhang, Q. and Baker, C.C. (2011) Evaluation of herbivore-induced plant volatiles for monitoring green lacewings in Washington apple orchards. *Biological Control*, 56, 98–105.

Kessler, A. and Baldwin, I.T. (2001) Defensive function of herbivore-induced plant volatile emissions in nature. *Science*, 291, 2141–2144.

Kessler, A. and Heil, M. (2011) The multiple faces of indirect defences and their agents of natural selection. *Functional Ecology*, 25, 348–357.

Khan, Z.R., James, D.G., Midega, C.A.O. and Pickett, C.H. (2008) Chemical ecology and conservation biological control. *Biological Control*, 45, 210–224.

Kvedaras, O.L., An, M., Choi, Y.S. and Gurr, G.M. (2010) Silicon enhances natural enemy attraction to pest-infested plants through induced plant defences. *Bulletin of Entomological Research*, 100, 367–371.

Lee, J.C. (2010) Effect of methyl salicylate based lures on beneficial and pest arthropods in strawberry. *Environmental Entomology*, 39, 653–660.

Lenardis, A.E., van Baren, C., Lira, D.L. and Ghersa, C.M. (2007) Plant-soil interactions in wheat and coriander crops driving arthropod assemblies through volatile compounds. *European Journal of Agronomy*, 26, 410–417.

Lou, Y.G. and Baldwin, I.T. (2003) *Manduca sexta* recognition and resistance among allopolyploid *Nicotiana* host plants. *Proceedings of the National Academy of Sciences of the United States of America*, 100, 14581–14586.

Lou, Y.G. and Cheng, J.A. (2003) Role of rice volatiles in the foraging behaviour of the predator *Cyrtorhinus lividipennis* for the rice brown planthopper *Nilaparvata lugens*. *BioControl*, 48, 73–86.

Lou, Y., Xiaoyan, H., Turlings, T.C.J., Cheng, J., Xuexin, C. and Gongyin, Y. (2006) Differences in induced volatile emissions among rice varieties result in differential attraction and parasitism of *Nilaparvata lugens* eggs by the parasitoid, *Anagrus nilaparvatae* in the field. *Journal of Chemical Ecology*, 32, 2375–2387.

Mallinger, R.E., Hogg, D.B. and Gratton, C. (2011) Methyl salicylate attracts natural enemies and reduces populations of soybean aphids (Hemiptera: Aphididae) in soybean agroecosystems. *Journal of Economic Entomology*, 104, 115–124.

Martin, N.A., Workman, P.J. and Hedderley, D. (2006) Susceptibility of *Scaptomyza flava* (Diptera: Drosophilidae) to insecticides. *New Zealand Plant Protection*, 59, 228–234.

McColloch, J.W. and Salmon, S.C. (1923) The resistance of wheat to the Hessian fly – a progress report. *Journal of Economic Entomology*, 16, 293–298.

Molleman, F., Drukker, B. and Blommers, L. (1997) A trap for monitoring pear psylla predators using dispensers with the synomone methyl salicylate. *Proceedings of Experimental and Applied Entomology N.E.V. Amsterdam*, 8, 177–182.

Mumm, R. and Dicke, M. (2010) Variation in natural plant products and the attraction of bodyguards involved in indirect plant defense. *Canadian Journal of Zoology*, 88, 628–667.

Orre, G.U.S. (2009) 'Attract and reward': combining a floral resource subsidy with a herbivore-induced plant volatile to enhance conservation biological control. Doctor of Philosophy thesis, Lincoln University, New Zealand.

Orre, G.U.S., Wratten, S., Jonsson, M. and Hale, R.J. (2010) Effects of an herbivore-induced plant volatile on arthropods from three trophic levels in *Brassicas*. *Biological Control*, 53, 62–67.

Ozawa, R., Arimura, G., Takabayashi, J., Shimoda, T. and Nishioka, T. (2000) Involvement of jasmonate and salicylate-related signaling pathways for the production of specific herbivore-induced volatiles in plants. *Plant Cell Physiology*, 41, 391–398.

Pare, P.W. and Tumlinson, J.H. (1996) Plant volatile signals in response to herbivore feeding. *Florida Entomologist*, 19, 93–103.

Pieterse, C.M.J. and Dicke, M. (2007) Plant interactions with microbes and insects: from molecular mechanisms to ecology. *Trends in Plant Science*, 12, 564–569.

Poelman, E.H., van Loon, J.J.A. and Dicke, M. (2008) Consequences of variation in plant defense for biodiversity at higher trophic levels. *Trends in Plant Science*, 13, 534–541.

Poelman, E.H., Oduor, A.M.O., Broekgaarden, C., Hordijk, C.A., Jansen, J.J., van Loon, J.J.A., van Dam, N.M., Vet, L.E.M. and Dicke, M. (2009) Field parasitism rates of caterpillars on *Brassica oleracea* plants are reliably predicted by differential attraction of *Cotesia* parasitoids. *Functional Ecology*, 23, 951–962.

Reynolds, O.L., Keeping, M.G. and Meyer, J.H. (2009) Silicon-augmented resistance of plants to herbivorous insects: a review. *Annals of Applied Biology*, 155, 171–186.

Sasamoto, K. (1958) Studies on the relation between silica content of the rice plant and insect pests. VI. On the injury of silicated rice plant caused by the rice-stem-borer and its feeding behaviour. *Japanese Journal of Applied Entomology and Zoology*, 2, 88–92.

Shimoda, T., Takabayashi, J., Ashira, W. and Takafuji, A. (1997) Response of a predatory insect, *Scolothrips takahashi* towards herbivore induced plant volatiles under laboratory and field conditions. *Journal of Chemical Ecology*, 23, 2033–2048.

Shulaev, V., Silverman, P. and Raskin, I (1997) Airborne signalling by methyl salicylate in plant pathogen resistance. *Nature*, 385, 718–721.

Simpson, M., Gurr, G.M., Simmons, A.T., Wratten, S.D., James, D.G. and Nicol, H. (2011a) Insect attraction to synthetic herbivore-induced plant volatile-treated field crops. *Agricultural and Forest Entomology*, 13, 45–57.

Simpson, M., Gurr, G.M., Simmons, A.T., Wratten, S.D., James, D.G., Leeson, G. and Nicol, H. (2011b) Attract and reward: combining chemical ecology and habitat manipulation to enhance biological control in field crops. *Journal of Applied Ecology*, 48, 580–590.

Simpson, M. (2011c) *Attract and reward: a novel approach to enhancing biological control of crop pests.* Doctor of Philosophy thesis, Charles Sturt University, Orange, Australia.

Szpeiner, A., Alejandra Martinez-Ghersa, M. and Ghersa, C.M. (2009) Wheat volatile emissions modified by top-soil chemical characteristics and herbivory alter the performance of neighbouring wheat plants. *Agriculture, Ecosystems and Environment*, 134, 99–107.

Takabayashi, J., Dicke, M., Takahashi, S., Posthumus, M.A. and Vanbeek, T.A. (1994) Leaf age affects composition of herbivore-induced synomones and attraction of predatory mites. *Journal of Chemical Ecology*, 20, 373–386.

Tamogami, S., Ralkwal, R. and Agrawal, G.K. (2008) Interplant communication: airborne methyl jasmonate is essentially converted into JA and JA-Ile activating jasmonate signaling pathway and VOCs emission. *Biochemical and Biophysical Research Communications*, 376, 723–727.

Thaler, J. (1999) Jasmonate-inducible plant defences cause increased parasitism of herbivores. *Nature*, 399, 686–688.

Thaler, J.S., Karban, R., Ullma, D.E., Boege, K. and Bostock, R.M. (2002) Cross-talk between jasmonate and salicylate plant defense pathways: effects on several plant parasites. *Oecologia*, 131, 227–235.

Turlings, T.C.J. and Ton, J. (2006) Exploiting scents of distress: the prospect of manipulating herbivore-induced plant odours to enhance the control of agricultural pests. *Current Opinion in Plant Biology*, 9, 421–427.

Turlings, T.C.J. and Tumlinson, J.H. (1992) Systemic release of chemical signals by herbivore-injured corn. *Proceedings of the National Academy of Sciences of the United States of America*, 89, 8399–8402.

Turlings, T.C.J., Wackers, F.I., Vet, L.E.M., Lewis, W.J. and Tumlinson, J.H. (1993) Learning of host-finding cues by hymenopterous parasitoids, in *Insect Learning* (eds D.R. Papaj and W.J. Lewis), Chapman & Hall, New York, pp. 51–78.

van den Boom, C.E.M., van Beek, T.A., Posthumus, M.A., de Groot, A.E. and Dicke, M. (2004) Qualitative and quantitative variation among volatile profiles induced by *Tetranychus urticae* feeding on plants from various families. *Journal of Chemical Ecology*, 30, 69–89.

Yan, Z.G. and Wang, C.Z. (2006) Wound-induced green leaf volatiles cause the release of acetylated derivatives and a terpenoid in maize. *Phytochemistry*, 67, 34–42.

Yu, H., Zhang, Y., Wu, K., Gao, X. and Guo, Y.Y. (2008) Field-testing of synthetic herbivore induced plant volatiles as attractants for beneficial insects. *Environmental Entomology*, 37, 1410–1415.

Application

USING DECISION THEORY AND SOCIOLOGICAL TOOLS TO FACILITATE ADOPTION OF BIODIVERSITY-BASED PEST MANAGEMENT STRATEGIES

M.M. Escalada and K.L. Heong

Biodiversity and Insect Pests: Key Issues for Sustainable Management, First Edition. Edited by Geoff M. Gurr, Steve D. Wratten, William E. Snyder, Donna M.Y. Read.

INTRODUCTION

In recent decades agricultural production, characterised by high inputs of fertiliser, water and pesticides, new crop varieties and other technologies, has become more knowledge- and information-intensive. For farmers to respond to opportunities that will improve their productivity, access to information is essential. This general comment applies also to farmer adoption of the biodiversity-related pest management strategies detailed in this book. Through up-to-date and relevant information, farmers can make informed decisions about events as they unfold each cropping season and assess the economic viability of control options available to them (Mulhall and Garforth, 2000). To be useful, information must be communicated to its intended beneficiaries in a way that is easy to understand. In many farming areas, access to information can be uncertain, partly due to differences in farmers' financial circumstances and ability to adopt new practices; often, a functioning extension-communication infrastructure is unavailable (Rivera, 1990; Swanson, 2008). Small farmers in particular are often not well served by the existing extension systems in developing countries. In some cases, information from applied research is not available or it is disseminated through communication channels, such as printed information bulletins, journals and online media, to which many farming households have limited or no access. In much of Asia and other developing regions including Africa (Khan *et al.*, chapter 16 of this volume) the farm household is a unit of production and consumption, in which all members work towards self-maintenance and sustainability. The farm household's success depends on access to and management of resources including labour, money, land, agricultural inputs (seeds, fertilisers and pesticides), food, and technological know-how (Meynen and Stephens, 1996).

Mulhall and Garforth (2000) argued that cost-effective ways of providing appropriate advice and information services to resource-poor farming households are needed. These are most effective when delivered through existing farmer and community groups in a readily understandable form through extension agents that also work to ensure adoption. These methods ensure that the developments in communication infrastructure in rural areas benefit resource-poor farmers. New communication media such as television, video players and the internet are often not easily accessible to disadvantaged farmers, thus any information on crop management and agricultural support services conveyed through these channels may not reach them.

In the absence of scientifically based technical information farmers often rely largely on their own knowledge, beliefs and perceptions. While there are strengths in indigenous knowledge systems, there are also weaknesses because 'what farmers don't know cannot help them' (Bentley, 1989, p. 25; DeWalt, 1994). For example, many farmers mistakenly believe that all insects are pests and must be killed. Furthermore, they think that pests are spontaneously generated by either the plant or the pesticides or fertilisers (Bentley and Rodriguez, 2001). When well-planned communication strategies are applied to correct such misperceptions, farmers' resource-management decisions and skills can be improved. Thus, discovering the key weaknesses in their knowledge base and decision-making is a vital first step in order to develop the appropriate intervention and communication strategy to introduce new information to reach and help the many millions of poor farmers in developing countries.

This chapter explores the use of decision theory and sociological tools to understand farmers' knowledge, attitudes and practices (KAP) and to facilitate adoption in biodiversity conservation. The following major section deals with the nature of farmer decision-making and the relevant theoretical framework. This is followed by a detailed account of a current project in which this knowledge is being used to develop and extend a biodiversity-based pest management strategy in multiple Asian countries.

FARMER DECISION-MAKING: NEW TECHNOLOGIES

Seeds and knowledge

To increase productivity, farmers are often confronted with two major sets of decisions; 'what' varieties to use for the season and 'how' to grow them (Figure 12.1) (Heong *et al.*, 2010). Varieties are developed through research to discover genes and understand their functions. Plant breeders then incorporate the genes through plant breeding processes into new varieties. The new seeds are then delivered to farmers through normal marketing channels or seed merchants. For example, many modern rice varieties are capable of yielding more than five tons per hectare and deliver

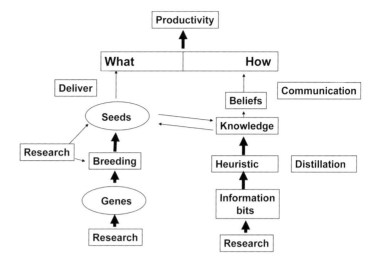

Figure 12.1 Seeds and knowledge: Farmers require differing but integrated approaches that synthesise and simplify innovative knowledge to improve decision-making and practice for productivity increase.

reasonable profits when the crop is well managed. However, most farmers obtain lower yields and profits although their input resources are adequate, probably because of inefficient management practices due to knowledge gaps (Siamwalla, 2001) and poor decision-making such as unnecessary insecticide sprays (Mumford and Norton, 1984).

What contributes to the farmers' knowledge gap is that most resource management information often ends up as research reports or scientific papers highly specific to a discipline, like entomology, plant pathology or agronomy. Such research information, although contributing to the scientific community, may not provide information that farmers can readily access. Much of the research-generated information is considered irrelevant to farmer circumstances as there is no effort to adapt and verify it for practical use (Byerlee and Alex, 1998). There is, therefore, potentially important information that is not being utilised by farmers. Most people use simple rules of thumb or heuristics in making decisions (Gigerenzer *et al.*, 1999). It is desirable, therefore, that information is synthesised and distilled into simple, testable and easy to communicate rules that can be used in decision-making and practice (Box 12.1).

Selective processes and biases

Heuristic rules may be thought of as 'rules of thumb' and their simplicity means they have the potential to

Box 12.1 Heuristics – improving pest management through easy-to-understand 'rules-of-thumb'

The rice leaffolder, *Cnaphalocrocis medinalis* (Guenée), damages leaves in the early crop stages but seldom causes yield loss because of plant compensation effects and natural biological controls that maintain the insect's population at non-damaging levels (Heong and Schoenly, 1998). Spraying to control this 'apparent pest problem' kills the natural enemies that suppress the insect, rendering the rice crop more vulnerable to subsequent invading adult leaffolders and planthoppers. A simple rule-of-thumb or 'heuristic' was therefore communicated to farmers in conflict with their prevailing belief that spraying was necessary, and they were encouraged to test the rule. The heuristic stated: 'In the first 30 days after transplanting (or 40 days after sowing), leaffolder control is not necessary'. The participatory experiments were carried out by 101 rice farmers in Leyte, Philippines. Although farmers' perceptions of pests and pesticide use were deeply entrenched, the simple experiment reduced their early-season insecticide applications and number of sprays. Farmers' attitudes towards leaf-feeding insects also changed. Besides apparently resolving any dissonance, the farmers' motivation for implementing the rule seemed mainly to relate to money saving and labour reduction.

exert a strong influence on farmers' behaviour. However, compared to the knowledge embodied within technologies such as seeds and pesticides, it is more difficult to transfer knowledge about a management practice to farmers as is often the case with biodiversity-based pest management. Difficulties include cognitive and psychological barriers such as selectivity and bias of both the disseminator and receiver of information (Schramm, 1973; Rogers, 1995). The communication of knowledge may be affected by the tendency for perceptions to be influenced by wants, needs, attitudes and other psychological factors (Krech and Crutchfield, 1971). Thus, individuals are selective about what information they respond to out of the messages with which they are constantly bombarded. Selection usually occurs at the subconscious level depending on whether new information and the perceptions that information elicits fit with strongly held attitudes, beliefs or behaviour (Schramm, 1973). In other words, an individual hears what he or she wants to hear.

People have the tendency to make judgements based on initial impressions and can be reluctant to change even when new (contradictory) information is presented. This is known as anchoring bias (Tversky and Kahneman, 1974). Unsupported information is insufficient to overcome a deeply held anchoring bias. Rather, messages must motivate by stressing the value of the knowledge or product and provide opportunities for farmers to put the knowledge or product into practical use. Box 12.1 provides a real example of a situation where encouragement of experimentation was found useful to overcome farmers' unease with new information and was based on changing practice with a simple 'rule of thumb'.

Decision theory – bounded rationality and heuristics

Studies on human judgement and choices have shown that prescriptive models such as the expected-utility theory (EUT), which deals with decisions under uncertainty, are unable to account for how people make decisions (Slovic *et al.*, 1977; Simon, 1978). Most people violate the prescriptive principles because decision-making is behavioural or cognitive in nature (Einhorn and Hogarth, 1981). Behavioural decision research is increasingly being used in fields such as public health management, business management and public policy management, making important contributions in the

design of services, information environments and decision systems (Payne *et al.*, 1992). The principles can be applied to quantify and understand farmers' decisions (Heong and Escalada, 1999).

In making resource management decisions, farmers are always faced with uncertainty in regard to factors such as rainfall and prices, and their knowledge is always limited in other ways (new products on the market or on developments that they have not heard about). Due to this limited knowledge, individuals are unable to make decisions that maximise outcomes but rather make decisions that suffice and satisfy. Simon (1956, 1982) termed this strategy 'satisficing' and this kind of decision-making 'bounded rationality'. As discussed above, individuals generally use heuristics in conditions of limited time, knowledge and computational capacities. However, heuristics that farmers have developed through experience and guesswork about possible outcomes may have inherent faults and biases. Research to understand farmers' current heuristics and reasons for their adoption (or lack of it) will help scientists and communication specialists frame alternative heuristics that improve outcomes. For instance, in the rice leaffolder example (Box 12.1), farmers spray insecticides to control the larvae (often called 'worms') because of the highly visible foliar damage. They strongly believe that the leaf damage will lead to yield loss and that the worms will multiply quickly and thus need to be killed immediately. These beliefs are likely to originate from farmers overestimating potential losses and a more general loss aversion. Such beliefs have a disproportionately higher influence on farmers' decisions if they are aware of a neighbour who has had a previous severe crop loss (even from a different pest species) and are in debt to the pesticide dealer.

Theoretical frameworks

The development of motivational campaigns and media materials to facilitate adoption of alternative heuristics is informed by theoretical frameworks such as behavioural decision-making theories (Einhorn and Hogarth, 1981), the theory of planned behaviour (TPB) (Ajzen, 1988) and the strategic extension campaign (SEC) framework (Adhikarya, 1994). The TPB asserts that the intention to behave in a particular manner is formed by the individual's attitude toward performing the behaviour, the social pressure they feel to perform the behaviour and their perception of the control they have in performing the behaviour. The theory has been applied to determine which factors influence individuals to act in certain ways and to identify better ways of effectively communicating the messages in campaigns relating to topics as diverse as health, breastfeeding, AIDS, anti-smoking, safety belt usage and anti-drugs (Rice and Paisley, 1981; Hornik, 1988; Adhikarya, 1994; Rogers, 1995; Escalada *et al.*, 1999; Singhal and Rogers, 2003). TPB helps to explain why some media campaigns have had limited success (Ajzen, 1988). Increasing knowledge alone does not change behaviour, whereas campaigns that also aim to change attitudes and perceptions of norms (accepted and sanctioned behaviour) produce better results (Ajzen, 1988; 1991; Rogers, 1995). Studies of behavioural intentions suggest that it is possible to predict the likelihood of the target audiences adopting desired practices. By understanding farmer behaviour, messages can be developed to more effectively modify their attitudes towards and perceptions of benefits of old and new practices. Farmers' attitudes towards and perception of their peers' response to their new behaviour may also be modified. On the other hand, within the SEC framework (Adhikarya, 1994), understanding the farmers' needs and problems is a prerequisite for formulating campaign objectives and developing a campaign strategy.

IMPLEMENTATION OF A BIODIVERSITY-BASED PEST MANAGEMENT STRATEGY

This section provides a detailed account of a current project in which theory is being used to inform practice in a current effort to develop and extend a biodiversity-based pest management strategy against rice pests in multiple Asian countries.

To facilitate the development of quality partnerships and local ownership of a communication campaign aimed at attitude and behaviour change, a multi-stakeholder participatory planning and review process involving research, extension, mass media, universities, NGOs and local governments, has been found to be useful (Heong *et al.*, 2010). This process comprises five phases (Figure 12.2) focusing on jointly identifying the problems, needs and opportunities, developing and evaluating intervention options and prototype materials, and developing hypotheses, instruments and data for research (see Snapp and Heong, 2003; Heong and Escalada, 2005 for more details).

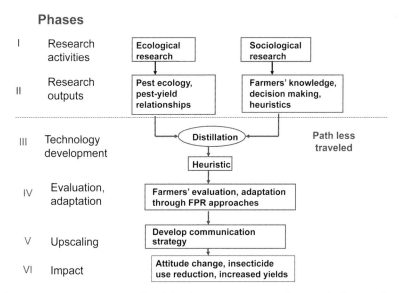

Figure 12.2 Pathways to impact – most research and development programmes focus on the first two phases. The remaining phases are often 'paths less travelled', but essential to change practice on a large scale.

Phase I: Research activities

In phase I the problem and the associated ecological and sociological issues are clearly identified. A scoping study (see Box 12.2) is conducted to achieve this level of understanding. Methods more commonly used in

---

Box 12.2 What is a scoping study?

The scoping study has been regarded as a type of literature review used to map relevant literature in the field of interest (Arksey and O'Malley, 2005). A scoping study examines broader dimensions of a problem often undertaken as a pre-project preliminary exercise. The scoping study has superseded the rapid rural appraisal or participatory rural appraisal (PRA) where a multidisciplinary team is commissioned to do a 'quick and dirty' review of the situation.

A scoping study is often done to focus on identifying the extent, nature and range of research and implementation issues related to a problem, to map the key concepts relevant to a research area and the main sources and types of evidence available. A review of available literature will help determine research gaps which a future larger study can address. A scoping study can be conducted as a stand-alone project where an area is complex or has not been reviewed comprehensively before (Mays et al., 2001).

As applied in development projects, the intent of a scoping study is to assess the magnitude, seriousness and intensity of the problem and the actions taken by the people concerned and affected by it. This is done by reviewing the literature, historical data, and reports, and collecting preliminary data (e.g. focus group discussions, observation, field visits, and interviews with key informants) to scope for research and implementation issues, to provide some understanding of the problem, and develop an integrated strategy or a set of recommendations to deal with the problem. Various tools are available that can be used in a scoping study. Historical profiles, problem tree, seasonal charts, discrimination profiles, and strengths-weaknesses-and-threats (SWOT) analysis are some examples. Collecting preliminary data will involve field visits, key informant interviews and a series of focus group discussions with stakeholders.

---

sociology and anthropology than agriculture are used to obtain data on what farmers know, their current practices and how they perceive a particular issue, such as non-crop vegetation, on-farm biodiversity or ecological engineering (Gurr et al., chapter 13 of this volume). These methods may include, for example, observation, formal and informal interviews, questionnaires and focus group discussions. The following sections outline recommendations for conducting farmer surveys.

Questionnaire pretest

When conducting farmer surveys an important first step is a questionnaire pretest. Pretesting involves interviewing a small group of respondents who are similar to the intended target group to determine their reactions to the draft questionnaire. This is done in order to determine

• the clarity of the wording and translation of the technical terms used
• whether the questions are in a logical sequence
• the adequacy of the response categories (e.g. where there is a multiple choice)
• the clarity of questionnaire instructions, and
• the estimated duration of the interview.

Results of the pretest are used to revise and refine the questionnaire and plan logistical arrangements for the fieldwork (see Box 12.3 for illustration). The pretest provides a means of catching and solving unforeseen problems in the use of the questionnaire, such as the phrasing and sequencing of questions. Linguistic and cultural differences complicate the task of questionnaire development, making pretesting indispensable.

Survey implementation

Once the questionnaire has been pretested, finalised and reproduced, the next step is to implement the field survey. Resources needed for the fieldwork include personnel, money and time. A field survey team is often composed of a survey coordinator, a field supervisor and interviewers. The survey coordinator is responsible for all aspects of the fieldwork – selection, training and deployment of interviewers. The field supervisor assists the survey coordinator in spot checking and monitoring the field interviews. Before they are fielded, interviewers are oriented on the purpose of the survey and trained on interviewing skills and how to conduct the interviews. Guided by the sampling plan and

Box 12.3 Questionnaire pre-testing, Hainan Island, China

A survey questionnaire on 'Conserving Arthropod Biodiversity and Ecosystem Services in Rice Environments of Hainan Island, China' was subjected to a pretest in Hong Qi town, Haikou on 20 November 2010. Findings led to six important changes in the subsequent, main survey.

1) It was discovered that farmers did not recognise a key pest problem in China, the brown planthopper (BPH). When the questions were framed, it was assumed that the BPH was recognised and known to all rice farmers in Asia. Yet, in Hainan this was not so. During the discussion, it was pointed out that perhaps it was a translation problem as Hong Qi farmers spoke the local Hainanese dialect and the Chinese translation would not be understood. It was agreed that each student interviewer would bring a vial containing BPH specimens to show to rice farmers during the survey.

2) Questions originally referred to the previous cropping season. However, the previous cropping season was badly affected by floods and farmers could not grow a rice crop. The wording was therefore changed to 'in the normal crop season'.

3) Farmers had difficulty estimating the number of days they spent and labour cost involved in sowing, so these questions were removed.

4) Farmers reported that they did not know the name of the pesticides they used as the pesticide shop mixed the chemicals for them. Again, the question asking for name of pesticide was deleted.

5) Farmers did not understand 'non-rice habitats'. It was agreed to reword questions to refer to 'other plants growing near the rice'.

6) The Chinese character that means 'to plant flowers' generally refers to ornamental flowers or landscape plants. Therefore it was suggested that the term 'beneficial flowers' be changed to 'flowering plants'.

professional agency staff may have. For example, in a survey of rice farmers' pest management perceptions and practices (Escalada and Heong, 1997a), it was observed that plant protection officers who had conducted the interviews tended to interpret rather than just record farmers' responses. Although many farmers reported that 'green worm' was their most important pest, this was recorded by interviewers as either army worm or rice bug based on their perception of what the term 'green worm' implied. In reality, farmers can use the term 'green worm' to refer to a range of leaf feeders such as rice leaffolders, cutworms, case worms and thrips.

To ensure efficient implementation of the survey and to minimise errors, interviewers should be supplied with a list of farmers to interview (the respondents), questionnaires, a map and pencils. Interviewers should be instructed to interview only the farmers on their list. Should the designated farmer be unavailable at the time of the interview, the interviewer should be told to schedule a return visit or select another farmer from a list of replacement respondents.

Interviewers unknown to respondents often need to be used. Respondents are often hesitant to give accurate and full information out of an implicit mistrust of strangers. To overcome this potential problem, when conducting field interviews, interviewers need to be able to quickly establish a rapport with respondents. Informing the respondent of the purpose of the survey and how it may benefit them may help to do this. At the end of the interview, interviewers should be instructed to check the interview schedule for completeness and thank the respondent for their help and cooperation.

Probing questions may be used to elicit additional information, expand an idea already expressed by the respondent, or clarify the respondent's response (Sedlack and Stanley, 1992). In many interview situations, respondents give vague replies such as 'okay' or 'good', which could mean different things. In such situations the interviewer should be coached to ask why it is 'okay' or 'good' and encourage respondents to give more specific answers. Probing questions are also necessary when respondents are asked to select from alternative answers, including the option of 'other', if full and meaningful data are to be collected. When respondents select 'other', the interviewer should ask for specific details.

Open-ended, probing questions usually require more than a 'yes' or 'no' answer and give the respondent the

respondent list, the interviewers locate the respondents, conduct the interviews, and check the completed questionnaires after the interview.

The interviewer is an important link in the survey chain so it is important that the interviewers selected are honest and objective. Our experience has shown that college students tend to be more objective interviewers because they do not have the inherent bias that

leeway to contribute to the data collected from their unique perspective. Some more generic follow-up questions to elicit more precise information include (Babbie, 2007; Krueger, 1988):

'Could you give an example?'
'In what way?'
'What do you mean?'
'Would you explain further?'
'Tell me a little more about it.'
'What do you mean when you said . . .'
'Tell me how it is so.'

Open-ended survey questions usually provide opportunities for probing, but the sequence of probe questions to ask would depend on the respondent's initial response (see Box 12.4).

Once the completed questionnaires have been reviewed, analysis of the data is necessary to present it in a useful form. Depending on the main objective of the survey, analysis can be relatively simple, employing descriptive statistics such as the percentage of respondents giving specific answers or listing the various ways in which farmers said they might utilise a new practice. For more complex surveys, particularly where the aim is to predict for the entire population from the results of the sample population, it is best that the data are encoded, processed and analysed using a statistical package. Ease of use, power, and cost are some of the important considerations in the choice of computer software for data analysis.

Analysis of the kind of data collected by the questionnaire method described first requires coding and tabulating the responses to questions. Coding involves sorting the qualitative (words) and the quantitative (numbers) data within the question responses and respondent information into specific categories. Tabulation records the numbers of types of responses in the appropriate categories. Statistical analysis may then be performed such as percentages, averages and appropriate tests of significance to compare, for example, respondents in different regions or in different treatment groups.

The meanings of the outcomes of the analysis must then be considered. In order to make the interpretation of the results accessible to others a survey report is compiled. The survey report outlines the research problem, data collection methods used, findings and conclusions.

Phase II: Research outputs

In Phase II, the baseline survey results are reviewed in a workshop to find out farmers' current attitudes and practices, and the potential to modify practices. The group then brainstorms for intervention ideas and develops a consensus on the scaling-up approach to use. Understanding the root causes as well as the direct causes of the problem is important. It is also important at this phase for all stakeholders to gain a common understanding of the various issues.

The following example illustrates how survey results are used to identify gaps in farmers' knowledge, attitudes and practices. In a current project led by the International Rice Research Institute, efforts are being made to reduce losses from planthopper pests (Gurr *et al.*, chapter 13 of this volume). An important aspect of this is managing rice pests in an ecological engineering manner. Achieving this involves using the information-gathering strategies described above to understand farmers' beliefs on how pest outbreaks are generated and how they manage planthoppers. Focus group discussions were conducted to develop the questionnaires. The surveys in Jin Hua, China (sample size = 327), Chainat, Thailand (sample size = 341) and Tien Giang, Vietnam (sample size = 1,009) showed

---

Box 12.4 Probing example

After establishing that farmers face pest problems and that leaf-feeding insects are important, interviewers might want to proceed to try to identify which species cause problems. Some of these questions would help to elicit the information needed:

• You said previously that *worms* are pest problems; can you tell me what they look like?
• Can you describe the colour?
• How big are they?
• Where do they live?
• At what stage of the crop do you see them?
• At what time of day do you see them?
• How many such insects do you often see on the rice crop?
• What are they doing to the crop?
• Can you show me these worms in your rice crop now?

Table 12.1 Farmers' beliefs on how intensive practices influence planthopper problems.

Practices that farmers believe would increase planthopper outbreaks	Percentage of farmers believing statement to be true		
	Jin Hua, China	Chainat, Thailand	Tien Giang, Vietnam
High seed rates	70.3	58.9	89.9
High fertiliser rates	77.1	55.4	81.0
Multiple crops of rice per year	46.2	73.0	74.5

that large proportions of farmers in China and Vietnam believed that high seed rates and high fertiliser inputs would tend to increase planthopper problems, while relatively fewer farmers in Thailand believed so (Table 12.1). In Thailand and Vietnam, farmers believed that growing multiple crops without a break/fallow period would increase planthopper problems, while in China fewer farmers believed that to be the case. This might be because triple cropping or growing seven crops in two years are normal practices in Thailand and Vietnam but less so in rice growing areas of China, much of which is temperate.

These surveys showed that most farmers believed intensive practices, such as high seed and fertiliser rates and growing rice continuously, would cause more planthopper problems and yet most continued to use these practices, probably because they perceived that higher inputs and more crops would result in higher incomes. Farmers in China and Vietnam seemed to have favourable attitudes towards resistant varieties; fewer did so in Thailand. Perhaps this partly explains why Thai farmers had used the same few varieties in the previous five years. A larger proportion of farmers in China favoured the use of insecticides than farmers in Thailand and Vietnam. In Vietnam, more farmers believed that insecticides caused planthopper problems and few believed that insecticides would increase yields; attitudes that favour reduced use of insecticides.

Phase II findings such as these are then used in phase III, 'technology development', where technical information is distilled into a heuristic to guide practice.

Phase III: Technology development

The role of heuristics in farmers' decision-making was discussed above. As an example of this, research results from the Ricehoppers project provide a cluster of three simple rules as follows.

• Flowers on the bunds (the earthen banks surrounding rice crops) provide food to attract bees and other beneficial insects.
• Some of the beneficial insects will help to control planthoppers invading crops, so insecticides are unnecessary.
• Insecticide application will kill the bees and other beneficial insects.

Although bees are not natural enemies of pests they serve as readily seen indicators of natural enemies. Compared to bees natural enemies are less familiar and recognisable and are usually small to minute in size. A heuristic can be developed from this cluster of rules: 'Flowers on bunds make the application of pesticide to control planthoppers unnecessary'.

Phase IV: Evaluation, adaptation and farmer participatory research

Farmer participatory research (FPR) involves encouraging farmers to test whether or not the principles being introduced via heuristics, such as that described above, are effective. FPR also allows farmers to adapt new technologies and spread the new knowledge to other farmers (Bunch, 1989). This is equivalent to providing samples to consumers for testing in marketing campaigns.

The advantages of FPR, which stimulates 'learning by doing', have been demonstrated in the spread of both traditional and recommended technologies (e.g. maize and cassava growing and soybean cultivation and utilisation in West Africa, soil conservation techniques in Cebu, Philippines, making contour ditches and planning Napier grass in Guatemala (Bunch, 1989; Reijntjes *et al.*, 1992)). In addition to adaptations of introduced innovations, farmers' experiments have evaluated new crop varieties and observed the results of new practices and procedures (e.g. use of diffused light storage of potatoes by farmers in Peru

(Reijntjes *et al.*, 1992; Rhoades, 1989). As tools for learning and demonstration, experimenting is also regarded as an important part of learning (Stoizenbach, 1992). Escalada and Heong (1997b) found changes in farmers' perception after evaluating conflicting information. See Box 12.5 for a more detailed example of the use of FPR.

Farmers are more likely to participate in an experiment if they perceive the source of information to be credible. In the case of the ricehoppers project involving ecological engineering to combat rice pests, information was from the International Rice Research Institute, widely regarded by farmers to be credible on rice-related issues. Credibility stems from a reputation

---

Box 12.5 Farmer experiments in the 'three reductions – three gains initiative' (Huan *et al.*, 2005)

Starting in the 1999–2000 wet and autumn–summer seasons, 30 volunteer farmers conducted experiments to evaluate the effects of reducing seed rates, fertiliser and pesticides in Vietnam. Farmers allocated a portion of their fields as the experimental area and the remainder as 'control'. For the experimental area, farmers were given guidelines in adjusting their seed rates and fertiliser rates, and told not to use insecticides in the first 40 days after sowing. Otherwise, farmers applied pesticides as needed. For the control area, farmers practised their normal routines. A simple system to record their inputs in the two plots was provided to facilitate data recording. Participating farmers were motivated to participate in the experiment through farmer meetings conducted by extension staff to discuss ways to reduce inputs and increase profits. No compensation of any form was provided as an incentive.

In 2001–2002, farmer participatory research was expanded to 920 farmers in Tân Lập Village, Tân Thạnh district, Long An Province, in collaboration with the

Cuu Long Delta Rice Research Institute (CLRRI). In the same season, farmers conducted 520 similar experiments in 8 provinces in the Mekong Delta. Results of the experiments were presented at a farmers' workshop in Tiền Giang province in March 2003. Another 446 demo fields were set up in 10 provinces in the Mekong Delta in 2002 and 600 in 6 coastal provinces in the Central region. In September 2002, another farmers' workshop was organised in Phu Yen Province.

The farmers' experiments demonstrated that seed, fertilisers, insecticides and fungicides can be reduced, resulting in higher profits (Table 12.2). After participating in this evaluation, most farmers modified their initial beliefs that reductions in seed and fertiliser rates would result in lower yields and profits. Farmers significantly increased their profits by an average of ~USD$58 ha^{-1} and ~$35 ha^{-1} in the two seasons, respectively. The highest contributions were from reduction in insecticide use, followed by reduction in fungicides and seed rates.

Table 12.2 Changes in farmers' seed rates, fertiliser and pesticide use in the summer–autumn season.

	Can Tho district		Tien Giang district	
	Pre-test	**Post-test**	**Pre-test**	**Post-test**
Seed rates (kg/ha)	275.1	210.6**	189.0	170.2**
Fertiliser use (kg/ha)				
Nitrogen	116.5	95.2**	105.9	100.0**
Phosphorus	62.4	54.5**	64.3	53.0**
Potassium	32.2	31.6	31.2	40.1**
Pesticide use (sprays/season)				
Insecticides	1.2	0.8**	2.27	1.95**
Fungicides	0.3	1.0**	0.49	0.23**
Herbicides	0.1	0.3**	0.05	0.01**

**indicates significant difference ($p < 0.01$) between pre- and post-tests.

for, or demonstration of, expertise in a given domain that stimulates perceptions of trustworthiness (Berlo *et al.*, 1970).

In the ecological engineering initiative in Vietnam, the farmers conducted simple field experiments in their own fields. Farmers were later invited to perform the 'experiment' in the Cai Be farming community in Tien Giang province. Using light traps, rice planting was synchronised and timed to be after the peak of immigrating planthopper adults. Bunds were also planted with nectar-producing flower species to attract parasitoids and insecticides were withheld in the early season (first 40 days after sowing). About 5,000 m of bunds were populated with thousands of locally appropriate nectar-producing plants. An important purpose of the participatory research was to encourage farmers to evaluate the values of such practices and see for themselves, and this was facilitated by a designating a large, nearby control area that was kept under normal management. The ecological engineering fields had no insecticide applications while the control fields were sprayed two or three times. Volunteer farmers were taught to observe bee populations as indicators of parasitism since parasitoids are too small to be seen readily and the concept of parasitism proved too difficult to explain. While the farmers grew the nectar-rich plants, the plant protection technicians assisted with insect counts. Such farmer evaluations can be powerful communication tools in their own right (see Box 12.5). As an illustration of this, as a consequence of the small-scale work at Cai Be, as much as 27,000 m of bunds were populated with flowers by the local farmers in the nearby district of Cai Lay (Plate 12.1).

Phase V: Upscaling – developing and launching a communication strategy

Phase V focuses on disseminating the message by developing and launching a communication strategy. This includes the use of pilot sites which may include areas outside the locations of the farmer participatory experiments. It is important to conduct this pilot project through partnership with local research, extension, mass media, government units, NGOs and other implementing agencies. The key stakeholders are invited to a 'message design workshop' where results from the baseline survey (phase II) and the farmer experiments (phase IV) are used to develop a communication strategy for scaling up the dissemination of

the message. The strategy includes the brand name of the initiative, the media to be used, prototype campaign materials, and an implementation plan to reach farmers and policy makers. For wide-scale dissemination of ecological engineering, for example, it is important to emphasise its benefits in the message to be communicated, such as: 1) Flowers in rice environments attract and support beneficial insects to protect rice from invading planthoppers, 2) insecticide use is reduced to avoid killing beneficial insects, and 3) profits are increased.

Developing an identity or 'brand' that is locally appropriate

In commerce, a brand name is a term, symbol, slogan or design, which is aimed to identify the goods or services of a company and to differentiate them from competitors. It conveys the concept, benefits and other pertinent information about the product or service. Similarly, in adoption of biodiversity-based pest management strategies, brand names and associated graphic images are used to refer to a product or innovation. For example: 'No early spray' (for stopping insecticide application for the rice leaffolder from 0–40 days after seeding) (Escalada and Heong, 1997b), 'Three reductions – three gains' (for reduced recommended seed rates, nitrogen fertiliser and insecticide application) (Huan *et al.*, 2008) and 'Minus-one element technique' (for a test that determines nutrient deficiency of the soil for lowland rice based on the principle that plants will show a physical reaction to limiting nutrients) (PhilRice, 2002). A message design workshop (Figure 12.3) can be useful in designing culturally appropriate materials. In particular, an innovation can be communicated better if it has a name that appropriately describes its attributes. Besides facilitating the promotion strategy, a brand name makes it easier to track down the product in impact assessment. In Vietnam, ecological engineering was brand named *Ruong Lua Bo Hoa* (rice fields with flower bunds). In China the equivalent in Mandarin was *Sheng Tai Gong Cheng*.

To reach thousands of farmers, it is important to use a combination of mass media including posters (Plate 12.2) as well as radio, television, video, billboards, leaflets, mobile phones, web-internet and interpersonal channels (demonstration farms, training, field visits). The prototype media materials for the 'rice fields with flower bunds' campaign in Vietnam consist of TV

Figure 12.3 Artist developing prototype materials during a Message Design Workshop.

Figure 12.4 Launching of Ecological Engineering Campaign in Tien Giang, Vietnam with Dr Bui Ba Bong, Vice Minister of Agriculture and Rural Development, Vietnam as keynote speaker.

broadcast videos, short radio drama episodes, billboards, posters and leaflets.

The campaign launch

The launching day is a high-profile event, which is officially graced by top-level national and provincial officials. Their involvement ensures the event receives extensive national coverage on radio, television and print media. In many cases, besides the launch in the capital city, another ceremony is held in the province where the project is to be implemented. The local event is aimed at enlisting the support and commitment of influential community members and at increasing the public prestige and credibility of the local extension workers who organised the launch in their district (Adhikarya and Posamentier, 1987). It is also an avenue to give recognition to researchers, extension staff and farmers who have contributed to research output and experimentation. Often, prizes are given to key farmers who participated in the experiments. The launching signals the beginning of the project and it is best held on an important day, such as World Environment Day, which is celebrated on June 5 each year. In Vietnam the Ecological Engineering Initiative was launched by the Vice Minister of Agriculture and Rural Development, Dr Bui Ba Bong (Figure 12.4).

Phase VI: Impact

Phase VI involves documenting the impact of activities in the preceding phases. A rigorous research frame-

work is required at this phase to quantify effects of the intervention. A management monitoring survey (MMS) is carried out about two months after the launch to enable the team to make adjustments as needed. Baseline data relating to farmers' initial and current beliefs, attitudes and practices are analysed and documented (see Box 12.6 for an example). A 'show and tell' press conference or workshop is another important event that can enhance adoption by other provinces and create multiplier effects. By involving policy-makers in these high-profile events, policy change that can favour widespread adoption is encouraged.

Managing multi-stakeholder participation

An effective multi-stakeholder partnership is essential to ensure success of the scaling out process. This is

Box 12.6 Summative evaluation of the 'three reductions – three gains' campaign (Huan *et al.*, 2008)

The 'three reductions – three gains', campaigns were launched in two provinces, Can Tho and Tien Giang in Vietnam. In both provinces, farmers' practices changed significantly. Their insecticide sprays reduced by 13–33% while their seed rates dropped about 10% and nitrogen rates by about 7%. The proportion of farmers using insecticides fell by about 11%. These practices were supported by modifications in attitudes that favoured high inputs.

Farmers who reported significant input reductions also changed their perception of yield loss. The campaigns in Can Tho and Tien Giang had significant multiplier effects. They stimulated several provincial governments as well as the Ministry of Agriculture and Rural Development to provide additional resources to reproduce the materials and campaign process for local use which eventually reached more than three million farmers in south and central Vietnam.

Table 12.2 shows that farmers' use of seeds, nitrogen and insecticides were significantly reduced between pre- and post-campaign surveys in both provinces. Farmers' attitude scores significantly changed. First the percentage of farmers who perceived that rice leaffolders were a serious cause of damage that needed to be sprayed in the early crop stages was significantly reduced and second the proportion of farmers who believed in the ability of the rice crop to recover from leaf damage was significantly increased.

tation plans. Thus, at the start of the project, a stakeholder analysis is useful to understand stakeholder relationships and implement processes to resolve them.

The participatory process requires the research team to work in partnership with various stakeholders. The major challenges to working in partnership include ensuring mutual beneficiality, avoiding unrealistic assumptions and expectations about the project and ensuring activities are carried out as planned. Among the rewards that can be derived from partnerships are:

• Mutual trust earned from partners
• Due recognition in knowledge products
• Participation in international workshops and meetings
• Opportunities for new collaboration in research
• Establishment of linkages with international and national organisations

Complementary evaluation tools can be used to determine the real response of partners to a partnership project. An indicator of a positive response includes gaining the partner's commitment to contribute matching funds or in-kind support which weaves the project into their agenda and creates a line item for budgeting – thereby establishing a commitment for implementation. Financial commitment to a project is likely to be encouraged by transparent budgeting, particularly providing estimates of partner matched funds (direct funds or in-kind (e.g. transport, equipment, labour)) being contributed by other partners. Other forms of support may also be listed. Budget planning carried out in a transparent manner will encourage matching contributions from research partners and can help leverage local resources. With counterpart funds put in, partners often work hard to make sure that the project succeeds. For example, cooperation from project partners can help leverage local resources to enhance campaign dissemination in a media campaign, thereby increasing returns on modest project investments. For instance the 'three reductions – three gains' project in Vietnam was supported by the central government and funds were allocated to help extend and eventually reached 80% adoption in some provinces (Heong *et al.*, 2010). Since in most Asian countries limited resources are allocated to agricultural extension, mass media, especially when implemented through multi-stakeholder partnerships, can be an effective option to communicate new technologies and information to farmers to motivate change.

achieved through use of a participative style of leadership to stimulate creative problem solving, to promote high morale, satisfaction, local ownership and commitment. Group decisions and supportive relationships based on mutual trust and respect are strongly emphasised in meetings and workshops.

To achieve large-scale diffusion of the biodiversity-based pest management heuristics, strong commitment and support of local government authorities and agencies is essential. It is necessary to recognise and satisfy the priorities of the local government as well as those of local implementing agencies. For instance, if the wages of the extension agents in the area are dependent on the sale of farm chemicals, the conflict of interest would significantly compromise implemen-

CONCLUSION

The sociological tools and phases described in the design and implementation of ecological engineering illuminated important principles. In order to elicit changes in farmers' practices it is first necessary to understand farmers' initial beliefs and how they make decisions. Then the relevant information from research may be distilled into easily communicated heuristics. For the heuristics to be adopted, farmers must be motivated to experiment or try out the approach by properly communicating the benefits.

It is currently rare in implementation programmes for information to be communicated as heuristics. This is far more important in extension efforts for biodiversity-based pest management strategies than in simpler technologies such as seeds of a new crop variety, a novel insecticide or a new piece of equipment. Farmers will not read books such as this one, nor have the ability to synthesise and apply to their farms the detailed ecological, agronomic, economic and other information. If greater attention and resources are directed to summarising key conclusions from the scientific literature, farmers can be presented with more appropriate information that motivates adoption.

ACKNOWLEDGEMENTS

The authors would like to thank Dr Ho Van Chien, Prof Zhu Zeng Rong and Mr Manit Luecha for assistance in carrying out the farm surveys. Dr Chien also organised the message design workshop and implemented the farmer participatory research in Vietnam. This work has been supported by the Asian Development Bank under the regional technical assistance grant (RETA 6489) to the International Rice Research Institute.

REFERENCES

Adhikarya, R. (1994) *Strategic extension campaign: a participatory oriented method of agricultural extension*, FAO of the UN, Rome.

Adhikarya, R., and Posamentier, H. (1987) *Motivating farmers for action: how strategic multi-media campaigns can help*. GTZ (Deutsche Gesellschaft für Technische Zusammenarbeit), Eschborn, Frankfurt.

Ajzen, I. (1988) *Attitudes, personality and behaviour*, Open University Press, Milton Keynes.

Ajzen, I. (1991) The theory of planned behavior. *Organization Behavior and Human Processes*, 50, 179–211.

Arksey, H. and O'Malley, L. (2005) Scoping studies: towards a methodological framework. *International Journal of Social Research Methodology*, 8, 19–32.

Babbie, E.R. (2007) *The practice of social research*, Wadsworth Publishing Company, Belmont, CA, pp. 246–247.

Bentley, J.W. (1989) What farmers don't know can't help them: the strengths and weaknesses of indigenous technical knowledge in Honduras. *Agriculture and Human Values*, 6, 25–31.

Bentley, J.W. and Rodriguez, G. (2001) Honduran folk entomology. *Current Anthropology*, 42, 285–301.

Berlo, D.K., Lemert, J.B., Robert, J. and Mertz, R.J. (1970) Dimensions for evaluating the acceptability of message sources. *Public Opinion Quarterly*, 33, 563–576.

Byerlee, D. and Alex, G.E. (1998) *Strengthening national agricultural research systems: policy issues and good practice*, World Bank Publications, Washington, DC.

Bunch, R. (1989) Encouraging farmers, experiments, in *Farmer first: Farmer innovation and agricultural research* (eds R. Chambers, A. Pacey and L.A. Thrupp), Intermediate Technology Publications, London, pp. 55–61.

DeWalt, B.R. (1994) Using indigenous knowledge to improve agriculture and natural resource management. *Human Organization*, 53, 123–131.

Einhorn, H.J. and Hogarth, R.M. (1981) Behavioural decision theory: processes of judgement and choice. *Annual Review of Psychology*, 32, 53–88.

Escalada, M.M and Heong, K.L. (1997a) Changing farmers' perceptions of pests through participatory experiments. *ILEIA Newsletter*, 13, 10–11.

Escalada, M.M and Heong, K.L. (1997b) Methods for research on farmers' knowledge, attitudes, and practices in pest management, in *Pest management of rice farmers in Asia* (eds K.L. Heong and M.M. Escalada), International Rice Research Institute, Manila, Philippines, pp. 1–24.

Escalada, M.M., Heong, K.L., Huan, N.H. and Mai, V. (1999) Communications and behavior change in rice farmers' pest management: The case of using mass media in Vietnam. *Journal of Applied Communication*, 83, 7–26.

Gigerenzer, G., Todd, P.M. and the ABC Research Group (1999) *Simple heuristics that make us smart*, Oxford University Press, New York.

Heong, K.L. and Escalada, M.M. (1999) Quantifying rice farmers' pest management decisions: beliefs and subjective norms in stem borer control. *Crop Protection*, 18, 315–322.

Heong, K.L. and Escalada, M.M. (2005) Scaling up communication of scientific information to rural communities. *Journal of Scientific Communication*, 4, 2–3.

Heong, K.L and Schoenly, K. (1998) Impact of insecticides on herbivore–natural enemy communities in tropical rice ecosystems, in *Ecotoxicology: pesticides and beneficial organisms* (eds P.T. Haskell and P. McEwen), Chapman & Hall, London, pp. 381–403.

Heong, K.L., Escalada, M.M., Huan, N.H., Chien, H.V. and Quynh, P.V. (2010) Scaling out communication to rural farmers; lessons from the 'Three Reductions, Three Gains' campaign in Vietnam, in *Research to impact: case studies for natural resource management for irrigated rice in Asia* (eds F.G. Palis, G. Singleton, M.C. Casimero and B. Hardy), International Rice Research Institute, Los Baños (Philippines), pp. 207–220.

Hornik, R.C. (1988) *Development communication: information, agriculture, and nutrition in the third world*, Longman, New York.

Huan, N.H., Thiet, L.V., Chien, H.V. and Heong, K.L. (2005) Farmers' evaluation of reducing pesticides, fertilizers and seed rates in rice farming through participatory research in the Mekong Delta, Vietnam. *Crop Protection*, 24, 457–464.

Huan, N.H., Chien, H.V., Quynh, P.V. *et al.* (2008) Motivating rice farmers in the Mekong Delta to modify pest management and related practices through mass media. *Journal of International Pest Management*, 54, 339–346.

Krech, D. and Crutchfield, R.S. (1971) Perceiving the world, in *The process and effects of mass communication* (eds W. Schramm and D.F. Roberts), University of Illinois Press, Urbana, pp. 235–264.

Krueger, R.A. (1988) *Focus groups: a practical guide for applied research*, Sage Publications, Beverly Hills.

Mays, N., Roberts, E. and Popay, J. (2001) Synthesising research evidence, in *Studying the organisation and delivery of health services: research methods* (eds N. Fulop, P. Allen, A. Clarke and N. Black), Routledge, London, pp. 188–220.

Meynen, C. and Stephens, A. (1996) Rural families and farm households in Asia and the Pacific: an overview, in *Rural families and household economies in Asia and the Pacific: report of a regional expert consultation*. RAP Publication (FAO), no. 1996/10/FAO, Bangkok. Regional Office for Asia and the Pacific, pp. 37–63.

Mulhall, A.E. and Garforth, C.J. (eds) (2000) *Equity implications for reforms in the financing and delivery of agricultural extension services*. Final Technical Report on research project R6470 to the Department for International Development. Reading. Agricultural Extension and Rural Development Department, University of Reading, UK.

Mumford, J.D. and Norton, G.A. (1984) Economics of decision making in pest management. *Annual Review of Entomology*, 29, 157–174.

Payne, J.W., Bettman, J.R. and Johnson, E.J. (1992) Behavioural decision research: a constructive processing perspective. *Annual Review of Psychology*, 43, 87–131.

PhilRice (2002) *Minus-one element technique: nutrient deficiency test made easy*. Rice Technology Bulletin No. 30, 2nd edn, Philippine Rice Institute, Nueva Ecija.

Reijntjes, C., Haverkort, B. and Waters-Bayer, A. (1992) *Farming for the future*, MacMillan Press, London.

Rhoades, R. (1989) The role of farmers in the creation of agricultural technology, in *Farmer first: farmer innovation and agricultural research* (eds R. Chambers, A. Pacey and L.A. Thrupp), Intermediate Technology Publications, London, pp. 55–61.

Rice, R.E. and Paisley, W.J. (1981) *Public communication campaigns*, Sage Publications, Beverly Hills.

Rivera, W.M. (1990) Trends and issues in international agricultural extension: the end of the beginning. *Journal of Extension Systems*, 6, 87–101.

Rogers, E.M. (1995) *Diffusion of innovations*, 4th edn, The Free Press, New York.

Schramm, W. (1973) *Men, messages and media: a look at human communication*, Harper & Row, New York.

Sedlack, R.G. and Stanley, J. (1992) *Social research: theory and methods*, Allyn and Bacon, Boston.

Siamwalla, A. (2001) *Study of rural Asia: the evolving roles of the state, private, and local actors in rural Asia*, ADB and Oxford University Press, Oxford.

Simon, H.A. (1956) Rational choice and the structure of environments. *Psychological Review*, 63, 129–138.

Simon, H.A. (1978) Rationality as process and as product of thought. *American Economic Review*, 86, 1–16.

Simon, H.A. (1982) *Models of bounded rationality*, MIT Press, Cambridge, MA.

Singhal, A. and Rogers, E. (2003) *Combating AIDS: communication strategies in action*, Sage Publications, New Delhi.

Slovic, P., Fishhoff, B. and Lichtenstein, S. (1977) Behavioural decision theory. *Annual Review of Psychology*, 28, 1–39.

Snapp, S. and Heong, K.L. (2003). Scaling up and out, in *Managing natural resource for sustainable livelihoods – uniting science and participation* (eds B. Pound, S. Snapp, C. McDougall and A. Braun), Earthscan Publications Ltd, London, pp. 67–83.

Stoizenbach, A. (1992) Farmers' experimentation: what are we talking about. *ILEIA Newsletter*, 9, 28–29.

Swanson, B. (2008) *Global review of good agricultural extension and advisory service practices*, FAO of the UN, Rome.

Tversky, A. and Kahneman, D. (1974) *Judgment under uncertainty: heuristics and biases. Science*, 185, 1124–1131.

Chapter 13

ECOLOGICAL ENGINEERING STRATEGIES TO MANAGE INSECT PESTS IN RICE

Geoff M. Gurr, K.L. Heong, J.A. Cheng and J. Catindig

Biodiversity and Insect Pests: Key Issues for Sustainable Management, First Edition. Edited by Geoff M. Gurr, Steve D. Wratten, William E. Snyder, Donna M.Y. Read.

INTRODUCTION

Rice production has a long history and more than two billion humans now depend upon this production system for their staple food. Around 96% of world rice production takes place in Asia, producing approximately 640 million tons of grain per annum (IRRI, 2011). Rice consumers and growers form the bulk of the world's poor so the stakes are high; any faltering of production can lead to civil unrest and potentially widespread starvation. Since the 'green revolution' of the 1970s, traditional production approaches have changed radically by the introduction of new rice varieties and hybrids and higher inputs of fertilisers and pesticides. Hybrid rice varieties have become very popular because of their high yield potential but they depend on heavy use of nitrogenous fertilisers. Although these technologies have led to increases in average rice yields, recent years have witnessed a resurgence of insect pests, especially planthoppers (Hemiptera: Delphacidae). Crop losses have resulted as pest populations develop resistance to widely used insecticides (Matsumura *et al.*, 2009) and adapt to overcome resistant variety traits (Horgan, 2009), a phenomenon often termed resistance 'breakdown' (Matteson, 2000). As an example of the impact of these effects, it is currently estimated that the world's largest rice producer, China, loses about a million tons of rice grain from planthopper outbreaks annually and in some years as much as 2.8 million tons (Heong and Hardy, 2009). Over recent years, pest damage has been amongst the most important factors leading to crop losses and price rises. Countries such as Vietnam, Brazil, India and Cambodia suspended exports in 2008 to prevent possible domestic shortages (Phoonphongphiphat, 2008). Outbreaks of delphacid pests and related virus diseases occurred in Thailand in 2009, 2010 and 2011. Combined with water shortages this has led the government to reduce production forecasts by 16% (Bangkok Post, 2010).

The factors outlined above imply an urgent need to rethink rice pest management. In response to this need the International Rice Research Institute organised an international conference in June 2008 to discuss new approaches, new techniques and management tools. The conference brought together scientists, agricultural directors and agribusiness representatives from Australia, Bangladesh, Cambodia, China, India, Indonesia, Japan, Korea, Laos, Malaysia, Philippines, Singapore, Taiwan, Thailand, the USA and Vietnam as well as the Food and Agriculture Organisation (FAO). The book that resulted from the conference (Heong and Hardy, 2009) recognised that new strategies are required. These strategies should be based on a better understanding of pest dynamics and employ ecologically based methods for enhancing biological control and reducing the adverse impacts on natural enemies of indiscriminate insecticide use. To a large degree, these strategies already exist in the form of proven technologies used in other crop systems (Gurr *et al.*, 2004).

Rice is usually grown in 'wet' or 'flooded' systems where water is used not only to support growth of the rice but also to suppress weeds and rodent pests. Irrigated rice production accounts for 75% of world production and is common in the Asian countries China, Japan, Indonesia, Vietnam and the Republic of Korea, as well as further afield in Egypt and the Senegal River Valley in Africa (IRRI, 2011). In this system, bunds (earthen levee banks) retain irrigation water so that rice grows in several centimetres of standing water. The second most important system for rice in terms of land area and yield is the rain-fed lowland production. This extends over about 60 million hectares and accounts for 20% of world rice production (IRRI, 2011). As in irrigated systems, in lowland rain-fed systems bunds are used to retain water. Additionally, bunds are sometimes used to make the best possible use of available water in rain-fed, upland rice production in Asia as well as Latin America and Africa. For obvious reasons, bunds are less relevant in areas subject to major flooding events where 'deep-water' rice is seeded before annual inundation. These special rice varieties are capable of growing up to 5 m tall in response to unregulated water levels. Deep-water rice is locally important but accounts for only a small proportion of world production, confined to the floodplains and river deltas of Bangladesh, the Irrawaddy of Myanmar, the Mekong of Vietnam and Cambodia and the Chao Phraya of Thailand in Asia, as well as the Niger of West Africa (IRRI, 2011).

The popular image of rice production is a tropical activity, but much of the world's rice is produced in non-tropical areas of Asia, such as Japan, Korea and much of China. Here, the winters can be severe with frosts and snowfall. Temperatures are too low for local survival of delphacids such as the rice brown planthopper and whitebacked planthopper. These pests migrate from tropical areas in the south each year. Although migration events can occasionally lead to crops being abruptly attacked by very large numbers of adult planthoppers, termed 'planthopper storms', the more normal situation is for modest numbers of delphacid adults to colonise a rice crop. A damaging level of infestation arises only if population increase is unchecked by natural enemies. In tropical areas, temperatures do not constrain local survival of pests and the growth of two or even three rice crops each year can facilitate pest increase unless checked by natural enemies.

The aim of this chapter is to assess the characteristics of rice production from the perspective of the potential for using biodiversity to contribute to more effective management of rice pests. As well as covering aspects of underlying ecological theory, the chapter considers the extent to which ecological pest management strategies have been researched in rice. This leads to a proposal for ecological engineering based on three 'planks': 1) reducing pesticide-induced mortality of natural enemies, 2) encouraging generalist natural enemies with organic matter inputs and 3) supporting parasitoids with nectar plants (including crop species) sown on the bunds. A current, multi-country project led by the International Rice Research Institute in which ecological engineering strategies are being assessed is also described. The principal taxonomic foci of the chapter are the planthopper species that currently constitute a serious and growing threat to production in Asia and the rice leaffolder (*Cnaphalocrocis medinalis* (Guenee) Lepidoptera: Pyralidae). The conspicuous foliar damage caused by *C. medinalis* larvae leads to early-season insecticide applications that disrupt biological control of various pests including planthoppers (Gurr *et al.*, 2011).

ECOLOGICAL BACKGROUND

Ecologically, the brown planthopper (*Nilaparvata lugens* (Stål)), whitebacked planthopper (*Sogatella furcifera* (Horvath)) and small brown planthopper (*Laodelphax striatellus* (Fallen)) can be viewed as essentially 'man-made pests'. These herbivores rarely assumed pest status prior to the 1960s but have now developed into serious threats to rice production all over Asia. This change is largely the result of an erosion of the natural mortality factors that previously checked population increase (Figure 13.1).

From a trophic, 'bottom-up' perspective, the traditional high-yielding and hybrid rice varieties that have

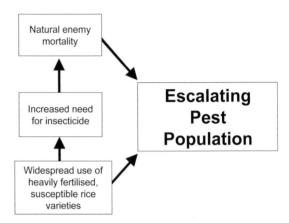

Figure 13.1 Escalation of rice planthopper pest populations as an ecological phenomenon: release of pest populations from 'bottom-up' control by widespread use of susceptible rice varieties with heavy nitrogenous fertiliser use has necessitated insecticide applications that kill natural enemies, so freeing pests from 'top-down' control.

been increasingly widely planted tend to have low resistance to delphacids. The whitebacked planthopper was a secondary insect pest of rice to the brown planthopper before the 1970s in China. However, when indica hybrid rice was planted in some rice areas in the 1980s, the numbers of whitebacked planthopper increased to the extent that it replaced the brown planthopper as the predominant pest (Sogawa *et al.*, 2009). Whitebacked planthopper species exhibited a major outbreak in 2000, extending over about 153,000 ha of winter–spring rice planted with Chinese hybrid rice in the Red River Delta. Since then it has steadily increased in significance as a rice pest in Asia (Thanh *et al.*, 2007).

'Green Revolution' varieties bred for resistance and released since the 1970s have major gene-based resistance that can be rendered ineffective by the well-known phenomenon of resistance 'breakdown' as a result of the pest's adaptation to plant resistance traits (Gallagher *et al.*, 1994). The use of pest-resistant varieties has faltered as a result of other factors inherent in the rice ecosystem. For example, on some (supposedly) resistant varieties, including IR26 and Utri Rajapan, brown planthopper growth rates increase when high levels of nitrogen fertiliser are applied (Cheng, 1971; Heinrichs and Medrano, 1985). Differences in nitrogen content between animal and plant tissues may be an important reason why herbivores are favoured by host plants with a high nitrogen content (Southwood, 1973; Altieri *et al.*, chapter 5 of this volume). As a consequence, nitrogen fertilisation influences not only rice plant nutrition and plant vigour but also the population dynamics of planthoppers (Denno *et al.*, 1994) by attracting immigrants and increasing population growth rate. Crops with nitrogen-enriched plants tend to favour brown planthopper population development (Lu and Heong, 2009), increasing feeding rate and honeydew secretion (Cheng, 1971; Sogawa, 1970). Brown planthoppers also probe less (Lu *et al.*, 2005; Sogawa, 1970), have higher survival rates, and exhibit greater population build-up on rice plants fertilised with nitrogen (Cheng, 1971; Preap *et al.*, 2001). Furthermore, they also produce more eggs (Preap *et al.*, 2001; Wang and Wu, 1991) and have a higher tendency for outbreaks (Hosamani *et al.*, 1986; Li *et al.*, 1996; Uhm *et al.*, 1985). The whitebacked planthopper exhibits similar responses (Hu *et al.*, 1986; Ma and Lee, 1996; Wu and Zhu, 1994). The overall effect of releasing pests from 'bottom-up' regulation is dramatically evident in data from China, where the rise in use of susceptible hybrid rice varieties has been accompanied by a dramatic increase in the area of land affected by whitebacked planthopper (Figure 13.2).

The need to protect the yield of these widely used and often heavily fertilised varieties of rice is one of several factors that have led to high levels of synthetic insecticide use. This is compounded by government subsidies for pesticide purchase (ostensibly in the interests of food security), and strong pesticide marketing and advertising, especially to exploit farmers' loss aversion with fears of losing 'face' or actual yield. This has led to pesticides being sold as fast-moving consumer goods (Box 13.1), literally alongside food and other household consumables in virtually every village store in many rice-growing countries. Often stores will facilitate sales by allowing credit to poor farmers until their rice is sold at the end of the season. The result has been that high levels of insecticide resistance are now commonly reported, particularly in Asia (Matsumura *et al.*, 2009). Moreover, the liberal use of insecticides leads to natural enemy mortality which, in turn, releases pest populations from 'top-down' control (Figure 13.1).

The foregoing account illustrates the need for the development and adoption of more sustainable pest management strategies. Biodiversity-related approaches have been the subject of much less study in rice than have insecticides and host-plant resistance but – as will become apparent – they offer good scope to contribute to future IPM in rice systems.

THREE PLANKS FOR ECOLOGICAL ENGINEERING IN RICE

Reducing the ecological fitness of pests

A wide range of technologies might be employed to reduce dependence on synthetic pesticides and manage

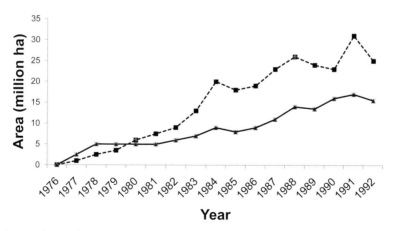

Figure 13.2 The historical rise in hybrid rice varieties (dashed line) has been accompanied by a dramatic increase in the area of land affected by the serious pest, the whitebacked planthopper (*Sogatella furcifera*) (solid line) in China (redrawn from data in Cheng, 2009).

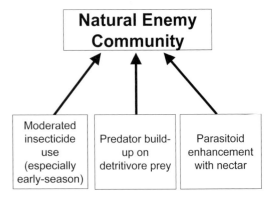

Figure 13.3 Reinstatement of 'top-down' control of rice planthopper pest populations by ecological engineering to 1) alleviate insecticide-induced mortality of natural enemies, 2) ensure early-season build-up of predators (mostly generalists) on detritivore prey and 3) maximise performance of parasitoids with nectar sources on bunds.

of 2.26 per 100 hills (a 'hill' is a clump of rice plants in a growing point) in early July reached 17,637 per 100 hills in late September. In contrast, late immigrants with a density of 172 per 100 hills in early September reached densities of only 5,836 per 100 hills. This showed the scope to manage planthopper pest numbers by delayed transplanting of rice (Cheng, 2009), but the following sections set out three more generally applicable strategies.

Plank 1: Moderated insecticide use – especially early in the season

Rice has a complex food web of herbivores and natural enemies (Barrion *et al.*, 1981). Exploiting this high level of biodiversity to improve pest suppression requires three strategies (Figure 13.3). The first is aimed at reducing the mortality of beneficial insects caused by insecticide applications to control early-season pests. The most important of these species, in terms of driving insecticide use, is the rice leaffolder (*C. medinalis*). Work aiming to set an action threshold for insecticidal control of this pest found that azinphos-ethyl and *Bt* gave poor results and even the best performing products (BPMC, endosulfan, and monocrotophos) gave only 53% control (Litsinger *et al.*, 2006). It was therefore concluded that an IPM approach was appropriate in order to exploit the activity of natural enemies and ability of rice to compensate for foliar damage by tiller-

rice pests. Reflecting this, there is a large literature on agronomic approaches such as altered cropping system (e.g. planting dates) and crop nutrient management. These approaches offer scope to influence pest populations via changes to the initial population size in a given area, fecundity and development pattern. Planthopper outbreaks are made possible by their high intrinsic capacity for increase, coupled with the release from mortality factors such as susceptible variety use and release from natural enemy impact. Managing initial populations – those that result from immigration into a new crop – is one way to reduce their ecological fitness. Field investigations and simulation studies indicate that early transplanting, an early immigration peak, and a high immigration rate all tend to favour brown planthopper outbreaks (Cheng *et al.*, 1990). In China, field investigations in 2006 showed that early immigrants with an initial density

ing (the production of additional stems). Unfortunately, early-season foliar damage from *C. medinalis* is very conspicuous so farmers often feel compelled to respond with insecticide application. This is despite Miyashita (1985) demonstrating that damage to 67% of leaves did not lead to significant yield loss. Further, a simulation model showed that normal field populations of leaffolders cause insignificant yield loss (Graf *et al.*, 1992). The major effect of these insecticide applications is to kill most members of the natural enemy community. This is because early-season applications are usually of broad-spectrum compounds such as synthetic pyrethroids, avermectins, chlorpyrifos and fipronil. This initial compromising of the natural enemy community and subsequent spraying prevents re-establishment of biological control for the rest of the growing season, locking the farmer into the 'pesticide treadmill' (van Den Bosch, 1978). When these insecticide applications can be avoided, effective biological control is encouraged via the 'integrated pest management (IPM) treadmill' (Tait, 1987) in which natural enemy communities remain intact and functional.

There is a wealth of empirical data associating insecticide use with pest problems in rice (Heong, 2009). This is illustrated by the within-season correlation between insecticide use and brown planthopper numbers reported by Way and Heong (1994) from work in the Philippines (Figure 13.4). On unsprayed rice brown planthopper began to multiply rapidly for about 20 days after the plants were transplanted but then rate of increase was reduced and then declined rapidly. Numbers continued to oscillate over the course of the growing season but never assumed densities above the damage threshold. In contrast, plants that were sprayed 30 days after transplanting exhibited increasingly severe pest infestation, up to a thousand times higher, despite repeated sprays.

Predators of *N. lugens* include the hemipterans *Orius tantillus* (Motschulsky) (Anthocoridae) (CABI, 2005) and *Cyrtorhinus lividipennis* (Reuter) (Miridae) (Luo and Zhuo, 1986; Guo *et al.*, 1994). A trombidiid mite was also reported to be frequently observed attacking nymph and adult *N. lugens* in India (Shankar and Baskaran, 1988). Spiders, too, are important predators of rice pests. *Ummelita insecticeps* (Bösenberg and Strand) (Erigoninae) consumes delphacids in rice in eastern China (Guo *et al.*, 1994) and the lycosid *Pirata subpiraticus* (Bösenberg and Strand) is considered a major predator of planthoppers, leafhoppers and *C. medinalis* (Li *et al.*, 2002). The latter species is

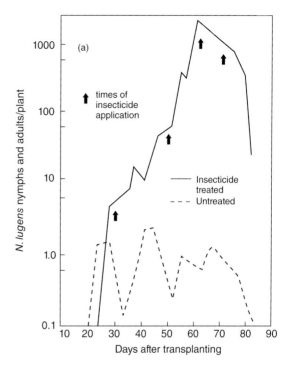

Figure 13.4 Comparison of brown planthopper (*Nilaparvata lugens*) numbers in insecticide-treated and unsprayed rice (from Way and Heong, 1994, reproduced with permission of Cambridge University Press via Copyright Clearance Centre).

numerically dominant in eastern China, accounting for between 68% and 96% of all spiders between May and September (Yu *et al.*, 2002). Studies of this spider in Hunan Province (central China) illustrated its appetite for whiteback planthopper, consuming 6–7 insects every 24 hours in the heading stage of early season rice and around 16 per day in the milk stage of rice (Wen *et al.*, 2003). Further information on the significance of spiders as predators of rice pests is given in the section below on build-up of generalist predators on non-pest prey.

Other evidence of the adverse impact of insecticide use on pest infestation is apparent in historical data from the farm at the International Rice Research Institute in the Philippines. There, it has been possible to introduce and adhere to a rigorous IPM system, and this has reduced insecticide use by more than 95% from 1994 levels, when almost 4 kg active ingredient

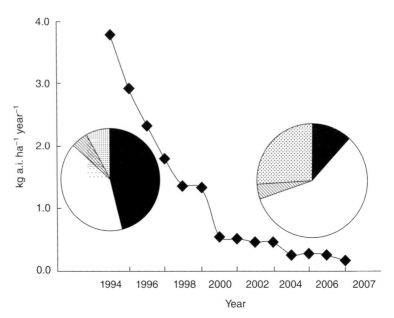

Figure 13.5 Reductions in insecticide use (trend line) on the International Rice Research Institute farm in the Philippines with changes in relative abundance of faunal groups from 1989 (left pie chart) and 2005 (right pie chart): reduced abundance of herbivores (black) and increased numbers of predators (white), parasitoids (hatched) and detritivores (speckled) (from Heong, 2009, reproduced with permission of IRRI).

was applied per hectare each year (Heong *et al.*, 2007). Prior to the implementation of IPM, 46% of the insects in the rice were herbivores (Figure 13.5). By 2005 insecticide inputs had been consistently low for several years and herbivores accounted for less than 12% of insects. Although such datasets may be viewed as unrepresentative of true production conditions on commercial farms, in fact there are cases of rice being grown economically with zero insecticide applications for many years in the Philippines (Box 13.2) and China (Box 13.3).

It is, nevertheless, unrealistic to promote a sudden cessation of insecticide applications to rice crops. Where rice is produced successfully with low or no insecticides it is often due to government-launched insecticide reduction campaigns and farmer training (Matteson, 2000) as well as a period of adjustment during which some 'trial and error' occurs. Importantly, however, these cases show that heavy insecticide use can be avoided and that there are natural mortality factors at play that can be exploited to suppress pest populations. Though maximising 'bottom-up' control by managing plant nutrition and host-plant

resistance is a significant factor (Figure 13.1), the remainder of this chapter covers ways in which 'top-down' control by natural enemies may be enhanced.

Plank 2: enhancement of generalist natural enemies – predator build-up on detritivore prey

The second plank of ecological engineering in rice is to ensure that an effective community of generalist predators establishes in the rice crop as early as possible in the growing season. The challenge to achieving this is that rice is often grown as a virtual monoculture over large areas of land and, prior to the young rice plants being planted out in temperate areas, fields are cultivated and flooded. This disturbance regime and the lack of vegetation (which may extend for weeks or months in cooler production areas) makes rice fields poor-quality habitat for beneficial insects. Therefore newly transplanted rice plants may be unprotected by natural enemies for several weeks before adequate numbers immigrate.

Box 13.2 Insecticide-free rice in the Philippines

Mr Sesinando Masajo, a rice farmer from Victoria, Laguna near Los Baños in the Philippines has grown rice profitably without insecticides for 39 years. The proximity of his farm to the headquarters of the International Rice Research Institute means it is a well-studied system (Islam and Heong, 1997) and many visitors to IRRI also visit there. Grain yields on his 28 ha farm are over 8 t/ha, significantly above the average level for the district. He relies upon natural enemies to prevent pest populations reaching damaging levels. Water buffalo are used to cultivate the land and these animals also enhance organic matter levels. Non-crop vegetation, most of it woody perennials, borders rice fields providing source vegetation for beneficial insects and spiders.

In temperate crops such as wheat, the ecological engineering strategy to alleviate this early-season paucity of natural enemies is to maximise the quality of overwintering habitat so that survival of predators is improved and large numbers are available to move into the crop as early as possible in the new season to check pest growth. Historically, overwintering habitat has taken the form of hedgerows but 'beetle banks' (Tillman *et al.*, chapter 19 of this volume; Thomas *et al.*, 1991) have also been developed to serve the same function for generalist natural enemies such as beetles and spiders. In rice agroecosystems, non-crop habitat is certainly important for various natural enemies of rice pests (Gurr *et al.*, 2011) but the aquatic nature of this crop lends itself particularly well to a complementary strategy: the build-up of generalist predators on non-pest prey supported by the detrital shunt (Figure 13.6) (Polis and Strong, 1996). This phenomenon operates when the predators of herbivores are not dependent solely on the presence of herbivores.

Rather, they are able also to consume detritivores, either directly via predation or indirectly by preying upon other species of predators that directly consume detritivores. In systems where this applies, the abundance of generalist natural enemies can be enhanced by this additional prey resource; for example web spiders in a forest system (Miyashita *et al.*, 2003). In agricultural systems the detrital shunt is an elegant solution to the problem of a natural enemy population otherwise being unable to grow until a pest population has reached a density where it provides sufficient prey. In the absence of a detrital shunt to allow early-season build-up of predators to suppress pests, the farmer is likely to spray the emerging pest population before it causes damage, unable to wait for biological control to take effect. If generalist predators are able to utilise abundant detritus feeders early in the season, they develop into a sentinel community of natural enemies that can respond immediately to migrating pests. Empirical evidence for this phenomenon operating in

Box 13.3 Insecticide-free rice in China

No insecticides have been used for over 10 years by Mr He and the other 100 farmer families at He Jia village near San Men in Eastern China. Bunds and field margins are well vegetated with flowering dicotyledons, grasses and woody plants, providing shelter and plant foods to natural enemies. Bunds also link fields with natural vegetation on nearby hills. Pig manure is used rather than synthetic fertiliser and this boosts organic matter in rice fields, supporting predator build-up on detritivore prey. In addition to lower production costs, rice from this village is sold for four times the price of conventional rice. (bottom picture by Zhu Zeng Rong; others, G.M. Gurr).

rice was evident in a study conducted in Indonesia by Settle *et al.* (1996). They compared composted cow manure fertilised plots of rice with control plots without organic matter supplementation. The results showed the organic matter increased the numbers of detritivores and predators in the mud, below the water surface, on the water surface and above the water, and had an impact on pest management (Figure 13.7a). Essentially, the early-season populations of generalist predators were supported by abundant detritus- and plankton-feeding arthropods and this detrital infusion gave predators a 'head start' on later-developing pest

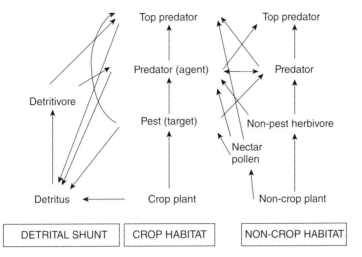

Figure 13.6 Stylised representation of trophic relationships in an agricultural system showing the importance of detritus-based and non-crop vegetation-based components.

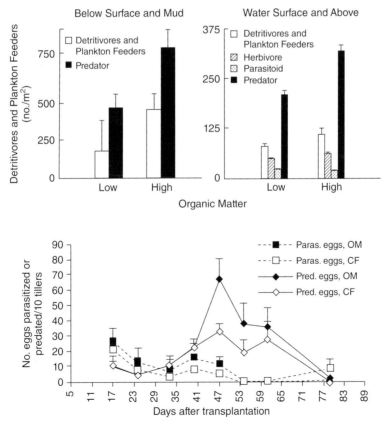

Figure 13.7 Cascading trophic effects of organic matter in irrigated rice: a) increased numbers of detritivores and plankton feeders and predators (from: Settle *et al.*, 1996); b) enhanced parasitism and predation of whitebacked planthopper (*Sogatella furcifera*) (OM = organic matter, CF = conventionally fertilised) (from Jiang and Cheng, 2004, reproduced with permission of Springer-Verlag Berlin/Heidelberg in the format Journal via Copyright Clearance Center).

populations. Similar work by Jiang and Cheng (2004) investigated biological control of whitebacked planthopper in China. Composted barnyard manure was compared with synthetic fertiliser at rates equivalent to the nutrient present in the manure to give an indication of the effects of organic matter without the confounding effect of plant nutrition being grossly different. This work was important not only in demonstrating changes in arthropod numbers but in measuring changes in ecological process rate (Figure 13.7b). Predation of whitebacked planthopper eggs was markedly increased 47 days after transplanting and thereafter. Direct observations of the activity of the spiders *P. subpiraticus* and *Pardosa pseudoannulata* (Bösenberg and Strand) (Araneae: Lycosidae) in Japanese rice fields has illustrated some of the underlying trophic relationships in the case of these important predators of rice delphacid pests (Ishijima *et al.*, 2006). Chironomids (Diptera) were important prey early in the rice-growing season when these detritivores comprised 20% and 50%, respectively, of the diet. One month later, the proportion of chironomids in the diet declined markedly, with delphacid pests constituting the major prey type. Similar trends were observed in Pakistan in detailed studies of the diets of locally important lycosid spiders (Tahir and Butt, 2009). In August the diet of all three spider species was dominated by Diptera prey whilst the following month, by which time planthopper prey were available in the crop, planthoppers were the chief prey type.

Plank 3: enhancement of specialist natural enemies by habitat manipulation – nectar plants on bunds

The strategies for encouraging specialist natural enemies – the third plank – are necessarily different from those described for generalists. Whilst generalist predators can be enhanced via the detrital shunt, this approach tends to have little effect on parasitoids (Figure 13.7b; Jiang and Cheng, 2004). Parasitoids tend to be more specialised, attacking hosts in the same genus, or at least order, and foraging in particular habitats. Consequently, they are unable to switch between detritivore and herbivore hosts so management practices based on other ecological mechanisms are required for parasitoids.

Dryinid (Hymenoptera) parasitoids are unusual in being carried with the adult host insects. In contrast, most other parasitoids attack eggs and nymphs and so are not dispersed by migrating adults. It is therefore critical for most types of parasitoids that habitat is provided for overwintering hosts. For mymarids in particular there is a wealth of evidence (see review by Gurr *et al.*, 2011) that parasitoids of delphacid pests can colonise rice crops after moving from overwintering habitat where they attack herbivore hosts that are non-pest or of minor pest significance. Non-rice habitat can also be important for certain generalist predators. For example, the spider *P. subpiraticus* overwinters in the dense foliage of the vegetable *Zizania caduciflora* (Turcz.) (Yu *et al.*, 2002). Pitfall trapping studies showed that large numbers of spiders move into rice fields from adjacent *Zizania* early in the rice season and move back to the *Zizania* before rice is harvested. Rubidium marking indicated dispersal up to 40 m into rice from a release point in the *Zizania*. This source/sink relationship between rice and other vegetation refuges is potentially supported by the bunds that surround fields in irrigated and most rain-fed rice production systems. The bunds form an intricate network through rice landscapes. These connect rice crops with any available source habitat, including natural and semi-natural vegetation fragments, riparian corridors and stands of perennial, woody crops grown for timber, fruits and nuts (Plate 13.1). The more general significance of bunds, as well as termite mounds, has been recognised as important in northeastern Thailand where they are considered biodiversity hotspots for soil macrofauna biodiversity (Chou Choosai *et al.*, 2009).

The network of bunds also offers an outstanding opportunity for a complementary strategy for enhancing the impact of parasitoids in rice: the use of flowering plants to provide floral and extrafloral nectar. Remarkably, although this conservation biological control strategy has been researched and deployed in many crop systems (Landis *et al.*, 2000; Gurr *et al.*, 2004; Wäckers, 2005; Wackers and van Rijn, chapter 9 of this volume) it has been the subject of very little research in the world's most important crop. The practicality of this approach is evident as the bunds are not solely used to control standing water and to allow human foot traffic around the network of fields. In some regions, secondary crops are commonly grown on the bunds, for example in Zhejiang Province, eastern China, where it is common to see soybean (Figure 13.8a). A logical extension of this accepted practice is to explore the use of plants that will provide nectar to parasitoids and potentially pollen to other

Figure 13.8 a) Traditional use of bunds surrounding rice for production of a secondary crop of soybean (*Glycine max*) in Zhejiang Province, eastern China; b) non-traditional, experimental use of sesame (*Sesamum indicum*) to provide nectar to parasitoids as well as serving as a valued secondary crop in Zhejiang Province, eastern China (photographs taken August 2010 by G.M. Gurr).

types of natural enemies that are known to benefit from this plant food, such as Syrphidae and Coccinellidae (Wäckers, 2005).

There has been some earlier recognition of the importance of bunds to rice pest management (Way and Heong, 1994) but its potential is far from fully realised (Gurr, 2009). A comprehensive review of the parasitoids of delphacid pests of rice (Gurr *et al.*, 2011) highlighted the scope to manipulate bund vegetation for enhancement of parasitoids of delphacid pests. The limited amount of work conducted on manipulation of bund vegetation for pest management illustrates the need for caution in terms of plant choice. In the Philippines, Marcos *et al.* (2001a) found that insect pests as well as natural enemies were more abundant and species richness was increased in rice paddies surrounded by bunds with vegetation than in paddies without this feature. An understanding of the ecological mechanisms at play and the quest for selectivity in the effects of companion plants (Baggen and Gurr,

1998) is, therefore, critical. An initial step towards this is evident in another Philippines study in which natural enemies were most abundant in bunds with only broadleaf as opposed to grassy weeds (Marcos *et al.*, 2001a; 2001b). Cowpea (*Vigna unguiculata* L.) crops were important reservoirs of the natural enemies. Other studies in China found that soybean (*Glycine max* L.) served the same function (Liu *et al.*, 2001). Importantly, therefore, there appears to be good scope for using crop species to provide nectar to parasitoids.

One such multiple-function plant is sesame (*Sesamum indicum* L.) (Pedaliaceae) (Figure 13.8b). Although this species is regarded as largely self-pollinating (Kinman and Martin, 1954), it is a recognised nectar and pollen source for beekeepers (Mc Gregor, 1976). Its flowers also attract various other insects (Langham, 1941). The nectar of sesame is reported to contain sucrose and alpha amino acids (Bahadur *et al.*, 1986). A more detailed analysis of the sugar composition is presented by Freeman *et al.* (1991), who found the relative proportions of sucrose: glucose: fructose to be 70.6: 28.8: 0.6. Such sucrose-rich nectars are generally considered typical of bee-pollinated flowers (Baker and Baker, 1983). The large (2–4 cm long), tubular flowers of sesame are consistent with this. Studies of the effect of sesame nectar on natural enemies and pests are now underway at the Zhejiang Academy of Agricultural Sciences in China. The delphacid parasitoid *Anagrus nilaparvatae* (Pang and Wang) (Hymenoptera: Mymaridae) is strongly attracted to the volatiles from this plant in Y tube olfactometer tests and longevity is enhanced to levels similar to those achieved with honey and water diet (Zhu Ping Yang, personal communication, August 2010). Whilst the nectar of sesame is likely to be accessible to most Hymenoptera with small bodies, the deep corolla of sesame flowers may preclude feeding by larger adult Lepidoptera – including major pests. Further, the moth's proboscis is unlikely to be long enough to access nectaries that are located at the base of the flowers (Abdalsalam and Al-Shebani, 2010).

Establishing sesame and other nectar-rich crops, and other flowering plants, on bunds is now being explored at a large scale in an Asian Development Bank-funded project led by the International Rice Research Institute (IRRI, 2010a; 2010b). Research sites have been established near the city of Guilin, southern China, by the Ministry of Agriculture, near the eastern Chinese city of Jinhua by the Zhejiang Academy of Agricultural Sciences (ZAAS) (Figure

Figure 13.9 Experimental comparison of the effects of vegetable crops and non-crop nectar plants in the margins of rice in a) Thailand (photograph taken March 2010 by G.M. Gurr); b) Vietnam (photograph taken December 2009 by G.M. Gurr).

13.8b), in Chainat north of Bangkok by the Thai Rice Department (Figure 13.9a) and in Cai Be near Ho Chi Minh City by the Plant Protection Department and Can Tho University in Vietnam (Figure 13.9b). Preliminary results from this study indicate enhanced densities of parasitoids and predators in the ecological engineering areas compared with nearby control areas (IRRI, 2010b). Densities of planthoppers have also been suppressed.

CONCLUSION

The three planks of ecological engineering elucidated in this chapter are based on an amalgam of sound ecology, successful precedents in other (usually temperate systems) and – most importantly – empirical evidence from the available research and evaluation in rice. Reducing pesticides inputs and promoting generalist natural enemies with enhanced organic matter inputs is likely to greatly alleviate the current problems of pest resurgence in rice. If these plank are also combined with a change of bund vegetation from being simply a structure to retain water, major advances could be made to help ensure food security for the large portion of the human population that is dependent on rice as their food staple. The bunds that are an integral system of irrigated and most rain-fed rice production offer scope to link rice fields with source vegetation that serves as natural enemy refuge as well as directly providing resources such as moderated microclimate, pollen and nectar to natural enemies. Farmers are likely to be receptive to this new approach if the bund plants are crop species from which a dual income is made.

ACKNOWLEDGEMENTS

This work was supported by the Asian Development Bank 13th RETA project 6489, coordinated by the International Rice Research Institute, Los Banos, Philippines. Donna Read is thanked for invaluable assistance in the preparation of this chapter.

REFERENCES

Abdalsalam, A.A. and Al-Shebani, Y.A. (2010) Phenological and productivity characteristics of sesame (*Sesamum indicum* L.) as affected by nitrogen rates under Sana'a conditions. *Journal of Plant Production*, 1, 251–264.

Baggen, L.R. and Gurr, G.M. (1998) The influence of food on *Copidosoma koehleri* (Hymenoptera: Encyrtidae), and the use of flowering plants as a habitat management tool to enhance biological control of the potato moth, *Phthorimaea operculella* (Lepidoptera: Gelechiidae). *Biological Control*, 11, 9–7.

Bahadur, B., Chaturvedi, A. and Swamy, N.R. (1986) Nectar types in Indian plants. *Proceedings of the Indian Academy of Sciences (Plant Sciences)*, 96, 41–48.

Baker, H.G. and Baker, I. (1983) A brief historical review of the chemistry of floral nectar, in *The biology of nectaries* (eds B. Bentley and T.S. Elias), Columbia University Press, New York, pp. 126–152.

Bangkok Post (2010) Second-crop output forecast cut by 16%. http://www.bangkokpost.com/business/economics/31763/second-crop-output-forecast-cut-by-16.

Barrion, A.T., Pantua, P.C., Bandong, J.P. *et al.* (1981) Food web of the rice brown planthopper in the Philippines. *International Rice Research Notes*, 6, 13–15.

CABI (2005) *Crop Protection Compendium*, 2005 edn, CAB International, Wallingford, www.cabicompendium.org/cpc.

Cheng, C.H. (1971) Effect of nitrogen application on the susceptibility in rice to brown planthopper attack. *Journal of Taiwan Agricultural Research*, 20, 21–30.

Cheng, J.A. (2009) Rice planthopper problems and relevant causes in China, in *Planthoppers: new threats to the sustainability of intensive rice production systems in Asia* (eds K.L. Heong and B. Hardy), International Rice Research Institute, Los Baños, pp. 157–177.

Cheng, J.A., Norton, G.A. and Holt, J. (1990) A systems analysis approach to brown planthopper control on rice in Zhejiang China. II. Investigation of control strategies. *Journal of Applied Ecology*, 27, 100–112.

Chou Choosai, C., Mathieu, J., Hanboonsong, Y. and Jouquet, P. (2009) Termite mounds and dykes are biodiversity refuges in paddy fields in north-eastern Thailand. *Environmental Conservation*, 36, 71–79.

Denno, R.F., Cheng, J.A., Roderick, G.K. and Perfect, T.J. (1994) Density related effects of the components of fitness and population dynamics of planthoppers, in *Planthopper: ecology and management* (eds R.F. Denno and T.J. Perfect), Chapman & Hall, London, pp. 257–281.

Freeman, C.E., Worthington, R.D. and Jackson, M.S. (1991) Floral nectar sugar compositions of some South and Southeast Asian species. *Biotropica*, 23, 568–574.

Gallagher, K.D., Kenmore, P.E. and Sogawa, K. (1994) Judicial use of insecticides deters planthopper outbreaks and extends the life of resistant varieties in Southeast Asian rice, in *Planthoppers: their ecology and management* (eds R.F. Denno and T.J. Perfect), Chapman & Hall, London, pp. 599–614.

Graf, B., Lamb, R., Heong, K.L. and Fabellar, L.T. (1992) A simulation model for the population dynamics of the rice leaffolders (Lepidoptera: Pyralidae) and their interactions with rice. *Journal of Applied Ecology*, 29, 558–570.

Guo, Y., Wang, N., Chen, J, Jiang, J., Hu, G. and Wu, J. (1994) Characteristics of the species richness and abundance of predators and preys in the arthropod community of three

rice ecosystems in China. *Chinese Journal of Biological Control*, 10, 157–161.

Gurr, G.M. (2009) Prospects for ecological engineering for planthoppers and other arthropod pests in rice, in *Planthoppers: new threats to the sustainability of intensive rice production systems in Asia* (eds K.L. Heong and B. Hardy), International Rice Research Institute, Los Baños, pp. 371–388.

Gurr, G.M., Wratten, S.D. and Altieri, M.A. (eds) (2004) *Ecological Engineering for Pest Management: Advances in Habitat Manipulation for Arthropods*, CABI Publishing, Wallingford.

Gurr, G.M., Liu, J., Read, D.M.Y. *et al.* (2011) Parasitoids of Asian rice planthopper (Hemiptera: Delphacidae) pests and prospects for enhancing biological control. *Annals of Applied Biology*, 158, 149–176.

Heinrichs, E.A. and Medrano, F.G. (1985) Influence of nitrogen fertilizer on the population development of brown planthopper. *International Rice Research Newsletter*, 10, 20.

Heong, K.L. (2009) Are planthopper problems caused by a breakdown in ecosystem services?, in *Planthoppers: new threats to the sustainability of intensive rice production systems in Asia* (eds K.L. Heong and B. Hardy), International Rice Research Institute, Los Baños, pp. 221–232.

Heong, K.L. and Hardy, B. (eds) (2009) *Planthoppers: new threats to the sustainability of intensive rice production systems in Asia*, International Rice Research Institute, Los Baños.

Heong, K.L., Manza, A., Catindig, J., Villareal, S. and Jacobsen, T. (2007) Changes in pesticide use and arthropod biodiversity in the IRRI research farm. *Outlooks in Pest Management*, October 2007, International Rice Research Institute, Los Baños.

Hettel, G. (2009) Bird's-eye views of an enduring rice culture. *Rice Today*, 7, 4–19.

Horgan, F. (2009) Mechanisms of resistance: a major gap in understanding planthopper-rice interactions, in *Planthoppers: new threats to the sustainability of intensive rice production systems in Asia* (eds K.L. Heong and B. Hardy), International Rice Research Institute, Los Baños, pp. 281–302.

Hosamani, M.M., Jayakumar, B.V. and Sharma, K.M. (1986) Sources and levels of nitrogenous fertilizers in relation to incidence of brown planthopper in Bhadra Project. *Current Research*, 15, 132–134.

Hu, J.Z., Lu, Q.H., Yang, J.S. and Yang, L.P. (1986) Effects of fertilizer and irrigation on the population of main insect pests and the yield of rice. *Acta Entomologica Sinica*, 29, 49–54.

IRRI (International Rice Research Institute) (2010a) Ricehoppers, a blog on the latest information and issues to manage rice planthopper problems. http://ricehoppers.net/about-2/

IRRI (International Rice Research Institute) (2010b) *Bringing about a sustainable agronomic revolution in rice production in Asia by reducing preventable pre- and postharvest losses* (RETA 6489). Annual Report submitted to the Asian

Development Bank (IRRI Ref. No.: DPPC2008-74). http://ricehoppers.net/wp-content/uploads/2010/04/dppc2008-74SemiAnnualRep0410Final.pdf

IRRI (2011) *Rice production and processing*. International Rice Research Institute, http://irri.org/about-rice/rice-facts/rice-production-and-processing.

Ishijima, C., Taguchi, A., Takagi, M., Motobayashi, T., Nakai, M. and Kunimi, Y. (2006) Observational evidence that the diet of wolf spiders (Araneae: Lycosidae) in paddies temporarily depends on dipterous insects. *Applied Entomology and Zoology*, 41, 195–200.

Islam, Z. and Heong, K.L. (1997) Letter to the editor: rice farming without insecticides: a farmer's long term experience. *The Bulletin, British Ecological Society*, 28, 259–262.

Jiang, M.X. and Cheng, J.A. (2004) Effects of manure use on seasonal patterns of arthropods in rice with special reference to modified biological control of whitebacked planthopper, *Sogatella furcifera* Horvath (Homoptera: Delphacidae). *Journal of Pest Science*, 77, 185–189.

Kinman, M.L. and Martin, J.A. (1954) Present status of sesame breeding in the United States. *Agronomy Journal*, 46, 24–27.

Landis, D.A., Wratten, S.D. and Gurr, G.M. (2000) Habitat management to conserve natural enemies of arthropod pests in agriculture. *Annual Review of Entomology*, 45, 175–201.

Langham, D.G. (1941) Natural and controlled pollination in sesame. *Journal of Heredity*, 35, 254–256.

Li, R.D., Ding, J.H., Wu, G.W. and Shu, D.M. (1996) *The brown planthopper and its population management*, Fudan University Press, Shanghai.

Li, J-Q., Shen, Z-R., Zhao, Z-M. and Luo, Y-J. (2002) Biology and ecology of the wolf spider *Pirata subpiraticus*. *Acta Ecologica Sinica*, 22, 1478–1484.

Litsinger, J.A., Bandong, J.P., Canapi, B.L. *et al.* (2006) Evaluation of action thresholds for chronic rice insect pests in the Philippines. III Leaffolders. *International Journal of Pest Management*, 52, 181–194.

Liu, G., Lu, Z., Tang, J. *et al.* (2001) Managing insect pests of temperate japonica rice by conserving natural enemies through habitat diversity and reducing insecticide use, in *Proceedings of the Impact Symposium on Exploiting Biodiversity for Sustainable Pest Management* (eds T.W. Mew, E. Borromeo and B. Hardy), Kunming, China, August 21–23, 2000, International Rice Research Institute, pp. 43–50.

Lu, Z. and Heong, K.L. (2009) Effects of nitrogen-enriched rice plants on ecological fitness of planthoppers, in *Planthoppers: new threats to the sustainability of intensive rice production systems in Asia* (eds K.L. Heong and B. Hardy), International Rice Research Institute, Los Baños, pp. 247–256.

Luo, X.N. and Zhuo, W.X. (1986) Studies on the relationships of population fluctuation between rice planthoppers and natural enemies and natural control effects. *Natural Enemies of Insects*, 8, 72–79 (Chinese with English abstract).

Lu, Z.X., Heong, K.L., Yu, X.P. and Hu, C. (2005) Effects of nitrogen nutrient on the behavior of feeding and oviposition of the brown planthopper, *Nilaparvata lugens*, on IR64. *Journal of Zhejiang University (Agriculture and Life Science)*, 31, 62–70.

Ma, K.C. and Lee, S.C. (1996) Occurrence of major rice insect pests at different transplanting times and fertilizer levels in paddy field. *Korean Journal of Applied Entomology*, 35, 132–136.

Marcos, T.F., Flor, L.B., Velilla, A.R. *et al.* (2001a) Relationships between pests and natural enemies in rainfed rice and associated crop and wild habitats in Ilocos Norte, Philippines, in *Proceedings of the Impact Symposium on Exploiting Biodiversity for Sustainable Pest Management* (eds T.W. Mew, E. Borromeo and B. Hardy), Kunming, China, August 21–23, 2000, International Rice Research Institute, pp. 23–24.

Marcos, T.F., Flor, L.B., Velilla, A.R. *et al.* (2001b) Exploiting biodiversity for sustainable pest management, in *Proceedings of the Impact Symposium on Exploiting Biodiversity for Sustainable Pest Management* (eds T.W. Mew, E. Borromeo and B. Hardy), Kunming, China, August 21–23, 2000, International Rice Research Institute, pp. 23–24.

Matsumura, M., Takeuchi, H., Satoh, M. *et al.* (2009) Current status of insecticide resistance in rice planthoppers in Asia, in *Planthoppers: new threats to the sustainability of intensive rice production systems in Asia* (eds K.L. Heong and B. Hardy), International Rice Research Institute, Los Baños, pp. 233–243.

Matteson, P.C. (2000) Insect pest management in tropical irrigated rice. *Annual Review of Entomology*, 45, 549–574.

Mc Gregor, S.E. (1976) Sesame, in *Insect Pollination of Cultivated Crop Plants*, United Stated Department of Agriculture. Online edition http://afrsweb.usda.gov/SP2UserFiles/Place/53420300/OnlinePollinationHandbook.pdf

Miyashita, T. (1985) Estimation on the economic injury level in the rice leafroller Cnaphalocrocis medinalis Guenée (Lepidoptera, Pyralidae). I. Relation between yield loss and injury of rice leaves at heading or in the grain filling period. *Japanese Journal of Applied Entomology and Zoology*, 29, 73–76.

Miyashita, T., Takada, M. and Shimazaki, A. (2003) Experimental evidence that aboveground predators are sustained by underground detritivores. *Oikos*, 103, 31–36.

Phoonphongphiphat A. (2008) *Thai rice export curbs would make no sense: analysts*. Reuters, http://www.reuters.com/article/idUSBKK17921220080425?pageNumber=1

Polis, G.A. and Strong, D.R. (1996) Food web complexity and community dynamics. *American Naturalist*, 147, 813–846.

Preap, V., Zalucki, M.P., Nesbitt, H.J. and Jahn, G.C. (2001) Effect of fertilizer, pesticide treatment, and plant variety on the realized fecundity and survival rates of brown planthopper, *Nilaparvata lugens*, generating outbreaks in Cambodia. *Journal of Asia-Pacific Entomology*, 4, 75–84.

Settle, W., Ariawan, H., Astuti, E.T. *et al.* (1996) Managing tropical rice pests through conservation of generalist natural enemies and alternative prey. *Ecology*, 77, 1795–1988.

Shankar, G. and Baskaran, P. (1988) Impact of the presence of parasites on the population of resident endosymbiotes in brown planthopper, *Nilaparvata lugens* (Stål) (Delphacidae: Homoptera). *Current Science*, 57, 212–214.

Sogawa, K. (1970) Studies on feeding habits of brown planthopper. I. Effects of nitrogen-deficiency of host plants on insect feeding. *Japanese Journal of Applied Entomology and Zoology*, 14, 101–106.

Sogawa, K., Liu, G. and Qiang, Q. (2009) Prevalence of whitebacked planthoppers in Chinese hybrid rice and whitebacked planthopper resistance in Chinese japonica rice, in *Planthoppers: new threats to the sustainability of intensive rice production systems in Asia* (eds K.L. Heong and B. Hardy), International Rice Research Institute, Los Baños, pp. 257–279.

Southwood, T.R.E. (1973) The insect/plant relationship – an evolutionary perspective, in *Insect-plant relationship* (ed F.V. Emden). Symposium of the Royal Entomological Society London, No. 6, Blackwell, Oxford, pp. 3–30.

Tait, E.J. (1987) Planning an Integrated Pest Management System, in *Integrated pest management* (eds A.J. Burn, T.H. Coaker and P.C. Jepson), Academic Press, London, pp. 198–207.

Tahir, H.M. and Butt, A. (2009) Predatory potential of three hunting spiders inhabiting the rice ecosystems. *Journal of Pest Science*, 82, 217–225.

Thanh, D.V., Dung, L.T., Thu, P.B. and Duong, N.T. (2007) *Management of rice planthopper in northern Vietnam*. Proceedings of International Workshop on 'Forecasting and Management of Rice Planthoppers in East Asia: Ecology and Genetics', December 4–5, 2007, Kumamoto, pp. 1–9.

Thomas, M.B., Wratten, S.D. and Sotherton, N.W. (1991) Creation of island habitats in farmland to manipulate populations of beneficial arthropods – predator densities and emigration. *Journal of Applied Ecology*, 28, 906–917.

Uhm, K.B., Hyun, J.S. and Choi, K.M. (1985) Effects of the different levels of nitrogen fertilizer and planting space on the population growth of the brown planthopper. *Research Report RDA*, 27, 79–85.

van Den Bosch, R. (1978) *The pesticide conspiracy*, University of California Press, Berkeley.

Wäckers, F.L. (2005) Suitability of (extra-)floral nectar, pollen, and honeydew as insect food sources, in *Plant-provided food for carnivorous insects* (eds F.L. Wäckers, P.J. van Rijn and J. Bruin), Cambridge University Press, Cambridge, pp. 17–74.

Wang, M.Q. and Wu, R.Z. (1991) Effects of nitrogen fertilizer on the resistance of rice varieties to brown planthopper. *Guangdong Agricultural Science*, 1, 25–27.

Way, M.J. and Heong, K.L. (1994) The role of biodiversity in the dynamics and management of insect pests of tropical irrigated rice – a review. *Bulletin of Entomological Research*, 84, 567–587.

Wen, D.D., He, Y-Y., Lu, Z-Y., Yang, H.M. and Wang, H-Q. (2003) A quantitative study of biomass flow in the rice-*Sogatella furcifera*-*Pirata subpiraticus* food chain using fluorescent substance tracing. *Acta Entomologica Sinica*, 46, 178–183.

Wu, L.H. and Zhu, Z.R. (1994) The relationship between rice leaf color and occurrence of rice diseases and insects and its mechanism. *Chinese Journal of Rice Science*, 8, 231–235.

Yu, X-P., Zheng, X-S., Xu, H-X., Lu, Z-X, Chen, J-M and Tao L-Y (2002) A study on the dispersal of lycosid spider, *Pirata subpiraticus* between rice and *Zizania* fields. *Acta Entomologica Sinica*, 45, 636–640.

Chapter 14

CHINA'S 'GREEN PLANT PROTECTION' INITIATIVE: COORDINATED PROMOTION OF BIODIVERSITY-RELATED TECHNOLOGIES

Lu Zhongxian, Yang Yajun, Yang Puyun and Zhao Zhonghua

Biodiversity and Insect Pests: Key Issues for Sustainable Management, First Edition. Edited by Geoff M. Gurr, Steve D. Wratten, William E. Snyder, Donna M.Y. Read.
© 2012 John Wiley & Sons, Ltd. Published 2012 by John Wiley & Sons, Ltd.

INTRODUCTION

China has a long recorded history of agriculture and it remains the foundation of the country's economy. Despite advances in technology and rising yields from its more than 100 million hectares of arable land, food security is critically important in China as a result of its large and still increasing human population (almost 1.4 billion) (FAO, 2011). One of the important factors affecting agricultural production is pest damage. In China, there are more than 1,700 pest species damaging crops (over 830 insect pests, over 720 diseases, over 60 weeds and over 20 rodent pests), and more than 100 of them cause heavy economic losses (Xia, 2010). On average, the area seriously affected by pest problems each year is between 400 and 467 million hectares, and between 60 and 90 million tons of losses are avoided by intervention (Xia, 2010).

The history of plant protection in China can be dated back to the Zhou Dynasty (about 2,000 years ago) when government officials were responsible for pest control (Pan, 1988). Botanical and mineral pesticides were used to control pests 1,800 years ago and by 200 years ago tobacco (with nicotine) had become an important material for controlling rice pests (Zhao, 1983). Nevertheless, in ancient China outbreaks of pests caused heavy losses because of limited technologies available for pest control. China proposed the concept of integrated pest control in 1953 (predating the 'integrated control concept' of Stern *et al.*, 1957 that led to integrated pest management (IPM)) and established a policy for integrated pest control (IPC) in 1975 (Guo, 1998). The principle of China's plant protection in the following years was integrated management with an emphasis on prevention, which is based on the widely accepted strategies of IPM.

In the past 30 years China has successfully extended IPM technologies. However, as a result of climate change, cultivar 'improvement', excessive use of agricultural chemicals and other biotic and abiotic factors, outbreaks of major agricultural pests have been more and more frequent in recent years (Xia, 2008). In order to meet the new challenges of pest control, China updated the concept of plant protection as 'Public Plant Protection, Green Plant Protection'. 'Public Plant Protection' gives recognition of the significance of social management and public service aspects of plant protection. This reflects the wider significance of public issues in agriculture and rural affairs. For example, plant quarantine and insecticide manage-

ment are administration issues whilst monitoring and surveillance of important pests needs the organisation and input of the government. 'Green Plant Protection' emphasises the support and safeguards needed to obtain high-yield, good-quality, ecologically sustainable agriculture. Achieving the aims of 'Green Plant Protection' requires research and implementation of appropriate technologies, a responsibility of the National Agro-Technology Extension and Service Centre (NATESC), Ministry of Agriculture (MOA). NATESC promotes the 'Professional and Standardised Management of Pests' (PSMP) (Box 14.1) to carry out the multiple strategies involved in the goal of 'Green Plant Protection'. This chapter focuses on the 'Green Plant Protection' element of China's current plant protection policy and illustrates the range of strategies that have been implemented.

BACKGROUND TO 'GREEN PLANT PROTECTION'

The preliminary concept of 'Green Plant Protection' was formed in the National Forum on Chinese Plant Protection in 2006. Scientists rethought the issues of Chinese plant protection and suggested that it should be based on the crop ecosystem, considering energy flow, economic factors, and ecological factors (Fan, 2006). 'Green' succinctly summarises the concept's aims in relation to ecologically sustainable, effective and economic agricultural systems.

Components of 'Green Plant Protection'

The implementation of the new concept 'Public Plant Protection, Green Plant Protection' and 'Law of the People's Republic of China on Agricultural Product Quality Safety' addressed public concerns about the safety of agricultural production in regard to the environment. The extension of green control technologies to farmers has been managed by NATESC. To 2009, 5.6 million hectares were managed using green pest control technologies, accounting for 15% of the area where pests occur and 10.4% of the area that exercised control for pests (Xia, 2011). Multiple strategies were used for pest control including physical-chemical attraction technology, biological control and ecological control methods. Each of these is expanded upon in following sections. The new strategies reduce pest

damage, protect the environment, increase yields and profits of grains, and contribute to the quality and safety of agricultural products (Xia, 2010).

Physical-chemical attraction technology

Most insect pests demonstrate forms of photo-taxis and chemo-taxis which can be exploited for pest management. The most popular methods of physical and chemical attractive technologies are frequency trembler grid lamp, ultraviolet lamp, colour sticky card, and pheromones (Plate 14.1; Wei *et al.*, 2008; Wang *et al.*, 2008; Yang *et al.*, 2010). In 2009, these methods were used in 4.5 million hectares (Zhao *et al.*, unpublished).

Biological control

Biological control is defined in this context as the effective utilisation of biological agents including living organisms and derivatives of living organisms to maintain the pest population below an economically significant threshold. With appropriate safeguards and proper implementation biological control is an environmentally sound and effective means for reducing or mitigating pest damage. Natural enemies including arthropods and microbes are important agents of biological control and their biodiversity in China is rich (Box 14.2). Predators and parasites have been widely used in biological control. Examples are predatory mites such as *Neoseiulus cucumeris* (Oudemans), predatory insects such as *Coccinella septempunctata* L. and parasitoids such as *Trichogramma* spp., *Microplitis* spp., *Encarsia formosa* (Gahan) and *Anastatus japonicus* (Ashmead). *Trichogramma* spp. have been produced in massive numbers and released in many Chinese provinces (Liu and Shi, 1996). The area involved recently reached 1 million hectares of cultivated land and 1 million hectares of forest (Liu and Shi, 1996). Entomopathogenic bacteria and fungi have also been employed in pest control. The well-known gram-positive bacterium *Bacillus thuringinensis* (Berliner) (*Bt*) has been widely used in pest control in China and almost every province has factories for production of *Bt* formulations. So far, six million hectares have been treated with *Bt* in China (Xiao *et al.*, 2008). The fungi *Beauveria bassiana* (Balsamo-Crivelli) and *Metarhizium anisopliae* (Metchnikoff) are used to control pine cater-

Box 14.2 Pest natural enemies: a biodiversity resource

The use of natural enemies is an important tactic in pest management. China is able to exploit a vast resource of various species of pest enemies. There are more than 1,000 species of enemies of rice pest, more than 960 species of enemies of corn pest and over 840 species of enemies of cotton pest. The most common enemies of pests are ladybeetles, wasps, spiders and entomogenous fungi. Pest enemies shown here are: a) *Apanteles* sp., b) *Anagrus* sp., c) *Zoophthora radicans* (Brefeld), d) spider (photos: a–c) Baoyu Han and d) Zhongxian Lu).

pillar moth (*Dendrolimus punctatus* (Walker)), European corn borer (*Ostrinia furnacalis* Guenée), the oriental migratory locust (*Locusta migratoria manilensis* (Meyen)) and other insects (Zeng *et al.*, 2008). Biological control was used on 82.6 million hectares in 2009, 88.04% of which was in grain crops such as rice, wheat, and corn; 11.1% was in cash crops such as cotton, vegetables, tea and fruit (Zhao *et al.*, unpublished data).

Ecological control

Ecological control as used here involves the manipulation of the environment to enhance biodiversity and regulate the ecological factors to balance the components of the ecosystem. It is, therefore, the approach of most relevance to this book. When ecological control, using the approaches described below, is successful, ecosystem functioning of top-down and bottom-up effects (see chapter 1) is maximised and yield losses caused by the damage of pests are reduced. Biodiversity is very important in ecosystem equilibrium; high biodiversity can enhance the stability of simple and disturbed agricultural systems.

In the production base of Shanghai Yuejing Modern Agriculture Co., the incidence of serious pests in rice fields was reduced by 50–70% by the use of ecological control. The importance of top-down control in this case is indicated by a twentyfold increase in natural enemy numbers compared to control paddy fields (Fan *et al.*, 2010). A second example, which has been the subject of more intense study, is the rice fields in Jinhua, Zhejiang province (Box 14.3) that were part of

---

Box 14.3 Demonstration of management of rice pests by increasing biodiversity

The demonstration of rice pest management in Jinhua aims to promote reduced applications of pesticides, to reduce the damage done by pests and to produce uncontaminated rice. Biodiversity was increased via ecological engineering involving plants that provided floral resources and alternative (non-pest) hosts for parasitoids and trap plants for stemborers. Combined with use of light traps and reduced fertiliser inputs, the need for insecticide was reduced with no adverse effect on rice yield. The illustration shows the noticeboard for the demonstration zone for the management of rice pests by ecological engineering in Jinhua, Zhejiang Province (managed in cooperation with International Rice Research Institute (IRRI)) (photo: Zhongxian Lu).

金华市水稻生态工程实验示范区

示范目标：示范区化学防治次数减少3次以上、化学农药使用量下降50%以上，水稻重大病虫危害损失总体控制在3%以下，稻米达到无公害标准。

技术措施：选用抗病虫品种；灯光、诱虫植物和性诱剂诱杀技术；田边留草和种植开花作物、保护和利用天敌；冬季种植绿肥、减少化肥使用量；水稻前期坚持不用或少用农药、全面放宽防治指标；优先选用生物农药、必要时选用选择性强、对天敌安全的化学农药；尽量选用农药单剂、实现农药轮换使用。

示范内容：水稻品种田间抗性评价；植物和节肢动物生物多样性；生物农药应用技术；开花作物对天敌种群增长的影响；肥料对害虫和天敌种群的影响；害虫抗药性监测；性诱剂、诱虫植物和杀虫灯对害虫的控制能力和对天敌种群的影响；优化农药防治策略。

建设单位：	农业部农业技术推广服务中心	浙江省农业科学院
	浙江省植物保护检疫局	金华市农业局
实施单位：	金华市植物保护站	金华寺平稻米专业合作社
技术依托：	国际水稻研究所（IRRI）	浙江省农业科学院植微所
	浙江大学	中国水稻研究所
资助项目：	亚洲发展银行ADB-IRRI基金项目	国家公益性行业（农业）科研专项
	部、省农作物病虫害绿色防控专项	

the 'ecological engineering' project featured in chapter 13 of this volume. On that site, floral diversity was enhanced by populating the bunds and non-rice habitats with sesame, Zizania and nectar-rich flowering plants. This non-rice habitat increased the density of parasitoids and predators and chemical insecticide applications were reduced by about 80% compared with control fields in 2010. The abundance of frogs, particularly *Rana limnocharis* (Boie) and *Rano nigromaculatta* (Hallowell), also increased substantially, but work to quantify the significance of these generalist predators in pest mortality is ongoing. Most significantly, yields from the ecological engineered fields were similar to those in control fields (10.01 tons per hectare compared with 10.03 tons per hectare), showing that productivity was not dependent on use of synthetic insecticides (Lu *et al.*, unpublished data).

NATESC has organised and established various ecological control demonstration zones and in 2009 these approaches were implemented on 2.9 million hectares in China. Its use extended over 769,200 hectares of cotton in Xinjiang, Henan, Hebei and Jiangxi provinces and in 692,000 hectares of rice fields in the central and southern areas of China (Zhao *et al.*, unpublished data).

New synthetic chemical and biotic insecticides

New synthetic chemicals differ from traditional synthetic chemicals. After the Second World War the chemical industry developed quickly, and traditional chemicals were widely utilised in pest control. However, use of these chemicals resulted in environment pollution, the '3Rs' (resistance, residues, and resurgence), and ecological destruction (Zhao, 1983). The traditional chemicals gained an unfavourable reputation. In China, 298,000 tons of pesticides were used in 2008, of which insecticides accounted for 46.98%, followed by herbicides (26.16%) and fungicides (25.49%) (Jin *et al.*, 2010). At the time of writing the Chinese government has banned 23 high-toxicity insecticides, such as DDT and methamidophos (MOA, 2002; 2007). Many pesticides have been banned from use in particular crops (e.g. fenvalerate banned for tea, daminozide for peanuts, methamidophos and fipronil for rice). Even so, many other pesticides have not been banned due to a lack of alternatives and these have negative effects on the environment. In order to be more 'green' the development of lower toxicity, more effective and 'environment-friendly' chemicals for pest control has been pursued. Growth regulators, hormones, botanicals, and other natural substances and their analogs are possible alternatives. The extracts of many plants have insecticidal properties. Toosendanin, an antifeedant limonoid from the bark of the trees *Melia toosendan* (Sieb. and Zucc.) and *Melia azedarach* L. (Meliaceae), was identified as potentially useful for control of pests such as the variegated cutworm, *Peridroma saucia* (Hübner) (Chiu, 1989; Chen *et al.*, 1995). A commercial botanical insecticide containing approximately 3% toosendanin as the active ingredient has now been produced in China (Koul, 2008). The use of plant compounds as novel insecticides is covered in more detail by Koul in chapter 6 of this volume.

CASE STUDIES

The 'Green Plant Protection' system aims to minimise pest damage to crops through the integrated use of various methods. NATESC established more than 210 demonstration zones in China between 2006 and 2009 in which 80% of the technologies employed for plant protection were 'green' and successfully controlled up to 90% of pests (Yang *et al.*, 2010). Some cases from this initiative are outlined below.

Locust management

The oriental migratory locust, *L. migratoria manilensis*, is a polyphagous insect listed as one of the most serious pests in ancient China since 707 BC (Wu, 1951). Outbreaks of this locust have caused great losses and have even gravely threatened food security and human lives. Outbreaks of this pest have occurred in the past two decades in northern China due to climate change-induced drought which favours oviposition and fast population growth of locusts (Lei and Wen, 2004). Every year large amounts of broad-spectrum chemical pesticides were used in locust control. This chemical use polluted the environment, caused health and safety issues and exacerbated the locust problems due to the loss of natural enemies (Zhu, 1999). In order to reduce the locust damage and avoid chemical problems, 'Green Plant Protection' was explored and applied to locust management.

Table 14.1 The relationship between reed vegetation characteristics and the density of pest locusts (data from Zhang *et al.*, 2009).

Reed status			Locust status		
Height (cm)	Density (/cm²)	Coverage (%)	No. of infested sites	Mean density (/m²)	Maximum density (/m²)
30–50	59	12	7	0.3	0.5
50–70	110	29	8	0.3	0.7
70–90	185	69	1	0.1	0.2
90–110	243	92	0	0	0
110–130	301	100	0	0	0
130–150	385	100	0	0	0
150–170	362	100	0	0	0
170–200	421	100	0	0	0

Vegetation is an important factor affecting locust populations. Suitable vegetation provides an environment compatible with the maintenance and propagation of locust enemies such as frogs and birds. Vegetation may also inhibit the migration of locusts, but the density of locusts varies depending on vegetation type (e.g. <0.01 per m² in cotton, <0.05 per m² in alfalfa and <0.1 per m² in winter dates) (Zhang *et al.*, 2006). Ji *et al.* (2007) found that vegetation with over 70% ground cover and at least 30% soil water content at 5 cm depth or greater than 3% soil salinity made ground unsuitable for *L. migratoria manilensis* oviposition. Further, although reeds (*Phragmites australis* Cavanilles) are among the most suitable foods for locust, the density of locusts was reduced when the height of the reeds was more than 90 cm and the coverage greater than 70% (Table 14.1; Zhang *et al.*, 2009). Although information on such biotic and abiotic effects on a given pest can be difficult to convert into practical pest management tactics, the potential for ecologically based strategies to be effective is apparent in medium-sized initiatives. Shiqiao, a suburb of Jining city in Shandong province of China, was frequently infested by locusts. More than 1,333 hectares were damaged annually, with the locust density ranging between 0.8 and 18.0 per m². In 2007, an ecological control demonstration zone was established in the area. Control measures were introduced, including cultivating vigorous reed beds in marshes and wasteland areas, reducing high-toxicity pesticides, increasing frog, bird and other predator populations and reducing the places for oviposition. These measures resulted in a reduction in the density of locusts and the damage they caused (Zhang *et al.*, 2009). In Funan county raising ducks in the locust-breeding zones reduced the density of locusts to less than 0.5 per m² (Zhang *et al.*, 2010). In a separate study of high prairie, Li (2010) found that chickens and ducks could effectively control locusts.

Biological agents may replace traditional chemical pesticides for locust control. A fungus, *M. anisopliae*, is effective against locusts and grasshoppers. Recent research showed that a moderate dose of a new strain of *M. anisopliae*, CQMa102 in oil miscible suspension resulted in high mortality in large-scale field trials and did not reduce the number of natural enemies. When aerially sprayed, the survival rate of locusts was lowered to about 10% at 11 and 14 days in the field cage and open field respectively (Peng *et al.*, 2008). Cheng *et al.* (2007) found that *Metarhizium* and the biopesticides Matrine® and Celangulin® were highly effective controls of oriental migratory locust and could replace the high-toxicity synthetic insecticides.

In 2010, the Chinese government organised and successfully implemented 'green' locust control. In Hebei Province 8,000 hectares of ecologically engineered sites were established using cotton, alfalfa and winter date crops. In the locust-prone zones of Weishan Lake and Hongze Lake in Jiangsu Province, poultry breeding and broad-leaved crops (which locusts dislike) were used. Additionally, 150 'professional locust control groups' were established in Shandong Province, eastern China. In 2010, ecological control techniques were employed on 120,000 hectares and biological control was used on 145,000 hectares involving the use of 129.8 tons of biological control agents (MOA, 2010).

Tea pest management

Tea, *Camellia sinensis* L., is a major cash crop in China. The area of tea planting is 18.6 million hectares which yielded 1.35 million tons of tea in 2009 (Yang, 2010). As living standards rise in China the demand for quality tea, free from the risk of pesticide residue, is increasing. As an indication of this trend, 20,000 tons of tea was grown without synthetic pesticides and fertilisers in 2008 (Jiang *et al.*, 2009). However, more than 1,300 pests of tea have been documented in the world; 900 of these occur in China and cause loss of tea yields (Hazarika *et al.*, 2009; Zhang, 2004). Neither high-quality tea nor ordinary tea production can use damaged bud-leaves due to their changed colour and poor taste. Nevertheless, the greatest threat for tea producers and international trade is consumers' and importers' concern about pesticide residues in tea (Yang, 2010). Therefore, 'Green Plant Protection' strategies are particularly important for pest control in tea.

Attraction technologies, biological and ecological control tactics are used to overcome concerns about pesticide residues in tea. Light traps are an important component of tea pest management. The black-light traps that emit near-UV light of 350 nm wave length or the Jiaduo® insect killer lamp have proven useful in tea fields for capturing adults of many lepidopterans (Qi *et al.*, 2005). Coloured sticky card traps have also been shown to be effective in control of pests of tea such as *Empoasca flavescens* Fabricius and spiny black whitefly (*Aleurocanthus spiniferus* (Quaintance)) (Plate 14.2). In an organic tea plantation, yellow sticky traps reduced the numbers of eggs laid on leaves by *E. flavescens* and *A. spiniferus* on average by 71.36% and 64.31%, respectively (Liu *et al.*, 2010).

Plant and insect chemical ecology is also playing a role. For example, natural enemies may be attracted by the release of a synomone (a semiochemical defined as beneficial to the emitter and recipient) by the plant when injured by an insect pest. These same compounds, such as methyl salicylic acid (MeSA), may be detected by 'eavesdropping' neighbouring plants such that the semiochemical serves as a pheromone (semiochemical that sends messages between individuals of the same species) that induces the defence of neighbouring plants (James *et al.*, chapter 11 of this volume). The pest may also release a kairomone (a semiochemical beneficial to the recipient) that also attracts natural enemies.

Figure 14.1 Intercropping in the tea plantation. Appropriate intercrop such as bamboo increased the abundance of diversity, reduced the damage of pests and enhanced the yields and quality of tea (photo: Baoyu Han).

Prudent intercropping is a good way to increase biodiversity within tea plantations and can also manipulate the spatial structure of the community (Figure 14.1). Ye *et al.* (2010) found that suitable intercrop species vary with the tea variety to the extent that a suitable intercrop and tea variety combination inhibited whitefly whilst an inappropriate combination promoted the pest. In tea plantations using Wuniuzao, a variety of tea, intercropping with waxberry, citrus, or snakegourd fruit, numbers of *A. spiniferus* were lower than in pure Wuniuzao tea plantations. However, when the tea variety Anjiebaicha was intercropped with snakegourd fruit the numbers of spiny whitefly was larger than in pure tea plantations (Ye *et al.*, 2010).

Vegetable pest management

Pest damage is one of the important factors affecting sustainable vegetable production. Some of the main pests of vegetables (e.g. cabbage, musk melon, tomato, and so on) are the diamondback moth (*Plutella xylostella* L.), whitefly (*Bemisia tabaci* (Gennadius)), beet armyworm (*Spodoptera exigua* (Hübner)), common cutworm (*Spodoptera litura* Fabricius) and cabbage white butterfly (*Artogeia rapae* L.) (Ma, 2001). Vegetables are a staple of the diet and 'Green Plant Protection' is aimed at ensuring their quality and that they are safe to eat.

The principle of photo-taxis in insect pests of vegetables has been exploited for effective pest management (Plate 14.3a, b). Pests vary in their sensitivity to colours. *Phyllotreta striolata* Fabricius is attracted to yellow and white, *Myzus persicae* (Sulzer) and *Liriomyza sativae* (Blanchard) to yellow and *P. xylostella* to green (Fu *et al.*, 2005). Field experiments have shown that the effectiveness of yellow sticky traps was correlated with the height, direction and shape of the card (Zhou *et al.*, 2003). Wang (2009) found blue sticky traps hung 15 cm above pepper plants was most effective for protecting the plants against pest damage.

Natural enemies have also been employed for pest management in vegetables. Shen *et al.* (2006) found that *A. rapae* could potentially be controlled by its natural enemies *Erigonidium graminicola* (Sundevall) and *Dyschiriognatha quadrimaculata* (Bösenberg and Strand) due to their being temporally and spatially synchronised with the eggs and larvae of *A. rapae*.

The use of sex pheromones as a 'Green Plant Protection' method is harmless to humans, livestock and the environment (Plate 14.3c). In general, the effectiveness of sex pheromone traps is related to densities of pests and traps. Field studies have shown that the strongest mating disruption of *P. xylostella* occured when 390 traps per hectare were employed and there were 4,515–6,000 individual moths per hectare (Zhong, 2008).

Another aspect of the use of biodiversity to combat pests is the identification and use of compounds from the plant kingdom that influence pest biology (Koul, chapter 6 of this volume). In an example of a Chinese study, Peng and Pang (2004) found that alcohol extracts of from the plants *Amaranthus retroflexus* L., *Rubia tinctorum* L., *Calystegia hederacea* (Wallich), *Scirpus wallichii* (Nees), and *Stepkania longa* (Lour.) repelled *P. xylostella*. The non-alkaloid extracts from *Tripterygium wilfordii* (Hook) had a strong antifeedant and growth inhibition effect on *P. xylostella* (Xu *et al.*, 2006). Zhong *et al.* (2005) investigated nine ecological measures to control *P. xylostella* based on cultural and biological control methods. In field studies, sex pheromones and the release of the egg parasite *Trichogramma confusum* Viggiani, supplemented with other biocontrol agents, was found to be an effective regime for the control of *P. xylostella*.

CONCLUSION

Pest damage is one of the main factors affecting crop production. The development of plant protection technologies aims to protect the economic and sustainable production of crops. Pests are components of the eco-system. Eliminating pests completely is rarely possible and – in the case of native species – may lead to disorder within food webs. The concept of China's 'Green Plant Protection' highlights a strategy for plant protection that is more ecologically rational than the use of synthetic pesticides.

Three examples of green protection implementation were outlined above. These employ specific technologies that range from well-developed methods that are well known overseas (e.g. sex pheromone traps and coloured sticky traps) to others that are less widely used internationally in field crop pest management (e.g. light traps) or are still in the process of being developed from biological phenomena into practicable pest management strategies. This broad field continues to be a very active area of applied research. Key factors in China's pursuit of these non-pesticidal approaches are the national-level 'Public Plant Protection, Green Plant Protection' policy and 'Law of the People's Republic of China on Agricultural Product Quality Safety' and the coordinating role of the NATESC in the Ministry of Agriculture. The Chinese government pays great attention to pest management and promotes the PSMP. With the ongoing creation of new technologies and concerted efforts to have these rapidly adopted over large areas, the outlook for food security in the world's most populous country is made more favourable.

ACKNOWLEDGEMENTS

The authors would like to thank Professor Cheng Jia'an, Zhejiang University, China and Dr G.M. Gurr, Charles Sturt University, Australia for their kind suggestions and critical revision on the draft. We are most grateful to Dr Han Baoyu of China Jiliang University, Mr Wu Jiangxing, the Director of Ningbo Plant Protection Station, and Mr Wang Guorong, the Deputy Director of Xiaoshan Plant Protection Station, China for providing invaluable pictures.

REFERENCES

Chen, W., Isman, M.B. and Chiu, S.F. (1995) Antifeedant and growth inhibitory effects of the *Limonoid toosendanin* and *Melia toosendan* extracts on the variegated cutworm, *Peridroma saucia*. *Journal of Applied Entomology*, 119, 367–370.

Cheng, Y.Q., Sun, Y.F., Xia, L., Sun, B. and Miao, H.Y. (2007) Evaluation of effect of the biocide and the pesticide stemming from botany eliminating *Locusta migratoria manilensis* Meyen. *Chinese Agricultural Science Bulletin*, 23, 262–264 (in Chinese with English abstract).

Chiu, S.F. (1989) Recent advances in research on botanical insecticides in China, in *Insecticides of plant origin* (eds J.T. Aranson, B.J.R. Philogene and P. Morand), American Chemical Society Symposium Series 387, Washington DC, pp. 69–77.

Fan, X.J. (2006) Speech on the meeting of plant protection of China. *China Plant Protection*, 26, 5–13 (in Chinese).

Fan, Y.X., Gu, H.P. and Xu, J.H. (2010) Extension and application of ecological control on organic rice pests. *Shanghai Agricultural Technology*, 3, 57–59 (in Chinese).

FAO (2011) *China*. Food and Agriculture Organization of the United Nations, http://www.fao.org/countries/55528/en/chn/ Accessed July 26, 2011.

Fu, J.W., Xu, D.M., Wu, W. and You, M.S. (2005) Preference of different vegetable insect pests to color. *Chinese Bulletin of Entomology*, 42, 532–533 (in Chinese with English abstract).

Guo, Y.Y. (1998) Review of China IPM research and prospect of its development in 21st century. *Plant Protection*, 24, 35–38 (in Chinese).

Hazarika, L.K., Bhuyan, M., and Hazarika, B.N. (2009) Insect pests of tea and their management. *Annual Review of Entomology*, 54, 267–284.

Ji, R., Yuan, H., Xie, B.Y., Li, Z. and Li, D.M. (2007) Oviposition site selection by *Locusta migratoria manilensis* in the coastal locust areas. *Chinese Bulletin of Entomology*, 44, 66–68 (in Chinese with English abstract).

Jiang, A.Q., Yu, Q.L., Jin, J.Z., Ying, H.J. and Liu, X. (2009) Development and domestic marketing strategies of organic tea in China. *China Tea*, 22–24 (in Chinese).

Jin F., Wang J., Shao H. and Jin M. (2010) Pesticide use and residue control in China. *Journal of Pesticide Science*, 35, 138–142.

Koul, O. (2008) Phytochemicals and insect control: an antifeedant approach. *Critical Reviews in Plant Sciences*, 27, 1–24.

Lei, Z.R. and Wen, J.Z. (2004) Advance in locust control with *Metarhizium anisopliae*. *China Plant Protection*, 30, 14–17 (in Chinese with English abstract).

Li, J.H. (2010) Experiments of locust control by raising chicken and ducks. *Xingjiang Technology of Farms and Land Reclamation*, 2, 56 (in Chinese).

Liu, S.S. and Shi, Z.H. (1996) Recent developments in research and utilization of *Trichogramma* spp. *Chinese Journal of Biological Control*, 12, 78–84 (in Chinese with English abstract).

Liu, F.J., Zeng, M.S., Wang, Q.S. and Wu, G.Y. (2010) Control effects of ecological attractive cards on the *Empoasca. flavescens* and *Aleurocanthus spiniferu*. *Tea Science and Technology*, 2, 4–6 (in Chinese).

Ma, K. (ed.) (2001) *An introduction to horticulture*, High Education Press, Beijing, pp. 211–218 (in Chinese).

MOA (2002) Announcement of Ministry of Agriculture of China, No. 199.

MOA (2007) Announcement of Ministry of Agriculture of China, No. 274.

MOA (2010) Fufillment of the Oriental Locust Control in summer 2010 in China, http://www.cropipm.natesc.gov.cn/Html/2010_07_19/104022_105116_2010_07_19_142121.html (in Chinese).

Pan, C.X. (1988) The development of integrated pest control in China. *Agricultural History*, 62, 1–12.

Peng, Y.F. and Pang, X.F. (2004) Studies on repelling action of extracts from 9 species of plants against *Plutella xylostella* L. *Shandong Agricultural Sciences*, 4, 40–43 (in Chinese with English abstract).

Peng, G.X., Wang, Z.K., Yin, Y.P., Zeng, D.Y. and Xia, Y.X. (2008) Field trials of *Metarhizium anisopliae* var. *acridum* (Ascomycota: Hypocreales) against oriental migratory locusts, *Locusta migratoria manilensis* (Meyen) in Northern China. *Crop Protection*, 27, 1244–1250.

Qi, S.C., Jun, L.J. and Zhong, X.H. (2005) Using the Jiaduo insect killer lamp to monitor and control tea pests. *Chinese Bulletin of Entomology*, 42, 324–325 (in Chinese with English abstract).

Stern, R.L., Smith, R.F., van den Bosch, R. and Hagen, K.S. (1957) The integrated control concept. *Hilgardia*, 29, 81–101.

Shen, Y., Wang, J.Y., Miao, Y. *et al.* (2006) Studies on the niche of *Artogeia rapae* and its predatory enemies. *Chinese Agricultural Science Bulletin*, 22, 437–439 (in Chinese with English abstract).

Wang, Z.M. (2009) Effects of blue attractive cards on pepper pests. *Agro-technology and Service*, 26, 70 (in Chinese).

Wang, X.M., Hua, L., Wei, J.L., Cheng, Y.Q. and Niu, Y.H. (2008) On black light lamp control efficacy to field orchard pest. *Agricultural Research in the Arid Areas*, 26, 253–256 (in Chinese with English abstract).

Wei, Q., Kong, D.S., Hui, X.H. and Zuo, X.F. (2008) Extension reports of the frequency trembler grid lamp applied in the green plant protection. *Modern Agricultural Technology*, 9, 98–100 (in Chinese).

Wu, F.L. (1951) *Locusts of China*, Yong Xiang Press, Shanghai, pp. 17–19 (in Chinese).

Xia, J.Y. (2008) Outbreaks of major agricultural pests and the control achievements in China. *China Plant Protection*, 28, 5–9 (in Chinese).

Xia, J.Y. (2010) Development and expectation of public and green plant protection. *China Plant Protection*, 31, 5–9 (in Chinese).

Xia, J.Y. (2011) Extension of the green plant protection technology. *Pesticide Today*, 3, 20–24 (in Chinese).

Xiao, Y., Huang, S.Q., Ding, J. and Shi, F.C. (2008) Microbial insecticides. *Bt. Science and Technology of Sichuan Agriculture*, 4, 44–45 (in Chinese).

Xu, H.X., Chen, J.M., Lu, Z.X., Chen, L.Z. and Yu, X.P. (2006) Insecticidal activities of the non-allkaloid extracts from *Tripterygium wilfordii* against *Plutella xylostella*. *Chinese Journal of Eco-Agriculture*, 14, 179–181 (in Chinese with English abstract).

Yang, J.F. (2010) *Chinese tea industry industrial research report (2010)*, Social Science Academy Press (China), Beijing (in Chinese).

Yang, P.Y., Xiong, Y.K., Yi, Z. and Shan, X.N. (2010) Progress and prospect of the demonstration of green plant protection. *China Plant Protection*, 30, 37–38 (in Chinese).

Ye, H.X., He, X.M. and Han, B.Y. (2010) Difference in influence of intercropping of tea plants with waxberry and citrus and snakegourd fruit plants respectively on numeral and spatial characteristics of population of citrus spiny whitefly. *Journal of Anhui Agricultural University*, 37, 183–188 (in Chinese with English abstract).

Zeng, Z., Sun, Y.J., Qian, R.H., Ding, X.Z. and Xia, L.Q. (2008) Present Situation and Prospect of Microbial Pesticide Research and Application in China. *Research of Agricultural Modernization*, 29, 252–256 (in Chinese with English abstract).

Zhao, S.H. (ed.) (1983) *Plant Chemical Protection*, China Agricultural Press, Beijing (in Chinese).

Zhang, S.M., Liu, J.X., Wang, G.S., Li, H.Q. and Zhang, Z.B. (2006) Study on the effects of ecological control on *Locusta migratoria manilensis*. *China Plant Protection*, 26, 7–10 (in Chinese with English abstract).

Zhang, J.H., Ruan, Q.Y., Yu, Y.L., Ge, Q.L. and Zhang, X.R. (2009) Effects of ecological factors in Binhu locust zone on the density of *Locusta migratoria manilensis*. *Modern Agricultural Technology*, 2, 85–86 (in Chinese).

Zhang, H.H. (2004) *China tea insect pests and their management*, Anhui Science and Technology Press, Beijing (in Chinese).

Zhang, T.W., Wang, J.D., Liu, F., Xiong, H.L. (2010) Locust control by ecological control and raising ducks in Funan county. *Anhui Agricultural Science Bulletin*, 16, 120–121 (in Chinese).

Zhong, P.S. (2008) Mating disruption effect of sexual trap on diamondback moth, *Plutella xylostella* (L.). *Journal of Changjiang Vegetables*, 10b, 66–68 (in Chinese with English abstract).

Zhong, P.S., Liang, G.W. and Zeng, L. (2005) Studies on the effects of ecological measures against diamondback moth (DBM), *Plutella xylostella* (L.). *Acta Agriculturae Universitatis Jiangxiensis*, 27, 417–421 (in Chinese with English abstract).

Zhou, F.C., Du, Y.Z., Sun, W., Yao, Y.L., Qin, J.Y. and Ren, S.X. (2003) Impact of yellow trap to sweetpotato whitefly *Bemisia tabaci* (Gennadius) in vegetable fields. *Entomological Journal of East China*, 12, 96–100 (in Chinese with English abstract).

Zhu, E.L. (1999) *Occurrence and management of oriental migratory locust in China*, China Agriculture Press, Beijing, pp. 3–38 (in Chinese).

Chapter 15

DIVERSITY AND DEFENCE: PLANT–HERBIVORE INTERACTIONS AT MULTIPLE SCALES AND TROPHIC LEVELS

Finbarr G. Horgan

INTRODUCTION

Over the past 10,000 years the enhancement of a few plant species through artificial selection has produced food and fibre crops that permitted human societies to grow and prosper. This enhancement, together with manipulation of crop habitat by the controlled addition of nutrients and water, the levelling of ground, and the clearing of weedy competitors, has made these same crops highly attractive for a few problematic insect herbivores. It appears that the very process of selecting plant species for large seed, heavy fruits and tubers, or fast growth rates, has, for many crops, all but left their ability to defend against herbivores in some distant ancestral past. This erosion of resistance may be due to the incompatibility of some defences (thorns, toxins, stinging trichomes) with desired crop traits or because of inherent trade-offs between defence and other plant functions (Simms, 1992; Rosenthal and Dirzo, 1997; Strauss *et al.*, 2002; Cipollini *et al.*, 2003). For decades scientists and breeders have tried to reverse this situation by seeking out plants (either varieties, wild relatives or crop ancestors) with high anti-herbivore resistance and by incorporating these into breeding programmes (Painter, 1951; Crute and Dunn, 1980; Jackson, 1997). This has produced varieties that are avoided by insect herbivores and sustain less damage compared to other varieties in the field (Crute and Dunn, 1980; Jackson, 1997).

Since the early 1990s, plants have been produced by genetically engineering crops to express anti-herbivore bacterial toxins, plant lectins and modified lectins or plant-derived protease inhibitors. One of these insecticidal traits, the production of *Bacillus thuringiensis* (*Bt*) toxins, is now perhaps the most geographically widespread crop anti-herbivore mechanism on the planet (Christou *et al.*, 2006). Transgenic anti-herbivore resistance in general has been the subject of a recent explosion in research articles, often at the expense of research into other areas of pest management (Figure 15.1). Nevertheless, transgenic crops have few advantages over conventionally bred crops and are often less attractive for agriculture because of public distrust, environmental concerns, and practical restrictions on their deployment (Benbrook, 2001; Chen *et al.*, 2011; Domingo and Bordonaba, 2011). Furthermore, both conventionally bred and genetically engineered crops can have associated physiological costs for the plants (Kalazich and Plaisted, 1991; Xia *et al.*,

2010) and are generally limited by the capacity of insects to adapt and overcome resistance (Gould, 1998; Rausher, 2001; Zhao *et al.*, 2005; Storer *et al.*, 2010).

The application of anti-herbivore resistance relies on a sound knowledge of interactions between plants and their environment and full consideration of the evolutionary dynamics occurring between plants and insect communities in farmers' fields. These dynamics are governed by interactions at multiple trophic levels (Price, 1986; van Emden, 1986; Gould, 1991). It is apparent that plants have evolved in diverse and intermittently stable ecosystems such that their defences have not only been determined by direct plant–herbivore coevolutionary responses, but have emerged under the influence of complex direct and indirect interactions with organisms at higher trophic levels, including predators, parasites and diseases (van Emden, 1986; Agrawal, 2000; Núñez-Farfán *et al.*, 2007). This chapter examines the nature and complexity of interactions between plants, herbivores and their natural enemies, and whether anti-herbivore resistance can sustainably protect and enhance biodiversity in modern agroecosystems, or conversely whether it might inadvertently reduce biodiversity. The chapter also examines how the natural enemies of herbivores can protect resistant crops against the selection of virulent herbivore populations and how host-plant resistance might be better integrated into sustainable strategies for crop protection and biodiversity conservation.

RESISTANCE, COEVOLUTION, AND THE RED QUEEN'S RACE

The identification and use of herbivore-resistant host plants dates back several centuries. Among the earliest successes were resistance to Hessian fly *Mayetiola destructor* (Say), in wheat, *Triticum* spp. (1792), resistance to woolly apple aphid, *Eriosoma lanigerum* (Hausmann) in apple, *Malus* spp. (1831), and resistance to grape phylloxera, *Daktulosphaira vitifoliae* (Fitch), in grapes, *Vitis* spp. (*c.* 1860) (Smith, 2005 and references therein). Interest in the potential use of host-plant resistance against herbivores has grown considerably since the early 1900s with two notable jumps during the 1960s and 1990s at the beginnings of the 'green' and 'gene' revolutions respectively. Despite the differences in methods between these 'revo-

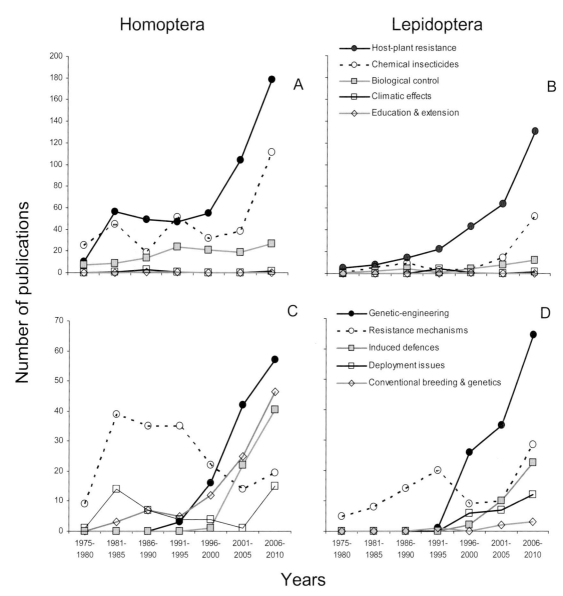

Figure 15.1 Number of scientific papers published between 1975 and 2010 concerning management of planthoppers (Homoptera) and caterpillars (Lepidoptera) on rice. A and B indicate research trends in five topics related to pest control/ management for Homoptera and Lepidoptera respectively; C and D indicate trends in five sub-topics related to host-plant resistance research for Homoptera and Lepidoptera respectively. Note the increased attention to molecular and genetic aspects of resistance since the 1990s and the absence of research into conventional resistance against caterpillars.

Functional categories

Examples of mechanisms

Direct

Indirect

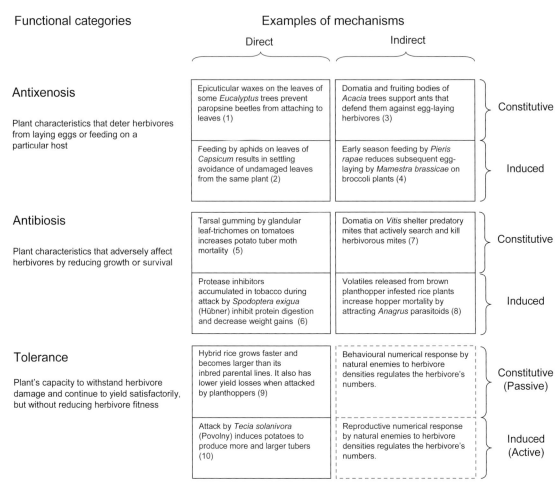

Figure 15.2 Functional categories and mechanisms of host-plant resistance and plant-insect interaction. Numbers in parentheses: 1 Edwards, 1982, 2 Prado and Tjallingii, 2007, 3 Janzen, 1966, 4 Poelman *et al.*, 2008, 5 Pelletier *et al.*, 2011, 6 Steppuhn and Baldwin, 2007, 7 Walter, 1996, 8 Lou *et al.*, 2005, 9 Horgan, unpublished, 10 Poveda *et al.*, 2010. Dashed-lines indicate a conceptual framework for understanding potential benefits of tolerance vis-à-vis natural pest regulation.

lutions', many of the basic underlying paradigms have been conserved (Smith, 2005). However, one important paradigm shift has been the separation of antixenosis (a plant's ability to deter herbivore attack) and antibiosis (a plant's ability to reduce the fitness of an attacking herbivore) from tolerance (a plant's ability to withstand herbivore attack by, for example, compensatory growth) as functional categories of plant–herbivore interaction (Strauss and Agrawal, 1999; Smith, 2005) (Figure 15.2). Painter (1951), one of the pioneers of host-plant resistance for agriculture, had originally regarded these as equal 'resistance mechanisms'. But 'mechanisms' are more correctly regarded as the complex chemical, physical and ecological processes underlying plant resistance and tolerance. As indicated in Figure 15.2, mechanisms can be constitutively or inductively expressed, and both resistance and tolerance can directly or indirectly affect target herbivores. Figure 15.2 indicates some of the diverse mechanisms underlying natural host-plant protection where each mechanism, whether direct or indirect, is ultimately determined by the genetics of the plant.

Until recently, resistant varieties generally each contained only a single major resistance gene. Hence early rice varieties, *Oryza sativa* L., with resistance to the brown planthopper, *Nilaparvata lugens* (Stål), contained either the *Bph1, bph2, Bph3* or *bph4* resistance genes (upper case denoting a dominant gene and lower case a recessive gene) (Alam and Cohen, 1998); sorghum, *Sorghum bicolor* L., resistant to greenbug, *Schizaphis graminum* (Rodani), contained either *Gb2, Gb3* or *Gb4* (Porter *et al.*, 1997); and wheat resistant to the Russian wheat aphid, *Diuraphis noxia* (Mordvilko), was based on nine *Dn*-genes, in several monogenic-resistant varieties (Haley *et al.*, 2004). Monogenic varieties of rice and sorghum lasted from two to five years under high herbivore pressure before the emergence of virulent (adapted) pest populations that were capable of feeding and reproducing on these plants to the same degree as on susceptible varieties (Porter *et al.*, 1997; Alam and Cohen, 1998) (Plate 15.1; Box 15.1). Meanwhile, whereas monogenic resistance to Russian wheat aphid initially appeared relatively stable, in 2004, a highly virulent population of the aphid emerged in Colorado, USA, capable of feeding on monogenic varieties with all but one of the nine *Dn*-genes available at that time (Haley *et al.*, 2004). Agricultural research institutes have generally responded to such virulence shifts by releasing new monogenic-resistant varieties with different resistance genes. This sequential release of resistant varieties following pest adaptation is often depicted as 'running the Red Queen's race' (Box 15.1). A recent study by Cambron *et al.* (2010) indicates that only about 25% of 33 genes for resistance to Hessian fly are still effective in the southern USA following the sequential release of several monogenic resistant varieties.

Because of the rapid pace of adaptation by insects to the large-scale deployment of monogenic resistant varieties, and with the advent of marker-assisted selection in the early 2000s, researchers have begun to include two or more resistance genes together in single crop varieties. This process has been called 'gene pyramiding' and is believed to broaden the spectrum of host-plant resistance, increase resistance strength, and consequently increase the durability of modern resistant varieties. This theory underlies the move from monogenic Bt-cotton, *Gossypium hirsutum*

Box 15.1 Plant breeding and the Red Queen's Race

The Red Queen hypothesis holds that for a given species, its effective environment is likely to include other species, such that an adaptation increasing the fitness of one species causes a decline in fitness of those species with which it interacts – these coevolutionary interactions give rise to continual natural selection for adaptation and counter-adaptation by interacting species (Stenseth and Smith, 1984). Therefore, in order to keep its place in the ecosystem, a species must continually adapt and change. This has been likened to the Red Queen character in Lewis Carroll's novel *Through the Looking Glass*, who explains to Alice:

> it takes all the running you can do, to keep in the same place. If you want to get somewhere else, you must run at least twice as fast as that!

The breeding and release of resistant host plants and the subsequent adaptation to these plants by target herbivores is a consequence of the Red Queen Race.

Two examples of such a sequence are indicated in tables A and B below.

A: Rice resistance to the brown planthopper

Variety	Released	Major gene	Breakdown
IR26	1973	*Bph1*	1975
IR36 + related	1976	*bph2*	1980s
IR56 + related	1982	*Bph3*	1990s

B: Sorghum resistance to greenbug

Variety	Released	Major gene	Breakdown
KS30, Shullu	1975	*Gb2*	1979
Capbam, P1220248	1985	*Gb3*	1990
Cargill 607E, Russian Pls	1990	*Gb4*	1992

L., which contains the Cry1Ac-gene alone, to pyramided plants containing two *Bt*-genes (Zhao *et al.*, 2005). In an experiment using *Bt*-transgenic broccoli, *Brassica oleracea* L., and synthetic populations of the diamondback moth, *Plutella xylostella* L., Zhao *et al* (2005) clearly demonstrated that pyramided transgenes can produce more durable resistance (Figure 15.3). Furthermore, by conducting a selection experiment where the proportions of monogenic-resistant, pyramided-resistant and susceptible-refuge plants were varied in greenhouse cages, the same authors indicated that the durability of pyramided *Bt*-plants was reduced where these occurred together with monogenic *Bt*-plants (Zhao *et al.*, 2005) (Figure 15.3). This example highlights one of the problems of deploying pyramided resistance in agricultural landscapes; in most cases, even with *Bt*-crops, it is difficult to eliminate monogenic-resistant varieties prior to deploying pyramided resistance.

The careful field deployment of anti-herbivore resistance varieties is essential to slow the pace of herbivore adaptation. Such deployment strategies are collectively termed 'resistance management' (MacIntosh, 2009). Prior to the advent of *Bt* crops, and despite the large-scale deployment of resistant varieties since the 1950s, little attention was given to resistance management (Figure 15.1) and many pests developed virulence against conventionally bred crop resistance (Porter *et al.*, 1997; Myint *et al.*, 2009; Cambron *et al.*, 2010). Several papers discuss the basic underlying principles of resistance management in detail (e.g. Gould, 1998; Rausher, 2001; MacIntosh, 2009). These principles are based on avoiding too widespread or long-term deployment of resistance and thus reducing selection pressure for virulence. This has resulted in the 'high dose/refuge' strategy that is widely proposed by suppliers of *Bt* crops (Gould, 1998; Rausher, 2001) and implemented at the point of seed sale under a supplier–client contract (MacIntosh, 2009). The 'high dose' refers to a strong level of resistance in the *Bt* plants that ensures high herbivore mortality. The 'refuge' refers to a 20% or greater area of susceptible host plants that reduces selection for resistance alleles because herbivores emerging from the refuge are expected to mate with any potential survivors from the *Bt* crop (Rausher, 2001). Overall, during the past two decades important lessons have been learned from the management of *Bt*-crop resistance in terms of the genetics and selection of herbivore populations, the design of refuges and the implementation of management actions. However, there are obvious logistical limitations to applying strict resistance management for the conservation of conventionally bred resistance, especially where publicly generated resistant lines and varieties are freely distributed and exchanged between research institutes or farmers (Teetes, 1994). The implementation of resistance management may also be more difficult in developing countries and among resource-poor farmers where the average size of farms is generally small (MacIntosh, 2009).

THE COSTS OF VIRULENCE UNDER THE INFLUENCE OF NATURAL ENEMIES

Herbivores that adapt to resistant plants may be constrained by fitness costs. In such cases, when the selection pressure is removed these virulent individuals should gain no advantage from their adaptations and will be out-competed by non-adapted conspecifics during intraspecific interactions. This is the rationale for removing selection pressure when virulence is high. For example, the sale of *Bt*-maize, *Zea mays* L., in Puerto Rico was temporarily stopped in response to high levels of virulence in *Spodoptera frugiperda* (Smith) in 2006 (Storer *et al.*, 2010). A similar removal of conventionally bred resistant varieties is difficult to achieve where herbivores have adapted to host-plant resistance, accumulating virulence against different genes over time (Myint *et al.*, 2009; Cambron *et al.*, 2010). One could argue that this indicates an absence of fitness costs associated with such virulence. However, the rarity of alleles for virulence against both conventional and transgenic resistance at the time of initial variety release suggests that virulence does indeed have costs. Some of these costs are mediated through natural enemies. Gassmann *et al.* (2009) found that the entomopathogenic nematodes *Steinernema riobrave* (Cabanillas) and *Heterorhabditis bacteriophora* (Poinar) reduced fitness in recessive and heterozygous strains respectively of *Bt*-resistant pink bollworm, *Pectinophora gossypiella* (Saunders). In the presence of natural enemies or competitors such fitness costs are expected to slow the rate of virulence build-up.

NATURAL ENEMIES: THE GUARDIANS OF RESISTANCE

In the context of biodiversity interactions, it is useful to consider the condition of those herbivores that

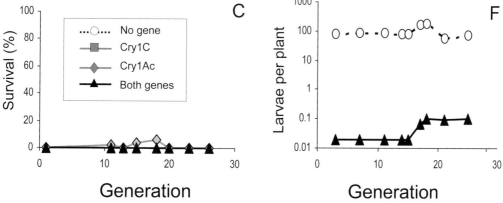

Figure 15.3 Adaptation by the diamondback moth (*Plutella xylostella*) to monogenic (Cry1Ac or Cry1C) and pyramided (containing both genes) *Bt*-broccoli over 30 generations. Greenhouse cages were set up with (A,D) 45% Cry1Ac + 45% pyramided + 10% refuge (susceptible) plants, (B,E) 45% Cry1C + 45% pyramided + 10% refuge plants, or (C,F) 90% pyramided + 10% refuge plants. The evolution of resistance to the plants is indicated in A, B and C. Moth densities on each plant type throughout the experiment are indicated in D, E and F. Graphs are redrawn from Zhao *et al.* (2005).

survive on resistant plants. Evidence suggests that initially (particularly during the first generations on a resistant host) survivors may be deformed, are generally small, and have a low fecundity. Furthermore, they are often lethargic compared to insects reared on susceptible varieties, and may actively seek improved food sources by regularly changing feeding locations (Gowling and van Emden, 1994; Alam and Cohen, 1998; Horgan et al., 2007). These physiological and behavioural changes can increase the vulnerability of herbivores to natural enemies on resistant plants. Longer development times of the vulnerable larvae and nymphs on resistant hosts may further increase exposure to natural enemies. There are several examples of higher predation rates on resistant compared to susceptible plants, including higher mortality of Russian wheat aphid due to parasitoids, Diaeretiella rapae (McIntosh), and lacewings, Chrysoperla plorabunda (Fitch), on resistant wheat (Farid et al., 1998; Messina and Sorenson, 2001) and higher predation of brown planthopper by a mirid bug, Cyrtorhinus lividipennis (Reuter) and spiders (e.g. Lycosa pseudoannulata (Boesenberg and Strand)) on resistant rice (Kartohardjono and Heinrichs, 1984). Evidence also suggests that the host plant can influence the herbivore's immune system and, therefore, its susceptibility to disease (Ojala et al., 2005). Therefore the potential for herbivores that survive on resistant plants to reproduce and establish new populations may be greatly reduced by their comparatively higher vulnerability to predators, parasitoids and diseases. This slows adaptation. The dynamics of herbivore–natural enemy interactions at the early stages of virulence build-up are complex. Alternative prey items can reduce the potential for natural enemies to prolong resistance. In a complex system where the 12-spotted lady beetle, Coleomegilla maculata (Lengi) attacks eggs of Colorado potato beetle, Leptinotarsa decemlineata (Say) and European corn borer, Ostrinia nubilalis (Hübner), as well as nymphs and adults of Myzus persicae (Sulzer), simulation models predict that the presence of corn borer eggs delayed Bt-resistance development in potato beetle by up to 40 generations. However, the presence of the aphid accelerated resistance evolution by 18 generations (Mallampalli et al., 2005). Gurr and Kvedaras (2010) give further examples of synergy between host-plant resistance and natural enemies in reducing herbivore population growth and indicate the potential to develop new strategies for biological control based on knowledge of such interactions.

Insecticides that reduce natural enemy densities, but allow the survival of herbivores, are expected to accelerate the adaptation of herbivores to resistance genes (Gallagher et al., 1994). Therefore, problems arise where pesticides are used on resistant plants. This can occur when farmers attempt to control species that are not the target of the specific host-plant resistance, or where farmers are unaware of the specific resistance that the plants possess. A further problem arises where the specific host-plant resistance causes increased mortality of natural enemies (Obrycki, 1986). This in effect will protect emerging virulent populations. For example, the parasitoid Copidosoma koehleri (Blanchard) makes fewer and shorter visits to plants with high densities of sticky leaf-trichomes, resulting in lower parasitism of potato tuber moth (Phthorimaea operculella (Zeller)) eggs on resistant plants (Baggen and Gurr, 1995; Gooderham et al., 1998). This creates an 'enemy-free space' for the moth and results in higher moth densities in crops with dense trichomes (such as tomatoes, Solanum lycopersicum L.) compared to those with fewer trichomes (such as potatoes, Solanum tuberosum L.) (Mulatu et al., 2004).

Several studies have examined the potentially negative effects of transgenic crops, particularly Bt crops, on natural enemies, which, if they occur, would accelerate resistance evolution. Literature on this subject has become extensive and is often polemic. A number of recent reviews and opinion pieces give interesting summaries of the extent, procedures and outcomes of this research (Lövei and Arpaia, 2005; Marvier et al., 2007; Shirai, 2007; Lövei et al., 2009; Shelton et al., 2009; Wolfenbarger et al., 2008; Gatehouse et al., 2011). Recently there has been a tendency to move away from simple bi-trophic studies of the effects of transgenic crops on non-target organisms to more complicated tri-tropic studies of their effects on natural enemies as mediated through the target herbivores. Many of these studies suggest zero effects of insecticidal proteins or protease inhibitors at higher trophic levels (Shirai, 2007); however, the choice of natural enemy can determine the outcome of the evaluation. For example, the effects of protease inhibitors on natural enemies, as mediated through the target herbivore, depend on the nature of the inhibitor expressed by the crop and whether the same or different proteases are used by the herbivore and its natural enemy (Gatehouse et al., 2011). The simple methods used in laboratory-based risk evaluation, and disagreement over the appropriateness of the statistical analyses employed, cast doubt on the meaning and value of small-scale and highly artificial risk-assessment procedures (Lövei and Arpaia, 2005; Lövei *et al.*, 2009; but see Shelton *et al.*, 2009). More recently, improved field-scale evaluations have been possible and have indicated positive effects of transgenic crops on certain predators compared to conventional pest control methods (Gatehouse *et al.*, 2011). However, it should be noted that 'conventional pest control' normally implies 'business as usual' with prophylactic insecticide inputs. A more appropriate comparison should include farms and crops employing more ecologically based pest management methods. In such cases the comparative effects of *Bt* crops on non-target biodiversity will have more useful meaning (Marvier *et al.*, 2007). Such comparative studies will also need to address mechanisms more carefully. For example natural resistance of hybrid cottonwoods (*Populus angustifolisa* (James) × *Populus fermontii* (Watson)) to leaf-galling aphid (*Pemphigus betae* (Doane)) affects species at multiple trophic levels and reduces the species richness and abundance of natural enemies through cascade effects (Dickson and Whitham, 1996). Such cascade effects are understandable and acceptable in the context of field crops. Concerns should arise where clear associations have been drawn between biodiversity loss and the deleterious toxic effects of transgenic crops on non-target beneficial species. In general, there is still a need to increase the complexity of risk-assessment hypotheses to reflect the multiple species interactions and governing processes in crop ecosystems.

Optimally, resistant crops, whether transgenic or conventionally bred, should work together with natural enemies to reduce pest densities and cause greater mortality to pests than when either is working alone (Price, 1986; van Emden, 1986; Bell *et al.*, 2001; Faria *et al.*, 2007; Gurr and Kvedaras, 2010). In some cases, the resistant plant might actually promote greater survival and efficiency of natural enemies (e.g. survival of the coccinellid predator, *Scymnus frontalis* (Fabricius) is higher on wheat resistant to the Russian wheat aphid compared to susceptible wheat. This is possibly because of lower levels of entrapment in rolled leaves produced after feeding by the aphid (Farid *et al.*, 1997)). In an interesting twist on the role of secondary pests, Faria *et al.* (2007) have related increased performance of corn leaf aphid, *Rhopalosiphum maidis* (Fitch) on *Bt* maize to higher parasitism of *Spodoptera littoralis* (Boisduval) by *Cotesia marginiventris* (Cresson). Honeydew

produced by the aphids allowed *C. marginiventris* to live longer and parasitise more caterpillars than on near-isogenic maize varieties (that is, the same varieties but without the Cry1Ab gene). Such improved survival and efficiency of natural enemies on resistant varieties could be particularly beneficial at low pest densities where, if natural enemies are to effectively slow down the build-up of virulent populations, predation and parasitism must be effective. This poses a problem where regulatory predators respond strongly to the density of a specific prey item or are specialised to feed on a narrow range of herbivore species, and suggests that the best guardians of resistance may be generalist predators such as spiders or ants, or omnivores such as crickets, which are buffered against low-density populations of the target species by using alternative food sources (Settle *et al.*, 1996). However, whether generalists are more efficient than specialist natural enemies is still debatable. Several laboratory and greenhouse studies have monitored insect adaptation to resistant varieties (Alam and Cohen, 1998; Zhao *et al.*, 2005) but surprisingly little is known of the early dynamics of resistance adaptation in farmers' fields where the build-up of virulence is influenced by communities of interacting generalist and specialist natural enemies and by farm management practices.

GENETIC DIVERSITY REDUCES CROP VULNERABILITY TO HERBIVORES

The overwhelming majority of host-plant resistance research has focused on interactions between specific plants, generally varieties or clones, and specific insect herbivores, usually model species from highly inbred laboratory colonies. With the ascent of modern molecular research tools, the focus has become increasingly narrow (Figure 15.1). This is unfortunate, since herbivore–plant interactions normally play out at landscape and regional levels where genetically diverse herbivore populations encounter an equally diverse range of crops and varieties. Some important insect pests undergo seasonal migrations that cover thousands of kilometres in a few days, during which their host varieties are likely to change considerably. It is well known that crop genetic diversity, even at field scales, can suppress pathogens, buffer against virulence (pathotype) development, and generate higher yields (Finckh *et al.*, 2000; Leung *et al.*, 2003; Newton *et al.*, 2009). Compared to the current gene-for-gene paradigm that permeates modern anti-herbivore resistance, this 'polygenic' or 'landscape' approach lies at the opposite extreme of the resistance research spectrum, and in the case of herbivores continues to receive minimal attention. Nevertheless, results from plant pathology have been impressive and a range of strategies has emerged to incorporate crop genetic diversity into cereal production in particular, with noted successes of variety mixtures against airborne diseases of wheat in the USA, against fungi in Germany, and against rice blast in China (Finckh *et al.*, 2000; Leung *et al.*, 2003).

Because of the rapid adaptation by insects to resistance genes, strategies have been proposed whereby resistant varieties are deployed together with susceptible or otherwise genetically different varieties either as seed mixes or refuge crops. This reduces selection pressure as described above, but also spreads the risk of herbivore damage and severe crop losses (Hajjar *et al.*, 2008). Furthermore, varietal mixes can improve the efficiency of natural enemies by increasing habitat favourability; for example, the seven-spotted lady beetle, *Coccinella septempunctata* L., is more highly attracted to mixed odours of barley, *Hordeum vulgare* L., than to odours from a single variety (Ninkovic *et al.*, 2011). Advances in marker-aided selection have allowed breeders to develop multilines as a means of achieving the benefits of seed mixtures, but without incurring problems due to variable yields, phenologies, or management requirements (e.g. harvesting) between the different varieties. Multilines are near-isogenic lines that differ in the anti-herbivore resistance genes they possess. Because they usually share about 90% of genes, they are generally morphologically indistinguishable and physiologically similar (Fujita *et al.*, 2010). One potential disadvantage of multilines is that they still depend on major resistance genes, although little is generally known about the exact function of these genes, and they are therefore potentially prone to cross-resistance adaptation by target herbivores.

Crop genetic diversity as a means of reducing vulnerability to herbivores may depend on the host's genetic background, rather than specific single-gene resistance sources. Pests and diseases are expected to adapt differentially to the diverse genetic backgrounds they encounter in their host plant, effectively delaying adaptation to any major resistance mechanisms. Observations of differential herbivore adaptation to host plants are apparent from traditional biotype studies, particularly where insects perform poorly on a novel host genotype, irrespective of the presence or absence of any major resistance genes (Box 15.2). This suggests that increasing the genetic diversity of crops could also reduce yield losses due to certain insect pests and may be a fruitful future research avenue for the sustainable management of herbivorous insects. Unfortunately, one worrisome trend in modern agriculture has been the narrowing of crop genetic diversity (Cavanagh *et al.*, 2008; Beeck *et al.*, 2008; Shu *et al.*, 2009). This potentially increases host-plant susceptibility by decreasing levels of differential adaptation. For example, since the early 1970s, a single wild-abortive cytoplasmic male sterile line (WA-CMS) has dominated Chinese hybrid rice development. Currently, about 20 million hectares of hybrid rice is planted in China and 3 million hectares in Southeast Asia, of which over 40% is derived from WA-CMS (Cheng *et al.*, 2007). Therefore, migratory insects such as planthoppers (Delphacidae: Homoptera) and leafrollers (Pyralidae: Lepidoptera) can find genetically similar, or the same, rice varieties throughout the course of their 1,000-km migrations, substantially depleting the protection normally afforded through differential adaptation. Recently, novel approaches to crop breeding have been initiated to address the erosion of crop genetic diversity: new genetic approaches including introgression libraries (ILs), multi-parent advanced generation intercross (MAGIC) populations, and association genetics have the potential to increase the diversity of sources of resistance and the range of genes and quantitative loci that protect crops against herbivores (Cavanagh *et al.*, 2008). Meanwhile, recurrent selection can potentially increase the genetic distances between the elite lines or varieties that emerge from crop breeding programmes. Recurrent selection refers to selecting for certain traits generation after generation with interbreeding of reselected plants (Beeck *et al.*, 2008). This could theoretically reduce crop vulnerability at landscape levels because of differential adaptation to genetically distinct varieties.

A resistant crop landscape might include a diversity of crops and interspersed natural vegetation, thereby reducing the extent of monoculture (and thus the dominance and availability of a few crop species) and providing habitat for natural enemies (Settle *et al.*, 1996; Bianchi *et al.*, 2010). As discussed above, increased crop genetic diversity will also reduce vul-

Box 15.2 Are herbivores choosy feeders?

Some insects feed and reproduce during several generations on a single host plant or variety. These might develop differential preferences for the specific plant type with which they are associated, creating a barrier to feeding on other plants with different genetic backgrounds. For example, in a study by Claridge and Den Hollander (1982) brown planthoppers from Australia had poor survival on the TN1 rice variety from Asia, but planthopper populations from the International Rice Research Institute (IRRI) in the Philippines had high survival on TN1 (A). Similarly, planthoppers collected on five rice varieties in Sri Lanka generally performed better when reared on the same variety from which they were collected, but often did poorly when moved to new varieties (Claridge et al., 1982) (B). Differential preferences in these cases were due to background plant genetics. These results suggest that, in a similar manner as with plant diseases, increasing crop genetic diversity could reduce planthopper densities.

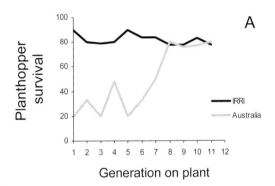

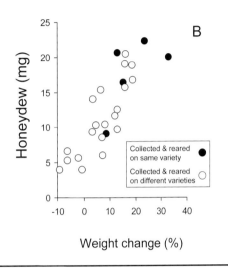

nerability to herbivores through differential adaptation. Crop heterogeneity may also be achieved at the landscape level through asynchronous crop planting. Even where plants are genetically homogenous, plant ontogeny can change the appearance and functional categories of standing crops (Boege et al., 2007). For herbivores with short generation times, it is conceivable that plants of different ages represent a barrier to inter-field dispersal. For example, in laboratory colonies, brown planthoppers reared successively on young rice plants become poorly adapted to older plants (personal observation). It is largely unknown how this might affect crop vulnerability. In fact, debate over the merits of asynchronous versus synchronous planting as a strategy for reducing crop vulnerability has not been resolved (Settle et al., 1996; Schoenly et al., 2010). It is likely that different herbivore species, and the viruses they transmit, will respond differently to cropping patterns.

PLANT DEFENCE WITH A LITTLE HELP FROM FRIENDS

Agroecosystems are generally more diverse than the planned crop components in which the farmer is interested. Crops have an associated flora and fauna that spills out from native habitat patches or has adapted to coexist with our crops. Some species, including most pests, but also some abundant natural enemies, are highly synanthropic (associated with human managed systems). These include many predatory carabid and staphylinid beetles and parasitoid wasps (Horgan and Myers, 2004; Nicholls et al., 2010).

Throughout evolution, natural enemies have been essential in protecting plants, and some plant species have gone to great lengths to enhance their protectors' populations or efficiency (Figure 15.2). Perhaps the most widely recognised example is that of the ant-acacias, *Acacia* spp. (Mimosoideae: Leguminosae) of Central America: these bear swollen thorns for ant (Pseudomyrmecinae: Formicidae) dwelling, as well as modified leaflet tips called Beltian bodies and enlarged foliar nectaries that provide food for the ants (Janzen, 1966). Similarly, dense foliar hairs or specialised tufts of hair, called domatia, on the leaf axils of several plants, including grapes and coffee, have been associated with the increased occurrence of predatory mites (Phytoseiidae: Acarina) that protect these plants against herbivores (Walter, 1996; Mineiro *et al.*, 2008). Defence mechanisms such as these are 'indirect' because they rely on other members of the ecological community.

Induced plant defences have gained considerable attention over the past 15 years, partly because of advances in molecular research tools such as micro arrays, and because of improved knowledge of plant biochemical pathways (Figure 15.1). In theory, induced defences will be less costly in terms of yield penalties than are constitutively expressed plant defences. Induced defences, both direct and indirect, are triggered by specific elicitors present in plant cells only at the time of herbivore attack. These may include saliva from the insect attacker or cell-wall components from damaged cells. Their effect is to alter the regulation of specific response genes resulting in the *de novo* synthesis of defensive chemicals through jasmonic acid (JA)- or salicylic acid (SA)-mediated biochemical pathways. The best-known examples of these include protease inhibitors that actively block herbivore digestion (Howe and Jander, 2008). Relatively less well studied are the induced volatile substances that indicate that the plant has been damaged by a specific herbivore. These can function to reduce subsequent attack by the same herbivore species (Figure 15.2). For example, many aphid species avoid settling on the undamaged leaves of their host plant after an initial inductive infestation on different leaves of the same plant (Prado and Tjallingii, 2007). Such induced volatiles can also reduce attacks by different herbivore species (e.g. the generalist herbivore *Mamestra brassicae* L. avoids broccoli plants that have been previously attacked by *Pieris rapae* L. (Poelman *et al.*, 2008)). Furthermore, several studies have shown that insect natural enemies locate their prey by responding to specific volatile cues and are not directly attracted to the prey species. For example, parasitism of brown planthopper eggs by *Anagrus nilaparvata* (Pang and Wang) and predation of eggs and nymphs by *C. lividipennis* is higher on JA-induced or herbivore-damaged plants than on control plants (Rapusas *et al.*, 1996; Lou *et al.*, 2005).

TOLERANCE: A SMALL PRICE FOR RELIABLE PROTECTION

The role of herbivory tolerance in plant defence has been overlooked by ecologists and agriculturalists alike until relatively recently (Strauss and Agrawal, 1999; Smith, 2005). Several studies suggest that tolerance is an inherent property of rapidly growing plants that is determined both by resource availability and plant genetics (Maschinski and Whitham, 1989; Cornelissen and Fernandes, 2008). Unlike resistance, tolerance is often a general response by the plant to a range of different herbivore species (Leimu and Koricheva, 2006). Tolerant plants do not affect the fitness and – importantly for sustainability – do not impose selection pressures on herbivores (Espinosa and Fornoni, 2006). Further, because tolerance allows low-density pest populations to persist without causing yield loss, specialist natural enemies are more likely to persist and offer a safeguard against instability of herbivore populations. As such, tolerance of low-level herbivore infestations may be an investment in plant protection that is indirect and potentially delayed. In Figure 15.2, an attempt has been made to separate tolerance into constitutive (or passive) and induced (or active) mechanisms. Constitutive tolerance is expressed throughout the plant's growth and is inherent to the plant's morphology and growth patterns, whereas induced mechanisms include physiological responses to damage such as increases in leaf photosynthetic activity or compensatory growth (Tiffin, 2000).

Further research is required to better elucidate the interactions between plant tolerance and regulatory ecosystem functions. Figure 15.2 proposes a conceptual framework that links plant tolerance and natural enemies through numerical responses to prey densities. In this framework, natural enemy populations that build up over time might be considered part of an induced regulation that is aided by tolerant plants.

Similarly, behavioural or numerical responses by natural enemies might be regarded as constitutive regulation that can be aided by behavioural cues such as honeydew excreted by planthoppers (Faria *et al.*, 2007). In this context, tolerance can be regarded as a plant's 'payment' for efficient regulation. In some cases this 'payment' may be in terms of plant biomass and not yield. In fact, plants can overcompensate in response to low levels of damage to produce even higher yields (Poveda *et al.*, 2010).

Tolerance does, however, have some potential disadvantages. Where natural enemies are absent or depleted, under excessive insecticide use or because a herbivore is invasive, pest populations on tolerant plants can build up to higher levels than would normally occur on relatively susceptible, comparatively low-tolerance varieties. Furthermore, because population density is often directly related to the extent of an outbreak area (Wallner, 1985), poor herbivore regulation on tolerant varieties could result in relatively large pest outbreaks. As fertiliser increases plant tolerance, excessive fertiliser applications can also potentially lead to larger pest outbreaks. The interactions between herbivores, natural enemies and tolerant crop varieties have so far received little research attention. Most crop varieties are never screened for tolerance, and damage or economic thresholds rarely consider varietal differences in tolerance levels. In terms of breeding for sustainable pest management, further attention should be focused on the merits and weaknesses of plant tolerance and on its optimal combination with resistance.

EXTENSION: A KEY COMPONENT OF HOST-PLANT RESISTANCE

Many research papers justify their focus on crop resistance with opening remarks on its utility in reducing the need for chemical insecticides. Consequently, host-plant resistance is expected to increase the diversity and abundance of natural enemies in crops and promote natural biological control. However, clear evidence that conventionally bred host-plant resistance has actually decreased insecticide use is extremely rare. Published studies on the effects of resistant crops on insecticide use have mainly been associated with *Bt* crops. Several studies have documented impressive reductions in insecticide use following the adoption by farmers of *Bt* crops (Christou *et al.*, 2006 and refer-

ences therein). However, *Bt* crops are carefully marketed through an effective extension system that plays a large role in their distribution, deployment and management. Therefore, comparative studies between adopter and non-adopter farms are often flawed by potentially confounding the effects of extension and the *Bt* technology per se (e.g. Heong *et al.*, 2005). In many cases, extension alone can produce similar reductions in insecticide use without the need for host-plant resistance (Benbrook, 2001; Escalada *et al.*, 2009). Meanwhile, as mentioned earlier, conventionally bred resistant lines and varieties are frequently absorbed into local crops without recognition of their resistance value, without guidelines for proper resistance management, or without informing farmers of their potential to decrease insecticide use (Teetes, 1994).

In the face of varying efforts and quality of extension, the main advantage of host-plant resistance may be limited to maintaining yields where pest management is weak or inefficient (Widawsky *et al.*, 1998). In effect, as stated by Pingali and Gerpacio (1997), insect- and disease-resistant varieties may serve only to reduce the productivity benefits and profitability of applying pesticides. Where resistance against resurgence and secondary pests is concerned, this 'business as usual' approach to crop management only prolongs the overuse of damaging insecticides while avoiding the consequences of depleted natural enemy populations. Evidence from the effects of *Bt* cotton supports the role of extension in reducing insecticide use on resistant varieties. For example, in some regions of China and India the failure of *Bt* cotton to reduce insecticide inputs has been associated with poor extension in remote areas and a proliferation of black-market varieties (Fitt, 2008). Similarly, *Bt* cotton in Australia has had a varying impact on insecticide use depending on the efficiency of extension services. According to Fitt (2008), the excessive use of insecticides on *Bt* cotton by some farmers, in spite of crop monitoring, is due to a poor understanding of the potential damage caused by different insects. As *Bt* crops receive less sociological and scientific scrutiny and as alternative seed options decrease, one might expect a future decline in extension efforts, which are expensive for both public and private institutions. This will lead to increased insecticide use on *Bt* crops, as it has for conventionally bred resistant varieties, particularly where farmers' pest-control decisions continue to be based on exaggerated perceptions of yield loss (Widawsky *et al.*, 1998; Rola and Pingali, 1993) and where the agrochemical industry continues to increase marketing efforts (Tilman et al., 2001). Therefore, although resistant crops can greatly improve ecosystem health and function, and promote biodiversity by reducing insecticide inputs, it appears that the extent of this reduction will depend on effective extension and education programmes, as well as cooperation from a reluctant agrochemical industry.

WHERE TO FROM HERE?

It is my hope that this chapter has stimulated more questions than it has provided answers. We have seen that crop genetics can determine the interactions between plants, herbivores and their natural enemies; however crop genetics should be regarded in a holistic manner: herbivores, predators and parasitoids alike respond to dozens of plant chemicals that are governed by several interacting genes, most of which are part of some poorly comprehended 'genetic background'. Decisions on crop genetics (genes, varieties and variety mixtures) can affect agroecosystem biodiversity at multiple trophic levels and drive population regulatory processes. On reviewing current research into pest management it becomes immediately apparent that since the early 1990s there has been a major shift toward narrow, molecular-based approaches for reducing crop vulnerability without regard to landscape and biodiversity issues. However, it should be clear that such 'gene-for-gene' approaches cannot be the only answer and, because of coevolution, are severely limited. As scientists we often look for solutions in the wrong places. For example, outbreaks of resurgence pests are a response to the overuse of pesticides, but science currently looks to host-plant resistance to 'solve the symptom'. It appears that very often the insect herbivores are less impressed with our new technologies than we are. The scientific community will need to restore the balance and focus also on landscapes, biodiversity and management issues and not only on molecular biology. One of the worrisome trends in modern agriculture is the erosion of crop diversity and of the genetic diversity of individual crop species. As a general rule, landscape homogeneity increases pest incidence. Increasing the genetic diversity of our crops, and not only the diversity of major resistance genes, is one way to reduce crop vulnerabil-

ity. But there is also a need to better understand and utilise the synergies between crop traits and natural biological control agents to promote natural pest regulation and slow herbivore adaptation to resistance. Above all, there is a need for crops and farms to become better integrated with nature and ecology: nature has provided all the solutions through negative feedbacks and regulatory processes, it is up to us to better understand and protect these important ecosystem functions.

REFERENCES

Agrawal A.A. (2000) Overcompensation of plants in response to herbivory and the by-product benefits of mutualism. *Trends in Plant Science*, 5, 309–313.

Alam, S.N. and Cohen, M.B. (1998) Durability of brown planthopper, *Nilaparvata lugens*, resistance in rice variety IR64 in greenhouse selection studies. *Entomologia Experimentalis et Applicata*, 89, 71–78.

Baggen, L.R. and Gurr, G.M. (1995) Lethal effects of foliar pubescence of solanaceous plants on the biological control agent *Copidosoma koehleri* Blanchard (Hymenoptera: Encyrtidae). *Plant Protection Quarterly*, 10, 116–118.

Beeck, C.P., Wroth, J.M., Falk, D.E., Khan, T. and Cowling, W.A. (2008) Two cycles of recurrent selection lead to simultaneous improvement in black spot resistance and stem strength in field pea. *Crop Science*, 48, 2235–2244.

Bell, H.A., Fitches, E.C., Down, R.E. *et al.* (2001) Effects of dietary cowpea trypsin inhibitor (CpTI) on the growth and development of tomato moth *Lacanobia oleracea* (Lepidoptera: Noctuidae) and on the success of the gregarious ectoparasitoid *Eulophus pennicornis* (Hymenoptera: Eulophidae). *Pest Management Science*, 57, 57–65.

Benbrook, C. (2001) Do GM crops mean less pesticide use? *Pesticide Outlook*, 12, 204–207.

Bianchi, F.J.J.A., Schellhorn, N.A., Buckley, Y.M. and Possingham, H.P. (2010) Spatial variability in ecosystem services: simple rules for predator-mediated pest suppression. *Ecological Applications*, 20, 2322–2333.

Boege, K., Dirzo, R., Siemens, D. and Brown, P. (2007) Ontogenic switches from plant resistance to tolerance: minimizing costs with age? *Ecology Letters*, 10, 177–187.

Cambron, S.E., Buntin, G.D., Weisz, R. *et al.* (2010) Virulence in Hessian fly (Diptera: Cecidomyiidae) field collections from the southern United States to 21 resistance genes in wheat. *Journal of Economic Entomology*, 103, 2229–2235.

Cavanagh, C., Morell, M., Mackay, I. and Powell, W. (2008) From mutations to MAGIC: resources for gene discovery, validation and delivery in crop plants. *Current Opinion in Plant Biology*, 11, 215–221.

Chen, Z.H., Chen, L.J., Zhang, Y.L. and Wu, Z.J. (2011) Microbial properties, enzyme activities and the persistence of exogenous proteins in soil under consecutive cultivation of transgenic cottons (*Gossypium hirsutum* L.). *Plant Soil Environment*, 57, 67–74.

Cheng, S.H., Zhuang, J.Y., Fan, Y.Y., Du, J.H. and Cao, L.Y (2007) Progress in research and development on hybrid rice: a super-domesticate in China. *Annals of Botany*, 100, 959–966.

Christou, P., Capell, T., Kohli, A., Gatehouse, J.A. and Gatehouse, A.M.R. (2006) Recent developments and future prospects in insect pest control in transgenic crops. *Trends in Plant Science*, 11, 301–308.

Cipollini, D., Purrington, C.B. and Bergelson, J. (2003) Costs of induced resistance. *Basic and Applied Ecology*, 4, 79–85.

Claridge, M.F. and Den Hollander, J. (1982) Virulence to rice cultivars and selection for virulence in populations of the brown planthopper, *Nilaparvata lugens*. *Entomologia Experimentalis et Applicata*, 32, 213–221.

Claridge, M.F., Den Hollander, J. and Furet, I. (1982) Adaptations of brown planthopper (*Nilaparvata lugens*) populations to rice varieties in Sri Lanka. *Entomologia Experimentalis et Applicata*, 32, 222–226.

Cornelissen, T. and Fernandes, G.W. (2008) Size does matter: variation in herbivory between and within plants and the plant vigor hypothesis. *Oikos*, 117, 1121–1130.

Crute, I.R. and Dunn, J.A. (1980) An association between resistance to root aphid (*Pemphigus bursarius* L.) and downy mildew (*Bremia lactucae* Regel) in lettuce. *Euphytica*, 29, 483–488.

Dickson, L.L. and Whitham, T.G. (1996) Genetically based plant resistance traits affect arthropods, fungi and birds. *Oecologia*, 106, 400–406.

Domingo, J.L. and Bordonaba, J.G. (2011) A literature review on the safety assessment of genetically modified plants. *Environment International*, 37, 734–742.

Edwards, P.B. (1982) Do waxes on juvenile *Eucalyptus* leaves provide protection from grazing insects? *Australian Journal of Ecology*, 7, 347–352.

Escalada, M.M., Heong, K.L., Huan, N.H. and Chien, H.V. (2009) Changes in rice farmers' management beliefs and practices in Vietnam: an analytical review of survey data from 1992 to 2007, in *Planthoppers: new threats to the sustainability of intensive rice production systems in Asia* (eds K.L. Heong and B. Hardy), International Rice Research Institute, Philippines, pp. 447–457.

Espinosa, E.G. and Fornoni, J. (2006) Host tolerance does not impose selection on natural enemies. *New Phytologist*, 170, 609–614.

Faria, C.A., Wäckers, F.L., Pritchard, J., Barrett, D.A. and Turlings, T.C.J. (2007) High susceptibility of *Bt* maize to aphids enhances the performance of parasitoids of lepidopteran pests. *PloS ONE 7*, e600.

Farid, A., Johnson, J.B. Quisenberry, S.S. (1997) Compatibility of a coccinellid predator with a Russian wheat aphid resistant wheat. *Journal of the Kansas Entomological Society*, 70, 114–119.

Farid A., Johnson J.B., Shafii B. and Quisenberry S.S. (1998) Tritrophic studies of Russian wheat aphid, a parasitoid, and resistant and susceptible wheat over three parasitoid generations. *Biological Control*, 12, 1–6.

Finckh, M.R., Gacek, E.S., Goyeau, H. *et al.* (2000) Cereal variety and species mixtures in practice with an emphasis on disease resistance. *Agronomie*, 20, 813–837.

Fitt, G.P. (2008) Have *Bt* crops led to changes in insecticide use patterns and impacted IPM?, in *Integration of insect-resistant genetically modified crops within IPM programs* (eds J. Romeis, A.M. Shelton and G. Kennedy), Springer, Dordrecht, pp. 303–328.

Fujita, D., Atsushi, Y. and Hideshi, Y. (2010) Development of near-isogenic lines carrying resistance genes to green rice leafhopper (*Nephotettis cincticeps* Uhler) with Taichung 65 genetic background in rice (*Oryza sativa* L.). *Breeding Science*, 60, 18–27.

Gallagher, K.D., Kenmore, P.E. and Sogawa, K. (1994) Judicial use of insecticides deters planthopper outbreaks and extends the life of resistant varieties in Southeast Asian rice *Planthoppers: their ecology and management* (eds R.F. Denno and T.J. Perfect), Chapman & Hall, London, pp. 599–614.

Gassmann, A.J., Fabrick, J.A., Sisterson, M.S. *et al.* (2009) Effects of pink bollworm resistance to *Bacillus thuringiensis* on phenoloxidase activity and susceptibility to entomopathogenic nematodes. *Journal of Economic Entomology*, 102, 1224–1232.

Gatehouse, A.M.R., Ferry, N., Edwards, M.G. and Bell, H.A. (2011) Insect-resistant biotech crops and their impacts on beneficial arthropods. *Philosophical Transactions of the Royal Society B*, 1569, 1438–1452.

Gooderham, J., Bailey, P.C.E., Gurr, G.M. and Baggen, L.R. (1998) Sub-lethal effects of foliar pubescence on the egg parasitoid *Copidosoma koehleri* and influence on parasitism of potato tuber moth, *Phthorimaea operculella*. *Entomologia Experimentalis et Applicata*, 87, 115–118.

Gould, F. (1991) The evolutionary potential of crop pests. *American Scientist*, 79, 496–507.

Gould, F. (1998) Sustainability of transgenic insecticidal cultivars: integrating pest genetics and ecology. *Annual Review of Entomology*, 43, 701–726.

Gowling, G.R. and van Emden, H.F. (1994) Falling aphids enhance impact of biological control by parasitoids on partially aphid-resistant plant varieties. *Annals of Applied Biology*, 125, 233–242.

Gurr, G.M. and Kvedaras, O.L. (2010) Synergizing biological control: Scope for sterile insect technique, induced plant defences and cultural techniques to enhance natural enemy impact. *Biological Control*, 52, 198–207.

Hajjar, R., Jarvis, D.I. and Gemmill-Heren, B. (2008) The utility of crop genetic diversity in maintaining ecosystem services. *Agriculture, Ecosystems and Environment*, 123, 261–270.

Haley, S.D., Peairs, F.B., Walker, C.B., Rudolph, J.B. and Randolph, T.L. (2004) Occurrence of a new Russian wheat aphid biotype in Colorado. *Crop Science*, 44, 1589–1592.

Heong, K.L., Chen, Y.H., Johnson, D.E., Jahn, G.C., Hossain, M. and Hamilton, R.S. (2005) Debate over a GM rice trial in China. *Science*, 310, 231–231.

Horgan, F.G. and Myers, J.H. (2004) Interactions between predatory ground beetles, the winter moth and an introduced parasitoid on the lower mainland of British Columbia. *Pedobiologia*, 48, 23–35.

Horgan, F.G., Quiring, D.T., Lagnaoui, A. and Pelletier, Y. (2007) Variable responses of tuber moth to the leaftrichomes of wild potatoes. *Entomologia Experimentalis et Applicata*, 125, 1–12.

Howe, G.A. and Jander, G. (2008) Plant immunity to insect herbivores. *Annual Review of Plant Biology*, 59, 41–66.

Jackson, M.T. (1997) Conservation of rice genetic resources: the role of the International Rice Genebank at IRRI. *Plant Molecular Breeding*, 35, 61–67.

Janzen, D.H. (1966) Coevolution of mutualism between ants and acacias in Central America. *Evolution*, 20, 249–275.

Kalazich, J.C. and Plaisted, R.L. (1991) Association between trichome characters and agronomic traits in *Solanum tuberosum* (L.) × *S. berthaultii* (Hawkes) hybrids. *American Potato Journal*, 68, 833–846.

Kartohardjono, A. and Heinrichs, E.A. (1984) Populations of the brown planthopper, *Nilaparvata lugens* (Stal) (Homoptera: Delphacidae), and its predators on rice varieties with different levels of resistance. *Environmental Entomology*, 13, 359–365.

Leimu, R. and Koricheva, J. (2006) A meta-analysis of tradeoffs between plant tolerance and resistance to herbivores: combining the evidence from ecological and agricultural studies. *Oikos*, 112, 1–9.

Leung, H., Zhu, Y., Revilla-Molina, I. *et al.* (2003) Using genetic diversity to achieve sustainable rice disease management. *Plant Disease*, 87, 1156–1168.

Lou, Y., Du, M., Turlings, T.C.J., Cheng, J. and Shan, W. (2005) Exogenous application of jasmonic acid induces volatile emissions in rice and enhances parasitism of *Nilaparvata lugens* eggs by the parasitoid *Anagrus nilaparvatae*. *Journal of Chemical Ecology*, 31, 1985–2002.

Lövei, G.L. and Arpaia, S. (2005) The impact of transgenic plants on natural enemies: a critical review of laboratory studies. *Entomologia Experimentalis et Applicata*, 114, 1–14.

Lövei, G.L., Andow, D.A. and Arpaia, S. (2009) Transgenic insecticidal crops and natural enemies: a detailed review of laboratory studies. *Environmental Entomology*, 38, 293–306.

MacIntosh, S.C. (2009) Managing the risk of insect resistance to transgenic insect control traits: practical approaches in local environments. *Pest Management Science*, 65, 100–106.

Mallampalli, N., Gould, F. and Barbosa, P. (2005) Predation of Colorado potato beetle eggs by a polyphagous ladybeetle in the presence of alternate prey: potential impact on resistance evolution. *Entomologia Experimentalis et Applicata*, 114, 47–54.

Marvier, M., McCreedy, C., Regetz, J. and Kareiva, P. (2007) A meta-analysis of effects of *Bt* cotton and maize on nontarget invertebrates. *Science*, 316, 1475–1477.

Maschinski, J. and Whitham, T.G. (1989) The continuum of plant responses to herbivory: The influence of plant association, nutrient availability, and timing. *American Naturalist*, 134, 1–19.

Messina, F.J. and Sorenson, S.M. (2001) Effectiveness of lacewing larvae in reducing Russian wheat aphid populations on susceptible and resistant wheat. *Biological Control*, 21, 19–26.

Mineiro, J.L.C., Sato, M.E., Raga, A. and Arthur, V. (2008) Population dynamics of phytophagous and predaceous mites on coffee in Brazil, with emphasis on *Brevipalpus phoenicis* (Acari: Tenuipalpidae). *Experimental and Applied Acarology*, 44, 277–291.

Mulatu, B., Applebaum, S.W. and Coll, M. (2004) A recently acquired host plant provides an oligophagous insect herbivore with enemy-free space. *Oikos*, 107, 231–238.

Myint, K.K.M., Yasui, H., Takagi, M. and Matsumura, M. (2009) Virulence of long-term laboratory populations of the brown planthopper, *Nilaparvata lugens* (Stål),and whitebacked planthopper, *Sogatella furcifera* (Horvath) (Homoptera: Delphacidae), on rice differential varieties. *Applied Entomology and Zoology*, 44, 149–153.

Newton, A.C., Begg, G.S. and Swanston, J.S. (2009) Deployment of diversity for enhanced crop function. *Annals of Applied Biology*, 154, 309–322.

Nicholls, J.A., Fuentes-Utrilla, P., Hayward, A. *et al.* (2010) Community impacts of anthropogenic disturbance. Natural enemies exploit routes in pursuit of invading herbivore hosts. *BMC Evolutionary Biology*, 10, e322.

Ninkovic, V., Al Abassi, S., Ahmed, E., Glinwood, R. and Pettersson, J. (2011) Effect of within-species plant genotype mixing on habitat preference of a polyphagous insect predator. *Oecologia*, 166, 391–400.

Núñez-Farfán, J., Fornoni, J. and Valverde, P.L. (2007) The evolution of resistance and tolerance to herbivores. *Annual Review of Ecology, Evolution and Systematics*, 38, 541–566.

Obrycki, J.J. (1986) The influence of foliar pubescence on entomophagous species, in *Interactions of plant resistance and parasitoids and predators of insects* (eds D.J Boethel and R.D. Eikenbary), John Wiley & Sons, Inc., New York, pp. 61–83.

Ojala, K., Julkunen-Tiitto, R., Lindström, L. and Mappes, J. (2005) Diet affects the immune defence and life-history traits of an Arctiid moth *Parasemia plantaginis*. *Evolutionary Ecology Research*, 7, 1153–1170.

Painter, H. (1951) *Insect resistance in crop plants*, McMillan Company, New York.

Pelletier, Y., Horgan, F.G. and Pompon, J. (2011) Potato resistance to insects. *Americas Journal of Plant Science and Biotechnology*, 5, 37–51.

Pingali, P.L. and Gerpacio, R.V. (1997) *Towards reduced pesticide use for cereal crops in Asia*, CIMMYT Economics working paper 97-04, CIMMYT, Mexico.

Poelman, E.H., Broekgaarden, C., van Loon, J.J.A. and Dicke, M. (2008) Early season herbivore differentially affects plant defence responses to subsequently colonizing herbivores and their abundance in the field. *Molecular Ecology*, 17, 3352–3365.

Porter, D.R., Burd, J.D., Shufran, K.A., Webster, J.A. and Teetes, G.L. (1997) Greenbug (Homoptera: Aphidae) biotypes: Selection by resistant cultivars or preadapted opportunists? *Journal of Economic Entomology*, 90, 1055–1065.

Poveda, K., Jímenez, M.I. and Kessler, A. (2010) The enemy as ally: herbivore-induced increase in crop yield. *Ecological Applications*, 90, 1787–1793.

Prado, E. and Tjallingii, W.F. (2007) Behavioral evidence for local reduction of aphid-induced resistance. *Journal of Insect Science*, 7, available online: insectscience.org/7.48.

Price, P.W. (1986) Ecological aspects of host plant resistance and biological control: Interactions among three trophic levels, in *Interactions of plant resistance and parasitoids and predators of insects* (eds D.J. Boethel and R.D. Eikenbary), John Wiley & Sons, Inc., New York, pp. 11–30.

Rapusas, H.R., Bottrell, D.G. and Coll, M. (1996) Intraspecific variation in chemical attraction of rice to insect predators. *Biological Control*, 6, 394–400.

Rausher, M.D. (2001) Co-evolution and plant resistance to natural enemies. *Nature*, 411, 857–864.

Rola, A.C. and Pingali, P.L. (1993) *Pesticides, rice productivity and farmers' health: an economic assessment*, IRRI, Manila.

Rosenthal, J.P. and Dirzo, R. (1997) Effects of life history, domestication and agronomic selection on plant defence against insects: evidence from maizes and wild relatives. *Evolutionary Ecology*, 11, 337–355.

Schoenly, K.G., Cohen, J.E., Heong, K.L., Litsinger, J.A., Barrion, A.T. and Arida, G. (2010) Fallowing did not disrupt invertebrate fauna in Philippine low-pesticide irrigated rice fields. *Journal of Applied Ecology*, 47, 593–602.

Settle, W.H., Ariawan, H., Astuti, E.T. *et al.* (1996) Managing tropical rice pests through conservation of generalist natural enemies and alternative prey. *Ecology*, 77, 1975–1988.

Shelton, A.M., Naranjo, S.E., Romeis, J. *et al.* (2009) Setting the record straight: a rebuttal to an erroneous analysis on transgenic insecticidal crops and natural enemies. *Transgenic Research*, 18, 317–322.

Shirai, Y. (2007) Nontarget effects of transgenic insecticidal crops: overview to date and future challenges. *Japanese Journal of Applied Entomology and Zoology*, 51, 165–186 (in Japanese).

Shu, A., Kim, J.H., Zhang, S. *et al.* (2009) Analysis of genetic similarity of japonica rice variety from different origins of geography in the world. *Scientia Agricultura Sinica*, 8, 513–520.

Simms, E.L. (1992) Costs of plant resistance to herbivory, in *Plant resistance to herbivores and pathogens: ecology, evolution, and genetics* (eds R.S. Fritz and E.L. Simms), The University of Chicago Press, Chicago, pp. 392–425.

Smith, C.M. (2005) *Plant resistance to arthropods: molecular and conventional approaches*, Springer, Dordrecht.

Stenseth, N.C. and Smith, J.M. (1984) Coevolution in ecosystems: Red Queen evolution or stasis? *Evolution*, 38, 870–880.

Steppuhn, A. and Baldwin, I.T. (2007) Resistance management in a native plant: nicotine prevents herbivores from compensating for plant protease inhibitors. *Ecology Letters*, 10, 499–511.

Storer, N.P., Babcock, J.M., Schlenz, M. *et al.* (2010) Discovery and characterization of field resistance to *Bt* maize: *Spodoptera frugipera* (Lepidoptera: Noctuidae) in Puerto Rico. *Journal of Economic Entomology*, 103, 1031–1038.

Strauss, S.Y. and Agrawal, A.A. (1999) The ecology and evolution of plant tolerance to herbivory. *Trends in Ecology and Evolution*, 14, 179–185.

Strauss, S.Y., Rudgers, J.A., Lau, J.A. and Irwin, R.E. (2002) Direct and ecological costs of resistance to herbivory. *Trends in Ecology and Evolution*, 17, 278–285.

Teetes, G.L. (1994) Adjusting crop management recommendations for insect resistant crop varieties. *Journal of Agricultural Entomology*, 11, 191–200.

Tiffin, P. (2000) Mechanisms of tolerance to herbivore damage: what do we know? *Evolutionary Ecology* 14, 523–536.

Tilman, D., Fargione, J., Wolff, B. *et al.* (2001) Forecasting agriculturally driven global environmental change. *Science*, 292, 281–284.

van Emden, H.F. (1986) The interactions of plant resistance and natural enemies: Effects on populations of sucking insects, in *Interactions of plant resistance and parasitoids and predators of insects* (eds D.J. Boethel and R.D. Eikenbary), John Wiley & Sons, Inc., New York, pp. 138–150.

Wallner, W.E. (1985) Factors affecting insect population dynamics: differences between outbreak and non-outbreak species. *Annual Review of Entomology*, 32, 317–340.

Walter, D.E. (1996) Living on leaves: mites, tormenta, and leaf domatia. *Annual Review of Entomology*, 41, 101–114.

Widawsky, D., Rozelle, S., Jin, S. and Huang, J. (1998) Pesticide productivity, host plant resistance and productivity in China. *Agricultural Economics*, 19, 203–217.

Wolfenbarger, L.L., Naranjo, S.E., Lundgren, J.G., Bitzer, R.J. and Watrud, L.S. (2008) *Bt* crop effects on functional guilds of non-target arthropods: a meta-analysis. *PLoS ONE 3*, e2118.

Xia, H., Chen, L., Wang, F. and Lu, B.R. (2010) Yield benefit and underlying cost of insect-resistance transgenic rice: Implication in breeding and deploying transgenic crops. *Field Crops Research*, 118, 215–220.

Zhao, J-Z., Cao, J., Collins, H.L. *et al.* (2005) Concurrent use of transgenic plants expressing a single and two *Bacillus thuringiensis* genes speeds insect adaptation to pyramided plants. *Proceedings of the National Academy of Sciences of the United States of America*, 102, 8426–8430.

Chapter 16

'PUSH–PULL' REVISITED: THE PROCESS OF SUCCESSFUL DEPLOYMENT OF A CHEMICAL ECOLOGY BASED PEST MANAGEMENT TOOL

Zeyaur R. Khan, Charles A.O. Midega, Jimmy Pittchar, Toby J.A. Bruce and John A. Pickett

Biodiversity and Insect Pests: Key Issues for Sustainable Management, First Edition. Edited by Geoff M. Gurr, Steve D. Wratten, William E. Snyder, Donna M.Y. Read.
© 2012 John Wiley & Sons, Ltd. Published 2012 by John Wiley & Sons, Ltd.

INTRODUCTION

Approximately 80% of the human population in sub-Saharan Africa (SSA) depends on agriculture for food, income and employment. However, the agricultural sector is characterised by very low productivity, which results in high poverty levels and undernourishment. Poverty is widespread with approximately 1.2 billion people living below the poverty line (projections following FAO, 2006 report). Cereals, including maize (*Zea mays* L.), sorghum (*Sorghum bicolor* (L.) Moench.), finger millet (*Eleusine coracana* (L.) Gaertn.) and rice (*Oryza sativa* L.), are the principal food and cash crops for millions of the poorest people in the predominantly mixed crop-livestock farming systems of the region (Polaszek, 1998). Unfortunately, grain yields in SSA are the lowest in the world at around 1t/ha (Jagtap and Abamu, 2003). The factors responsible for the exceptionally low cereal productivity include biotic constraints (principally lepidopterous stemborers and parasitic striga weeds), and abiotic constraints (land degradation and poor soil fertility) (Lal, 1988; Nwilene *et al.*, 2008).

Stemborers, the larvae of moths (Lepidoptera), seriously limit cereal production in the region (Khan and Pickett, 2004). Heavy yield losses, 20 to 40% of the potential output, are attributed to 4 of the 17 species of stemborers which infest cereal crops in SSA, with the indigenous *Busseola fusca* (Fuller) and the invasive *Chilo partellus* (Swinhoe) being the most destructive (Kfir *et al.*, 2002). Stemborers are difficult to control, largely because of the cryptic and nocturnal habits of the adult moths and the protection provided to immature stages by the stem of the host crop (Ampofo *et al.*, 1986; Kfir *et al.*, 2002). Chemical pesticides are the main method of stemborer control recommended to farmers by the governments' ministries of agriculture. However, these can be environmentally unfriendly and unsustainable, uneconomical and impractical for most resource-poor farmers (Kfir *et al.*, 2002) and a direct threat to beneficial arthropods (Epstein *et al.*, 2000).

The ravages caused by parasitic witchweeds in the genus *Striga* (commonly known as striga) negatively affect lives of over 100 million people in SSA and infest over 40% of arable land in the savanna region, causing an estimated annual loss of US$7 to $13 billion (Lagoke *et al.*, 1991). There are over 20 species of striga in the region (Gethi *et al.*, 2005). Of these, *Striga hermonthica* (Del.) Benth. is the most destructive in cereals and infests 24% of maize crop area in SSA

(Pickett *et al.*, 2010). Infestations by striga have resulted in the abandonment of much arable land by farmers in Africa (Khan, 2002), with the problem being more widespread and serious in areas where both soil fertility and rainfall are low (Oswald, 2005). Unfortunately, subsistence farmers in the region must weed out striga, because most of the damage to the host crop occurs before Striga plants emerge above the soil, (Press, *et al.*, 2001) but this is an ineffective and labour-intensive activity, mainly delegated to women and children.

Recommended control methods to reduce striga infestation include heavy applications of nitrogen fertiliser, crop rotation, use of imidazolinone herbicide resistant maize (IR-maize) together with a seed treatment of this herbicide, use of trap crops and chemicals to stimulate suicidal seed germination, hoeing and hand pulling, herbicide application and the use of resistant or tolerant crop varieties (Oswald, 2005). The effectiveness of all these methods, including the most widely practised hand hoe weeding, is seriously limited by the reluctance of farmers to accept them, for biological and socio-economic reasons (Lagoke *et al.*, 1991). As a result most farmers do not attempt to control striga beyond the usual weeding as an agronomic practice, and striga continues to cause very serious reductions in yields of staple cereals (Parker, 2009). The seedbank of Striga can stay dormant in the soil for over 20 years. This suggests a need for an approach that aims at reduction of the seed bank (Westerman *et al.*, 2007).

The most serious abiotic constraint to cereal production is poor soil fertility, which has also been identified by farmers. It results from poor inherent fertility status, together with high pressure on the land and poor management practices. Furthermore, soils are degrading rapidly due to their low buffering capacity and the inability of small-scale farmers to invest in soil fertility management. These soils are hardly able to sustain acceptable maize yields. Nitrogen (N) and phosphorus (P) are the major limiting nutrients. Lack of adequate soil management also negatively affects the soil organic matter pool that is responsible for a series of production and environmental service functions essential for sustainable crop production. The use of inorganic fertilisers to supply plant nutrients by resource-poor farmers is limited (Jama *et al.*, 1998). Although many soil fertility restoration technologies such as cereal–legume rotations, biomass transfers, farmyard manures, tree-fallowing and green manure cover crop exist that can redress this situation, their potential for use is greatly influenced by land size, capital and labour constraints (Sanchez *et al.*, 1997).

THE PUSH–PULL SYSTEM

Reducing the yield losses caused by stemborers and striga and improving soil fertility through qualitatively improved management strategies was anticipated to significantly increase cereal production, resulting in better nutrition and purchasing power for many producers of these crops. Thus the International Centre of Insect Physiology and Ecology (*icipe*) (www.icipe.org), Rothamsted Research (www.rothamsted.bbsrc.ac.uk) and East African national collaborative partners have developed a sustainable solution that simultaneously addresses these major constraints to cereal production. This innovative conservation agricultural platform technology, called 'push–pull' (www.push-pull.net), provides integrated pest, weed and soil management through efficient use of natural resources to increase farm productivity. This chapter takes this particularly successful example of using aspects of biodiversity to manage an insect pest problem in an integrated manner and considers the factors that led to its widespread adoption and high level of impact. Further, this case illustrates the need for ongoing research once a successful system has been introduced. This is to realise the potential of the system by making it responsive to changes (such as a new plant disease) and allow upscaling and use over a wider range of environmental conditions. Finally, the impacts of climate change are considered.

THE PUSH–PULL CROPPING INNOVATION

The push–pull system comprises specifically selected intercrops and trap crops grown in a mixed cropping system (Cook *et al.*, 2007). These companion plants release semiochemical repellents and attractants that manipulate the distribution and abundance of stemborers and beneficial insects for management of stemborer pests as well as releasing root exudates that suppress striga (Khan and Pickett, 2004; Khan *et al.*, 2010). The main cereal crop is intercropped with molasses grass (*Melinis minutiflora* P. Beav.), or desmo-

Figure 16.1 A farmer's field in Kuria district in Kenya, planted with the push–pull technology. Maize is intercropped with desmodium (*Desmodium uncinatum*) for management of parasitic striga weeds, stemborers and soil fertility. Napier grass (*Pennisetum purpureum*), a trap plant for stemborers, is planted around the border of the field (not seen here). Both Napier grass and desmodium provide quality fodder for livestock. The technology is highly appropriate, sustainable and economical for the resource-poor smallholder African farmers.

dium [*Desmodium uncinatum* (Jacq.) DC. or *Desmodium intortum* (Mill.) Urb.], which are repellent to gravid stemborer moths (push) (Figure 16.1) while Napier grass (*Pennisetum purpureum* Schumacher), planted as a border crop around the main crop, simultaneously attracts the stemborers so acts as a trap plant (pull) (Cook *et al.*, 2007; Hassanali *et al.*, 2008; Khan *et al.*, 2010). The intercrop also attracts parasitic wasps which are natural enemies of the stemborer (Khan *et al.*, 1997). Such a system requires a complete understanding and exploitation of the chemical ecology of plant–plant and plant–insect interactions.

While the push–pull system was being developed specifically for the control of cereal stemborers in smallholder maize production in Kenya (Khan and Pickett, 2004), it was discovered that use of desmodium as an intercrop had further benefits in regard to suppression of striga (Khan *et al.*, 2002; Pickett *et al.*, 2010) and improvement of soil fertility (Khan *et al.*, 2006a). Molasses grass is therefore a preferred intercrop for stemborer control in highland areas where striga is not present, while desmodium is a preferred intercrop where both striga and stemborers occur. The system is well suited to African socio-economic conditions because it includes multiple cropping and is based on locally available plants, and not expensive external and seasonal inputs (Khan et al., 2010).

In addition to providing protection against stemborers and the parasitic striga weeds, the companion plants in the push–pull technology also provide high-quality animal fodder (Khan and Pickett, 2004). Furthermore, since both companion plant species are perennials, push–pull conserves soil moisture, and improves soil health and beneficial biodiversity (Khan et al., 2006a; Midega et al., 2008). Using the push–pull approach, cereal yields are at least doubled (Khan et al., 2008a). Table 16.1 outlines how the push-pull system addresses the various constraints of smallholder farming.

Table 16.1 How push–pull technology addresses major constraints facing smallholder farmers in SSA.

Major constraints	How push–pull addresses the constraints
Low soil fertility	Increased nitrogen fixation by the intercrop, desmodium (Khan et al., 2006a)
Degraded land	Controls soil erosion; increased organic matter and soil physical properties (Khan et al., 2006a)
Parasitic striga weed	Striga control by the intercrop, striga seed depletion (Khan et al., 2008a; 2008b)
Stemborer pests	Effective stemborer control by companion plants, and activity of natural enemies (Khan et al., 2008a; Midega et al., 2008; 2009)
Moisture stress	Soil moisture conservation, improved water holding capacity by intercrops (Khan et al., 2002; 2006a)
Low crop yields	Increased cereal yields (maize from 1 to 3.5 t/ha; sorghum 0.8 t to 2 t/ha; millet 0.4 t to 0.8 t/ha; upland rice) (Khan et al., 2006b; 2008a; Midega et al., 2010; Khan et al., 2010)
Shortage of livestock fodder	All-year-round quality fodder from the trap and intercrop plants leading to improved milk production (Khan and Pickett, 2004; Khan et al., 2008c)
Loss of biodiversity	Increased abundance and diversity of beneficial organisms (Khan et al., 2006a; Midega et al., 2008)
Shortage of labour	Reduced labour requirement for land preparation and weed control (Khan et al., 2008d)

CHEMICAL ECOLOGY OF THE PUSH–PULL SYSTEM

In the push–pull system, the interactions with pests and weeds are based on semiochemicals released by the companion plants. During the development of the system, efforts were made to elucidate the underlying mechanisms underpinning the effects of the companion plants on stemborer pests and striga weed for maintaining system sustainability. This illustrates a more general message that understanding the biological basis for the ecological interactions between components of biodiversity is important. Semiochemicals are the principal mediators of insect responses to plants. Suitable host plants release attractive semiochemicals (or specific ratios of semiochemicals) which attract insects (Nordlund and Lewis, 1976; Dicke and Sabelis, 1988; Bruce et al., 2005), while semiochemicals associated with non-host taxa deter them (Pettersson et al., 1994).

Stemborer control by the companion plants

Semiochemicals involved in stemborer control were described in our earlier publication, Khan and Pickett (2004). The trap plants such as Napier grass produce attractive semiochemicals, notably octanal, nonanal, naphthalene, 4-allylanisole, eugenol, and linalool. Napier grass also produces significantly higher amounts of these attractive compounds than maize and sorghum. Moreover, emission varies over the day–night cycle so that it is over 100 times higher during the first two hours of darkness (Birkett et al., 2006; Chamberlain et al., 2006), the period when stemborer moths are most active (Päts, 1991). A similar response is also seen in maize and sorghum, but the increase is 10 times less than in Napier grass. Napier grass is highly preferred over maize for oviposition by both C. partellus and B. fusca (van den Berg, 2006; Khan et al., 2006c; 2007a). Nevertheless, high larval mortality rates result (approximately 80% (Khan et al., 2006c)) because Napier grass is nutritionally poor, produces sticky sap that immobilises stemborer larvae and harbours a high diversity of predatory arthropods (Khan et al., 2006a; Midega et al., 2008).

Plants often produce stress-induced volatile organic compounds (VOCs) in response to insect attacks, called herbivore-induced plant volatiles (HIPVs), such as (E)-ocimene and (E)-4,8-dimethyl-1,3,7-nonatriene (DMNT) (see review by Khan et al., 2008e and

James *et al.*, chapter 11 of this volume). These HIPVs protect the producer-plant in two ways: they prevent further colonisation by the herbivore, and also attract natural enemies of the herbivores, principally parasitic wasps, the activity of which lowers herbivore populations. The intercrops, molasses grass and desmodium constitutively produce similar HIPVs which repel stemborer moths and attract parasitic wasps (Khan *et al.*, 1997; 2000; Midega *et al.*, 2009). The DMNT in particular was demonstrated to be responsible for the increased parasitoid foraging in the plots intercropped with molasses grass (Khan *et al.*, 1997). The effectiveness of the push–pull system in reducing pest levels and increasing stemborer parasitism rates has been demonstrated extensively under farmers' field conditions (Khan *et al.*, 1997; 2000; 2008a; Midega *et al.*, 2005; 2009).

Striga control by allelopathic effects of desmodium

Striga is so well adapted to its environment and integrated with its hosts that it will only germinate in response to specific chemical cues present in root exudates of these hosts or highly specific non-host plants. The interaction between host and parasite begins with the secretion of secondary metabolites from the roots of the host (and some false non-hosts) that induce the germination of the parasite's seeds (Matusova *et al.*, 2005). Because the seeds of striga contain only low levels of energy and nutrient reserves, they cannot survive for more than a few days after germination unless they reach a host root and a xylem connection is established. We made a serendipitous discovery that striga development was significantly suppressed in the fields intercropped with desmodium for stemborer control. *Desmodium* spp. are efficient nitrogen-fixing legumes; for example *D. uncinatum* can fix up to 110 kg N/ha/yr (Henzel *et al.*, 1966), and because they have thick canopies, they provide soil cover thereby smothering the weeds. While these attributes of desmodium contributed to suppression of striga emergence, a stronger and much more efficient allelopathic effect was observed with the desmodium root exudates (Khan *et al.*, 2002). The phytochemical profiles of desmodium root exudates and root extracts revealed a complex array of plant secondary compounds (Hooper *et al.*, 2009). In subsequent studies, it was found that the root exudates of *D. uncinatum* contain novel flavonoid and isoflavonoid compounds, some of which induced

premature germination of striga seeds before host roots had been contacted, while others interfered with attachment of striga roots to those of its hosts (Tsanuo *et al.*, 2003; Khan *et al.*, 2008b). Isolated fractions from the root exudates of *D. uncinatum* containing uncinanone B induced germination of striga seeds and fractions containing uncinanone C inhibited radical growth (Tsanuo *et al.*, 2003). This provided the first example of a potential allelopathic mechanism to prevent striga parasitism.

Efforts to identify and characterise other constituent compounds in the root exudates of *D. uncinatum* have continued, with recent characterisation of another key post-germination inhibitor, isoschaftoside, as well as other *C*-glycosylflavones from a more polar fraction of the root exudates and solvent extracts (Pickett *et al.*, 2007; Hooper *et al.*, 2009; 2010). Indeed, isoschaftoside was found to be the main compound in the most potent fraction inhibiting growth of germinated *S. hermonthica* radicles, present at biologically active concentrations of 10–100 nM. Intercropping with desmodium, with its combined effects of germination stimulants and post-germination inhibitors, thus provides a novel means of continual *in situ* reduction of the striga seed bank in the soil, even in the presence of graminaceous host plants. Moreover, field data show continual reduction in striga seed bank in the plots intercropped with desmodium but the seed bank steadily increasing in the maize monoculture plots (Khan *et al.*, 2008b). There are also significant reductions in the number of emerged striga in the plots intercropped with *D. uncinatum* (Khan *et al.*, 2008a), and other *Desmodium* spp. (Khan *et al.*, 2006b; 2006d; Midega *et al.*, 2010). Intercropping with desmodium thus increases the options available to farmers in different socio-economic strata and agroecological zones in Africa.

NEW DEVELOPMENTS

Extending the push–pull strategy to other cereal crops

In the semi-arid tropics of SSA sorghum and finger millet are important food and cash crops and provide livestock feed and firewood (Ahmed *et al.*, 2000). They are grown under severe moisture and nutrient stress, in areas where it is difficult to grow other food grains such as maize, and where the resource base of the farmer is low (Gebremedhin *et al.*, 2000). They are often referred to as poor man's crops or orphan crops. Both are indigenous to Africa with outstanding nutritional qualities. Finger millet in particular has nutritional qualities superior to rice and is equivalent to wheat (Madhavi *et al.*, 2005). They also have superior storage qualities, enabling safe storage for several years. Thus they are a traditional component of farmers' risk-avoidance strategies in drought-prone areas where harsh climatic conditions limit the production of other crops. Production of these crops is similarly constrained by striga, stemborers and poor soil fertility.

Once the push–pull technology had been developed and implemented to address pest, weed and soil fertility constraints in maize, demand arose from farmers for a similar approach for sorghum and millet cropping systems. Therefore, *icipe* and partners initiated studies to address this demand by evaluating different *Desmodium* spp. for use in these crops. This was necessary as *D. uncinatum* that had been used as the principal maize intercrop was only suited to moderate agroecologies with annual rainfall of at least 700 mm and temperatures of up to 30 °C. For sorghum and finger millet cultivation, *D. intortum* was found to be suitable as it withstands drought conditions better and wilts less (Ostrowski, 1966). It also has a relatively higher nitrogen-fixing ability, over 300 kg N/ha per year under optimum conditions (Whitney, 1966). On-station and on-farm studies showed intercropping sorghum and finger millet with drought-tolerant *D. intortum* resulted in highly significant reductions in striga emergence and the proportion of stemborer-damaged plants relative to the corresponding monoculture plots (Khan *et al.*, 2006b; Midega *et al.*, 2010). Moreover, grain yields were significantly higher in the intercropped plots than in the monoculture, which led to higher economic returns compared to the monoculture plots (Midega *et al.*, 2010). Farmers in drier areas of western Kenya are currently implementing this approach for management of striga, stemborer and soil fertility and exploiting the fodder and commercial value of the companion plants.

In recent years rice has gained in importance in SSA as a food and cash crop, largely because of changing lifestyles and diets. In some regions of Africa non-irrigated upland rice is becoming a popular crop. For example, in Uganda, NERICA (New Rice for Africa) cultivars are promoted. One of the most important biotic constraints on effective production of NERICA varieties is striga, particularly *S. hermonthica*. Recent research shows that intercropping rice with *D. uncinatum* effectively suppresses *S. hermonthica*, resulting in

significant yield increases (Khan *et al.*, 2010; Pickett *et al.*, 2010).

Integration of edible beans into the push–pull technology

Smallholder farmers in the region expressed concern that intercropping with desmodium hindered them from interplanting cereals with edible legumes, particularly beans (*Phaseolus vulgaris* L.), an important source of protein for resource-poor farmers. Therefore efforts were made to address this as it negatively affected technology adoption amongst some farming communities. Farmer surveys in the region revealed that farmers planted beans either between the rows of maize, in the same holes with maize or in between maize plants within a row (Figure 16.2). Studies were therefore conducted to evaluate effects of integrating beans in the maize–desmodium intercrops. These studies showed that integration of beans in the maize–desmodium intercrops and the planting arrangement, either with maize in the same or different holes, did not compromise the striga and stemborer control efficacy of desmodium (Khan *et al.*, 2009). The practice did, however, increase labour and total variable costs, with these being significantly higher in plots with both crops in different holes than in the same hole. However, the revenue accrued from the yields, gross benefits and benefit:cost ratios remained the same as in maize–desmodium intercrops. It was concluded that integrating beans into the desmodium intercrop provided an

Figure 16.2 Integrating edible beans (*Phaseolus vulgaris*) into the push-pull strategy provides an additional food crop, a protein source to the farmers. Adoption rates of the push–pull technology have increased significantly since farmers are incorporating beans in their push–pull plots.

additional crop, a protein source, to the farmers but yielded the same economic benefits (Khan *et al.*, 2009). Adoption rates of the push–pull technology have increased significantly since bean cultivation was made possible and farmers are incorporating beans in their push–pull plots. It is advised that maize and beans be planted in different holes, where labour is not limiting, as planting both crops in the same hole poses a challenge of competition for moisture and nutrients which are often limiting factors in most areas of SSA experiencing the other three main cereal production constraints.

ECONOMICS OF THE PUSH–PULL TECHNOLOGY

Other than the biological and logistical issues detailed above, economic issues too have proven critical to the adoption of the push–pull system. Smallholder farmers in SSA are often risk-averse and do not take up new agricultural technologies unless there are clearly demonstrated agronomic and economic benefits that accrue direct to themselves. In their general practices these farmers practise multiple cropping, such as an intercrop of cereals and legumes, as a means of increasing crop yields per unit area of land, and also as a security should one crop fail (Abate *et al.*, 2000). The push–pull technology thus fits within their cropping systems. The technology has been subjected to several rigorous economic analyses with a view to establishing the extent of its profitability. These analyses show that it yields higher economic returns than farmers' conventional practices of intercropping maize with edible beans and maize monoculture (Khan *et al.*, 2008d). In western Kenya, for example, the technology yielded significantly higher gross revenues than the conventional farmer practices involving maize–bean intercropping and maize monoculture. Indeed, these two conventional practices yielded negative gross benefits during some of the cropping seasons, meaning the cost of production was higher than the revenue obtained from the produce (Khan *et al.*, 2008d). Returns to land and labour were also significantly higher with the push–pull technology than the maize–bean intercropping and maize monoculture, with a cost:benefit analysis showing returns by a factor of 2.2 in push–pull relative to 0.8 the maize monoculture system (Khan *et al.*, 2001; 2008d). These findings show unequivocally that the use of push–pull technology is economically efficient at the farm level.

Attempts have also been made to compare economic performance of the push–pull technology against other striga and soil fertility management approaches in western Kenya. These include rotations with promiscuous (nodulating with various fast-growing *Rhizobium* spp.) soybean varieties and green manure crops, and imidazolinone-resistant maize with seed treatment (IR-maize). The economic performance of these technologies, both with and without fertiliser, was compared with that of push–pull based on marginal analysis using a multi-output, multi-period model. Results showed that push–pull was significantly more profitable than the other approaches (De Groote *et al.*, 2010a). A profit function was estimated with the discounted benefits as dependent variables, and the different technologies as independent variables. The extra-discounted benefit over six seasons to push–pull was estimated at US$1,218/ha, markedly higher than the other striga control technology combinations estimated at $356/ha and below (De Groote *et al.*, 2010a).

The profitability of the push–pull technology stems from higher grain yields, attributable to the effective control of striga and stemborer (Khan *et al.*, 2000, 2001; 2002; 2006b; Midega *et al.*, 2005) (Figure 16.3), and improvements in soil fertility through nitrogen fixation by desmodium (Whiteman, 1969; Khan *et al.*, 2006a), moisture retention by desmodium (Khan *et al.*, 2002) and provision of additional organic matter from desmodium foliage (as suggested by Midega *et al.*, 2005). Additionally, the companion plants are valuable fodder crops, either consumed by the farmer's own livestock or sold for cash, with the sale of desmodium seeds generating additional income for the farmers. In addition to these benefits farmers have reported increased milk production (Khan *et al.*, 2008c); also reported in an independent review by Fischler (2010). Moreover, total costs of production fall significantly in the push–pull plots from the second cropping season as the companion plants are perennial, thus requiring a one-time only expenditure on planting materials and labour for establishment, and reduced ploughing and weeding (Khan *et al.*, 2008d). These factors have resulted in significant improvements in economic returns to the farmers. An impact assessment study estimated that in Kenya the push–pull technology generates total annual additional gross benefit in the range of US$2 to 3 million (reported again in an independent review by Fischler (2010)). The technology thus contributes significantly to reducing the vulnerability of farm families by ensuring higher yield and

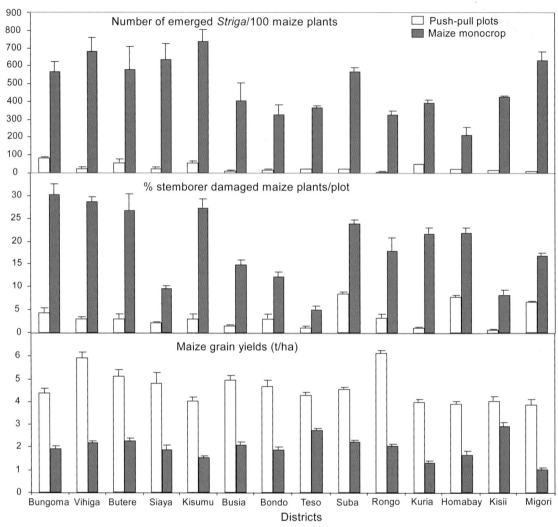

Figure 16.3 Mean number of emerged *Striga hermonthica* per plot, proportion of stemborer-damaged plants per plot, and average maize grain yields (t/ha) from maize monocrop and push–pull plots in different districts in western Kenya. Means represent data averages from 30 farmers' fields per district over two cropping seasons (long and short rains 2009). In all districts, mean number of *S. hermonthica* emerged per 100 maize plants and percent stemborer damaged maize plants were significantly higher in the maize monocrop than push–pull plots while maize grain yields were significantly higher in the push–pull plots ($p < 0.0001$).

economic stability in areas where rain-fed agriculture is the main source of income and employment.

Farmers practising the technology are able to harvest on average 3.5 t/ha of maize from plots that previously yielded below 1 t/ha (Khan *et al.*, 2008a). Additionally, the technology provides a 'springboard' for diversifying farmers' income streams, including sale of surplus

produce, fodder and desmodium seed. The technology thus strengthens individual farmers, creates both on-farm and off-farm rural employment opportunities while enhancing trade, and helps to develop cereal and livestock value chains. It contributes to the overall improvement in livelihoods of entire communities, poverty alleviation, and attainment of the Millennium

Development Goals as smallholder disposable income is injected into the local economy. Less discernible but equally important are the positive impacts that the technology has on the socio-economic well-being of the communities, such as development of social capital, gender protection and food security.

FARMERS' PERCEPTIONS OF THE PUSH–PULL TECHNOLOGY

Smallholder farmers in SSA often mention striga, stemborers and soil fertility as major constraints to efficient cereal production in the region. However, there is often low uptake of technologies developed to alleviate these constraints because of both biological and socio-economic reasons, and overall weaknesses in the extension delivery methods. Adoption of agricultural technologies by farmers is influenced by a range of factors, including perceptions of the severity of the farming constraints, efficacy, profitability and compatibility of a given technology with their production systems, and the clarity with which the new knowledge and information is communicated (Adesina and Baidu-Forson, 1995; Boahene et al., 1999; Emechebe et al., 2004).

Khan et al. (2008c) assessed perceptions of adopters and non-adopters of the push–pull technology and their influence on its adoption in western Kenya. Stemborers and striga weeds were seen as severe maize production constraints by all farmers interviewed. The majority of adopters cited a number of motivations for adopting the technology, including to control stemborers and striga weed, improve soil health and increase farm productivity. The adopters reported reduced infestation by the pests, improvement in soil fertility, significant increases in maize grain yields, and improved fodder and milk productivity. Farmers who had not yet adopted the technology but attended farmers' field days rated push–pull as significantly superior to the farmers' own practices on all attributes, indicating that they perceived it as an effective technology for the control of stemborers and striga, improved soil fertility and increased maize productivity.

In addition to push–pull technology, soybean and crotalaria rotations, and IR-maize seed have been developed to help alleviate the three constraints mentioned above. Evaluations of different combinations of these approaches by smallholder farmers in western Kenya revealed that push–pull combinations were generally preferred, with a combination of push–pull,

IR-maize and fertiliser being most favoured (De Groote et al., 2010b). The push–pull technology was also more appreciated by women than men, while other technologies were gender-neutral. Older farmers were also more likely to prefer push–pull and crotalaria with fertiliser. In spite of the companion plants in the push–pull technology being fodder plants rather than for human consumption, livestock ownership had no effect on technology preferences (De Groote et al., 2010b).

UPSCALING OF THE PUSH–PULL TECHNOLOGY

The various benefits derived by the farmers from adoption of the technology suggest that push–pull could contribute to livelihood improvement of smallholder farmers who depend on agricultural land as their main resource base and its productivity as the single most important source of livelihood. This partly explains the increase in uptake from a few farmers to over 45,000 farmers in East Africa (Khan et al., 2010) (Figure 16.4). This has been achieved through a combination of factors, including (a) the deployment of dissemination pathways catering for different socio-cultural contexts and literacy levels of farmers, (b) multi-level collaboration with research institutions, national extension networks and NGOs, and farmer groups, and (c) extension efforts underpinned by a robust scientific base and continuous technical support by scientists and the national extension staff to ensure the correct implementation of the technology.

Access to information about an agricultural technology with a demonstrated efficacy is one of the key factors determining uptake. Such access can be facilitated by different information channels including mass media, information bulletins, field days, technical support and farmer field schools (Khan et al., 2008c). The different communication channels and learning tools are effective at different stages of adoption decision-making (Khan et al., 2008c).

Technology transfer of the push–pull technology was facilitated by a series of activities (Khan et al., 2008c), including information bulletins (brochures, detailed practical manuals on how to plant push–pull), mass media (radio programme in local languages, Tembea na majira, a Kiswahili term meaning 'keep up with modern trends', and newspaper articles), farmer-to-farmer learning methods (field days, farmer teachers, farmer field schools and enactment of drama), training by specialised extension staff, and

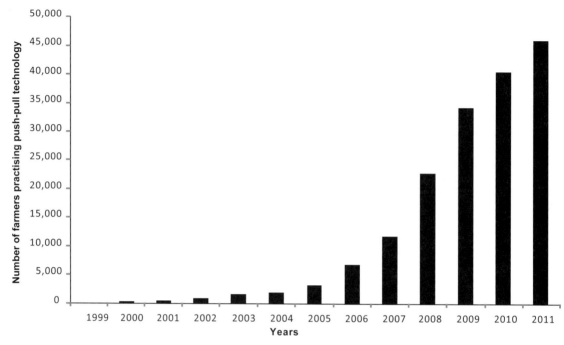

Figure 16.4 Number of farmers who have progressively adopted the push–pull technology from 1997 to June 2011 in East Africa (1997–2011).

public meetings. Farmer field days and farmer teachers were found to be very effective in disseminating the technology (Amudavi *et al.*, 2009a; 2009b), and confirmed the premise that peer influence is most effective for getting farmers to adopt new technology.

Extension was also facilitated through collaboration with national extension organisations and NGOs for wider dissemination of the push–pull technology. Collaboration with Heifer Project International, an international NGO, has facilitated the integration of livestock production with the push–pull technology thereby allowing efficient utilisation of the technology's fodder components.

CHALLENGES TO THE PUSH–PULL TECHNOLOGY

Napier stunt disease

Napier grass, the principal trap crop for stemborer control in the push–pull technology, is the main source of fodder of smallholder dairy systems in SSA. There has been increased commercialisation of smallholder dairies and uptake of the push–pull technology in

eastern Africa in recent years, which has led to intensified cultivation of Napier grass in the region. However, these production systems are challenged by a disease of Napier grass that is new to the region, commonly referred to as Napier stunt disease. The disease, caused by a phytoplasma belonging to the 16SrXI group (Jones *et al.*, 2004; 2006), becomes apparent in regrowth after cutting or grazing of Napier grass. Its symptoms include severe stunting, profuse tillering, and yellowing of the plants. Often the whole stool is affected with complete loss in yield and eventual death (Jones *et al.*, 2004; 2006). There is thus an urgent need for a management approach for the disease as recent results indicate that some of the grasses in the typical agroecosystems in the region such as star grass, *Cynodon dactylon* (L.) Pers, are alternative hosts of the Napier stunt phytoplasma (Obura *et al.*, 2010). Moreover, scientists from *icipe* have shown that the disease can be successfully transmitted to cereal crops, notably maize, sorghum and rice, while pearl millet is an effective host plant for the vector (unpublished report). As a first step in developing a management approach for the disease, *icipe* and partner institutions have identified *Maiestas* (=*Recilia*) *banda* Kramer (Hemiptera: Cicadellidae) as the vector of the disease in Kenya

(Obura et al., 2009). Efforts are ongoing to develop an integrated management approach for the disease whose components will include use of disease-resistant Napier grass varieties, management of the vector and phytoplasma alternative host plants, and diversified use of alternative fodder grasses thereby addressing the problem of intensified use of Napier grass.

Shortage of desmodium seeds

The shortage of desmodium seeds is an important limiting factor for much wider uptake of the push-pull technology in sub-Saharan Africa (SSA). In a recent impact assessment report (Fischler, 2010), lack of desmodium seeds within reach of farmers was reported as an important constraint by 30% of farmers. Due to current high demand of the push-pull technology and unavailability of desmodium seed in market, desmodium prices have remained high, about US$ 30 per kg. ICIPE has collaborated with a private seed company to produce desmodium seed through contract production with smallholder farmer groups. The strategy was to enable a public-private partnership in which contracted smallholder farmers produced seed for the seed company which cleaned the seed, ensured its quality (germination and viability), packaged and branded the seed, and promoted and distributed it through a network of private agro-stockists. The volume of desmodium seed produced under this arrangement remains far short of the amount of seed required to fulfill the demands of tens of thousands of farmers. The seed company limited production because of perceived high risk and low profitability compared to other seed brands.

Two informal pathways have emerged: (1) vegetative propagation among smallholder farmers and (2) community-based seed production initiatives. Although vegetative mechanism could evolve into an important mechanism for local diffusion of the 'push-pull' technologies among resource-limited farming communities, large quantities of seed are still required, through both formal and informal systems, for massive expansion of the technology in SSA.

Climate change

Climate change is anticipated to have far-reaching effects on sustainable agricultural development in SSA. Indeed, the magnitude and speed of climate change over major crop areas in Africa has been predicted by calculating the percentage overlap between historical (1960–2002) crop-growing season temperature and the projected 2025, 2050 and 2075 values over reported crop area. Results indicate that growing season temperature at any given maize-growing region in Africa will overlap on average 58% with its historical observations by 2025, 14% by 2050, and 3% by 2075 (Burke et al., 2009). This suggests that within two decades the growing season average temperature will be hotter than any year in historical experience for four years out of 10 for most of the African maize area, growing to nearly nine out of 10 by 2050 and nearly 10 out of 10 in 2075 (Burke et al., 2009). Millet and sorghum are expected to be similarly affected. Rainfall is expected to become progressively more unpredictable. Increasing atmospheric temperature and incidences of flood and drought will result in progressively more serious land degradation and increased pest and weed pressure, increased incidences of crop failure and general increases in food and nutritional insecurity for resource-poor farmers in many parts of SSA. To adapt to these adverse conditions, the resource constrained smallholder farmers will need to move to more drought-tolerant cereal crops, such as sorghum and millet, and small ruminants, such as dairy goats, for dairy production. Push–pull technology works effectively with these crops, giving highly significant increases in grain yields (Khan et al., 2006b; Midega et al., 2010), while still delivering quality fodder for livestock and soil improvement benefits. However, the trap and intercrop components are limited by rainfall and temperature. Therefore, to improve cereal and livestock productivity in dry areas, and to ensure the technology continues to positively impact food security in the region over the longer term, new drought-tolerant trap and intercrop plants need to be identified and incorporated. These should have correct chemistry in terms of stemborer attractancy for the trap component and stemborer repellence and striga suppression, and ability to improve soil fertility and soil moisture retention, for the intercrop component. In addition, they should provide other ecosystem services such as biodiversity improvement and conservation and organic matter improvement, with an overall resilience to withstand the climate change effects.

WIDER DEPLOYMENT OF THE PUSH–PULL TECHNOLOGY

Wider scale adoption of push–pull technology to various agroecologies will be achieved by ensuring that, first and foremost, it is sustainable and fully adapted to the increasingly hot and dry conditions in arid and semi-arid zones. Towards this end, efforts are being made to adapt the technology to climate change effects by identifying and selecting drought-tolerant companion plants suitable to arid and semi-arid environments. Associated with these is a need for sufficient attention to strategically important crops in drier agroecologies such as sorghum and finger millet, capacity-building of national research and extension agencies, and multi-stakeholder collaboration with organisations, including partnership with non-government organisations (NGOs) and seed companies serving farmers in the region. Secondly, the current push–pull technology needs to be extended as widely as possible, while integrating it with animal husbandry knowledge, in the target countries. A critical mass of adopters would allow its horizontal diffusion within and beyond the target sites (Khan *et al.*, 2008c; Amudavi *et al.*, 2009a; 2009b) and where necessary adapt and optimise the technology transfer pathways to local conditions. Thirdly, desmodium accessibility and availability should be ensured in the target areas by combining commercial production by seed companies, community-based seed production and distribution systems, and farmer groups trained on vegetative propagation of desmodium using vines. Most importantly, there is a need for strategic partnerships and linkages, concerted resource mobilisation and efficient utilisation to up- and out-scale the technology.

Push–pull is a platform technology and, therefore, there is need for multiple stakeholder forums beyond research and development to also address the domains of policies and integration, and the effects that these have on the productivity, profitability, and sustainability of agriculture. As the technology is optimised for and extended to different target areas, there is a continuous need to ensure the technology is economically viable; is compatible with the target population's farming systems; offers sound natural resource and environmental management; offers additional opportunities and is relevant to the farming systems. There is also a need to improve accessibility and efficiency of input and output markets, as well as value-chain development, of smallholder cereal, legume and livestock products.

Furthermore, policy support by national governments and international stakeholders is needed to integrate research and upscaling (and out-scaling) processes (see Box 16.1).

Box 16.1 Policy support needed to integrate research and upscaling (and out-scaling) processes

a) Promotion of cross-disciplinary research, development of innovation system approaches, and multi-institutional collaboration;

b) Capacity-building for project teams (in mutual learning and knowledge sharing), farmers (in skills acquisition and organisational capacity), and scientists in national institutions (in adequate funding, real collaboration and action research orientation);

c) Development of information and knowledge management capacity, and wide dissemination of the findings of proven research work, underpinned by sound scientific bases, and

d) Continuous learning by all involved stakeholders, participatory monitoring and evaluation, and a systemic approach to impact assessment to track programme progress towards overall goals, precisely identify the needs for mid-course adjustments, and document the returns on project/technology investment.

FUTURE OUTLOOK

Exploiting early herbivory alert in the push–pull technology

Maize plants produce compounds such as (*E*)-ocimene and (*E*)-4,8-dimethyl-1,3,7-nonatriene during damage by herbivorous insects (Turlings *et al.*, 1990; 1995). These HIPVs function as attractants for parasitic wasps, the activity of which lowers pest populations (see review by Khan *et al.*, 2008e and James *et al.*, chapter 11 of this volume). However, these natural enemies are often ineffective in reducing pest damage in farmers' fields because the 'cry for help' comes after damage has already been inflicted on the plants and therefore activity of the natural enemies does not prevent yield losses. Previous studies have shown that egg deposition induces production of defence VOCs in some plant species (Hilker and Meiners, 2006). Moreover, we have shown recently that an African forage grass, *Brachiaria brizantha* (Hochst. EX A. Rich.), suppresses emission of (*Z*)-3-hexenyl acetate, a VOC used in host location by other stemborers, as a result of oviposition by stemborer moths thus preventing further oviposition (Bruce et al., 2010). This causes a qualitative change in the odour profile which is important as the ratio of volatiles emitted by a plant is used in host recognition by insects (Bruce et al., 2005). Larval parasitoids were also attracted to the VOC blend of the grass after stemborer oviposition. Breeding has been known to result in loss of certain traits found in the 'primitive' relatives of the crops. Therefore this finding opens up a new opportunity for development of new crop protection strategies at the earliest stage of herbivore attack, oviposition. The cereal varieties which can naturally defend themselves against herbivory are urgently needed in SSA for resource-poor farmers. Studies have thus been initiated to identify and select 'smart' cereal varieties with the early herbivory defence trait. Incorporating these varieties will enhance effectiveness of the push–pull technology by initiating biological control of stemborers at oviposition, the earliest stage of attack.

Exploiting plant–plant communication in the push–pull technology

Chemical communication also occurs between plants (Pickett et al., 2003). Plants respond to HIPVs produced by neighbouring plants that have been attacked. For example, when cotton is attacked by herbivorous mites, the rate of oviposition of mites is reduced on seedlings exposed to HIPVs from infested plants. Predatory mites are also attracted to uninfested seedlings exposed to the HIPVs (Bruin et al., 1992). Pioneering studies, conducted by icipe and Rothamsted Research chemical ecology groups, showed that intact plants such as molasses grass can produce the same compounds used as HIPVs in maize without being attacked by pests (Khan et al., 1997). More recently, we have shown that molasses grass volatiles induce defence responses in intact neighbouring maize plants (Khan et al., unpublished data). These volatiles thus serve as airborne signals that induce resistance in the neighbouring, unharmed plants. In addition to induced plant defence, plants are also 'primed' to respond more quickly or aggressively to future biotic or abiotic stress (Bruce et al., 2007). This condition may be initiated in response to an environmental cue that reliably indicates an increased probability of encountering a biotic stress such as HIPVs. Cotton is known to produce HIPVs in response to insect attack (Loughrin et al., 1994; Röse et al., 1996). In cotton the HIPV cis-

jasmone causes dramatic induction of direct and indirect defence compounds, such as (E,E)-4,8,12-trimethyltridecaca-1,3,7,11-tetraene (TMTT), which cause a reduction in colonisation by sucking insects (M. Birkett et al., unpublished data). cis-Jasmone can also activate defence in other plants (Birkett et al., 2000; Bruce et al., 2008; Moraes et al., 2008; 2009; Matthes et al., 2010). Research building on these findings is now being undertaken to identify companion plants with the ability to induce defence against insect attack in cereal crop varieties. These companion plants will then be used in intercropping arrangements within the push–pull technology to enhance natural plant defence.

Biotechnological options with desmodium for striga control

There is good evidence that desmodium root exudate chemistry is the source of the allelopathic effects on striga (Khan et al., 2002; 2008b; Tsanuo et al., 2003; Hooper et al., 2009; 2010). Desmodium, however, is a cattle fodder legume and therefore smallholder farmers would find intercropping with legumes for human consumption, such as cowpea and beans, more appealing. Root exudates of legumes for human consumption have been demonstrated to stimulate significant levels of striga germination through strigolactones (Matusova et al., 2005). These food legumes also possess the biosynthetic pathway for flavone biosynthesis, which comprises most of the desired biochemical pathway to C-glycosylflavones, which are crucial to the striga-inhibitory biosynthetic pathways (Hooper et al., 2009). However, these legumes have not proven to be as potent suppressors of striga as desmodium (Khan et al., 2002; 2007b). Some understanding has been gained of the secondary metabolism involved in the mechanism by which desmodium suppresses striga, and therefore there is now a possibility of transferring the relevant biochemical traits for striga suppression into the edible legumes. To be able to transfer the allelochemical pathway into food legumes there is a need to identify the active molecules and the genes responsible for the superior controlling effects of desmodium; efforts are being made to isolate these (Tsanuo et al., 2003; Hamilton et al., 2009). The active ingredients in desmodium root exudates will be fully characterised and compared with the exudates from traditional food legumes. If similar traits are found in the food legumes, conventional breeding could then be used to select and enhance the traits in the food legumes or, alternatively, if these traits are lacking, genes could be transferred directly from desmodium into the food legumes via genetic modification. In the longer term, it may be possible to transfer the same traits to cereal crops through heterologous gene expression, principally to the open pollinated varieties, rather than hybrids (Pickett *et al.*, 2010), for ease of accessibility to the smallholder farmers.

CONCLUSIONS

The push–pull technology is an excellent example of how basic science can be exploited to alleviate the major crop production constraints of pests, weeds and low soil fertility among smallholder farmers in Africa. The environmental benefits of the technology include soil and moisture conservation, improved soil health, enhanced biodiversity through eliminating pesticide usage, increased soil cover and organic matter rendering ecological services such as carbon sequestration. It also offers a platform upon which other livelihood improvement approaches such as livestock husbandry can be integrated, and provides an opportunity for the smallholder farmers to enter into the cash economy through the sale of grain and fodder surpluses and desmodium seed. The technology thus ensures sustained contribution of the cereal–livestock production farming system to attainment of food security and poverty alleviation in the region. An independent impact assessment report by Fischler (2010) based on participatory beneficiary assessment estimated that over 45,000 households are benefiting from the technology in East Africa. Household sizes in SSA range from 7 to 10 people, which suggests that more than 350,000 people are benefiting from push–pull. There is potential to extend push–pull to a much larger number of farmers in most of the regions of SSA, and to adapt it to withstand increasingly hot and dry conditions associated with climate change. With knowledge already acquired on the mechanisms by which the companion plants alleviate the constraints, particularly striga control by desmodium, there is now an opportunity to transfer these traits into food legumes, or to cereals themselves. Furthermore, discoveries relating to early herbivory alert and plant signalling represent opportunities for further enhancing the effectiveness of the technology, extending its appeal to

a range of farmer profiles in different agroecologies throughout SSA, and improving inbuilt long-term sustainability components.

From the foregoing, it is apparent that push–pull technology represents an effective and sustainable ecologically based pest management tool for low-input agriculture that incorporates elements from farmers' own practice of polycropping with agricultural intensification that exploits available biodiversity. Pest management programmes aimed at smallholder farmers should incorporate these tenets and be based on a clear understanding of the biological basis for the ecological interactions between components of biodiversity. Equally essential is understanding of the scientific mechanisms underpinning pest management in such technologies for maintaining sustainability. Moreover, the technology should offer multiple benefits to the farmers, with every biodiversity component having commercial value, or at least social value. Finally, the farmer must have complete ownership of the technology, apply it rigorously, and sufficiently understand it so as not to depart from essential practices within the technology.

REFERENCES

Abate T., van Huis, A. and Ampofo, J.K.O. (2000) Pest management strategies in traditional agriculture: an African perspective. *Annual Review of Entomology*, 45, 631–659.

Adesina, A.A. and Baidu-Forson, J.J. (1995) Farmers' perceptions and adoption of new agricultural technology: evidence from analysis in Burkina-Faso and Guinea, West Africa. *Agricultural Economics*, 13, 1–9.

Ahmed, M.M., Sanders, J.H. and Nell, W.T. (2000) New sorghum and millet cultivar introduction in sub-Saharan Africa: Impacts and research agenda. *Agricultural Systems*, 64, 55–65.

Ampofo, J.K.O., Saxena, K.N., Kibuka, J.G. *et al.* (1986) Evaluation of some maize cultivars for resistance to the stemborer *Chilo partellus* (Swinhoe) (Lepidoptera: Crambidae) in western Kenya. *Maydica*, 31, 379–389.

Amudavi, D.M., Khan, Z.R., Wanyama, J.M. *et al.* (2009a) Assessment of technical efficiency of farmer teachers in the uptake and dissemination of push–pull technology in Western Kenya. *Crop Protection*, 28, 987–996.

Amudavi, D.M., Khan, Z.R., Wanyama, J.M. *et al.* (2009b) Evaluation of farmers' field days as a dissemination tool for push-pull technology in western Kenya. *Crop Protection*, 28, 225–235.

Birkett, M.A., Campbell, C.A.M, Chamberlain, K. *et al.* (2000) New roles for *cis*-jasmone as an insect semiochemical and in plant defense. *Proceedings of the National Academy of Sciences of the United States of America*, 97, 9329–9334.

Birkett, M.A., Chamberlain, K., Khan, Z.R. *et al.* (2006) Electrophysiological responses of the lepidopterous stemborers *Chilo partellus* and *Busseola fusca* to volatiles from wild and cultivated host plants. *Journal of Chemical Ecology*, 32, 2475–2487.

Boahene K., Snijders T.A.B. and Folmer H. (1999) An integrated socioeconomic analysis of innovation adoption: the case of hybrid cocoa in Ghana. *Journal of Policy Modeling*, 21, 167–184.

Bruce, T.J.A., Wadhams, L.J. and Woodcock, C.M. (2005) Insect host location: a volatile situation. *Trends in Plant Science*, 10, 269–274.

Bruce, T.J.A., Matthes, M.C., Napier, J.A. *et al.* (2007). Stressful 'memories' of plants: evidence and possible mechanisms. *Plant Science*, 173, 603–608.

Bruce, T.J.A., Matthes, M.C., Chamberlain, K. *et al.* (2008). *Cis*-Jasmone induces *Arabidopsis* genes that affect the chemical ecology of multitrophic interactions with aphids and their parasitoids. *Proceedings of the National Academy of Sciences USA*, 105, 4553-4558.

Bruce, T.J.A., Midega, C.A.O., Birkett, M.A. *et al.* (2010) Is quality more important than quantity? Insect behavioural responses to changes in a volatile blend after stemborer oviposition on an African grass. *Biology Letters*, 6, 314-317.

Bruin, J., Dicke, M. and Sabelis, M.W. (1992) Plants are better protected against spider mites after exposure to volatiles from infested conspecifics. *Experientia*, 48, 525–529.

Burke, M.B., Lobell, D.B. and Guarino, L. (2009) Shifts in African crop climates by 2050, and the implications for crop improvement and genetic resources conservation. *Global Environmental Change*, 19, 317–325.

Chamberlain, K., Khan, Z.R., Pickett, J.A. *et al.* (2006) Diel periodicity in the production of green leaf volatiles by wild and cultivated host plants of stemborer moths, *Chilo partellus* and *Busseola fusca*. *Journal of Chemical Ecology*, 32, 565–577.

Cook, S.M., Khan, Z.R. and Pickett, J.A. (2007) The use of push-pull strategies in integrated pest management. *Annual Review of Entomology*, 52, 375–400.

De Groote, H., Vanlauwe, B., Rutto, E. *et al.* (2010a) Economic analysis of different options in integrated pest and soil fertility management in maize systems of Western Kenya. *Agricultural Economics*, 41, 471–482.

De Groote, H., Rutto, E., Odhiambo, G. *et al.* (2010b) Participatory evaluation of integrated pest and soil fertility management options using ordered categorical data analysis. *Agricultural Systems*, 103, 233–244.

Dicke, M. and Sabelis, M.W. (1988) Infochemical terminology: based on cost-benefit analysis rather than origin of compounds. *Functional Ecology*, 2, 131–139.

Emechebe, A., Ellis-Jones, J., Schulz, S. *et al.* (2004) Farmers' perception of the *Striga* problem and its control in northern Nigeria. *Experimental Agriculture*, 40, 215–232.

Epstein, D.L., Zack, R.S., Brunner, J.F. *et al.* (2000) Effects of broad spectrum insecticides on epigeal arthropod biodiversity in Pacific Northwest apple orchards. *Environmental Entomology*, 29, 340–348.

FAO (2006) *The state of food security in the world 2006: Eradicating world hunger – taking stock ten years after the World Food Summit*. Food and Agriculture Organisation of the United Nations, Viale delle Terme di Caracalla, 00153, Rome.

Fischler, M. (2010) *Impact assessment of push-pull technology developed and promoted by icipe and partners in eastern Africa*, Intercooperation, Switzerland, *icipe* Science Press, Nairobi.

Gebremedhin, W., Goudriaan, J. and Naber, H. (2000) Morphological, phonological and water-use dynamics of sorghum varieties (*Sorghum bicolor*) under *Striga hermonthica* infestation. *Crop Protection*, 19, 61–68.

Gethi, J.G., Smith, M.E., Mitchell, S.E. *et al.* (2005) Genetic diversity of *Striga hermonthica* and *Striga asiatica* populations in Kenya. *Weed Research*, 45, 64–73.

Hamilton, M.L., Caulfield. J.C., Pickett, J.A. *et al.* (2009) C-glucosylflavonoid biosynthesis from 2-hydroxynaringenin by *Desmodium uncinatum* (Fabaceae). *Tetrahedron Letters*, 50, 5656–5659.

Hassanali, A., Herren, H., Khan, Z.R. *et al.* (2008) Integrated pest management: the push-pull approach for controlling insect pests and weeds of cereals, and its potential for other agricultural systems including animal husbandry. *Philosophical Transactions of the Royal Society B*, 363, 611–621.

Henzel, E.F., Fergus, I.F. and Martin, A.E. (1966) Accumulation of soil nitrogen and carbon under a *Desmodium uncinatum* pasture. *Australian Journal of Experimental Agriculture and Animal Husbandry*, 6, 157–160.

Hilker, M. and Meiners, T. (2006) Early herbivore alert: insect eggs induce plant defense. *Journal of Chemical Ecology*, 32, 1379–1397.

Hooper, A.M., Hassanali, A., Chamberlain, K. *et al.* (2009) New genetic opportunities from legume intercrops for controlling *Striga* spp. parasitic weeds. *Pest Management Science*, 65, 546–52.

Hooper, A.M., Tsanuo, M.K., Chamberlain, K. *et al.* (2010) Isoschaftoside, a C-glycosylflavonoid from *Desmodium uncinatum* root exudate, is an allelochemical against the development of *Striga*. *Phytochemistry*, 71, 904–908.

Jagtap, S.S. and Abamu, F.J. (2003) Matching improved maize production technologies to the resource base of farmers in a moist savanna. *Agricultural Systems*, 76, 1067–1084.

Jama, B., Buresh, R.J. and Place, F.M. (1998) Sesbania tree fallows on phosphorus- deficient sites: maize yield and financial benefit. *Agronomy Journal*, 90, 717–726.

Jones, P., Devonshire, B.J., Holman, T.J. *et al.* (2004) Napier grass stunt: a new disease associated with a 16SrXI group phytoplasma in Kenya. *Plant Pathology*, 53, 519.

Jones, P., Arocha, T., Zerfy, J. *et al.* (2006) A stunting syndrome of Napier grass in Ethiopia is associated with a 16SrIII Group phytoplasma. *New Disease Reports*, 13, 42.

Kfir, R., Overholt, W.A., Khan, Z.R. *et al.* (2002) Biology and management of economically important lepidopteran cereal stemborers in Africa. *Annual Review of Entomology*, 47, 701–731.

Khan, Z.R. (2002) Cover crops, in *Encyclopedia of pest management* (ed. D. Pimentel), Markel Dekker, Inc., USA, pp. 155–158.

Khan, Z.R. and Pickett, J.A. (2004) The 'push–pull' strategy for stemborer management: a case study in exploiting biodiversity and chemical ecology, in *Ecological engineering for pest management: advances in habitat manipulation for arthropods* (eds G.M. Gurr, S.D. Wratten and M.A. Altieri), CABI, Wallington, pp. 155–164.

Khan, Z.R., Ampong-Nyarko, K., Chilishwa, P. *et al.* (1997) Intercropping increases parasitism of pests. *Nature*, 388, 631–632.

Khan, Z.R., Pickett, J.A., van den Berg, J. *et al.* (2000) Exploiting chemical ecology and species diversity: stemborer and *Striga* control for maize and sorghum in Africa. *Pest Management Science*, 56, 957–962.

Khan, Z.R., Pickett, J.A., Wadhams, L.J. *et al.* (2001) Habitat management strategies for the control of cereal stemborers and *Striga* in maize in Kenya. *Insect Science and Applications*, 21, 375–380.

Khan, Z.R., Hassanali, A., Overholt, W. *et al.* (2002) Control of witchweed *Striga hermonthica* by intercropping with *Desmodium* spp., and the mechanism defined as allelopathic. *Journal of Chemical Ecology*, 28, 1871–1885.

Khan, Z.R., Hassanali, A. and Pickett, J.A. (2006a) Managing polycropping to enhance soil system productivity: a case study from Africa, in *Biological Approaches to Sustainable Soil Systems* (ed. N. Uphoff), CRC Press, pp. 575–586.

Khan, Z.R., Midega, C.A.O., Pickett, J.A. *et al.* (2006b) Management of witchweed, *Striga hermonthica*, and stemborers in sorghum, *Sorghum bicolor*, through intercropping with greenleaf desmodium, *Desmodium intortum*. *International Journal of Pest Management*, 52, 297–302.

Khan, Z.R., Midega, C.A.O., Hutter, N.J. *et al.* (2006c) Assessment of the potential of Napier grass (*Pennisetum purpureum*) varieties as trap plants for management of *Chilo partellus*. *Entomologia Experimentalis et Applicata*, 119, 15–22.

Khan, Z.R., Pickett, J.A., Wadhams, L.J. *et al.* (2006d) Combined control of *Striga* and stemborers by maize–*Desmodium* spp. intercrops. *Crop Protection*, 25, 989–995.

Khan, Z.R., Midega, C.A.O., Wadhams, L.J. *et al.* (2007a) Evaluation of Napier grass (*Pennisetum purpureum*) varieties for use as trap plants for the management of African stemborer (*Busseola fusca*) in a 'push–pull' strategy. *Entomologia Experimentalis et Applicata*, 124, 201–211.

Khan, Z.R., Midega, C.A.O., Hassanali, A. *et al.* (2007b) Assessment of different legumes for the control of *Striga hermonthica* in maize and sorghum. *Crop Science*, 47, 728–734.

Khan, Z.R., Midega, C.A.O., Amudavi, D.M. *et al.* (2008a) On-farm evaluation of the 'push-pull' technology for the control of stemborers and striga weed on maize in western Kenya. *Field Crops Research*, 106, 224–233.

Khan, Z.R., Pickett, J.A., Hassanali, A. *et al.* (2008b) *Desmodium* species and associated biochemical traits for controlling *Striga* species: present and future prospects. *Weed Research*, 48, 302–306.

Khan, Z.R., Amudavi, D.M., Midega, C.A.O. *et al.* (2008c) Farmers' perceptions of a 'push–pull' technology for control of cereal stemborers and striga weed in western Kenya. *Crop Protection*, 27, 976–987.

Khan, Z.R., Midega, C.A.O., Njuguna, E.M. *et al.* (2008d) Economic performance of 'push-pull' technology for stemborer and striga weed control in smallholder farming systems. *Crop Protection*, 27, 1084–1097.

Khan, Z.R., James, D.G., Midega, C.A.O. *et al.* (2008e) Chemical ecology and conservation biological control. *Biological Control*, 45, 210–224.

Khan, Z.R., Midega, C.A.O., Wanyama, J.M. *et al.* (2009) Integration of edible beans (*Phaseolus vulgaris* L.) into the push–pull technology developed for stemborer and striga control in maize-based cropping systems. *Crop Protection*, 28, 997–1006.

Khan, Z.R., Midega, C.A.O., Bruce, T.J.A. *et al.* (2010) Exploiting phytochemicals for developing a 'push–pull' crop protection strategy for cereal farmers in Africa. *Journal of Experimental Botany*, 61, 4185–4196.

Lagoke, S.T.O., Parkinson, V. and Agunbiade, R.M. (1991) Parasitic weeds and control methods in Africa, in *Combating striga in Africa* (ed. S.K. Kim), Proceedings of an International Workshop organized by IITA, ICRISAT and IDRC, August 22–24, 1988. IITA, Ibadan, Nigeria, pp. 3–14.

Lal, R. (1988) Soil degradation and the future of agriculture in sub-Saharan Africa. *Journal of Soil and Water Conservation*, 43, 444–451.

Loughrin, J.H., Manukian, A., Heath, R.R. *et al.* (1994) Diurnal cycle of emission of induced volatile terpenoids by herbivore-injured cotton plants. *Proceedings of the National Academy of Sciences of the United States of America*, 91, 11836–11840.

Madhavi, L.A., Venkateswara, K.R. and Dashavantha, V.R. (2005) Production of transgenic plants resistant to leaf blast disease in finger millet (*Eleusine coracana* (L.) Gaertn.). *Plant Science*, 169, 657–667.

Matthes, M.C., Bruce, T.J.A., Ton, J. *et al.* (2010) The transcriptome of *cis*-jasmone-induced resistance in *Arabidopsis thaliana* and its role in indirect defence. *Planta*, 232, 1163–1180.

Matusova, R., Rani, K., Verstappen, W.A. *et al.* (2005) The strigolactone germination stimulants of the plant-parasitic *Striga* and *Orobanche* spp. are derived from the carotenoid pathway. *Plant Physiology*, 139, 920–934.

Midega, C.A.O., Khan, Z.R., van den Berg, J. and Ogol, C.K.P.O. (2005) Habitat management and its impact on maize stemborer colonization and crop damage levels in Kenya and South Africa. *African Entomology*, 13, 333–340.

Midega, C.A.O., Khan, Z.R., van den Berg, J. *et al.* (2008) Response of ground dwelling arthropods to a 'push–pull' habitat management system: spiders as an indicator group. *Journal of Applied Entomology*, 132, 248–254.

Midega, C.A.O., Khan, Z.R., van den Berg, J. *et al.* (2009) Non-target effects of the 'push–pull' habitat management strategy: Parasitoid activity and soil fauna abundance. *Crop Protection*, 28, 1045–1051.

Midega, C.A.O., Khan, Z.R., Amudavi, D.M., Pittchar, J. and Pickett, J.A. (2010) Integrated management of *Striga hermonthica* and cereal stemborers in finger millet (*Eleusine coracana* (L.) Gaertn.), through intercropping with *Desmodium intortum*. *International Journal of Pest Management*, 56, 145–151.

Moraes, M.C.B., Birkett, M.A., Gordon-Weeks, R. *et al.* (2008) *cis*-Jasmone induces accumulation of defence compounds in wheat, *Triticum aestivum*. *Phytochemistry*, 69, 9–17.

Moraes, M.C.B., Laumann, R.A., Pareja, M. *et al.* (2009) Attraction of the stink bug egg parasitoid *Telenomus podisi* to defence signals from soybean activated by treatment with *cis*-jasmone. *Entomologia Experimentalis et Applicata*, 131, 178–188.

Nordlund, D.A. and Lewis, W.J. (1976) Terminology of chemical-releasing stimuli in intra-specific and inter-specific interactions. *Journal of Chemical Ecology*, 2, 211–220.

Nwilene, F.E., Nwanze K.F. and Youdeowei A. (2008) Impact of integrated pest management on food and horticultural crops in Africa. *Entomologia Experimentalis et Applicata*, 128, 355–363.

Obura, E., Midega, C.A.O., Masiga, D. *et al.* (2009) *Recilia banda* Kramer (Hemiptera: Cicadellidae), a vector of Napier stunt phytoplasma in Kenya. *Naturwissenschaften*, 96, 1169–1176.

Obura, E., Masiga, D., Midega, C.A.O. *et al.* (2010) First report of a phytoplasma associated with Bermuda grass white leaf disease in Kenya. *New Disease Reports*, 21, 23.

Ostrowski, H. (1966) Tropical pastures for the Brisbane district. *Queensland Agricultural Journal*, 92, 106–116.

Oswald, A. (2005) *Striga* control technologies and their dissemination. *Crop Protection*, 24, 333–342.

Parker, C. (2009) Observations on the current status of *Orobanche* and *Striga* problems worldwide. *Pest Management Science*, 65, 453–459.

Päts, P. (1991) Activity of *Chilo partellus* (Lepidoptera: Pyralidae): eclosion, mating and oviposition time. *Bulletin of Entomological Research*, 81, 93–96.

Pettersson, J., Pickett, J.A., Pye, B.J. *et al.* (1994) Winter host component reduces colonization by bird cherry-oat aphid, *Rhopalosiphum padi* (L.) (Homoptera: Aphididae), and other aphids in cereal fields. *Journal of Chemical Ecology*, 20, 2565–2574.

Pickett, J.A., Rasmussen, H.B., Woodcock, C.M., *et al.* (2003) Plant stress signalling: understanding and exploiting plant-plant interactions. *Biochemical Society Transactions*, 31, 123–127.

Pickett, J.A., Khan, Z.R., Hassanali, A. *et al.* (2007) Chemicals involved in post-germination inhibition of *Striga* by *Desmodium*: opportunities for utilizing the associated allelopathic traits, in *Integrating new technologies for striga control: towards ending the witch-hunt* (eds G. Ejeta and J. Gressel), World Scientific, Singapore, New Jersey, London, pp. 61–70.

Pickett, J.A., Hamilton, M.L., Hooper, A.M. *et al.* (2010) Companion cropping to manage parasitic plants. *Annual Review of Phytopathology*, 48, 161–177.

Polaszek, A. (1998) *African cereal stemborers: economic importance, taxonomy, natural enemies and control*, CABI, Wallingford.

Press, M. C., Scholes, J. D., Riches, C. R., (2001). Current status and future prospects for management of parasitic weeds (*Striga* and Orobanche). In: Riches, C. R. (Ed.), *The World's Worst Weeds. Proceedings BCPC/Monograph Series no 77*, 71–90.

Röse, U.S.R., Manukian, A., Heath, R.R. *et al.* (1996) Volatile semiochemicals released from undamaged cotton leaves: a systemic response of living plants to caterpillar damage. *Plant Physiology*, 111, 487–495.

Sanchez, P.A., Shepherd, K.D., Soule, M.J. *et al.* (1997) Soil fertility replenishment in Africa: an investment in natural resource capital, in *Replenishing soil fertility in Africa* (eds R.J. Buresh *et al.*), *Soil Science Society of America*. Special Publication No. 51, Madison, pp. 1–46.

Tsanuo, M.K., Hassanali, A., Hooper, A.M. *et al.* (2003) Isoflavanones from the allelopathic aqueous root exudates of *Desmodium uncinatum*. *Phytochemistry*, 64, 265–273.

Turlings, T.C.J., Scheepmaker, J.W.A., Vet, L.E.M. *et al.* (1990) How contact experiences affect preferences for host-related odours in the larval parasitoid *Cotesia marginiventris* (Cresson) (Hymenoptera: Braconidae). *Journal of Chemical Ecology*, 16, 1577–1589.

Turlings, T.C., Loughrin, J.H., McCall, P.J. *et al.* (1995) How caterpillar damaged plants protect themselves by attracting parasitic wasps. *Proceedings of the National Academy of Sciences of the United States of America*, 92, 4169–4174.

van den Berg, J. (2006) Oviposition preference and larval survival of *Chilo partellus* (Lepidoptera: Pyralidae) on Napier grass (*Pennisetum purpureum*) trap crops. *International Journal of Pest Management*, 52, 39–44.

Westerman, P.R., van Ast, A., Stomph, T.J. and van der Werf, W. (2007) Long term management of the parasitic weed *Striga hermonthica*: Strategy evaluation with a population model. *Crop Protection*, 26, 219–227.

Whiteman, P.C. (1969) The effects of close grazing and cutting on the yield, persistence and nitrogen content of four tropical legumes with Rhodes grass at Samford, south-eastern Queensland. *Australian Journal of Experimental Agriculture and Animal Husbandry*, 9, 287–294.

Whitney, A.S. (1966) Nitrogen fixation by three tropical forage legumes and the utilization of legume-fixed nitrogen by their associated grasses. *Herbage Abstracts*, 38, 143.

Chapter 17

USING NATIVE PLANT SPECIES TO DIVERSIFY AGRICULTURE

Douglas A. Landis, Mary M. Gardiner and Jean Tompkins

Biodiversity and Insect Pests: Key Issues for Sustainable Management, First Edition. Edited by Geoff M. Gurr, Steve D. Wratten, William E. Snyder, Donna M.Y. Read.
© 2012 John Wiley & Sons, Ltd. Published 2012 by John Wiley & Sons, Ltd.

INTRODUCTION

The human activity of agriculture has pervasive influences on the health of the Earth's ecosystems. At present, over one-third of Earth's ice-free land surface is devoted to agriculture and grazing systems (Foley *et al.*, 2007) and the continuing intensification of agriculture is resulting in losses of biodiversity and ecosystem function (Flynn *et al.*, 2009; Geiger *et al.*, 2010). Given continued expansion of human populations and the increasing expectation of higher standards of living, agriculture is likely to continue to extend its footprint on the globe. In this context, we explore whether there are ways of sustaining agricultural productivity while also enhancing the range of benefits that humans obtain from the biodiversity in such working landscapes.

The benefits that humans obtain from biodiversity have been termed 'ecosystem services' (Gillespie and Wratten, chapter 4 of this volume; Daily *et al.*, 1997). These benefits can be categorised into four broad groups: supporting services such as soil formation and nutrient cycling that maintain the basic productivity of ecosystems; provisioning services such as the food, fuel and fibre that we harvest from them; regulating services like flood prevention, pollination and pest suppression that influence system functioning (see Box 17.1); and cultural services including the recreational, aesthetic and spiritual values that humans derive from the world around us (Millennium Ecosystem Assessment, 2005). All of these services are clearly vital to human well-being and provide a framework for evaluating the impacts of differing land management strategies.

Beneficial arthropods (insects, spiders and mites) provide key ecosystem services in all agricultural landscapes. These arthropod-mediated ecosystem services (Isaacs *et al.*, 2009) include: decomposition and nutrient recycling, pollination, pest suppression and even aesthetic and spiritual values. While the precise value of these services is difficult to measure, one attempt has placed the value of insects to the US at over $57 billion per year (Losey and Vaughan, 2006). Those authors estimate the value of crop pollination and pest suppression alone at more than $7 billion per year to US agriculture. Thus, maintaining these services is clearly a priority.

Entomologists have long recognised that the arthropods which provide ecosystem services in agriculture depend on a variety of plant resources in their environment. For example, pollinators require a constant supply of floral resources over their lifetime. Likewise, predatory and parasitic insects that help to control crop pests frequently require access to pollen and nectar resources to maintain and enhance their life span and fecundity (Wackers and van Rijn, chapter 9 of this volume). Research aimed at managing plant communities in and around agricultural fields to enhance the abundance and activity of beneficial insects has been an area of active research for several decades and the subject of numerous papers, books and reviews (Pickett and Bugg, 1998; Landis *et al.*, 2000; Gurr *et al.*, 2003; 2004).

In this chapter we focus on the use of native plants to enhance beneficial arthropods. First, we briefly review the major concepts and theories relevant to topic. Next we provide key examples of the real-world use of biodiversity for pest management. Finally, we conclude with some thoughts on the future of this exciting area of applied research.

MAJOR CONCEPTS

Plants provide resources for beneficial insects

Disturbances within agroecosystems such as tillage, pesticide application and crop harvesting can negatively impact the survival of beneficial insects and the supply of arthropod-mediated ecosystem services within crop fields. To provide these critical services, beneficial insects require resources including shelter and overwintering sites, alternative hosts or prey and nectar and pollen (Landis *et al.*, 2000; Jonsson *et al.*, 2008; Lundgren, 2009). The introduction of flowering plants into agroecosystems can provide these key resource needs and increase the diversity of beneficial

---

Box 17.1 Pest suppression services

Pest control is an ecosystem service that can benefit human welfare through its moderation of crop damage, increasing both the quantity and quality of yields.

insects in agricultural systems including; field crops, vegetables, tree fruit and turf (Altieri and Whitcomb, 1979a; Patt *et al.*, 1997; Pickett and Bugg, 1998; Chaney, 1998; Braman *et al.*, 2002; Gurr *et al.*, 2003; Fitzgerald and Solomon, 2004; Lee and Heimpel, 2005; Begum *et al.*, 2006). These issues are explored in detail in chapter 1 of this volume, but the rest of this section provides a concise overview of the three ways in which flowering plant resources can be critical to the stability of beneficial insect communities in agroecosystems.

Shelter and overwintering habitat

Harvest of annual crops and removal of plant debris from agricultural fields results in large areas of bare ground and reduces overwintering habitat for beneficial insects. Nearby non-crop vegetation can, however, provide some valuable overwintering habitat (Gámez-Virués *et al.*, chapter 7 of this volume; Scherber *et al.*, chapter 8 of this volume; Jonsson *et al.*, 2008; Geiger *et al.*, 2009). The establishment of flowering resource plants within agroecosystems can further enhance natural enemy overwintering. Such habitats can also provide refugia from adverse conditions and predators during the growing season and supply nesting sites for stem-nesting pollinators. In addition, an unaltered soil surface within floral resource plantings can provide an undisturbed soil and litter layer to support overwintering of natural enemies and pollinators at or below the soil surface (Altieri *et al.*, chapter 5 of this volume). The presence of these sites near crop fields can promote early-season colonisation of fields by emerging natural enemies and result in enhanced pest suppression primarily by generalist predators such as Carabidae, Staphylinidae and Araneae which are active in the spring (Symondson *et al.*, 2002; Geiger *et al.*, 2009). The ability of native pollinators to nest within non-crop habitats near crop fields can also enhance the frequency of pollinator visitation and the supply of pollination services (Ricketts, 2004).

Alternative hosts and prey

Access to alternative hosts and prey has been shown to alter the distribution, dispersal frequency, longevity and fecundity of natural enemies (Eubanks and Denno, 1999; Tylianakis *et al.*, 2004; Lee and Heimpel, 2005; Jacometti *et al.*, 2010). Non-crop habitats surrounding

agroecosystems can also enhance biological control services by providing locally supported natural enemy populations to nearby crop fields (Thies *et al.*, 2003; Gardiner *et al.*, 2009). The presence of flowering plants enhances biological control services by supplying alternative hosts or prey early in the season, allowing populations of natural enemies to establish prior to the arrival of pest populations in nearby crops (Landis *et al.*, 2000). These plants may support a specific host necessary for the survival of a parasitoid, or a diversity of potential prey to sustain generalist predators. For generalist predators, mixed plantings are likely to support a diversity of alternative insect prey throughout the growing season. For example, Fiedler and Landis (2007a) found that a mixed planting of native perennial plants can support a diversity of alternative hosts and prey for natural enemies including aphids, leafhoppers and caterpillars.

Pollen and nectar

Many parasitoids and predators require non-host or prey food sources such as pollen and nectar (Wackers and van Rijn, chapter 9 of this volume; Lundgren, 2009). The sugars in floral nectar are commonly used by natural enemies as an energy source (Jervis and Kidd, 1986; Heimpel *et al.*, 2004). Moreover, providing this resource can increase the longevity of natural enemies (Wäckers, 2001) and the supply of biological control services (Jonsson *et al.*, 2008; Wäckers *et al.*, 2008; Jonsson *et al.*, 2010). For example, availability of flowering buckwheat (*Fagopyrum esculentum* Moench) and alyssum (*Lobularia maritima* (L.) Desv.) has been shown to enhance parasitism of the light-brown apple moth (*Epiphyas postvittana* Walker) an important pest of grapes (Berndt and Wratten, 2005). Landis *et al.* (2000) note that substantial work has also examined the use of flowering plant resources by predators such as aphidophagous syrphids. For example, borders of the annual plant *Phacelia tanacetifolia* Benth. in cabbage systems have been shown to increase syrphid numbers and decrease aphid populations. Pollinators also benefit from additional pollen and nectar resources within agroecosystems. Flowering plants provide bees with the bulk of their nutrition (Michener, 2007). Often, a pollinating species is active in the environment either before or after a target crop has bloomed (Tuell *et al.*, 2008). Therefore, the incorporation of flowering plant species to provide resources throughout the life cycles of key pollinators will sustain pollination services within an agroecosystem.

Importantly, not all plant species supply equivalent alternative food resources (Wäckers and van Rijn, chapter 9 of this volume). Nectar can vary in sugar concentration and composition (Tompkins *et al.*, 2010) and this may affect the attractiveness of plants to natural enemies or pollinators as well as the benefits they gain from consuming particular nectar resources. In addition to sugars, beneficial insects acquire a diversity of amino acids, lipids, vitamins, proteins, minerals, organic compounds and secondary plant metabolites from pollen and nectar resources (Baker and Baker, 1983; Tompkins *et al.*, 2010). Therefore, the potential benefits as well as any possible negative impacts (i.e. plant toxins or deterrents) produced by resource plants should be evaluated to maximise the benefits of flowering plant addition (Tompkins *et al.*, 2010).

A history of non-native plants for habitat management

The incorporation of flowering plants into agricultural systems as a habitat management practice can provide the important resources to beneficial insects addressed in the preceding sections, enhancing natural enemy activity in agroecosystems. Interestingly, most habitat management projects have focused on a small number of non-native annual or biennial plants. Fiedler *et al.* (2008) note the repeated use of alyssum (*L. maritima*), borage (*Borago officinalis* L.), coriander (*Coriandrum sativum* L.), dill (*Anethum graveolens* L.), buckwheat (*F. esculentum*), faba bean (*Vicia faba* L.) and phacelia (*P. tanacetifolia*) in habitat management programmes. The use of annual plants such as these species has several benefits: the seed is readily available and cost-effective, the plants are easy to cultivate in strips within annual crops or in borders alongside agricultural fields, and they produce prolific floral displays and can be planted in different locations on-farm each year (Fiedler *et al.*, 2008). However, the use of non-native annuals has limitations as well. Growers incur costs annually, as they must be purchased and planted each year. In addition, these plants do not provide overwintering habitat if removed following crop harvest, they do not enhance native biodiversity and in some cases non-native species have been found to become invasive (Isaacs *et al.*, 2009).

Box 17.2 The diversity of native plants

Currently, only seven annual flowering plant species are common in the literature as nectar resources for parasitoid wasps. These plants have had some success in improving biological control; ranging from improving parasitoid fitness in the laboratory, to reducing pest populations in the field. They are, however, often deployed outside their native range. Considering that there are more than 200,000 other flowering plant species on Earth, there is large scope for utilising other species with the potential to enhance parasitoid fitness.

The selection of native plants to support beneficial insects

Given the potential negatives associated with the use of non-native plants for habitat management, Isaacs *et al.* (2009) present a rationale supporting the use of native plants (see Box 17.2). First, plants native to a given region are locally adapted. Frequently, these native species will have lower water, nutrient and pest control requirements within their native range compared with non-native species (Isaacs *et al.*, 2009). Second, native perennials can create habitats that provide resources to beneficial insects year-round. In addition to supporting natural enemy and pollinator populations, native perennial plant habitats also supply additional ecosystem services and environmental benefits by enhancing wildlife habitat, stabilising soils and reducing agricultural runoff. Third, the use of native plants directly supports their conservation by encouraging their propagation and reintroduction into agroecosystems where they have been largely eliminated. Fourth, establishing these permanent habitats can be cost-effective, as native perennial plants require little management after establishment and can persist within agroecosystems for decades (Isaacs *et al.*, 2009). Finally, in many cultures native plants have important medicinal, religious and aesthetic values (Fiedler *et al.*, 2008). For example, the reincorporation of native plant species may support the production of indigenous medicines or culturally important arts and crafts.

Table 17.1 Selected examples of the use of native plants to enhance ecosystem services in agricultural landscapes (see text for discussion and references).

Example	Location	Concept
Insectary hedgerows	California, USA	Native shrubs to enhance pest suppression and ecological function in intensive agriculture
Greening Waipara	North Canterbury, New Zealand	Restoring native vegetation to vineyards for the provision of multiple ecosystem services
Revegetation by design	Queensland, Australia	Replacement of weedy and pest-supporting vegetation with native plants to enhance natural enemies
Reincorporating prairies	Midwestern USA	Restoring multiple ecosystem services by reincorporation of native prairie vegetation into annual agriculture

How to screen native plants for habitat management

Given the potential of native plants in attracting and sustaining natural enemy and pollinator populations and providing several additional ecosystem services, it is critical to select the best species to use in a given agricultural context. Fiedler and Landis (2007a; 2007b) and Fiedler *et al.* (2008) caution that to maximise the benefits of habitat management, plant species must be selected carefully. First, it is important to consider plant architecture. Plants with deep, narrow corollas may not support natural enemies that cannot reach pollen and nectar resources (Jervis *et al.*, 1993; Orr and Pleasants, 1996; Wäckers *et al.*, 1996). Second, there is a need to evaluate the phenology of target pests, natural enemies, pollinators and the cropping system. Pollen and nectar resources need to present at the correct time to attract and support beneficial insects and arthropod-mediated ecosystem services (Jervis *et al.*, 1993; Orr and Pleasants, 1996; Colley and Luna, 2000). Third, it is important that the selected native plants support alternative hosts and prey, but not herbivores which will attack the focal crop (Baggen and Gurr, 1998; Baggen *et al.*, 1999; Lee and Heimpel, 2005; Winkler *et al.*, 2005; Lavandero *et al.*, 2006). Finally, consideration must be given to the regional context of the habitat management project and select plant species which are currently present in, or adapted to, the growing conditions present (Altieri and Whitcomb, 1979b; Foster and Ruesink, 1984; Idris and Grafius, 1995; Nicholls *et al.*, 2001).

CASE STUDIES

Some classic examples of the use of native plants to enhance beneficial insects include the use of 'beetle banks' (Collins *et al.*, 2002; MacLeod *et al.*, 2004) and 'sown weed strips' (Nentwig *et al.*, 1998) in Western Europe. Here we focus on several contemporary efforts to use native plants to enhance multiple ecosystem services (Table 17.1).

Insectary hedgerows

Historically, hedgerows have been defined as linear habitats containing bushes, shrubs or trees that border farm fields or lanes. Originally managed to confine animals or form boundaries, in agricultural landscapes hedgerows also serve as windbreaks and frequently harbour higher biodiversity than the surrounding farmlands (Bugg *et al.*, 1998). Unfortunately, with increased agricultural intensification (mechanisation and chemical use), hedgerows have disappeared from many agricultural landscapes resulting in a loss of potential ecosystem services. However, researchers and land managers in California have been re-inventing hedgerows to serve multiple purposes in modern agricultural landscapes.

Agriculture in California's Central Valley can be highly intensive, leaving little habitat for native plants and animals. Beginning in the 1990s researchers began to investigate the potential to diversify these systems to support beneficial insects, suppress weeds, improve water quality and provide habitat for wildlife

(Bugg *et al.*, 1998). One of the concepts that emerged from this work was the idea of establishing 'insectary hedgerows' along field borders or riparian areas, consisting primarily of native perennial shrubs, forbs and grasses (Long *et al.*, 1998). Early work focused on selecting appropriate plant species for various soil types, establishment techniques and studies of the pest and beneficial insects that utilised the habitats. To determine how far beneficial insects visiting the hedgerows moved into adjacent crop fields, researchers used a marking technique using rubidium, a naturally occurring element. When sprayed on hedgerow plants a solution of rubidium chloride was incorporated into plant pollen and nectar. Insects feeding on these resources developed elevated rubidium levels that could be detected by flame emission spectrophotometry. Using this method they documented that many natural enemies moved 75 m from the hedgerows into crop fields, while some such as green lacewings were found as far as 137 m away (Long *et al.*, 1998).

Research on hedgerows has also investigated the relative attractiveness of native grasses to pest and beneficial insects. For example the native stinkbug, *Euschistus conspersus* Uhler, a pest of tomato, is frequently found in weedy roadside vegetation but seldom in stands of native grasses. Replacement of weedy vegetation with native grasses has been proposed as a means of reducing overwintering habitat for the stinkbug and supporting its natural enemies (Ehler *et al.*, 2002). Combining both shrub and grasses in hedgerows has also been investigated. For example, hedgerows consisting of 3 m-wide strips of native perennial grasses planted on one or both sides of a row of native shrubs (Figure 17.1), attracted a wide variety of beneficial insects. Moreover, such hedgerows always attracted higher ratios of beneficial insects than pests (Figure 17.2). This suggests that such habitats could serve as a replacement for weedy vegetation and support higher levels of beneficial insects on California farms (Morandin *et al.*, in press).

Finally, in an analysis of farmer attitudes, Brodt *et al.* (2009) identified the key social, economic and agronomic factors that influence hedgerow adoption. Farmers cited ability to attract beneficial insects, increased wildlife habitat, increased dust control and increased shade for livestock among the key benefits they perceive from hedgerows. Among the key constraints reported were cost – native plant hedgerows cost about US$12 per linear metre to establish and maintain for the first three years (Long and Anderson,

2010) – fear of harbouring pests and lack of certainty about the benefits. They concluded that both increased research to define the benefits and the availability of cost-sharing programmes were keys to increased adoption.

Greening Waipara

The Waipara Valley, a wine-producing region of North Canterbury, New Zealand was originally covered in extensive forests, woodlands and shrublands. However, early Polynesian burning and more recent European farming have resulted in the loss of most native biodiversity and its replacement by pasture and crops (Molloy, 1969; Meurk, 2008) and most recently vineyards. It is estimated that only 1% of indigenous vegetation remains in Waipara (C. Meurk, personal communication, 2011).

Today there is a growing awareness that biodiversity in these agricultural lowlands is at risk and that action by private landowners is required to reduce potential losses (Lee *et al.*, 2008; Moller *et al.*, 2008; Landcare Research, 2009). Partly in response to this awareness and partly due to sustainable production opportunities, Waipara landowners began to collaborate with other interest groups to reincorporate native vegetation into the agricultural landscape. By 2005, this effort had evolved into the Greening Waipara project (http://bioprotection.org.nz/greening-waipara)   and included the collaboration of the Waipara winegrowers association, local council and community members, international financiers, scientists from Lincoln University and the Crown Research Institute Landcare Research. The project proposed that the establishment of native New Zealand plants within vineyard properties was a form of ecological engineering (Gurr *et al.*, 2004) aimed at enhancing ecosystem services (Costanza *et al.*, 1997; Daily *et al.*, 1997) with tangible benefits for improved sustainability of the area's wine production (see Box 17.3).

Potential benefits for growers included marketing and eco-tourism opportunities, erosion management, filtration of winery effluent and the enhanced biological control of pests. This latter service, which was the initial link between the winegrowers and Lincoln University researchers, became a key focus of the project. Previous research undertaken in 2005 had shown that non-native buckwheat (*F. esculentum*) could be deployed within Waipara vineyards to enhance the biological

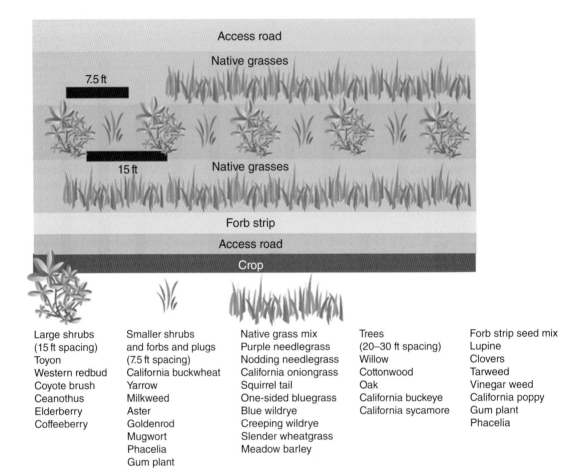

Large shrubs
(15 ft spacing)
Toyon
Western redbud
Coyote brush
Ceanothus
Elderberry
Coffeeberry

Smaller shrubs
and forbs and plugs
(7.5 ft spacing)
California buckwheat
Yarrow
Milkweed
Aster
Goldenrod
Mugwort
Phacelia
Gum plant

Native grass mix
Purple needlegrass
Nodding needlegrass
California oniongrass
Squirrel tail
One-sided bluegrass
Blue wildrye
Creeping wildrye
Slender wheatgrass
Meadow barley

Trees
(20–30 ft spacing)
Willow
Cottonwood
Oak
California buckeye
California sycamore

Forb strip seed mix
Lupine
Clovers
Tarweed
Vinegar weed
California poppy
Gum plant
Phacelia

Figure 17.1 Suggested design criteria for California insectary hedgerows showing relative spatial relationships of native grasses, forbs and shrubs, and listing of potential species. Reprinted with permission from Long and Anderson, 2010.

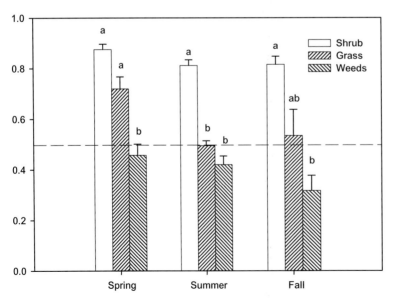

Figure 17.2 Ratio of beneficial (selected predators and parasitoids) to total insect abundance (beneficials + key herbivores) in four planted hedgerows and adjacent weedy areas over two years during the growing season, Yolo County, California. Different letters over bars indicate significant differences in proportions (p < 0.05). Reprinted with permission from Morandin *et al.*, in press.

Box 17.3 Consumers matter

As part of the Greening Waipara project, visitors to vineyards were encouraged to walk native plant trails and provide their opinions on the use of native plants in sustainable wine production. Once introduced to the trails, the many visitors were interested in walking them, and of those who did, the vast majority felt it added to their experience of the winery. Surveys indicated that after walking the trails, visitors felt more connected to the winery after and some were more likely to buy wine from the vineyard based on the experience. Additional studies will be required to determine if such activities actually increase consumer loyalty to a wine-growing region or to particular vineyards.

control of the light-brown apple moth (*E. postvittana*) (Scarratt, 2005). The Greening Waipara project built upon this work to investigate whether native New Zealand plants could be used to enhance vineyard natural enemies.

In 2007, 14 native ground covering plant species were tested under vines at a North Canterbury field site to assess their potential to improve vineyard pest man-

agement (Plate 17.1). The screening of these plants considered their growth and survival, floral characteristics, ability to host alternative prey and the effect of their floral resources upon both natural enemy and pest longevities in the laboratory (Tompkins, 2010). After considering agronomic factors, three of the 14 plant species, creeping pohuehue (*Muehlenbeckia axillaris* (Hook. f.) Walp.), shore cotula (*Leptinella dioica* Hook. f.) and bidibid (*Acaena inermis* Hook. f.) were identified as having potential to enhance pest control in vineyards. The Greening Waipara project disseminated these findings to growers via regular newsletters (available at: http://bioprotection.org.nz/greening-waipara), actively promoting the use of native plants, not only for pest control but also for various other ecosystem services. Grower workshops, newspaper articles, tasting room brochures, internet resources and a news story on national television in 2005 are examples of other media the project has employed to enhance the awareness and adoption of these relatively new concepts. While awareness is now high, adoption of protocols requires further work and will then likely depend on the effective and efficient transfer of knowledge between the scientists and growers. Warner (2006), who studied the adoption of conservation biological control protocols by Californian wine-growers, found that adoption was improved where a multi-year relationship between the growers, their organisation and extension scientists existed. Given the strong collaboration built through the Greening Waipara project, potential exists for widespread adoption of research proven protocols. The importance of effective extension methods is further explored by Escalada and Heong (chapter 12 of this volume).

Revegetation by design

Initiated in 2003 the 'Revegetation by Design' project, based in Queensland, Australia and facilitated by Horticulture Australia Ltd, promotes the replacement of weedy areas within horticultural systems with native plants. This is intended to achieve several benefits for growers, including long-term weed management, improved farm profitability, enhanced biodiversity and improved invertebrate pest management. It was proposed that this latter benefit would be achieved through two avenues. Firstly the native plants would replace weedy species which were known to be key hosts to pest thrips (Thysanoptera: Thripinae) and would themselves be poor hosts to these pests. Pest thrips can cause extensive feeding damage to vegetable crops and are known vectors of the Tomato Spotted Wilt Virus (TSWV). In the Northern Adelaide Plains thrips cost the horticultural industry around A $25 million annually (Taverner *et al.*, 2006). Consequently, removing the surrounding reservoir of pests that may damage vegetable crops could significantly reduce this cost. Secondly, the native plants may provide resources (shelter, nectar, alterative prey or pollen) to support natural enemies of vegetable pests which could move into the crop to provide early-season pest control services.

Research by CSIRO's Ecosystem Sciences Division found that native vegetation could indeed help vegetable farmers manage invertebrate pests by replacing weeds which hosted the pests (Schellhorn *et al.*, 2010) and by supporting populations of natural enemies, particularly hoverflies and parasitoids (Schellhorn, 2007; Stephens *et al.*, 2006). Trials in the Northern Adelaide Plains horticultural region investigated the use of weedy vegetation and native plants by four pest thrips (*Frankliniella occidentalis* (Pergande), *Frankliniella schultzei* Trybom, *Thrips tabaci* Lindeman, and *T. imaginis* Bagnall). Nineteen exotic and 12 native plant taxa were sampled between March 2003 and May 2004. While the exotic plants were found to harbour high densities of the pest thrips, all 12 native plant species demonstrated low levels of thrips (Schellhorn *et al.*, 2010). Further research compared parasitic wasp assemblages sampled from native plants with those from weedy exotic plants. This work found that both plant types supported abundant and species-rich wasp assemblages, including many species considered to be important biological control agents of vegetable pests (Stephens *et al.*, 2006). Overall, while weedy plants may provide resources to biological control agents they are not selective in their provisions and also support pests. The native Australian vegetation, however, offers not only the potential to enhance biological control agents but also the ability to reduce pest reservoirs by failing to be suitable as pest hosts. These studies provide evidence that establishing native Australian vegetation in areas traditionally covered by weeds offers growers a means to improve their management of vegetable pests.

While further work remains to clarify the extent to which particular native plants would reduce pest damage to vegetable crops, the concept has been met with widespread grower and industry support. A guidebook made available to growers of the Northern Adelaide Plains in 2006 provides information on the project's concept and gives detailed plans of how to implement the 'Revegetation by Design' approach (Taverner *et al.*, 2006). Tailoring the plans to meet the needs of various vegetable systems, which are faced with different pest assemblages, is made possible by the guide detailing the invertebrate assemblages supported by each of the proposed native plant species compared to a common Brassica weed, *Rapistrum rugosum* (L.). Invertebrates are grouped as either pest, beneficial or non-target, giving the grower an idea of which native plant would be most appropriate given their pest management concerns (Table 17.2). Costs of establishing and maintaining the native plants are also provided so that growers can assess the economic viability of adopting the project's concept. This guidebook, alongside presentations to interest groups at various conferences and workshops as well as online publications, has further developed the awareness of the concept amongst growers. The next step to the project will be to demonstrate that this on-farm manipulation of non-crop vegetation in favour of native Australian plants significantly disadvantages pest populations to such an extent that pest control costs are reduced.

Reincorporating prairies

The tallgrass prairie once formed the dominant vegetation type over vast areas of the Midwestern US and parts of Canada. However, due in large part to its conversion to field crop agriculture, less than 1% of the original prairie remains (Packard and Mutel, 2005). Virtually all of these remnant patches are highly isolated, limiting their long-term viability. Given this situation, there is intense interest in prairie conservation and restoration and many groups are investigating ways to reincorporate prairie vegetation into Midwestern US landscapes (see Box 17.4).

In 2003, Michigan State University initiated a collaborative project to investigate the potential use of native prairie plants to support beneficial insects in agricultural landscapes. Partnering with commercial native plant producers they compared the attractiveness of native and exotic plants to both natural enemies and pollinating bees (Fiedler and Landis, 2007a; Tuell *et al.*, 2008). In total they measured the abundance of natural enemies on the flowers of 43 species of native prairie and savanna species and compared them to

Table 17.2 Comparison of pest thrips found on the common weed, giant mustard (*Rapistrum rugosum* (L.)), in comparison to selected native plants. Reprinted with permission from Taverner *et al.*, 2006.

Plant Species		Thrips Species			
Common Name	**Species Name**	**Western flower thrips (V)** *Frankliniella occidentalis*	**Tomato thrips (V)** *Frankliniella schultzei*	**Onion thrips (V)** *Thrips tabaci*	**Plague thrips (Native)** *Thrips imaginis*
Giant mustard weed	*Rapistrum rugosum*	••	••	••	••••
Windmill grass	*Chloris truncata*	•	nil	•	•••
Wallaby grass	*Austrodanthonia linkii*	nil	•	nil	•••
Black-head grass	*Enneapogon nigricans*	nil	•	•	•••
Kangaroo grass	*Themeda triandra*	nil	nil	nil	••••
Berry saltbush*	*Atriplex semibaccata*	nil	nil	nil	•••
Grey coastal saltbush	*Atriplex cinerea*	•	•	•	•••
Marsh saltbush	*Atriplex paludosa*	nil	nil	nil	••
Small leafed bluebush	*Maireana brevifolia*	••	•	•••	••
Fragrant saltbush*	*Rhagodia parabolica*	•	•	nil	••
Fleshy saltbush	*Rhagodia crassifolia*	nil	nil	••	••
Ruby saltbush	*Enchylaena tomentosa*	nil	nil	nil	••
Elegant wattle*	*Acacia victoriae*	•	•	•	••
Tallerack	*Eucalyptus tetragona*	•	•	•	••
Muntries*	*Kunzea pomifera*	•	nil	•	•

(V) Indicates that the thrips transmits Tomato Spotted Wilt Virus. *Tested free of Tomato Spotted Wilt Virus.

LEGEND

Not detected	nil
Rare	•
Occasional	••
Common	•••
Numerous	••••

Averages of thrips derived from sampling September–December 2005

- All Pest thrips that transmit Tomato Spotted Wilt Virus (TSWW) were found on the Giant mustard weed, but they were not always seen on native plants.
- Overall, native plague thrips was the most abundant thrips, but they were not as important a pest because they do not transmit TSWW.
- Plant species showing 'nil' means no thrips of this type were seen throughout the entire sampling period.
- The 'best bet' to replace weeds would be plant species that have 'nil' or 'rare' thrips that transmit TSWW.

Box 17.4 A win–win for agriculture and the environment

The incorporation of native plants into agricultural landscapes for ecosystem service enhancement is a relatively new endeavour. It has the potential to provide mutually beneficial outcomes for both food production and conservation; two factions traditionally in conflict.

---

five species of non-native plants that have been widely recommended as attractive to natural enemies. They found that the abundance of beneficial insects on native plants frequently equalled or exceeded those on recommended non-native plants in bloom at similar times (Figure 17.3). Based on their results they developed a list of 26 species that were highly attractive to beneficial insects, forming a set of plants that bloomed over the entire summer and would establish under a variety of growing conditions (Figure 17.4) (Fiedler *et al.*, 2008). Many of these species are now being used in ongoing research and demonstration plots as well as being incorporated into state programmes to promote wildlife habitat. For example, researchers are currently using sets of these plants to investigate the potential for native plant habitats to enhance pest control in blueberry, grape and cherry productions systems in Michigan (B. Blauw and R. Isaacs, personal communication).

US federal agencies are also promoting the use of native plants to enhance wildlife conservation and improve ecosystem services in agricultural landscapes. In Michigan, the US Department of Agriculture – Natural Resource Conservation Service (USDA-NRCS) currently has three programmes that support landowners in establishing native prairie plants for conservation purposes. The largest is the Conservation Reserve Program (CRP), which sponsors multiple programmes that provide technical advice and cost-sharing for landowners to establish a variety of habitats including those based on native prairie plants (Plate 17.2). Two additional USDA-NRCS programmes: the State Acres for Wildlife (SAFE) program and the Conservation Reserve Enhancement Program (CREP), feature the use of native species as designated in their 'eligible wildflower lists'.

Other Midwestern states in the USA are exploring creative ways to reincorporate prairie plants into conventional agriculture (Schulte *et al.*, 2008). The Science-based Trials of Rowcrops Integrated with Prairies (STRIPS) experiment at the Neal Smith National Wildlife Research Center in Iowa is a unique example (http://www.nrem.iastate.edu/research/STRIPs). In this project, a multi-disciplinary group of scientists are exploring how integration of perennial prairie habits at different levels influences multiple ecosystems services at the watershed level. They are testing the hypothesis that even relatively small levels of perennial habitat (e.g. 10–20% of the total watershed) might result in disproportionately large increases in biodiversity and improvements in ecosystems functioning, particularly for water, nutrient and carbon cycling.

Finally, the increased interest in the potential of cellulosic biofuels has created an opportunity for potentially large-scale reestablishment of prairie habitats in the Midwestern US (Tilman *et al.*, 2006). Recent work conducted as part of the US Department of Energy's Great Lakes Bioenergy Research Center confirms that perennial grassland-based biofuel crops (e.g. switchgrass and mixed prairie grasses and forbs) support greater abundance of insect natural enemies and pollinators than the current major biofuel crop, corn (Gardiner *et al.*, 2010) and that this translates into increased pest suppression services within the biofuel crop (Werling *et al.*, 2011). The occurrence of more diverse and perennial habitats in agricultural landscapes is known to increase the probability of effective biological control (Bianchi *et al.*, 2006) and this is predicted to be the case in the addition of perennial biofuel crops (Landis and Werling, 2010).

CONCLUSION

The intensification of farming – with associated losses of biodiversity and ecological function – has heightened concerns about the sustainability of agricultural ecosystems in many regions of the world. Understanding the value of native plants in terms of their provision of ecosystem services presents opportunities for their reincorporation into agricultural landscapes. The case studies presented here provide evidence that native plants can indeed improve pest management and also point to multiple other ecosystem services that may accrue from the establishment of these

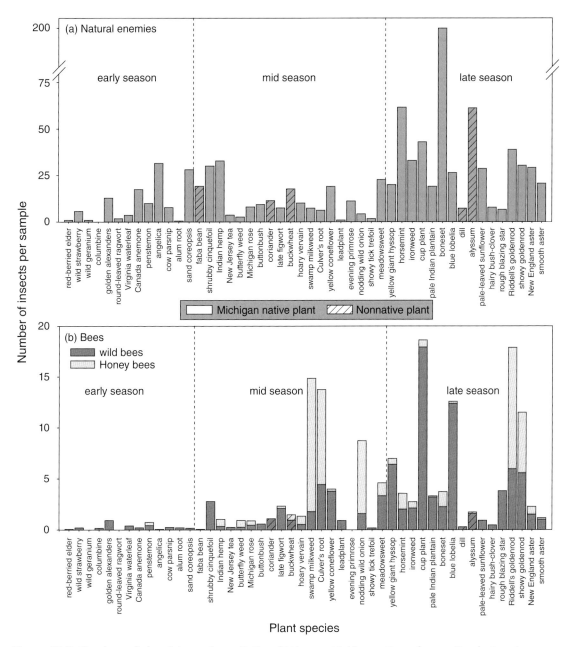

Figure 17.3 Abundance of a) predators and parasitoids and b) bees on Michigan native and non-native plant species grown in a common garden experiment during the early, mid and late growing season 2006. Data from Fiedler *et al.*, 2007b and Tuell *et al.*, 2008.

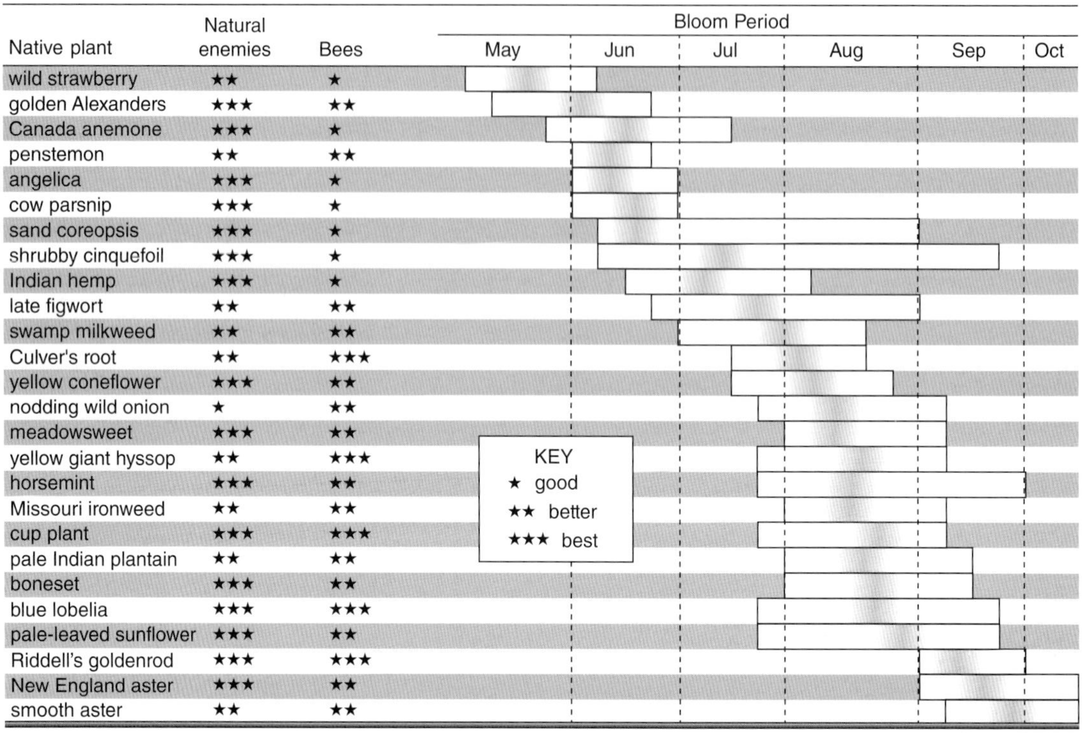

Figure 17.4 Twenty-six of the Michigan native plants most attractive to natural enemies and bees showing their overall (white bar) and peak (shaded portion of bar) bloom periods and relative suitability for attracting beneficial insects. Data from Fiedler *et al.*, 2007b and Tuell *et al.*, 2008.

plants. These include benefits to soil and water conservation, wildlife and the maintenance of culturally important natural resources. This is significant because, while improvements in pest management may be attractive to growers, the addition of other ecosystem services may make it far more likely that society at large will support their implementation.

In spite of the benefits, the case studies also reveal that the cost of installing such habitats, and uncertainty about the economic payback, continues to present significant barriers to more widespread adoption. A common factor in successful efforts is that the active collaboration of growers, industry groups, researchers and government agencies is often required. Interestingly, the concept of native plant reintroduction is extending beyond traditional agroecosystems. The use of native plants as a bioenergy source is a growing research area. The incorporation of native plants into urban ecosystems where vacant lands are being re-purposed by local residents to produce food crops is also underway (see Box 17.5 and Shrewsbury and Leather (chapter 18 of this volume)). Through continued efforts, this diverse field of research will continue to elucidate the ecological and economic benefits of native plants in supporting ecosystem services throughout our working landscapes.

Box 17.5 Habitat management in the city?

In many US cities economic decline and foreclosures have created vacant properties within city limits. Residents are actively re-purposing some of this land to produce food crops. Just as in traditional agricultural landscapes, food production in the city relies on the activity of beneficial arthropods. The use of floral resources and other habitat management practices to enhance biological control and pollination services, as well as additional provisioning, supporting, regulating and cultural ecosystem services within urban agroecosystems is a rapidly growing field of research.

REFERENCES

Altieri, M.A. and Whitcombm W.H. (1979a) Potential use of weeds in the manipulation of beneficial insects. *HortScience*, 14, 12–18.

Altieri, M.A. and Whitcomb, W.H. (1979b) Manipulation of insect populations through seasonal disturbance of weed communities. *Protection Ecology*, 1, 185–202.

Baggen, L.R. and Gurr, G.M. (1998) The influence of food on *Copidosoma koehleri* (Hymenoptera : Encyrtidae), and the use of flowering plants as a habitat management tool to enhance biological control of potato moth, *Phthorimaea operculella* (Lepidoptera: Gelechiidae). *Biological Control*, 11, 9–17.

Baggen, L.R., Gurr, G.M. and Meats, A. (1999) Flowers in tri-trophic systems: mechanisms allowing selective exploitation by insect natural enemies for conservation biological control. *Entomologia Experimentalis et Applicata*, 91, 155–161.

Baker, H.G. and Baker, I.B. (eds) (1983) *Floral nectar sugar constituents in relation to pollinator type*, Scientific and Academic Editions, Claremont, CA.

Begum, M., Gurr, G.M., Wratten, S.D., Hedberg, P.R. and Nicol, H.I. (2006) Using selective food plants to maximize biological control of vineyard pests. *Journal of Applied Ecology*, 43, 547–554.

Berndt, L.A. and Wratten, S.D. (2005) Effects of alyssum flowers on the longevity, fecundity and sex ratio of the leaf-roller parasitoid *Dolichogenidea tasmanica*. *Biological Control*, 32, 65–69.

Bianchi, F., Booij, C.J.H. and Tscharntke, T. (2006) Sustainable pest regulation in agricultural landscapes: a review on landscape composition, biodiversity and natural pest control. *Proceedings of the Royal Society B*, 273, 1715–1727.

Braman, S.K., Pendley, A.F. and Corley, N. (2002) Influence of commercially available wildflower mixes on beneficial arthropod abundance and predation in turfgrass. *Environmental Entomology*, 31, 564–572.

Brodt, S., Klonsky, K., Jackson, L., Brush, S.B. and Smukler, S. (2009) Factors affecting adoption of hedgerows and other biodiversity-enhancing features on farms in California, USA. *Agroforestry Systems*, 76, 195–206.

Bugg, R.L., Anderson, J.H., Thomsen, C.D. and Chandler, J. (1998) Farmscaping in California: managing hedgerows, roadside and wetland plantings, and wild plants for biointensive pest management, in *Enhancing biological control: habitat management to promote natural enemies of agricultural pests* (eds C.H. Pickett and R.L. Bugg), University of California Press, Berkeley, pp. 339–374.

Chaney, W. (1998) Biological control of aphids in lettuce using in-field insectaries, in *Enhancing biological control: habitat management to promote natural enemies of agricultural pests* (eds C.H. Pickett and R.L. Bugg), University of California Press, Berkeley, pp. 73–83.

Colley, M.R. and Luna, J.M. (2000) Relative attractiveness of potential beneficial insectary plants to aphidophagous hoverflies (Diptera: Syrphidae). *Environmental Entomology*, 29, 1054–1059.

Collins, K.L., Boatman, N.D., Wilcox, A., Holland, J.M. and Chaney, K. (2002) Influence of beetle banks on cereal aphid predation in winter wheat. *Agriculture, Ecosystems and Environment*, 93, 337–350.

Costanza, R., dArge, R., deGroot, R. *et al.* (1997) The value of the world's ecosystem services and natural capital. *Nature*, 387, 253–260.

Daily, G.C., Alexander, S., Ehrlich, R.R. *et al.* (1997) Ecosystem services: benefits supplied to human societies by natural ecosystems. *Issues in Ecology*, 2, 1–18.

Ehler, L.E., Pease, C.G. and Long, R.F. (2002) Farmscape ecology of a native stinkbug in the Sacramento Valley. *Fremontia*, 30, 3–4.

Eubanks, M.D. and Denno, R.F. (1999) The ecological consequences of variation in plants and prey for an omnivorous insect. *Ecology*, 80, 1253–1266.

Fiedler, A.K. and Landis, D.A. (2007a) Attractiveness of Michigan native plants to arthropod natural enemies and herbivores. *Environmental Entomology*, 36, 751–765.

Fiedler, A.K. and Landis, D.A. (2007b) Plant characteristics associated with natural enemy abundance at Michigan native plants. *Environmental Entomology*, 36, 878–886.

Fiedler, A.K., Landis, D.A. and Wratten, S.D. (2008) Maximizing ecosystem services from conservation biological control: the role of habitat management. *Biological Control*, 45, 254–271.

Fitzgerald, J.D. and Solomon, M.G. (2004) Can flowering plants enhance numbers of beneficial insect habitat for management of Lepidoptera pests? *Journal of Economic Entomology*, 13, 664–668.

Flynn, D.F.B., Gogol-Prokurat, M., Nogeire, T. *et al.* (2009) Loss of functional diversity under land use intensification across multiple taxa. *Ecology Letters*, 12, 22–33.

Foley, J.A., Monfreda, C., Ramankutty, N. and Zaks, D. (2007) Our share of the planetary pie. *Proceedings of the National Academy of Sciences*, 104, 12585–12586.

Foster, M.A. and Ruesink, W.G. (1984) Influence of flowering weeds associated with reduced tillage in corn on a black cutworm (Lepidoptera, Noctuidae) parasitoid, *Meteorus rubens* (Nees von Esenbeck). *Environmental Entomology*, 13, 664–668.

Gardiner, M.M., Landis, D.A., Gratton, C. *et al.* (2009) Landscape diversity enhances biological control of an introduced crop pest in the north-central USA. *Ecological Applications*, 19, 143–154.

Gardiner, M.M., Tuell, J.K., Isaacs, R., Gibbs, J., Ascher, J.S. and Landis, D.A. (2010) Implications of three model biofuel crops for beneficial arthropods in agricultural landscapes. *BioEnergy Research*, 3, 6–19.

Geiger, F., Bengtsson, J., Berendse, F. *et al.* (2010) Persistent negative effects of pesticides on biodiversity and biological control potential on European farmland. *Basic and Applied Ecology*, 11, 97–105.

Geiger, F., Wackers, F. and Bianchi, F.J.J.A. (2009) Hibernation of predatory arthropods in semi-natural habitats. *BioControl*, 54, 529–535.

Gurr, G.M., Wratten, S.D. and Luna, J.M. (2003) Multifunction agricultural biodiversity: pest management and other benefits. *Basic and Applied Ecology*, 4, 107–116.

Gurr, G., Wratten, S.D. and Altieri, M.A. (2004) *Ecological engineering for pest management: advances in habitat manipulation for arthropods*, CABI Publishing, Wallingford.

Heimpel, G.E., Lee, J.C., Wu, Z.S., Weiser, L., Wäckers, F. and Jervis, M.A. (2004) Gut sugar analysis in field-caught parasitoids: adapting methods originally developed for biting flies. *International Journal of Pest Management*, 50, 193–198.

Idris, A.B., and Grafius, E. (1995) Wildflowers as nectar sources for *Diadegma insulare* (Hymenoptera: Ichneumonidae), a parasitoid of diamondback moth (Lepidoptera: Yponomeutidae). *Environmental Entomology*, 24, 1726–1735.

Isaacs, R., Tuell, J., Fiedler, A., Gardiner, M. and Landis, D. (2009) Maximizing arthropod-mediated ecosystem services in agricultural landscapes: the role of native plants. *Frontiers in Ecology and the Environment*, 7, 196–203.

Jacometti, M., Jorgensen, N. and Wratten, S. (2010) Enhancing biological control by an omnivorous lacewing: Floral resources reduce aphid numbers at low aphid densities. *Biological Control*, 55, 159–165.

Jervis, M.A. and Kidd, N.A.C. (1986) Host-feeding strategies in hymenopteran parasitoids. *Biological Reviews*, 61, 395–434.

Jervis, M.A., Kidd, N.A.C., Fitton, M.G., Huddleston, T. and Dawah, H.A. (1993) Flower-visiting by hymenopteran parasitoids. *Journal of Natural History*, 27, 67–105.

Jonsson, M., Wratten, S.D., Landis, D.A. and Gurr, G.M. (2008) Recent advances in conservation biological control of arthropods by arthropods. *Biological Control*, 45, 172–175.

Jonsson, M., Wratten, S.D., Landis, D.A., Tompkins, J.M.L. and Cullen, R. (2010) Habitat manipulation to mitigate the impacts of invasive arthropod pests. *Biological Invasions*, 12, 2933–2945.

Landcare Research (2009) *Annual report 2009 part 1: the economy and the environment*, Landcare Research Manaaki Whenua, Lincoln.

Landis, D.A. and Werling, B.P. (2010) Arthropods and biofuel production systems in North America. *Insect Science*, 17, 1–17.

Landis, D.A., Wratten, S.D. and Gurr, G.M. (2000) Habitat management to conserve natural enemies of arthropod pests in agriculture. *Annual Review of Entomology*, 45, 175–201.

Lavandero, B., Wratten, S.D., Didham, R.K. and Gurr, G.M. (2006) Increasing floral diversity for selective enhancement of biological control agents: a double-edged sward? *Basic and Applied Ecology*, 7, 236–243.

Lundgren, J.G. (2009) *The relationships of natural enemies and non-prey foods*, Springer Publishers, Dordrecht.

Lee, J.C. and Heimpel, G.E. (2005) Impact of flowering buckwheat on Lepidopteran cabbage pests and their parasitoids at two spatial scales. *Biological Control*, 34, 290–301.

Lee, W.G., Meurk, C.D. and Clarkson, B.D. (2008) Agricultural intensification: whether indigenous biodiversity? *New Zealand Journal of Agricultural Research*, 51, 457–460.

Long, R.F. and Anderson, J. (2010) *Establishing hedgerows on farms in California*. UC ANR publication number 8390, URL: http://anrcatalog.ucdavis.edu/pdf/8390.pdf

Long, R.F., Corbett, A., Lamb, C., Reberg-Horton, C., Chandler, J. and Stimmann, M. (1998) Movement of beneficial insects from flowering plants to associated crops. *California Agriculture*, 52, 23–26.

Losey, J.E. and Vaughan, M. (2006) The economic value of ecological services provided by insects. *Bioscience*, 56, 311–323.

MacLeod, A., Wratten, S.D., Sotherton, N.W. and Thomas, M.B. (2004) 'Beetle banks' as refuges for beneficial arthropods in farmland: long-term changes in predator communities and habitat. *Agricultural and Forest Entomology*, 6, 147–154.

Meurk, C.D. (2008 Vegetation of the Canterbury plains and downlands. In: *The natural history of Canterbury*. Canterbury University Press, Canterbury, pp. 195–229.

Michener, C.D. (2007) *Bees of the world*, Johns Hopkins University Press, Baltimore.

Millennium Ecosystem Assessment (2005) *Ecosystems and human well-being: biodiversity synthesis*, World Resources Institute, Washington, DC.

Moller, H., MacLeod, C., Haggerty, J. *et al* (2008) Intensification of New Zealand agriculture: implications for biodiversity. *New Zealand Journal of Agricultural Research*, 51, 253–263.

Molloy, B.P.J. (1969) Recent history of the vegetation, in *The natural history of Canterbury* (ed. G.A. Knox), A.H. and A.W. Reed, Wellington, pp. 340–360.

Morandin, L., Long, R.F., Pease, C.G. and Kremen, C. (in press) Hedgerows enhance beneficial insects in agricultural systems. *California Agriculture*, 65.

Nentwig, W., Frank, T. and Lethmayer, C. (1998) Sown weed strips: artificial ecological compensation areas as an important tool in conservation biological control, in *Conservation biological control* (ed. P. Barbosa), Academic Press, San Diego, pp. 133–153.

Nicholls, C.I., Parrella, M. and Altieri, M.A. (2001) The effects of a vegetational corridor on the abundance and dispersal of insect biodiversity within a northern California organic vineyard. *Landscape Ecology*, 16, 133–146.

Orr, D.B. and Pleasants, J.M. (1996) The potential of native prairie plant species to enhance the effectiveness of the *Ostrinia nubilalis* parasitoid *Macrocentrus grandii*. *Journal of the Kansas Entomological Society*, 69, 133–143.

Packard, S. and Mutel, C.F. (2005) *The tallgrass restoration handbook: for prairies, savannas, and woodlands*, Island Press, Washington.

Patt, J.M., Hamilton, G.C. and Lashomb, J.H. (1997) Impact of strip-insectary intercropping with flowers on conservation biological control of the Colorado potato beetle. *Advances in Horticultural Science*, 11, 175–181.

Pickett, C.H. and Bugg, R.L. (eds) (1998) *Introduction: enhancing biological control-habitat management to promote natural enemies of agricultural pests*, University of California Press, Berkeley.

Ricketts, T.H. (2004) Tropical forest fragments enhance pollinator activity in nearby coffee crops. *Conservation Biology*, 18, 1262–1271.

Scarratt, S.L. (2005) Enhancing the biological control of leafrollers (Lepidoptera: Tortricidae) using floral resource subsidies in an organic vineyard in Marlborough, New Zealand. Unpublished Doctor of Philosophy thesis, Lincoln University, Lincoln.

Schellhorn, N.A. (2007) *Native vegetation to enhance biodiversity, beneficial insects and pest control in horticulture systems,*

Final Report 2007 Horticulture Australia, Limited No. VG06024.

Schellhorn, N.A., Glatz, R.V. and Wood, G.M. (2010) The risk of exotic and native plants as hosts for four pest thrips (Thysanoptera: Thripinae). *Bulletin of Entomological Research*, 100, 501–510.

Schulte, L.A., Asbjornsen, H., Atwell R. *et al.* (2008) *A targeted conservation approach for improving environmental quality: multiple benefits and expanded opportunities*, Extension Bulletin PMR 1002, Iowa State University, Ames.

Stephens, C.J., Schellhorn, N.A., Wood, G.M. and Austin, A.D. (2006) Parasitic wasp assemblages associated with native and weedy plant species in an agricultural landscape. *Australian Journal of Entomology*, 45, 176–184.

Symondson, W.O.C., Sunderland, K.D. and Greenstone, M.H. (2002) Can generalist predators be effective biocontrol agents? *Annual Review of Entomology*, 47, 561–594.

Taverner, P., Wood, G., Jevremov, D. and Doyle, B. (2006) *Revegetation by design: guidebook*, SARDI, Adelaide, SA. URL: www.sardi.sa.gov.au/__data/assets/pdf_file/0008/44945/reveg_by_design_guidebook.pdf

Thies, C., Steffan-Dewenter, I. and Tscharntke, T. (2003) Effects of landscape context on herbivory and parasitism at different spatial scales. *Oikos*, 101, 18–25.

Tilman, D., Hill, J. and Lehman, C. (2006) Carbon-negative biofuels from low-input high-diversity grassland biomass. *Science*, 314, 1598–1600.

Tompkins, J.M.L. (2010) *Ecosystem services provided by native New Zealand plants in vineyards*. Doctor of Philosophy Thesis Lincoln University, Lincoln. Available at: URL: http://researcharchive.lincoln.ac.nz/dspace/handle/10182/3077

Tompkins, J.M.L., Wratten, S.D. and Wäckers, F.L. (2010) Nectar to improve parasitoid fitness in biological control: Does the sucrose:hexose ratio matter? *Basic and Applied Ecology*, 11, 264–271.

Tuell, J.K., Fiedler, A.K., Landis, D. and Isaacs, R. (2008) Visitation by wild and managed bees (Hymenoptera: Apoidea) to eastern US native plants for use in conservation programs. *Environmental Entomology*, 37, 707–718.

Tylianakis, J.M., Didham, R.K. and Wratten, S.D. (2004) Improved fitness of aphid parasitoids receiving resource subsidies. *Ecology*, 85, 658–666.

Wäckers, F.L. (2001) A comparison of nectar- and honeydew sugars with respect to their utilization by the hymenopteran parasitoid *Cotesia glomerata*. *Journal of Insect Physiology*, 47, 1077–1084.

Wäckers, F.L., Bjornsen, A. and Dorn, S. (1996) A comparison of flowering herbs with respect to their nectar accessibility for the parasitoid *Pimpla turionellae*. *Proceedings of the Section Experimental and Applied Entomology of the Netherlands Entomological Society*, 177–182.

Wäckers, F.L., van Rijn, P.C.J. and Heimpel, G.E. (2008) Honeydew as a food source for natural enemies: making the best of a bad meal? *Biological Control*, 45, 176–184.

Warner, K.D. (2006) Extending agroecology: grower partici-pation in partnerships is key to social learning. *Renewable Agriculture and Food Systems*, 21, 84–94.

Werling, B., Meehan, T.D., Robertson, B., Gratton, C. and Landis, D. (2011) Biocontrol potential varies with changes in biofuel-crop plant communities and landscape perenni-ality. *Global Change Biology-Bioenergy*, doi: 10.1111/j.1757-1707.2011.01092.x

Winkler, K., Wäckers, F.L., Stingli, A. and van Lenteren, J.C. (2005) *Plutella xylostella* (diamondback moth) and its parasitoid *Diadegma semiclausum* show different gustatory and longevity responses to a range of nectar and honey-dew sugars. *Entomologia Experimentalis et Applicata*, 115, 187–192.

Chapter 18

USING BIODIVERSITY FOR PEST SUPPRESSION IN URBAN LANDSCAPES

Paula M. Shrewsbury and Simon R. Leather

Biodiversity and Insect Pests: Key Issues for Sustainable Management, First Edition. Edited by Geoff M. Gurr, Steve D. Wratten, William E. Snyder, Donna M.Y. Read.

INTRODUCTION

Rapid urbanisation has become a phenomenon of great concern because of dramatic changes in land use patterns and the global decline of biodiversity (Shochat *et al.*, 2006; Anton *et al.*, 2010). The transition from natural areas or agricultural lands to cities and suburbs brings buildings, roads, parking lots, waterways and green spaces (McDonnell and Pickett, 1990; McKinney, 2002; Faeth *et al.*, 2005; Grimm *et al.*, 2008; Raupp *et al.*, 2010). Worldwide human population growth and movement from rural to urban areas is predicted to result in over 84% of humans being concentrated in urban areas by 2050 (United Nations, 2009). As urbanisation continues, so too will changes to land use patterns with associated changes to biota and ecosystem function (McKinney, 2002; Smith *et al.*, 2006a). There are biota shifts from natural communities of plants and animals to species capable of exploiting human-altered systems associated with collections of ornamental plants within various forms of urban green space and remnants of natural habitats (McIntyre, 2000; McKinney, 2002; Zerbe *et al.*, 2003; Faeth *et al.*, 2005; Shochat *et al.*, 2006; Raupp *et al.*, 2010; Snep and Opdam, 2010). These shifts in biota and other aspects of urbanisation often result in plant and animal populations that are considered pests of urban environments. These include those that affect living plants such as many insect and mite species, structural pests such as termites and some beetles, and pests of sanitary concern such as rodents, houseflies, cockroaches, house finches, pigeons and mosquitoes. This chapter focuses on the relationships between urbanisation and arthropods associated with living plants as this is where most research efforts relating to biodiversity have been concentrated.

Of particular interest is how features and properties common to urban landscapes affect the biodiversity (richness and abundance) of arthropod communities and the potential for herbivorous arthropods to outbreak and become pests. Urban landscapes consist of contrived plant communities resulting from the widespread use of alien plants, impervious surfaces (e.g. hardscapes), simplified designs with reduced vegetation diversity and structural complexity. Common maintenance practices such as inputs of fertilisers, water and pesticides (McDonnell and Pickett, 1990; McIntyre *et al.*, 2001; Faeth *et al.*, 2005; McKinney, 2002; 2008; Grimm *et al.*, 2008; Raupp *et al.*, 2010), and increased nitrogen inputs from pollution (Dohmen *et al.*, 1984) provide opportunities for herbivorous insect and mite populations to increase (Raupp *et al.*, 2010; Sattler *et al.*, 2010). For example, human-altered systems experience change in bottom-up forces such as increased host plant quality or reduced plant defences and greater plant productivity; disruption of top-down forces such as reductions in natural enemy abundance and species richness; changes in microhabitats such as the creation of heat islands; and the development of matrixes that disrupt movement and colonisation of herbivores and natural enemies (McIntyre *et al.*, 2001; McKinney, 2008; Raupp *et al.*, 2010). These unique features of urbanised areas result in biodiversity that differs from natural systems, or those of other land use types, that can differentially affect ecosystem function (McKinney, 2002; Faeth *et al.*, 2005; Shochat *et al.*, 2006; Marussich and Faeth, 2009).

This chapter explores the importance and role of biodiversity as it affects ecosystem services within urban landscapes and the effects of urban habitat biodiversity on arthropod community dynamics. Moreover, the manipulation of habitat biodiversity as a pest management approach for insects and mites in urban landscapes is examined. We discuss mechanisms explaining how variation in habitat biodiversity influences insect or mite suppression in urban landscapes, along with case studies where manipulation of habitat biodiversity was implemented as a pest management approach. We also describe varying types of habitat biodiversity found in urban landscapes, and discuss the link between biodiversity and the ecosystem services it provides or could potentially provide in mitigating negative effects of urbanisation. Gaps in knowledge relating to the use of habitat biodiversity for pest management are identified in the context of future research opportunities.

BIODIVERSITY IN URBAN LANDSCAPES

Biodiversity, conservation and ecosystem services

Habitat biodiversity in urban landscapes relates to levels of vegetation diversity and structural complexity. Vegetation diversity is a measure of the species richness, spatial array and temporal overlap of plants (Andow, 1991). Structural complexity is the amount of vegetation within the three-dimensional space (e.g. vegetation strata) of the habitat (Shrewsbury and

Raupp, 2000). Moreover, urban habitat biodiversity varies along the urbanisation gradient which spans the inner city to suburban and rural locales. Habitat biodiversity can be altered by the removal of plants as they are replaced by impervious surfaces, by adding plants, often non-native plants that replace native species, or by increasing primary productivity (plant biomass) through inputs of chemicals (Smith *et al.*, 2006a; McKinney, 2002; 2008; Raupp *et al.*, 2010). A common pattern in urban landscapes is a reduction in habitat biodiversity, especially at high levels of urbanisation (e.g. urban core with over 50% impervious surface) (McKinney, 2002; 2008). Alternatively, features of urbanisation may alter landscapes in ways that increase habitat biodiversity, especially at intermediate levels of urbanisation (e.g. suburban areas with 20–50% impervious surface and with significant amounts of ornamental horticulture (Figure 18.1) (McKinney, 2002; 2008) or in urban desert environments (Hope *et al.*, 2003; Faeth *et al.*, 2005).

Urban habitat biodiversity plays an important, although not exclusive, role in maintaining urban ecosystem function. Consequently, studies have examined the effect of urbanisation on habitat biodiversity and its cascading effects on ecosystem function, and measures to restore habitat biodiversity and its benefits to ecosystem services (Faeth *et al.*, 2005; Smith *et al.*, 2006a; McKinney, 2008; Goddard *et al.*, 2009; Sattler *et al.*, 2010). More common are studies on increasing biodiversity for conservation purposes, and then its role in relation to ecosystem services (McKinney, 2002; Goddard *et al.*, 2009). In addition, studies have exam-

ined the influence of vegetation diversity and structural complexity in urban landscapes on herbivore and natural enemy trophic dynamics in the context of conservation biological control and habitat manipulation aimed at preventing pest insect and mite outbreaks (Raupp *et al.*, 2001a; Frank and Shrewsbury, 2004; Shrewsbury *et al.*, 2004; Shrewsbury and Raupp, 2006). Others have examined outbreaks of herbivorous arthropods to elucidate patterns and underlying mechanisms influencing outbreaks (Hanks and Denno, 1993; Tooker and Hanks, 2000).

Types of urban habitat biodiversity

Urban landscapes are highly variable, but almost all have substantial areas of green space, although the range of land area in green space is great. Green spaces are the major source of urban biodiversity. In Europe, the average land area covered by green space is approximately 19% of urban areas, New Zealand averages 36% (Mathieu *et al.*, 2007); and South African cities on average contain 11% green space (McConnachie *et al.*, 2008). Urban green spaces cover a range of habitats, from large formal parks and botanic gardens, to recreational parks, smaller residential gardens, residential landscaping schemes, street trees, school grounds, green roofs, green walls, roadside verges and traffic islands, to natural remnants of urban forests and woodlands (Sadler *et al.*, 2010). Green spaces and their associated biodiversity provide a variety of benefits to people, conservation of nature, and ecosystem function, but the extent of these benefits can be affected by the nature of the vegetation in these habitat patches, which may be relatively natural (Figure 18.2a) or highly managed (Figure 18.2b).

Benefits of urban green space for people, conservation and ecosystem services

Green spaces within the urban environment, be they natural remnants or manmade features, are very important as they create better living conditions for the human population and provide a number of direct and indirect benefits (Box 18.1; Niemelä, 1999; Chiesura, 2004; Wolf, 2004; Matsuoka and Kaplan, 2008). It has been argued that urban plant communities are different from plant communities in other areas such as agricultural or forested land (Duguay *et al.*, 2007).

Figure 18.1 A suburban residential landscape demonstrating high levels of vegetation diversity and structural complexity (photo by Paula Shrewsbury).

Figure 18.2 Urban roundabouts can add habitat diversity using a) a more natural or b) a more managed landscape design (photos by Simon Leather).

Therefore, urban plant communities should be conserved as they provide unique habitat. Moreover, the green spaces found in urban environments are extremely important sources of biological diversity and play a critical role in the conservation of species assemblages and maintaining ecosystem services (Gaston *et al.*, 2007; Goddard *et al.*, 2009). Consequently the ecological value of these areas should be routinely taken into account in urban planning (Czechowski,

1982; Zerbe *et al.*, 2003; Gaston *et al.*, 2007; Colding and Folke, 2009; Goddard *et al.*, 2009). Many cities throughout the world have been designed with green space in mind, although the underlying premise for green space has mainly focused on the benefits it has for people. Until recently there has been little attempt by urban planners or policy-makers to include conservation of biodiversity, and even less so the enhancement of ecosystem services, in urban planning (Wolf, 2004; Hunter and Hunter, 2008; Goddard *et al.*, 2009). In an effort to promote this approach, Wolf (2004) discusses the economic valuation of ecosystem services provided by green space biodiversity where an economic value is projected of the benefits (direct and indirect) to people and the environment. Based on modelling of air pollution, stormwater mitigation and energy impacts, annual values of urban forest services were estimated. For example, an urban ecosystem analysis of the Washington DC, US area found that tree cover had reduced stormwater storage costs by US\$4.7 billion and generated annual air quality savings of \$49.8 million (Wolf, 2004). Unfortunately, the analysis did not make reference to the economic valuation of pest suppression services of biodiversity.

More information is, however, needed about the best methods to maximise green spaces for species conservation and maintenance of ecosystem services, although it is possible to suggest ways in which ecological principles might benefit nature conservation in cities and other urban areas (Snep and Opdam, 2010). Green spaces such as domestic gardens artificially elevate habitat biodiversity (Weller and Ganzhorn, 2004; Breuste *et al.*, 2008), and mitigate the detrimental effects of urbanisation on animal biodiversity by

---

Box 18.1   Urban green space and its associated biodiversity provide benefits to:

People	Conservation of nature	Ecosystem function
Aesthetic appeal	Maintenance of biodiversity of plants and animals	Reduced likelihood of pest outbreaks
Emotional and physical well being	Habitat and food resources	Enhanced biological control
Increased housing values	Corridors for movement	Improved water infiltration, reduced runoff and erosion
Recreation and social interaction		Reduced pollution
Improved ecosystem services		Reduced heat island effects

preserving or creating habitat and food resources, and retaining corridors for movement through the urban matrix (Zapparoli, 1997; Smith *et al.*, 2006a). Traditionally, nature conservation has focused on preserving native vegetation or conserving and restoring wildlife habitats (Snep and Opdam, 2010). Specific examples of conserved remnant woodlands within the boundaries of cities include Northeram Woods in Bracknell, UK (Jones, 2010) and Helyar Woods in New Brunswick, NJ, USA (Rutgers, 2010), to name just two. These habitats are understandably rare and although desirable and worthy of preservation and enhancement, not likely to play a large role in increasing urban biodiversity overall or in providing wide-ranging ecosystem services (Marussich and Faeth, 2009).

Enhanced ecosystem service provisioning is more likely to come from the deliberate and planned provision of urban green spaces amongst the areas used by human inhabitants either as dwelling places or work areas (e.g. roadside verges, traffic islands, parks and gardens). For example, management should be of green spaces and their constituent biodiversity (e.g. management of collections of gardens across neighbourhoods or cities) to form interconnected networks of biodiversity, and urban planners need to consider the role of networks of domestic gardens (Goddard *et al.*, 2009). Their value if managed as a network would be greater than a random diversity of gardens or green spaces within urban landscapes. Management of green spaces as a network would probably be done at the city or town level or through community associations.

A characteristic of urban environments is the extensive use of flowering plants to enhance the aesthetic appeal of working and leisure space (Gaston *et al.*, 2005a; 2005b). These plants play a vital function in providing habitat and food for the animals that use urban areas as refugia (Smith *et al.*, 2006a; 2006b; Davies *et al.*, 2009). Vegetation diversity and structural complexity has been found to be positively correlated with invertebrate biodiversity (Raupp *et al.*, 2001b; Mullen *et al.*, 2003; Helden and Leather, 2004; 2005; Shrewsbury and Raupp, 2006; Smith *et al.*, 2006a) and the biodiversity of birds (Fernandez-Juricic, 2000), emphasising the importance of maintaining a high diversity of plants for the benefit of all animals. Phytophagous invertebrates are an important food resource for higher trophic levels and thus the presence, diversity and health of plants in urban green spaces is vital to maintain overall levels of biodiversity (Dearborn and Kark, 2010).

URBAN HABITAT BIODIVERSITY AND ARTHROPOD COMMUNITIES

Patterns and mechanisms

Arthropod communities and the abundance of individual species vary in urban green spaces along the urbanisation gradient and between urban landscapes and other land use types (McIntyre, 2000; Smith *et al.*, 2006a; 2006b; McKinney, 2008; Raupp *et al.*, 2010). Reduced habitat biodiversity may disrupt trophic dynamics, alter the abundance, diversity, and composition of animal and plant communities and impair ecosystem services such as suppression of insect and mite populations (McIntyre *et al.*, 2001; Faeth *et al.*, 2005; Shochat *et al.*, 2006; Smith *et al.*, 2006a, b; Grimm *et al.*, 2008; McKinney, 2002; 2008; Haddad *et al.*, 2009; Christie *et al.*, 2010; Raupp *et al.*, 2010). For example, urbanised areas with reduced levels of habitat biodiversity correlate with greater abundance of herbivorous insects and mites that often reach pest levels (Balder *et al.*, 1999; McIntyre, 2000; Schneider *et al.*, 2000; Shrewsbury and Raupp, 2000; Shochat *et al.*, 2006; Shrewsbury and Raupp, 2006; Smith *et al.*, 2006a; Raupp *et al.*, 2010). Many herbivorous arthropods found on woody plants achieve significantly greater abundance in urban landscapes than in natural areas (reviewed in Raupp *et al.*, 2010). Alternatively, high levels of habitat biodiversity may maintain, restore or positively impact trophic dynamics, increase diversity of arthropod communities, support native, alien, and even rare species, and improve ecosystem services such as those that regulate herbivorous arthropod abundance (e.g. biological control services) (Gibson, 1998; McIntyre *et al.*, 2001; Helden and Leather, 2004; Shochat *et al.*, 2006; Shrewsbury and Raupp, 2006; Smith *et al.*, 2006a; b; McKinney, 2008; Haddad *et al.*, 2009; Raupp *et al.*, 2010). Increasing the biodiversity of urban habitats by habitat manipulation to specifically provide key ecological resources that attract and retain natural enemies within the habitat with the goal of suppressing pest populations is a form of conservation biological control (Gurr *et al.*, 2000). In addition to conserving natural enemies, habitat manipulation can also reduce the impact of pests via resource concentration effects (e.g. reducing the concentration of any given host plant, making it difficult for herbivores to find their hosts) (Gurr *et al.*, 2000).

Other studies have inferred, and in some cases empirically elucidated, mechanisms underlying relationships between habitat biodiversity and arthropod richness and abundance. Diverse ecosystems support greater species richness which has been found to stabilise ecosystems (Naeem and Li, 1997; Folke et al., 2004; Sattler et al., 2010). Therefore, reduction in biodiversity may lead to a decoupling of fundamental ecological processes that result in arthropod outbreaks (Balder et al., 1999; Schneider et al., 2000; Shochat et al., 2006; Shrewsbury and Raupp, 2000; 2006). Understanding these dynamics provides fundamental knowledge that will assist in managing urban biodiversity with the objective of restoring ecological processes and preventing pest outbreaks (Niemelä, 1999). For example, in highly urbanised areas (cities) where the proportion of impervious surface is high, and vegetation diversity is low and structurally simple, arthropod species richness and some arthropod herbivore populations are reduced (Smith et al., 2006a, b; McKinney, 2008). However, other arthropod herbivore populations become elevated (Hanks and Denno, 1993; Kahn, 1988; Balder et al., 1999; Shrewsbury and Raupp, 2000). Taxa that thrive in urban landscapes, 'urban exploiters', increase in abundance whereas other taxa less suited to the environment decline and may disappear (Shochat et al., 2006; Smith et al., 2006a). Additionally, reduced natural enemy abundance and/or richness may allow certain herbivore populations to erupt (Hanks and Denno, 1993; Shrewsbury and Raupp, 2006; Smith et al., 2006a, b). Altered bottom-up forces such as improved host plant quality or reduced plant defences (Herms and Mattson, 1992; Speight et al., 1998), greater accessibility to resources, or higher plant productivity may also favour populations of herbivores (Raupp et al., 2010). Urban matrixes may disrupt movement and dispersal of arthropods (Thomas, 1989). In some cases, however, trees in urban and ornamental plantings were found to have less damage than those in natural forests in the US. In this case dispersal into urban habitats was limited and survival rates of herbivorous insects were lower in those environments (Nuckols and Connor, 1995).

Top-down regulation of arthropods

A growing number of studies have found that habitat biodiversity influences top-down regulation of herbivorous arthropods by their natural enemies in urban landscapes. In Berlin, broad city streets lined with Tilia spp. supported spectacularly higher numbers of spider mites, including Eotetranychus tiliarium (Hermann), than Tilia growing on narrower side streets where there were gardens, parks or natural areas that resulted in higher biodiversity (Balder et al., 1999; Schneider et al., 2000). Greater spider mite abundance was attributed to the relative rarity of predators in trees along wide boulevards compared with side streets where vegetation surrounded trees (Balder et al., 1999; Schneider et al., 2000). The native holly leafminer, Phytomyza ilicicola Loew, was 10 times more abundant on holly in urban landscapes than in forests (Kahn, 1988). Lower rates of parasitism and predation in urban sites were suggested as the mechanism underlying this pattern (Kahn and Cornell, 1989). Populations of green apple aphid, Aphis pomi DeGeer, were significantly greater on crab apples growing along motorways in Basle, Switzerland compared with crab apples some 200 m away from motorways where predation levels were significantly higher (Braun and Flückiger, 1984). High densities of pine needle scale, Chionaspis pinifoliae (Fitch), on pines was attributed to low diversity and abundance of natural enemies in urban landscapes, nurseries, and Scots pine plantations in the US (Eliason and McCullough, 1997; Tooker and Hanks, 2000). Armoured scale, Pseudaulacaspis pentagona (Targioni), on mulberry trees was three times more abundant on trees in mesic urban landscapes than on mulberry trees edging nearby forest remnants, due to the difference in abundance of natural enemies (Hanks and Denno, 1993). In urban habitats in China, larch aphids achieve higher densities than in rural habitats due to high light conditions and reduced natural enemy abundance (Fang et al., 1983). Clearly, reductions in habitat biodiversity may alter trophic dynamics and contribute to important reductions in the abundance and richness of natural enemies and biological control services they provide.

Mechanisms underlying top-down regulation

Several studies demonstrate that urban landscapes with greater vegetation diversity or complexity support greater abundance (Hanks and Denno, 1993; Frank and Shrewsbury, 2004; Shrewsbury et al., 2004; Rebek et al., 2005; Shrewsbury and Raupp, 2006; Smith et al., 2006a) or richness of natural enemies, especially generalist predators (Tooker and Hanks, 2000; Smith et al., 2006b). For example, predatory centipedes and parasitic and predatory wasps (vespids) were more abundant in habitats with lower levels of impervious surface and greater numbers of trees, respectively (Smith *et al.*, 2006a). Native plant richness, which is often greatest at intermediate levels of urbanisation, was positively correlated with the abundance and species richness of solitary bees and hoverflies (pollinators and predators) (Smith *et al.*, 2006a; 2006b). This is not always the case, however. Small-scale additions of native flowering plants to urban gardens in New York City, US, failed to increase beneficial insect (bees, butterflies and predatory wasps) richness (Matteson and Langellotto, 2010). Spider richness (Smith *et al.*, 2006b) and abundance (Shrewsbury and Raupp, 2006) was greatest in gardens and residential landscapes, respectively, with greater structural complexity of the vegetation. A study of trophic dynamics of arthropod communities found that predatory ground insects controlled herbivores in urban areas (urban desert remnant and mesic yard) more effectively compared with outlying desert sites (Marussich and Faeth, 2009).

Studies of urban and other land use types have demonstrated increased habitat diversity provides alternative food resources, such as pollen and nectar, and a range of alternative prey for generalist predators (Landis *et al.*, 2000; Langellotto and Denno, 2004; Shrewsbury and Raupp, 2006). Urban habitats with greater biodiversity provided more consistent sources of food such as alternative prey or floral resources for generalist predators (Hanks and Denno, 1993; Tooker and Hanks, 2000; Frank and Shrewsbury, 2004; Shochat *et al.*, 2004; Shrewsbury *et al.*, 2004;

Shrewsbury and Raupp, 2006) and provided refuge for parasitoids (Frankie *et al.*, 1992). Sustaining an abundance of resident predators provides a mechanism for reducing herbivore outbreaks in diverse and complex habitats (Shrewsbury and Raupp, 2006). A review by Langellotto and Denno (2004) further supports these hypotheses. They found that increasing vegetation structure resulted in significant increases in natural enemy abundance in seven of nine arthropod guilds examined. Along these same lines a series of studies were conducted in the UK to examine the value of domestic urban gardens for the conservation of species richness and abundance and their benefits to ecosystem function (Smith *et al.*, 2006a; 2006b; Gaston *et al.*, 2007). The most important determinant of invertebrate diversity in urban gardens was structural complexity of the vegetation – the presence of trees, large shrubs, and hedges (Box 18.2; Smith *et al.*, 2006a; 2006b; Gaston *et al.*, 2007).

In a test of mechanisms underlying patterns in herbivorous arthropod abundance, residential landscapes with high vegetation complexity were compared to those with low complexity (Shrewsbury and Raupp, 2006). Landscapes with complex vegetation had greater abundance and retention of generalist predators, especially a hunting spider, *Anyphaena celer* (Hentz), than did landscapes with low complexity. Greater abundance of predators was credited to more abundant alternative prey which attracted and retained spiders in complex landscapes. Stronger top-down pressure in complex landscapes reduced the

---

Box 18.2 Enhanced urban habitat biodiversity and its benefits

Urban biodiversity	Benefits
Vegetation diversity • Increase plant species richness, spatial array, and temporal overlap • Add flowering plants to provide season-long resources • Incorporate native plant species	Increased abundance and richness of arthropods (all trophic guilds) • Increased food resources (prey, floral resources) for natural enemies • Increased habitat (ex. refuge) • Increased natural enemy abundance and/or richness • Favourable matrixes for natural enemy movement and colonisation
Structural complexity • Increase amount of vegetation in all vegetation strata (ground cover, herbaceous plant, shrub, understory tree, and overstory tree layers)	Enhanced biological control by natural enemies

abundance of a key pest, the azalea lace bug, *Stephanitis pyrioides* (Scott). Elevated populations of lace bugs were observed only in landscapes with low levels of complexity despite the fact that host quality was superior in complex landscapes (Shrewsbury and Raupp, 2006) suggesting that top-down regulation was more important than bottom-up in this system.

Effect of increased abundance of alien plants on arthropod dynamics

Another feature of urban landscapes that relates to habitat diversity is the composition of plant species relative to plant origin. The proportion of alien plants in landscapes increases with increasing levels of urbanisation (Owen, 1983; McDonnell and Pickett, 1990; Pysek, 1998; McIntyre, 2000; McKinney, 2002; Smith *et al.*, 2006c), thus adding to a complex matrix that includes endemic arthropods encountering alien hosts and exotic arthropods encountering endemic hosts (Raupp *et al.*, 2010). The impact of these phenomena on the diversity and abundance of herbivores and their natural enemies in urban settings is unclear; evidence variously indicates negative, positive, and no effects (McIntyre, 2000; Tallamy, 2004; Smith *et al.*, 2006a; 2006b; Gaston *et al.*, 2007; McKinney, 2008; Burghardt *et al.*, 2009; Matteson and Langellotto, 2010; Raupp *et al.*, 2010).

On one hand, an increase in alien plants may be detrimental to the arthropod community. For example, it is predicted that alien plants should support fewer herbivores than natives (Leather, 1986; Keane and Crawley, 2002; Tallamy, 2004) because many herbivores are restricted to hosts with which they have coevolved (Jaenike, 1990). Supporting this hypothesis Burghardt *et al.* (2009) found the richness and abundance of Lepidoptera in suburban landscapes dominated by native plants to be three and four times greater, respectively, than the richness and abundance of Lepidoptera in similar landscapes but dominated by alien plants (although these results are confounded because the native landscapes had greater plant species richness). This suggests that at least for some taxa of herbivorous arthropods, abundance and richness may be reduced in urban landscapes dominated by alien plants. Further confounding this phenomenon of reduced herbivorous arthropods is the fact that many alien plants used in urban landscapes have been selected specifically for their resistance to key pests

(Raupp *et al.*, 1992; Herms, 2002; Tallamy, 2004), partially explaining reduced richness and abundance of some herbivore taxa in urban landscapes. If this is indeed the case, then urban landscapes dominated by alien plants would be herbivore-poor communities and be likely to support a reduced natural enemy community (Tallamy, 2004). Consequently, those herbivore species that are adapted to alien-dominated landscapes are more likely to reach outbreak densities due to the relaxation of top-down regulation.

Although many herbivores may be restricted to a narrow range of coevolved hosts (Janieke, 1990), others are not. There are examples of native herbivores switching to non-native hosts, especially hosts in the same family or genera of their native host (Gaston *et al.*, 2007; Gandhi and Herms, 2010; Raupp *et al.*, 2010). For instance, some native generalist herbivores readily incorporate introduced plants into their diet (Owen, 1983; Keane and Crawley, 2002; Agrawal and Kotanen, 2003; Tallamy, 2004; Gaston *et al.*, 2007). Tent caterpillars, *Malacosoma* spp., fall webworm, *Hyphantria cunea* (Drury), and evergreen bagworm, *Thyridopteryx ephemeraeformis* (Haworth), are classic examples of North American folivores that readily include aliens from several plant families into their diets (Raupp *et al.*, 2010). A possible mechanistic explanation for this pattern is found in the hypothesis of 'defence free space' (Gandhi and Herms, 2010). If alien host plants are not well defended (e.g. chemically) from endemic herbivores owing to lack of a coevolutionary history, then release from bottom-up regulation may be just as important as relaxation of top-down suppression, allowing herbivores to flourish in 'defence free space' (Gandhi and Herms, 2010). To add to these complex interactions, many species of plants are regionally native and have limited natural geographic ranges. Within their natural range they are often pest-free; however, many are extensively planted outside their range and become susceptible to pests. Further research is needed to elucidate the impact of alien and endemic plants, and exotic and native arthropods, on the relative strength of top-down and bottom-up regulation and the occurrence of pest outbreaks in urban landscapes.

The addition of alien plants can also lead to an increase in plant species richness along portions of the urban gradient (Zerbe *et al.*, 2003), especially where alien plants are being added at a faster rate than natives are removed. This accumulation may positively affect arthropod richness or abundance (Shochat *et al.*,

2006; Smith *et al.*, 2006a; 2006b; McKinney, 2008; Goddard *et al.*, 2009). For example, species richness of phytophagous and predatory mites on trees lining avenues and populating parks was greater than richness on trees in wooded suburbs of Como, Italy. On avenues and in parks, the diverse combination of resident native and alien trees supported greater arthropod diversity of herbivores and natural enemies (Rigamonti and Lozzia, 1999). Further support for this hypothesis comes from studies that examined the ecological contribution of domestic gardens in urban landscapes in Sheffield, UK (Smith *et al.*, 2006a; 2006b; Gaston *et al.*, 2007). Gardens with few species of native plants were just as rich in invertebrates as those with many species of native plants (Smith *et al.*, 2006a; 2006b). Proposed explanations for this pattern are that a majority of garden arthropods are predators, parasitoids, detritivores or pollinators, which usually do not have a close association with a particular host plant species. Further, many garden herbivores are generalists that are adapted to consume a variety of plants. Moreover, many plants from North America, Europe, Asia and the UK are closely related. In the Sheffield garden studies, 87% of the alien plants in the gardens were in the same families as native British species; moreover half had native relatives in the same genus (Smith *et al.*, 2006c; Gaston *et al.*, 2007). Studies concluded that vegetation structure and diversity are more important for maintaining arthropod species richness than plant origin (Smith *et al.*, 2006a; 2006b; Gaston *et al.*, 2007). Along these lines the more current school of thought is to base conservation and land management decisions on whether species are beneficial or detrimental to biodiversity, ecological services, human health and the economy, rather than species origin (Davis *et al.*, 2011).

CASE STUDIES OF URBAN HABITAT MANIPULATION AND PEST SUPPRESSION

A number of studies have been conducted on the implementation of habitat manipulation and conservation measures with the goal of increasing the success rate of biological control in urban landscapes. The addition of flowering plants to urban green space to enhance the services of natural enemies is a practice that can readily be implemented by many gardeners, and has the added benefit of increasing the aesthetic beauty of the urban habitat. This approach has been found to be effective in suppressing some herbivorous arthropods but not others. Questions surrounding this approach include the following. What type and how many species of flowers need to be added? Are native flowers better than aliens? How large does the flower patch have to be? How close do flowers have to be to landscape plants that are pest-prone to provide protection? What are the actual mechanisms that lead to pest suppression?

Several studies have addressed one of more of these questions. Experimental field plots were planted with flowering forbs at various densities and distances from shrubs infested with bagworms, *T. ephemeraeformis* (Ellis *et al.*, 2005). Parasitism rates increased by 71% on shrubs surrounded by forbs compared with shrubs with no surrounding forbs. Higher densities of forbs increased parasitism of bagworms relative to shrubs with low density or no forbs around them (Ellis *et al.*, 2005). Flowering forbs had a local effect on host-searching behaviour of female parasitoids, encouraging them to parasitise bagworms in the immediate vicinity. In a similar study, Shrewsbury *et al.* (2004) added two species of herbaceous plants that provided season-long bloom and variation in floral architecture, Shasta daisy (*Chrysanthemum* sp.) and coriander (*Coriandrum sativum* Linn.), to landscape beds containing azaleas (*Rhododendron* sp.) that were infested with azalea lace bug, *S. pyrioides*. Lace bugs on azaleas with flowering plants adjacent to them had lower survival than lace bugs on azaleas without flowering plants. Flowers increased alternative prey and natural enemy abundance, which translated into reduced survival of the pest insect (Shrewsbury *et al.*, 2004). Similarly, Rebek *et al.* (2005) found that natural enemies in general, and spiders and parasitic wasps in particular, were usually more abundant in beds of *Euonymus fortunei* (Turcz.) surrounded by flowers than beds with no flowers. Four species of flowering perennials (*Trifolium repens* L., *Euphorbia epithymoides* L., *Coreopsis verticillata* L. var. 'Moonbeam', and *Solidago canadensis* L. var. 'Golden Baby') were used. Interestingly, Rebek *et al.* (2005) concluded that increased abundance of natural enemies was probably due to vegetative characteristics of the plants rather than the actual floral resources. Further, Rebek *et al.* (2006) found that populations of the armoured scale, *Unaspis euonymi* (Comstock), on *E. fortunei* were reduced due to parasitism, regardless of the flower treatment, relative to initial population densities.

Figure 18.3 A conservation strip of flowering plants and ornamental grass borders a golf course fairway to conserve natural enemies and increase biological control (photo by Steven Frank, North Carolina State University, USA).

In golf courses, a different type of urban green space, Frank and Shrewsbury (2004) added beds of flowering plants (referred to as conservation strips) to golf course roughs about 4 m from the fairway:rough ecotone (Figure 18.3). Conservation strips ranged from 3 by 8 m to 4 by 16 m in size and plants included alyssum, *Lobularia maritima* (L.), coreopsis, *C. verticillata*, and switchgrass, *Panicum virgatum* L. These plants provided season-long blooms and variation in floral architecture and vegetative structure. The abundance and taxa of predators and predation of black cutworm, *Agrotis ipsilon* (Hufnagel), were measured at increasing distances from the conservation strip into the fairway (predator abundance at 0, 2, 4, 6, 8 and 12 m; cutworm predation at 0, 2, 6 and 10 m). Predator and alternative prey abundances were greatest in turfgrass with conservation strips compared to areas of the fairway without such strips. Abundance, however, varied with taxon and distance from the conservation strip. Predation of cutworms was greater in fairways adjacent to conservation strips, even out to the furthest distance of 10 m. This did not, however, correlate with predator abundance, suggesting natural enemy abundance may not be the most reliable estimate of the effects of habitat manipulations. Frank and Shrewsbury (2004) concluded that by increasing biodiversity, conservation strips could be an important tool towards conservation biological control of golf course pests. These studies demonstrate that in some cases enhancing urban habitat biodiversity can restore ecosystem function and lead to increased suppression of herbivorous pests by natural enemies in urban landscapes. Exact mechanisms underlying enhanced biological control are, however, not always clear.

USE OF HABITAT MANIPULATION FOR PEST MANAGEMENT IN URBAN LANDSCAPES

More recently with increased information, education and media coverage on global change, much of society worldwide has recognised the need for humans to develop more environmentally sensitive approaches for managing pests in landscapes. Also relevant in urban landscapes is the high potential for human contact as pesticides are applied to plants and turfgrass in recreation areas and parks where people walk and visit, and yards and gardens where children and pets play and adults entertain. All of these factors have led to societal pressures and desires to reduce or not use pesticides in urban landscapes and a generally increased interest in using ecologically based pest management practices such as biological control (Landis *et al.*, 2000), but only if the price differential is not prohibitive (Jetter and Paine, 2004).

To the authors' knowledge, there are no surveys or statistics on the prevalence of habitat manipulation practices to increase biodiversity of urban landscapes to suppress pest insects and mites. In general, research and implementation of habitat manipulation, however, have increased over recent decades (Landis *et al.*, 2000). This approach, however, requires fundamental knowledge of the biology and ecology of the urban landscape, the plants and arthropods within the landscape, their interactions, and other factors that influence these interactions. There is also a lack of economic valuation of the benefits that practitioners might achieve using habitat manipulation (Cullen *et al.*, 2008). Therefore, the adoption of habitat manipulation in urban landscapes has not been quick or widespread.

Generally, urban green spaces have long been managed either for biodiversity conservation purposes, especially public areas and a few wild gardens, or for pest control in domestic gardens and allotments, as evidenced by the plethora of books available for gardeners (e.g. Buczacki and Harris, 2005). Residential gardeners, until fairly recently, were more likely to turn to the spray gun than the natural enemy for help with their pest problems, and even today the advice given on some popular gardening programmes such as *Gardener's Question Time* (BBC Radio 4 in the UK) tends to focus on which pesticide to use rather than on how to encourage biological control, although this varies from country to country, e.g. *Gardening Australia* (Australian Broadcasting Corporation) very much encourages an organic approach. That said, pioneering work in the 1980s (Gaugler, 1988), has now made it an easy task to obtain ready-to-apply packets of entomopathogenic nematodes for pest control in domestic gardens. Ladybird beetles and Delta pheromone traps to prevent codling moth damage to domestic fruit trees, sprays containing the insect killing bacteria *Bacillus thuringiensis* (Bt), and sticky bands to prevent ants from protecting aphids are now commonly seen for sale in garden centres around the developed world. People are encouraged to avoid killing ladybird beetles and honey bees, but most still perceive vespids as dangerous intruders and ground beetles either go unnoticed or are regarded as pests when seen. The general level of biological knowledge in urban dwellers is, however, depressingly low (Leather and Quicke, 2010), reflecting a general tendency seen since the 1970s as children lost touch with nature and the human population in general developed an over-reliance on technology (Louv, 2005). This state of affairs seems unlikely to improve dramatically in the near future (Randler *et al.*, 2007; Leather and Quicke, 2009). There are, nevertheless, many garden enthusiasts, master gardeners (university extension volunteers) and practitioners who are informed and strive for ecologically based pest management. One of the authors of this chapter, P. Shrewsbury, an Extension Specialist in the eastern US, receives an ever-increasing number of requests to discuss and train landscape practitioners and garden hobbyists on topics such as landscape ecology, conservation biological control, the use of landscape design and flowering plants to prevent pest outbreaks, and sustainable and ecologically based pest management.

Further examples exist of practitioners implementing habitat manipulation practices to manage green space and biodiversity for pest suppression, or using conservation and good stewardship for the same end. For example, there has been a rapid increase in the number of examples where introduced pests have resulted in catastrophic loss of trees in urban landscapes. This environmental and aesthetic devastation is also combined with the extraordinary costs to cities for the removal and replacement of dead trees (Raupp *et al.*, 2006). The best-known example was the introduction of Dutch Elm Disease, *Ophiostoma ulmi* (Buisman), into the US. More current introductions into the US include Asian long-horned beetle, *Anoplophora glabripennis* (Motschulsky), and emerald ash borer, *Agrillus planipennis* Fairmaire, both of which continue to result in further catastrophic loss of trees

and the increasing costs associated with these introductions. Inventories of urban street trees have been conducted in several major US cities and urban planners have come to realise the importance of increasing the diversity of tree species in urban forests to reduce the potential for future catastrophic losses (Raupp *et al.*, 2006). In Middleton, Wisconsin, city planners have implemented an 8-year removal / 10-year replacement plan of ash (*Fraxinus* spp.), which make up 28% of their street and park trees, in response to emerald ash borer. The plan consists of pre-emptive removals of ash and replacement with a high diversity of tree species (Libby and Wegner, 2010). Increased tree diversity will enhance the quality of the city's urban forest and reduce the likelihood of future catastrophic pest outbreaks. Many city parks participate in 'Alternatives to Pesticides Program' (NCAP, 2010) and public school systems in many US states are mandated to implement IPM practices (Mertz *et al.*, 2002). These programmes have components of biological control and conservation. Similarly, golf courses throughout the world have prioritised conservation so that ecological design and management practices have resulted in insect conservation and maintenance of ecosystem function (Audubon International, 2010). At the other end of the spectrum, there are still many managers of green spaces, urban planners and decision-makers who have done little to prioritise the planning and development of urban green space and biodiversity for enhancement of the ecosystem service of biological control, or for that matter ecosystem function in general. To move forward will require further integration of efforts by scientists, practitioners, economists, and urban planners and policy-makers (Gaston *et al.*, 2007; Hunter and Hunter, 2008; Colding and Folke, 2009; Goddard *et al.*, 2009).

FUTURE RESEARCH DIRECTIONS: INCREASING THE USE OF URBAN BIODIVERSITY TO IMPROVE PEST SUPPRESSION

This chapter has reviewed key aspects of biodiversity in urban settings from the perspective of suppressing the impact of pests, chiefly via enhancement of natural enemies. The available knowledge allows a set of guidelines to be proposed that exploit the benefits of vegetation structure and diversity, the presence of flowering plants, and the optimal spatial arrangement of

Box 18.3   Research-based recommendations for using urban habitat biodiversity for pest suppression

Although many questions remain, the research reviewed in this chapter allows some recommendations to be made on practices to enhance habitat biodiversity and probably suppress herbivorous insects and mites in urban landscapes:

• Encourage greater vegetation diversity and increased vegetation overall with its associated structure to provide resource benefits to the widest range of invertebrates.

• Individual domestic garden size does not seem to be as important as the addition of vegetation structure in general.

• Green spaces should be managed as networks since they can form interconnected tracts of habitat and constitute much of the urban matrix, especially public green space.

• Addition of flowering plants should be encouraged for their alternative food resources and structure. Plants should be selected to provide season-long bloom and a variety of floral architecture and vegetative structure.

• Flowering plants should be as close to pest-prone plants as possible, but distances of 10 m have resulted in increased predation.

• Plant species richness in general appears to be more important than native plant richness, at least at the local scale of urban green spaces, and should be encouraged.

2. More research is necessary on urban trophic dynamics and mechanistic underpinnings as influenced by urban biodiversity and measures to enhance biological control services of arthropods.

3. Knowledge of optimal patch sizes, qualities and configurations for green space is needed to enhance pest suppression. This knowledge should be linked to the needs of individual taxa and communities of natural enemies in general.

4. Knowledge of arthropod movement and green space connectivity must be established to design and manage green spaces within cities for optimal benefits.

5. Translational research must be implemented and evaluated in urban landscapes and accompanied by outreach and education to demonstrate habitat manipulation practices and the associated economic and environmental benefits (e.g. reduction in pest outbreaks and plant damage; reduced pesticide inputs; reduced costs; conservation and maintenance of ecosystem function).

6. Development of economic valuation models for conservation biological control services in urban landscapes is necessary to encourage adoption of conservation tactics and strategies.

7. Ultimately, greater collaborative efforts by scientists, practitioners, economists, sociologists, urban planners and policy-makers are needed to accelerate discovery, implementation and adoption.

ACKNOWLEDGEMENTS

The authors thank Michael Raupp for reviewing and providing input on this chapter and Nancy Harding for her editorial assistance.

vegetation (Box 18.3). Notwithstanding these guidelines, there remain significant gaps in knowledge relating to the use of habitat biodiversity for pest management still exist and are prevalent. These include answers to basic research questions and applied research, and education to further the implementation of conservation biological control. Topics are identified in the context of future research opportunities below.

1. The trophic dynamics between alien and indigenous plants and exotic and native insects are complex and need to be elucidated, especially in the context of increased dominance of alien plants in urban landscapes. The focus should be on the ecosystem function of the species rather than origin.

REFERENCES

Agrawal, A.A. and Kotanen, P.M. (2003) Herbivores and the success of exotic plants: a phylogenetically controlled experiment. *Ecology Letters*, 6, 712–715.

Andow, D.A. (1991) Vegetational diversity on arthropod population response. *Annual Review of Entomology*, 36, 561–586.

Anton, C., Young, J., Harrison, P.A. *et al.* (2010) Research needs for incorporating the ecosystem service approach into EU biodiversity conservation policy. *Biodiversity Conservation*, 19, 2979–2994.

Audubon International (2010) *Golf and the environment.* URL: http://www.auduboninternational.org/ge.html

Balder, H., Jäckel, B. and Pradel, B. (1999) Investigations on the existence of beneficial organisms on urban trees in Berlin, in *Proceedings of the international symposium on urban tree health* (eds M. Lemattre, P. Lamettre and F. Lemaire), International Society for Horticultural Science, Brugge, pp. 189–195.

Braun, S. and Flückiger, W. (1984) Increased population of aphid *Aphis pomi* at a motorway. Part 2 – The effect of drought and deicing salt. *Environmental Pollution Series A, Ecological and Biological,* 36, 261–270.

Breuste, J., Niemelä, J. and Snep, R.P.H. (2008) Applying landscape ecological principles in urban environments. *Landscape Ecology,* 23, 1139–1142.

Buczacki, S. and Harris, K. (2005) *Collins guide to the pests, diseases and disorders of garden plants,* Collins, London.

Burghardt, K., Tallamy, D.W. and Shriver, G. (2009) Impact of native plants on bird and butterfly biodiversity in suburban landscapes. *Conservation Biology,* 23, 219–224.

Chiesura, A. (2004) The role of urban parks for the sustainable city. *Landscape and Urban Planning,* 68, 129–138.

Christie, F.J., Cassis, G. and Hochuli, D.F. (2010) Urbanization affects the trophic structure of arboreal arthropod communities. *Urban Ecosystems,* 13, 169–180.

Colding, J. and Folke, C. (2009) The role of golf courses in biodiversity conservation and ecosystem management. *Ecosystems,* 12, 191–206.

Cullen, R., Warner, K.D., Jonsson, M. and Wratten, S.D. (2008) Economics and the adoption of conservation biological control. *Biological Control,* 45, 272–280.

Czechowski, W. (1982) Occurrence of carabids (Coleoptera, Carabidae) in the urban greenery of Warsaw according to the land utilization and cultivation. *Memorabilia Zoologica,* 39, 3–108.

Davies, Z.G., Fuller, R.A., Loram, A., Irvine, K.N., Sims, V. and Gaston, K.J. (2009) A national scale inventory of resource provision for biodiversity within domestic gardens. *Biological Conservation,* 142, 761–771.

Davis, M.A., Chew, M.K., Hobbs, R.J. *et al.* (2011) Don't judge species on their origin. *Nature,* 474, 153–154.

Dearborn, D.C. and Kark, S. (2010) Motivations for conserving urban biodiversity. *Conservation Biology,* 24, 432–440.

Dohmen, G.P., McNeill, S. and Bell, J.N.B. (1984). Air pollution increases *Aphis fabae* pest potential. *Nature,* 307, 52–53.

Duguay, S., Eigenbrod, F. and Fahrig, L. (2007) Effects of surrounding urbanization on non-native flora in small forest patches. *Landscape Ecology,* 22, 589–599.

Eliason, E.A. and McCullough, D.G. (1997) Survival and fecundity of three insects reared on four varieties of Scotch pine Christmas trees. *Journal of Economic Entomology,* 90, 1598–1608.

Ellis, J.A., Walter, A.D., Tooker, J.F. *et al.* (2005) Conservation biological control in urban landscapes: Manipulating parasitoids of bagworm (Lepidoptera: Psychidae) with flowering forbs. *Biological Control,* 34, 99–107.

Faeth, S.H., Warren, P.S., Schochat, E. and Marussich, W.A. (2005) Trophic dynamics in urban communities. *BioScience,* 55, 399–407.

Fang, S.Y., Zhong, H. and Ling, Y.M. (1983) An investigation on larch aphids in the botanical garden of Heilongjiang Province. *Journal of North Eastern Forestry Institute, China,* 11, 36–41.

Fernandez-Juricic, E. (2000) Bird community composition patterns in urban parks of Madrid: the role of age, size and isolation. *Ecological Research,* 15, 373–383.

Folke, C., Carpenter, S., Walker, B. *et al.* (2004) Regime shifts, resilience, and biodiversity in ecosystem management. *Annual Review of Ecology, Evolution and Systematics,* 35, 557–581.

Frank, S.D. and Shrewsbury, P.M. (2004) Effect of conservation strips on the abundance and distribution of natural enemies and predation of *Agrotis ipsilon* (Lepidoptera: Noctuidae) on golf course fairways. *Environmental Entomology,* 33, 1662–1672.

Frankie, G.W., Morgan, D.L. and Grissell, E.E. (1992) Effects of urbanization on the distribution and abundance of the cynipid gall wasp, *Disholcaspis cinerosa,* on ornamental live oak in Texas, USA, in *Biology of insect-induced galls* (eds J.D. Shorthouse and O. Rohfritsch), Oxford University Press, New York, pp. 258–279.

Gandhi, K.J.K. and Herms, D.A. (2010) Direct and indirect effects of invasive insect herbivores on ecological processes and interactions in forests of eastern North America. *Biological Invasions,* 12, 389–405.

Gaston, K.J., Smith, R.M., Thompson, K. and Warren, P.H. (2005a) Urban domestic gardens (II): experimental tests of methods for increasing biodiversity. *Biodiversity and Conservation,* 14, 395–413.

Gaston, K.J., Warren, P.H., Thompson, K. and Smith, R.M. (2005b) Urban domestic gardens (IV): the extent of the resource and its associated features. *Biodiversity and Conservation,* 14, 3327–3349.

Gaston, K.J., Cush, P., Ferguson, S. *et al.* (2007) Improving the contributions of urban gardens for wildlife: some guiding propositions. *British Wildlife,* 18, 171–177.

Gaugler, R. (1988) Ecological considerations in the biological control of soil inhabiting insects with entomopathogenic nematodes. *Agriculture, Ecosystems and Environment,* 24, 351–360.

Gibson, C.W.D. (1998) Brownfield: red data. The values artificial habitats have for uncommon invertebrates. *English Nature Research Report No. 273.* English Nature, Peterborough.

Goddard, M.A., Dougill, A.J. and Benton, T.G. (2009) Scaling up from gardens: biodiversity conservation in urban environments. *Trends in Ecology and Evolution,* 25, 90–98.

Grimm, N.B., Faeth, S.H., Golubiewski, N.E. *et al.* (2008) Global change and the ecology of cities. *Science,* 319, 756–760.

Gurr, G.M., Wratten, S.D. and Barbosa, P. (2000) Success in conservation biological control of arthropods, in *Biological control: measures of success* (eds G.M. Gurr and S.D. Wratten), Kluwer Academic, Dordrecht, pp. 105–132.

Haddad, N.M., Crutsinger, G.M., Gross, K., Haarstad, J., Knops, J.M.H. and Tilman, D. (2009) Plant species loss decreases arthropod diversity and shifts trophic structure. *Ecology Letters*, 12, 1029–1039.

Hanks, L.M. and Denno, R.F. (1993) Natural enemies and plant water relations influence the distribution of an armored scale insect. *Ecology*, 74, 1081–1091.

Helden, A.J. and Leather, S.R. (2004) Biodiversity on urban roundabouts – Hemiptera, management and the species-area relationship. *Basic and Applied Ecology*, 5, 367–377.

Helden, A.J. and Leather, S.R. (2005) The Hemiptera of Bracknell as an example of biodiversity within an urban environment. *British Journal of Entomology and Natural History*, 18, 233–252.

Herms, D.A. (2002) Strategies for deployment of insect resistant ornamental plants, in *Mechanisms and deployment of resistance in trees to insects* (eds M.R. Wagner, K.M. Clancy, F. Lieutier and T.D. Paine), Kluwer Academic, Dordrecht, pp. 217–237.

Herms, D.A. and Mattson, W.J. (1992) The dilemma of plants: to grow or defend. *Quarterly Review of Biology*, 67, 283–335.

Hope, D., Gries, C., Zhu, W.X. *et al.* (2003) Socioeconomics drive urban plant diversity. *Proceedings of the National Academy USA*, 100, 8788–8792.

Hunter, M.R. and Hunter, M.D. (2008) Designing for conservation of insects in the built environment. *Insect Conservation and Diversity*, 1, 189–196.

Jaenike, J. (1990) Host specialization in phytophagous insects. *Annual Review of Entomology*, 21, 243–273.

Jetter, K. and Paine, T.D. (2004) Consumer preferences and willingness to pay for biological control in the urban landscape. *Biological Control*, 30, 312–322.

Jones, E.L. (2010) Factors affecting the diversity and abundance of roadside invertebrates and plants in urban areas. Doctor of Philosophy thesis, Imperial College London.

Kahn, D.M. (1988) Population ecology of an insect herbivore: Native holly leafminer, *Phytomyza ilicicola*. Doctor of Philosophy thesis, University of Delaware.

Kahn, D.M. and Cornell, H.V. (1989) Leafminers, early leaf abscission, and parasitoids: a tritrophic interaction. *Ecology*, 70, 1219–1226.

Keane, R.M. and Crawley, M.J. (2002) Exotic plant invasions and the enemy release hypothesis. *Trends in Ecology and Evolution*, 17, 164–170.

Landis, D.A., Wratten, S.D. and Gurr, G.M. (2000) Habitat management to conserve natural enemies of arthropod pests in agriculture. *Annual Review of Entomology*, 45, 175–201.

Langellotto, G.A. and Denno, R.F. (2004) Responses of invertebrate natural enemies to complex-structured habitats: a meta-analytical synthesis. *Oecologia*, 139, 1–10.

Leather, S.R. (1986) Insect species richness of the British Rosaceae: the importance of host range, plant architecture, age of establishment, taxonomic isolation and species-area relationships. *Journal of Animal Ecology*, 55, 841–860.

Leather, S.R. and Quicke, D.J.L. (2009) Where would Darwin have been without taxonomy? *Journal of Biological Education*, 43, 51–52.

Leather, S.R. and Quicke, D.J.L. (2010) Do shifting baselines in natural history knowledge threaten the environment? *Environmentalist*, 30, 1–2.

Libby, H. and Wegner, M. (2010) *Emerald ash borer and Middleton's plan*. URL: http://www.ci.middleton.wi.us/City/Departments/Lands/Forestry/pdf/Middleton_EAB_PowerPoint_PDF.pdf

Louv, R. (2005) *Last Child in the Woods*, Algonquin Books, New York.

Marussich, W.A. and Faeth, S.H. (2009) Effects of urbanization on trophic dynamics of arthropod communities on a common desert host plant. *Urban Ecosystems*, 12, 265–286.

Mathieu, R., Freeman, C. and Aryal, J. (2007) Mapping private gardens in urban areas using object-oriented techniques and very high-resolution satellite imagery. *Landscape and Urban Planning*, 81, 179–192.

Matsuoka R.H. and Kaplan R. (2008) People needs in the urban landscape: analysis of landscape and urban planning contributions. *Landscape and Urban Planning*, 84, 7–19.

Matteson, K.C. and Langellotto, G.A. (2010) Small scale additions of native plants fail to increase beneficial insect richness in urban gardens. *Insect Conservation and Diversity* Online Early, doi:10.1111/j.1752-4598/2010.00103.x.

McConnachie, M.M., Shackleton, C.M. and McGregor, G.K. (2008) The extent of public green space and alien plant species in 10 small towns of the sub-tropical thicket biome, South Africa. *Urban Forestry and Urban Greening*, 7, 1–13.

McDonnell, M.J. and Pickett, S.T.A. (1990) Ecosystem structure and function along urban–rural gradients: an unexploited opportunity for ecology. *Ecology*, 71, 1232–1237.

McIntyre, N.E. (2000) Ecology of urban arthropods: a review and a call to action. *Annals of Entomological Society of America*, 93, 825–835.

McIntyre, N.E., Rango, J., Fagan, W.F. *et al.* (2001) Ground arthropod community structure in a heterogeneous urban environment. *Landscape and Urban Planning*, 52, 257–274.

McKinney, M.L. (2002) Urbanization, biodiversity, and conservation. *BioScience*, 52, 883–890.

McKinney, M.L. (2008) Effects of urbanization on species richness: a review of plants and animals. *Urban Ecosystems*, 11, 161–176.

Mertz, T., Shrewsbury, P.M., Raupp, M.J. and Crow, E. (2002) *Integrated Pest Management in Schools: Plant Selection and Care*. University of Maryland, Cooperative Extension Service and Maryland Department of Agriculture, Bulletin 364.

Mullen, K., Fahy, O. and Gormally, M. (2003) Ground flora and associated arthropod communities of forest road edges in Connemara, Ireland. *Biodiversity and Conservation*, 12, 87–101.

Naeem, S. and Li, S.B. (1997) Biodiversity enhances ecosystem reliability. *Nature*, 390, 507–509.

NCAP (2010) *Pesticide-free places and urban IPM*. Northwest Coalition for Alternatives to Pesticides (URL: http://www.pesticide.org/Our%20Work/pesticide-free-parks/

Niemelä, J. (1999) Ecology and urban planning. *Biodiversity and Conservation*, 8, 119–131.

Nuckols, M.S. and Connor, E.F. (1995). Do trees in urban or ornamental plantings receive more damage by insects than trees in natural forests? *Ecological Entomology*, 20, 253–260.

Owen, J. (1983) Effects of contrived plant diversity and permanent succession on insects in English suburban gardens, in *Urban entomology: interdisciplinary perspectives* (eds G.W. Frankie and C.S. Koehler), Praeger, New York, pp. 395–422.

Pysek, P. (1998) Alien and native species in Central European urban floras: a quantitative comparison. *Journal of Biogeography*, 25, 155–163.

Randler, C., Höllwarth, A. and Schaal, S. (2007) Urban park visitors and their knowledge of animal species. *Anthrozoös*, 20, 65–74.

Raupp, M.J., Koehler, C.S. and Davidson, J.A. (1992) Advances in implementing integrated pest management for woody landscape plants. *Annual Review of Entomology*, 37, 561–585.

Raupp, M.J., Holmes, J.J., Sadof, C.S., Shrewsbury, P.M. and Davidson, J.A. (2001a) Effects of cover spray and residual pesticides on scale insects and natural enemies in urban forests. *Journal of Arboriculture*, 27, 203-213.

Raupp, M.J., Shrewsbury, P.M., Holmes, J.J. and Davidson, J.A. (2001b) Plant species diversity and abundance affects the number of arthropod pests in residential landscapes. *Journal of Arboriculture*, 27, 222–229.

Raupp, M.J., Buckelew Cumming, A. and Raupp, E.C. (2006) Street tree diversity in Eastern North America and its potential for tree loss to exotic borers. *Journal of Arboriculture*, 32, 297–304.

Raupp, M.J., Shrewsbury, P.M. and Herms, D.A. (2010) Ecology of herbivorous arthropods in urban landscapes. *Annual Review of Entomology*, 55, 19–38.

Rebek, E.J., Sadof, C.S. and Hanks, L.M. (2005) Manipulating the abundance of natural enemies in ornamental landscapes with floral resource plants. *Biological Control*, 33, 203–216.

Rebek, E.J., Sadof, C.S. and Hanks, L.M. (2006) Influence of floral resource plants on control of an armored scale pest by the parasitoid *Encarsia citrina* (Craw.) (Hymenoptera: Aphelinidae). *Biological Control*, 37, 320–328.

Rigamonti, I.E. and Lozzia, G.C. (1999) Injurious and beneficial mites on urban trees in Northern Italy, in *Proceedings of the international symposium on urban tree health* (eds M. Lemattre, P. Lamettre and F. Lemaire), International Society for Horticultural Science, Brugge, pp. 177–182.

Rutgers Gardens (2010) *History*. The State University of New Jersey URL: http://rutgersgardens.rutgers.edu/history.html

Sadler, J., Bates, A., Hale, J. and James, P. (2010) Bringing cities alive: the importance of urban green spaces for people and biodiversity, in *Urban Ecology* (ed. K.J. Gaston), Cambridge University Press, Cambridge, pp. 230–260.

Sattler, T., Duelli, P., Obrist, M.K., Arlettaz, R. and Moretti, M. (2010) Response of arthropod species richness and functional groups to urban habitat structure and management. *Landscape Ecology*, 25, 941–954.

Schneider, K., Balder, H., Jackel, B. and Pradel, B. (2000) Bionomics of *Eotatranychus tiliarum* as influenced by key factors, in *International symposium on plant health in urban horticulture* (eds G.F. Backhaus, H. Balder and E. Idczak), Parey Buchverlag, Berlin, pp. 102–108.

Shochat, E., Stefanov, W.L., Whitehouse, M.E.A. and Faeth, S.H. (2004) Urbanization and spider diversity: influences of human modification of habitat structure and productivity. *Ecological Applications*, 14, 268–280.

Shochat, E., Warren, P.S., Faeth, S.H. McIntyre, N.E. and Hope, D. (2006) From patterns to emerging processes in mechanistic urban ecology. *Trends in Ecology and Evolution*, 4, 186–191.

Shrewsbury, P.M. and Raupp, M.J. (2000) Evaluation of components of vegetational texture for predicting azalea lace bug, *Stephanitis pyrioides* (Heteroptera: Tingidae), abundance in managed landscapes. *Environmental Entomology*, 29, 919–1026.

Shrewsbury, P.M. and Raupp, M.J. (2006) Do top-down or bottom-up forces determine *Stephanitis pyrioides* abundance in urban landscapes? *Ecological Applications*, 16, 262–272.

Shrewsbury, P.M., Lashomb, J.H., Hamilton, G.C., Zhang, J., Patt, J.M. and Casagrande, R.A. (2004) The influence of flowering plants on herbivore and natural enemy abundance in ornamental landscapes. *International Journal of Ecology and Environmental Sciences*, 30, 23–33.

Smith, R.M., Gaston, K.J., Warren, P.H. and Thompson, K. (2006a) Urban domestic gardens (VIII): environmental correlates of invertebrate abundance. *Biodiversity and Conservation*, 15, 2515–2545.

Smith, R.M., Warren, P.J., Thompson, K. and Gaston, K.J. (2006b) Urban domestic gardens (VI): environmental correlates of invertebrate species richness. *Biodiversity and Conservation*, 15, 2415–2438.

Smith, R.M., Thompson, K., Hodgson, J.G., Warren, P.J. and Gaston, K.J. (2006c) Urban domestic gardens (IX): composition and richness of the vascular plant flora, and implications for native biodiversity. *Biological Conservation*, 129, 312–322.

Snep, R. and Opdam, P. (2010) Integrating nature values in urban planning and design, in *Urban Ecology* (ed.

K.J. Gaston), Cambridge University Press, Cambridge, pp. 261–286.

Speight, M.R., Hails, R.S., Gilbert, M. and Foggo, A. (1998) Horse chestnut scale (*Pulvinaria regalis*) (Homoptera: Coccidae) and urban host tree environment. *Ecology*, 79, 1503–1513.

Tallamy, D.W. (2004) Do alien plants reduce insect biomass? *Conservation Biology*, 18, 1689–1692.

Thomas, C.D. (1989) Predator–herbivore interaction and the escape of isolated plants from phytophagous insects. *Oikos*, 55, 291–298.

Tooker, J.F. and Hanks, L.M. (2000) Influence of plant community structure on natural enemies of pine needle scale (Homoptera: Diaspididae) in urban landscapes. *Environmental Entomology*, 29, 1305–1311.

United Nations (2009) *World Urbanization prospects: the 2009 Revision*, UN Department of Economic and Social Affairs',

New York, URL: http://esa.un.org/unpd/wup/Documents/WUP2009_Highlights_Final.pdf

Weller, B. and Ganzhorn, J.U. (2004) Carabid beetle community composition, body size, and fluctuating asymmetry along an urban-rural gradient. *Basic and Applied Ecology*, 5, 193–201.

Wolf, K.L. (2004) Economics and public value of urban forests. *Urban Agriculture Magazine*, 13, 31–33.

Zapparoli, M. (1997) Urban development and insect biodiversity of the Rome area, Italy. *Landscape and Urban Planning*, 38, 77–86.

Zerbe, S., Maurer, U., Schmitz, S. and Sukopp, H. (2003) Biodiversity in Berlin and its potential for nature conservation. *Landscape and Urban Planning*, 62, 139–148.

Chapter 19

COVER CROPS AND RELATED METHODS FOR ENHANCING AGRICULTURAL BIODIVERSITY AND CONSERVATION BIOCONTROL: SUCCESSFUL CASE STUDIES

P.G. Tillman, H.A. Smith and J.M. Holland

Biodiversity and Insect Pests: Key Issues for Sustainable Management, First Edition. Edited by Geoff M. Gurr, Steve D. Wratten, William E. Snyder, Donna M.Y. Read.

INTRODUCTION

In modern agricultural systems mechanical cultivation and chemical pesticides are used for crop production, restricting diversity and promoting landscapes dominated by large monocultures. Through the use of equipment such as harrows and mowers, large portions of the biomass are often removed and/or tilled annually, thereby forcing the growth process to start over. Herbicides are used to manage weeds, and fertilisers are used to foster rapid, lush growth of the crop. The prevailing insect pest control strategy in these agricultural systems is application of toxic agrochemicals. Such prophylaxis 'insurance' approaches can lead to biological control failure or a least a reduction in effectiveness as a result of the direct and indirect effects of pesticides, tillage, cultivation, lack of nectar and pollen sources, scarcity of hosts and lack of shelter and hibernation, mating and oviposition sites (Corbett and Rosenheim, 1996; Landis *et al.*, 2000; Heimpel and Jervis, 2005). In the absence of vital resources, colonisation by predatory species is often much lower than that by herbivores (Altieri and Whitcomb, 1979; Thies and Tscharntke, 1999), resulting in the failure of predators and parasitoids to control pests as they begin colonising crops (Landis *et al.*, 2000). Long-term solutions to escalating economic and environmental consequences of combating pests in agricultural crops can be achieved by restructuring and managing agroecosystems in ways that enhance agricultural diversity to increase biocontrol and other ecological services for pest management.

One of the most important aspects of enhancing biodiversity in agricultural systems involves the provision of resources for natural enemies of pest insects and insect pollinators. Interestingly, many of the habitats incorporated into agricultural systems for enhancing natural enemies are multifunctional, for they can provide other ecological benefits such as conserving wildlife, protecting water quality and reducing erosion and runoff (Leidner and Kidwell, 2000; Thomas *et al.*, 2001; SWCS, 2006; Triplett and Dick, 2008; FAO, 2010). Understanding the ecology of insect pests and their natural enemies in agroecosystems is essential in creating and designing habitats for enhancing agricultural biodiversity for pest suppression. It is important to have a clear understanding of what resources are needed and how specific habitats can successfully provide these resources. Strategic placement, in time and space, of a multifunctional habitat in an agricultural system may also be essential for successfully increasing biocontrol and other ecological services for pest management. For example, the southern green stinkbug (*Nezara viridula* L.) is a generalist feeder that exhibits edge-mediated dispersal from peanut into cotton at the common boundary of the two crops in peanut–cotton farmscapes (Tillman *et al.*, 2009). Addition of a habitat of sorghum along this boundary apparently enhances biocontrol of this pest by the adult fly *Trichopoda pennipes* F. (Tachinidae) (Tillman, 2006). Strategic establishment of a corridor composed of 65 flowering plant species enhanced predator colonisation and abundance on adjacent organic vineyards by providing timely circulation and dispersal of predators into the centre of the field (Nicholls *et al.*, 2001).

In this chapter, we present three examples of establishing a habitat in an agricultural system at the right time and location for enhancement of agricultural biodiversity and conservation biocontrol. These include use of cover crops and conservation tillage in cotton fields, provision of insectary plants within lettuce fields and establishment of beetle banks within cereal fields. For each of these examples we cover the driving forces that led to the introduction of a biodiversity-based pest management system, discuss the development of the habitat and evaluate its effectiveness and uptake, while providing some information on the economics.

COVER CROPS AND CONSERVATION TILLAGE IN COTTON IN GEORGIA, USA

Cotton, *Gossypium hirsutum*, is a fibre, feed and food crop. The fibre of cotton is used to make thousands of products including T-shirts, sheets, towels, etc. US textile mills spun over 3.5 million bales of cotton in 2010 (NCC, 2011), enough cotton fibre to make over 1 billion pairs of jeans. About 70% of the harvested crop is composed of the seed, which is crushed to separate its three products – oil, meal and hulls. Cotton seed oil is a common component of many food items, used primarily as a cooking oil, shortening and salad dressing. The oil is used extensively in the preparation of such snack food as crackers, cookies and chips (crisps). The meal and hulls are used as livestock, poultry and fish feed.

Traditionally cotton is one of the most pest-plagued, and thus one of the most pesticide-treated, commodities. Many of the pesticides used in conventional cotton production can adversely impact human and animal health and the environment both directly and indirectly. Indeed, a study conducted by researchers at the Technical University of Lódz in Poland has shown that hazardous pesticides applied during cotton production can sometimes be detected in cotton clothing (EJF, 2007). In California, the leaves, stems, and short fibres of cotton known as 'gin trash' can contain concentrated levels of pesticide residue, making it illegal to feed this cotton by-product to livestock (Maan and Beam, 2009). Highly toxic insecticides used to control cotton insect pests can kill their natural enemies which may lead to a resurgence of the pests or outbreaks of secondary pests. For example, heavy outbreaks of beet armyworms (*Spodoptera exigua* (Hübner) can be generated by insecticide treatments used to suppress the plant bug (*Lygus Hesperus* Knight) in cotton (Eveleens *et al.*, 1973). Biodiversity-based pest management systems are needed to reduce use of pesticides in cotton production.

Georgia is one of the leading cotton-producing states in the US; over 1.3 million bales were harvested there in 2010 (Williams, 2011). In Georgia cotton is planted in early summer and harvested in the fall. Since the eradication of the boll weevil (*Anthonomus grandis* Boheman), larvae of the heliothines, the tobacco budworm (*Heliothis virescens* F. and the cotton bollworm (*Helicoverpa zea* (Boddie), and nymphs and adults of the stinkbugs, the southern green stinkbug, the brown stinkbug (*Euschistus servus* (Say), and the green stinkbug (*Chinavia hilaris* (Say), have been the two major pest complexes causing economic damage to cotton in this state.

Many farmers in Georgia became increasingly interested in conservation tillage as research efforts began demonstrating that it could work with cover cropping to improve soil quality and long-term soil productivity, reduce soil erosion, promote beneficial insects and provide greater agroecosystem stability (Blumberg and Crossley, 1982; McPherson *et al.*, 1982; Sprague and Triplett, 1986; Triplett, 1986; Guthrie *et al.* 1993). Thus, from the mid-1990s grower-driven research was conducted in the southern region of the state to determine the effects of a legume, crimson clover (*Trifolium incarnatum* L.), and a grass, rye (*Secale cereal* L.), cover crop in a conservation-tillage system on the populations of insect pests and their natural enemies in non-*Bt* cotton. In these experiments, primarily the heliothines reached economic threshold (i.e. pest level at which a control measure should be applied to prevent economic damage), which is a 5% infestation

of first instars on cotton plants. Either equal or fewer insecticide applications were needed for control of these pests in the cover crop/conservation-tilled fields compared to the conventional ones (Ruberson *et al.*, 1995; Lewis *et al.*, 1996; Ruberson *et al.*, 1997).

Building on the work of previous researchers, a group of research scientists funded by SARE (Sustainable Agriculture Research and Education) initiated a two-year cover/conservation tillage on-farm experiment in non-*Bt* cotton in Georgia in the fall of 2000 (Tillman *et al.*, 2004). The main insect research goal was to develop an early-season habitat for natural enemies of pests that would promote biological control of these pests in cotton and minimise the need for insecticides without sacrificing yield. The five cover crop treatments were 1) rye (standard grass cover crop), 2) crimson clover (standard legume cover crop), 3) a legume cover crop mix of balansa clover (*Trifolium michelianum* Savi), crimson clover and hairy vetch (*Vicia villosa* Roth.), 4) a combination of the legume cover crop mix plus rye and 5) no cover crop (Figure 19.1). The legume cover crop mix of an early- (balansa clover), mid- (crimson clover) and late- (hairy vetch) spring flowering legume was used to extend the availability of the habitat and the provision of nectar to insect pollinators in the field beyond what could be attained using a single legume species. For the legume and rye treatment, alternating strips of the legume mix and rye were planted to combine the benefits of a legume habitat and nitrogen fixation with the enhanced biomass production of rye. The rye and legume mix were planted so that a strip of rye grew in the centre of the future cotton row, and strips of the legume mix grew between the rows. All cover crops were killed approximately three weeks before planting the cotton. Legume cover crops were strip-killed by applying an herbicide in a 46–53 cm-wide strip of cover crop in the centre of the future cotton row. The tallness of rye made it difficult to maintain row patterns in this cover crop, and so it was broadcast-killed. While planting cotton, the soil was strip-tilled in the band of dead legume cover crop or in the centre of the row for the dead rye. In control fields with no cover crop, conventional tillage practices were used for cotton production.

The heliothine complex was the only group of insect pests that caused economic damage to cotton in both years of the study. Cotton bollworms and tobacco budworms only cause damage to cotton in the larval (worm) stage. First instars feed on plant terminals and

Figure 19.1 Winter cover crops used to promote biological control in cotton: a) rye, (b) crimson clover, c) legume mixture, and d) legume mixture + rye (K.J. Graham).

small squares (buds) and may sometimes destroy the terminal bud, which results in branching of the plant. Later instars move into lower squares, blooms and then bolls (fruit). These larvae burrow into squares and bolls, often hollowing them out. A single larva feeds on 6 to 7 squares and 2 to 3 bolls during its developmental period. Injured squares are often shed. Bolls that are fed upon do not produce cotton fibre or seed.

Big-eyed bugs, *Geocoris punctipes* (Say), pirate bugs, *Orius insidiosus* Say and red imported fire ants, *Solenopsis invicta* Buren, preyed on eggs and small larvae of these pests in both the cover crops and the cash crop. In the spring, the flowers of each legume species produced nectar and were visited frequently by bees and other insect pollinators. Also, predator populations built up in the cover crops, especially in the crimson

clover and legume cover crop mix. Later, conservation tillage of these legume cover crops allowed a live strip of cover crop to remain between crop rows to serve as a habitat and food source for natural enemies until the cotton crop was established. Density of *G. punctipes* in crimson clover on the last sampling date was statistically similar to density of this predator in cotton on the first sampling date in 2001 (Figure 19.2) and in the legume cover crop mix in both years of the study. In 2002, density of *G. punctipes* was significantly higher in the first cotton sweeps compared to the last crimson clover sweeps. Evidently, intercropping cotton in strips of cover crop resulted in the relay of *G. punctipes* from these cover crops onto cotton.

Even though both nymphs and adults of *G. punctipes* are predatory, they fed on leaves of each cover crop

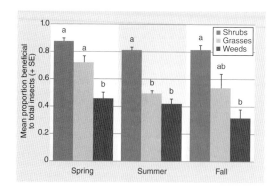

Figure 19.2 Seasonal occurrence of *Geocoris punctipes* in conservation-tillage cotton with crimson clover winter cover crop in 2001. Least squares means are not significantly different between the last sampling date in the cover crop and the first sampling date in cotton (one-tailed *t*-statistics, $P > 0.05$).

species and cotton and may have fed on the nectar produced by the flowers of the legumes and the extrafloral nectaries of hairy vetch. *Geocoris punctipes* does not require nectar for nymphal development and adult longevity when given abundant prey, although nectar helps them to survive in the absence of prey (De Lima and Leigh, 1984). Also, predation of heliothine eggs by *G. punctipes* is similar on cotton with or without extrafloral nectaries (Thead *et al.*, 1985). Thus, provision of prey is likely to have played a more significant role than nectar provision in the success of the early-season habitat in enhancing this natural enemy in cotton. Apparently, a reduction in tillage conserved the habitat of red imported fire ants. *Geocoris punctipes* density in cotton generally was higher in crimson clover fields and on some occasions in the legume-mix fields compared to control fields, and fire ant density was higher in conservation-tilled fields compared to control fields. These results suggest that the early-season build-up of these natural enemies and conservation of within-field habitat translated into higher numbers of predators in cotton fields with cover crop/conservation tillage systems than in control cotton fields.

The number of heliothine eggs on cotton was similar across cover crop treatments over the growing season. Nevertheless, the need for insecticides to manage heliothine larvae was reduced by one to two applications in conservation-tilled fields with winter cover crops compared to control fields. These results indicated that the early-season build-up of natural enemies in

the cover crops and conservation of habitat of natural enemies subsequently resulted in reduction in heliothine damage in conservation-tillage cotton with these cover crops compared to conventional-tillage cotton without cover crops (Box 19.1). Yields of seed cotton (i.e., cotton with lint and seed before cleaning) for cover crop/conservation-tilled fields were either higher, particularly for fields with a legume cover crop, or comparable to those for control fields. In summary, cover crops and conservation tillage promoted biological control of these pests in cotton and minimised the need for insecticides without reduction in yield. Detailed economic analyses were not done, but cost of insect control was reduced in cover crop/conservation-tilled fields without incurring economic loss through lower yields. Further benefits of this approach are the decrease in soil erosion, increase in soil organic matter, and a reduction in insecticide contamination of the environment.

In an earlier study, Bugg *et al.* (1991) determined that *G. punctipes* occurred in high densities on subterranean clover (*Trifolium subterraneum* L.), and there was evidence that the high densities observed amid dying mulches translated into greater predation of fall armyworm (*Spodoptera frugiperda* (J.E. Smith)) egg masses on cantaloupe foliage. Likewise, Ruberson *et al.* (1995) suggested that the high abundance of red imported fire ants in a crimson clover/strip-tilled field was responsible for the reduction in pest populations in this field relative to the conventionally tilled field without the cover crop.

According to the Conservation Technology Information Center's National Crop Residues Management Survey (2010), conservation tillage accounted for 41.5% of US planted crop area in 2008, compared with 26% in 1990. Widespread adoption of genetically engineered *Bt* germplasm has limited the usefulness of cover crops for managing heliothines in production of conventional cotton. However, this pest complex still ranked second in yield reduction for conventional cotton in the state in 2010 (Williams, 2011). Also, heliothines are major economic pests in organic cotton in Georgia (Tillman *et al.*, 2008). Analysis of available data collected by an Organic Trade Association survey of US organic cotton producers determined that 4,343 ha of organic cotton were planted in the US in 2009, an increase of 26% over the previous year (OTA, 2010). Because the use of *Bt* cotton is prohibited in organic production of cotton (SAN, 2007), organic growers may benefit economically by using cover crops

Box 19.1 Benefits of a crimson cover crop in conservation-tillage cotton

Crimson clover provided a habitat for predator populations to build up in the spring and provided nectar
for bees and other pollinators and parasitoids while
increasing nitrogen in the soil, improving soil quality
and reducing runoff and soil erosion.

Photo: K.J. Graham

Management of the cover crop strategically placed a
strip of habitat in the cotton field so that the heliothine
egg predator, G. punctipes, could relay from crimson
clover to cotton.

Photo: H. Pilcher

The decreased need for insecticides to manage heliothine larvae in crimson clover cotton compared to
control cotton indicated that the early-season build-
up of predators in the cover crop and relay of predators from the cover crop onto cotton resulted in a
reduction in heliothine larval damage to cotton.

Treatment	No. times economic threshold for heliothine pests exceeded
Control	2.0a
Crimson clover	0.75b

Means are not significantly different between cover
crop treatments (one-tailed t-statistics, $P > 0.05$).

and conservation tillage practices, and new Global
Positioning System (GPS) technology is available to
help growers establish and maintain row patterns in
cover crop fields. So, in Georgia, cover crops are beginning to be utilised in organic production of corn (*Zea
mays* L.) and soybean (*Glycine max* L. Merr.), and
research has been initiated to incorporate cover crops
and conservation tillage in organic cotton and peanut
(*Arachis hypogaea* L.) production (J. Tescher, personal
communication, 2010).

CONSERVATION BIOLOGICAL CONTROL BY SYRPHID LARVAE OF *NASONOVIA RIBISNIGRI* AND OTHER APHIDS IN ORGANICALLY GROWN LETTUCE ON THE CENTRAL COAST OF CALIFORNIA

Almost 80% of the lettuce consumed in the USA
is grown in California (NASS, 2010) where it is a
billion-dollar industry (CDFA, 2008). In 2008, over

101,175 ha of lettuce were planted in California
(NASS, 2010) with 8% grown using certified organic
methods (ERS, 2008). Over 70% of the lettuce produced in California is grown in the Central Coast
region, primarily in the highly productive Salinas
Valley, which runs through the heart of Monterey
County (Monterey County, 2009). Both leaf and head
lettuce varieties are produced on the Central Coast and
sold as whole heads in cartons, as trimmed hearts and
in bagged salad mixes. Lettuce is planted in Monterey
County from January to August, and it is harvested
from April to December (Smith *et al.*, 2009).

A new pest

In 1998, *Nasonovia ribisnigri* (Mosley), a new invasive
pest of lettuce, locally referred to as the lettuce aphid
or red aphid, became established in the Salinas Valley
(Chaney, 1999). Originating in Europe, *N. ribisnigri* has
become established in Asia, the Middle East and North

and South America (Blackman and Eastop, 2000). *Nasonovia ribisnigri* is greenish-orange to pink in colour and establishes dense colonies in the inner leaves of the lettuce head, making it unmarketable (Liu, 2004). Conventional growers have a range of insecticides available to suppress *N. ribisnigri* populations, including systemic insecticides such as neonicotinoids. Suppressing incipient *N. ribisnigri* infestations is crucial in conventional lettuce production because chemical control has limited efficacy once the infestation is protected by outer lettuce leaves, and the crop has outgrown the window of protection provided by at-plant neonicotinoid treatments. While specific thresholds have not been established, growers run the risk of having their lettuce crop rejected for sale if *N. ribisnigri* is detected at even very low levels.

From 1998 to 2001, management of *N. ribisnigri* was problematic for organic lettuce growers because attempts by these growers to suppress *N. ribisnigri* infestations using available insecticides, such as insecticidal soap, were ineffective (Colfer, 2004). Some, but not all, organic growers were planting strips of flowering plants, also known as insectary plants, in their lettuce fields to attract beneficial insects. During these initial years, growers and university personnel observed that, often in the absence of any pest management intervention, fields of organic lettuce that were infested with *N. ribisnigri* early in the season were largely aphid-free and marketable by the harvest date (Chaney and Smith, 2005). By 2001, University of California Farm Advisor William Chaney, Ramy Colfer of Mission Organics and other organic growers had concluded that aphidophagous syrphid larvae were playing a major role in the suppression of *N. ribisnigri* in organic lettuce fields on the Central Coast of California (Colfer, 2004). However, a small percentage of lettuce fields remained unmarketable at harvest time due to aphid infestation, leading growers and researchers to ask if there were ways to enhance the activity of syrphids and make aphid suppression in organic lettuce more predictable.

Syrphid flies belong to the family Syrphidae and are also referred to as hoverflies or flower flies. Many species of syrphids have predatory larvae that feed on aphids (Bugg *et al.*, 2008). Adult syrphids are not predaceous. Both male and female syrphids require pollen for gametogenesis (Chambers, 1988). For energy, syrphid adults exploit floral nectar and in some cases honeydew, a sugar-rich material excreted by aphids and other Hemiptera.

Insectary plantings

Once the connection between syrphid larvae and *N. ribisnigri* suppression seemed clear, interest in evaluating the best species and density for insectary plantings intensified among organic growers and university researchers on the Central Coast (Chaney and Smith, 2005). Insectary crops are plants that provide resources such as nectar and pollen to predators and parasitoids. Growers incorporate insectary crops into fields with the aim of enhancing the pest-suppression activity of natural enemies (Landis *et al.*, 2000). One of the first promoters of insectary plantings on organic farms in California was Robert 'Amigo' Cantisano (UCSC, 2010). Cantisano recalls that before the national demand for organic produce rose in the 1990s and product quality standards increased, there was limited interest in insectary plantings among the Central Coast's traditional organic vegetable growers (R. Cantisano, personal communication, 2010). According to Cantisano, the arrival of *N. ribisnigri* in the Salinas Valley 'changed everyone's attitude' towards the importance of insectary plantings (R. Cantisano, personal communication, 2010).

Among the many insectary plants that have been evaluated on California's Central Coast, sweet alyssum (*Lobularia maritima* L. (Desv.)) has proven to be among the most attractive to syrphid adults and the easiest to incorporate into field production (Chaney, 1998; Colfer, 2004; Chaney *et al.*, 2006; Bugg *et al.*, 2008). Damage by flea beetles to alyssum has led some Central Coast organic growers to incorporate phacelia (*Phacelia tanacetifolia* Bentham) and buckwheat (*Fagopyrum esculentum* M.) in their insectary plantings (Ramy Colfer, personal communication). Dhani-ya coriander (*Coriandrum sativum* L.), a cultivar of coriander which flowers quickly, is also highly prized by organic growers on the Central Coast for the apparent abundance and diversity of syrphid adults that it attracts. Some growers use 'good bug blends' – mixtures that include clovers, herbaceous plants and grasses – as insectary crops in their organic lettuce production in the Salinas Valley.

Both the ratio of insectary crop to lettuce and the insectary intercropping pattern vary among organic farms. Organic growers on the Central Coast dedicate up to 9% of lettuce fields to insectary plantings, with 5% considered the norm (Tourte *et al.*, 2009). Chaney (1998) demonstrated that densities of beneficial insects were higher and aphids were lower within an

11 m range of insectary strips planted in lettuce than at greater distances from insectary strips. He recommended one insectary strip roughly every 33 m. Colfer (2004) documented a threefold increase in syrphid oviposition on romaine lettuce adjacent to (0.6–3 m) alyssum versus romaine 16 m from alyssum. However, the concentration of syrphid eggs primarily near insectary plantings is brief, for syrphid eggs and larvae soon become distributed across lettuce fields (Colfer, 2004).

It is important to keep in mind that syrphid adults are strong fliers and that syrphid oviposition is stimulated by the presence of or proximity to aphids and compounds associated with them, regardless of immediate availability of floral resources (Chandler, 1968a; 1968b; Shonouda et al., 1998; Verheggen et al., 2008). The Salinas Valley is characterised by organic farms, roadsides, rangeland and riparian habitat that harbours abundant natural flowering vegetation. Depending on the specific agricultural landscape in which a field is located, naturally occurring floral resources may contribute as much as planted insectaries to the suppression of N. ribisnigri by syrphids. Some organic growers have benefited from the activity of syrphids without intercropping insectary plants in their lettuce fields, presumably because the floral resources near the lettuce field are sufficient to enhance syrphid activity.

The syrphid 'team'

Fourteen species of aphidophagous syrphids have been reared from commercial organic romaine fields in and around the Salinas Valley (Smith and Chaney, 2007; Smith et al., 2008). Four species predominate: *Toxomerus marginatus* (Say), *Platycheirus stegnus* (Say), *Sphaerophoria sulfuripes* (Thomson) and *Allograpta obliqua* (Say). *Toxomerus marginatus* and *S. sulfuripes* comprised 39% and 13%, respectively, of over 1,000 syrphids reared from several farms during an intensive seven-month survey in 2005 (Smith and Chaney, 2007). While these two species were recovered from moderately and highly infested romaine lettuce fields, it is noteworthy that their eggs and larvae were also recovered in significant numbers from romaine fields in which aphid densities were too low to be of concern to the grower. For example, *T. marginatus* and *S. sulfuripes* were collected from a variety of romaine that is highly resistant to *N. ribisnigri* but which supports low populations of the potato aphid, *Macrosiphum euphorbiae* (Thomas). From a biological control perspective, it

is significant that there are species in the syrphid complex that are present in fields where aphid populations are at sub-economic levels. By contrast, *P. stegnus*, which comprised 27% of the syrphids reared in 2005, was collected primarily from highly infested fields. While other syrphid species collected from organic romaine lay eggs singly or in groups of two or three, *P. stegnus* oviposits clusters of parallel, contiguous eggs. It was not uncommon to encounter clusters of five to seven eggs, and a single cluster of 18 eggs was recovered. This egg-laying behaviour suggests that *P. stegnus* is adapted to take advantage of plants with high aphid densities.

Weekly whole-plant samples from multiple organic romaine farms in 2005 and 2006 revealed that peak densities of five to nine syrphid larvae per romaine head one to two weeks before harvest are not uncommon (Smith and Chaney, 2007; Smith et al., 2008). These peak syrphid densities consistently coincided with the crashing of aphid populations in the romaine field. Replicated field trials were carried out in 2007 to determine the effect on aphid populations of experimentally removing syrphid larvae (Smith et al., 2008). The organically approved formulation of spinosad (Entrust®) was applied once a week for five weeks prior to harvest in 7.6 m plots of romaine to suppress syrphid larvae. Spinosad is an effective insecticide for suppressing certain dipterous larvae, and pilot studies had determined that it suppresses syrphid larvae without affecting aphid populations (W.E. Chaney, personal communication). Where Entrust was applied, syrphid larval populations were suppressed, and romaine was unmarketable at harvest because of aphid infestation (Smith et al., 2008). In plots where Entrust was applied, the highest average whole plant syrphid larva density was 2.84 ± 0.58 (SEM). In untreated romaine, the highest average whole plant densities of syrphid larvae ranged from 2.75 ± 0.58 to 9.08 ± 0.58, depending on the site (Smith et al. 2008). In untreated plots, where syrphids were allowed to persist, romaine was marketable at harvest (Figure 19.3).

Other natural enemies

Parasitic wasps are not considered important in the suppression of *N. ribisnigri* because the pest colonises the interior of the lettuce head, where it is largely protected from these natural enemies. Infections by the entomogenous fungus *Pandora* spp. commonly sup-

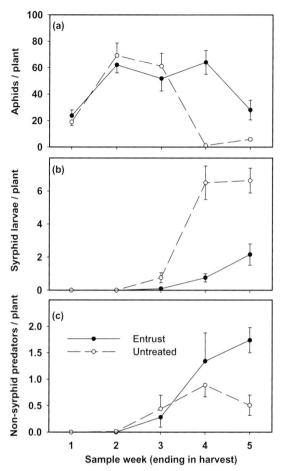

Figure 19.3 The effect of suppressing syrphid larvae with Entrust, an organically approved insecticide, in an organic romaine field in Hollister, California. The graphs illustrate three key aspects of aphid suppression by syrphids: 1) When syrphids are suppressed, lettuce is unmarketable because of aphid infestation a), 2) Syrphids can reach peak densities of ~6 larvae or more per romaine head before harvest b) and 3) Non-syrphid predator densities overall are very low c). Data represent average insect densities per plant (±SEM) (from: Smith *et al.*, 2008).

dwarf spiders (Linyphiidae: Araneae), big-eyed bugs (*Geocoris* spp.: Lygaeidae), minute pirate bugs (*Orius* spp.: Anthocoridae), green lacewings (*Chrysopa* and *Chrysoperla* spp.: Chrysopidae), brown lacewings (*Hemerobius* spp.: Hemerobiidae), rove beetles (Staphylinidae) and predatory thrips (Thysanoptera) (Smith and Chaney, 2007; Smith *et al.*, 2008). Between 61% and 97% of romaine plants collected from five organic farms in 2005 contained either syrphid eggs or larvae; the range for plants containing non-syrphid predators was 16–60%, depending on the field (Smith and Chaney, 2007). Syrphid larvae comprised between 85% and 96% of all predators collected from whole-plant samples at most research sites in 2006 (Smith *et al.*, 2008). Non-syrphid predators were consistently found at much lower densities than syrphid larvae. Except for dwarf spiders, no other predators besides syrphids were found in every field.

Organic romaine growers dedicate on average 5% of crop area to insectary plantings to ensure the marketability of the remaining 95% of the crop; the cost of alyssum seed is roughly US$10.00/ha (Tourte *et al.*, 2009). Organic leaf lettuce growers who apply insecticides to manage caterpillars and aphids spend on average an additional $290/ha on pest management. The estimate for pest management in conventionally grown romaine hearts is $1100/ha (Smith *et al.*, 2009). In recent years, the Dutch plant breeding company Rijks Zwaan has produced varieties of lettuce that are highly resistant to *N. ribisnigri* (van Helden *et al.*, 1995). These varieties have enabled some organic lettuce growers to reduce the area that they dedicate to insectary plantings (Phil Foster, personal communication, 2010). However a biotype of *N. ribisnigri* that is able to survive on resistant lettuce varieties has already been identified in Europe (Rijks Zwaan, 2010).

Summary

Within the diverse species complex of syrphids suppressing aphids in the Salinas Valley, the predominant species apparently complement each other by exploiting distinct predatory guilds. *Toxomerus marginatus* and *S. sulfuripes* will oviposit in fields that have very low numbers of aphids as well as in more infested fields, while there is evidence that *P. stegnus* is specifically adapted to colonise fields with high aphid populations (Smith and Chaney, 2007). At least 11 additional syrphid species contribute to aphid suppression. Syrphid

press *N. ribisnigri* and other aphids in lettuce during the early months of production, which overlap with the final winter rains in the Salinas Valley (S. Koike, personal communication, 2010).

Other predators found in organic romaine in the Salinas Valley include ladybird beetles (Coccinellidae),

larvae operate with a high degree of efficiency in the inner leaves of the lettuce head where other natural enemies are apparently less effective. Among predators in organic lettuce, only syrphid larvae reach high densities in the crucial weeks immediately before harvest. When these larvae are killed, the crop is unmarketable because of aphid infestation (Smith *et al.*, 2008). The suppression of *N. ribisnigri* by naturally occurring syrphid species on California's Central Coast is a noteworthy example of an invasive pest of a high-value fresh market crop being effectively managed by endemic natural enemies (Box 19.2). This is a contemporary example of the effectiveness of 'new associations' in biological control (Hokkanen and Pimentel, 1989). The break-even costs for California's organic lettuce growers depend on market price and

yield per hectare, which can vary considerably from year to year (Tourte *et al.*, 2009). Therefore it is difficult to put a dollar value on syrphid predation. However, it seems unlikely that California's organic growers could supply the bulk of the nation's demand for organic lettuce without the aphid suppression services provided by syrphid flies.

BEETLE BANKS

Background

In the European Union (EU) 30% of the farmed landscape is devoted to the production of cereal crops for human consumption and animal feed. Of this, winter

---

Box 19.2 Conservation biological control of the lettuce aphid in organic lettuce

Syrphid larvae (top right) feed on the lettuce aphid and other aphids. Organic growers enhance the activity of these natural enemies by providing floral resources (nectar and pollen) to adult syrphids (bottom right) in lettuce fields with in-field plantings of alyssum (below) and other insectary plants.

Photo: W.E. Chaney

Photo: W.E. Chaney

Photo: H.A. Smith

wheat (sown in autumn) accounts for 46% of cereal production. The main insect pests are cereal aphids, some of which invade the crop in the autumn, and because they transmit damaging viruses (e.g. barley/cereal yellow dwarf viruses), most crops are prophylactically treated with insecticides unless they are late-sown, which reduces the chance of aphid infestation. Aphids also invade crops in the summer, causing yield loss (Mann *et al.*, 1991), whilst the honeydew encourages sooty moulds near harvest (Poehling *et al.*, 2007). In Western and Central Europe, *Sitobion avenae* F. (grain aphid), *Metopolophium dirhodum* (Walker) (rose-grain aphid) and *Rhopalosiphum padi* L. (bird cherry-oat aphid) are the predominant pest species, but *Schizaphis gramminum* (Rondani) (greenbug) and *Diuraphis noxia* (Kurdjumov) (Russian wheat aphid) which are more typical of warmer climates, are spreading as Northern Europe experiences warmer winters (Poehling *et al.*, 2007).

The development of beetle banks

A programme of research was initiated in the 1980s in the UK by S. Wratten at Southampton University and the Game Conservancy Trust (now Game and Wildlife Conservation Trust) aimed at developing an IPM system for cereal crops. Exclusion studies had identified that generalist predators, mainly carabid and staphylinid beetles, were capable of contributing to cereal aphid control (Edwards *et al.*, 1979; Chiverton, 1986), especially early in the spring when predator:prey ratios were high and before aphidophagous species were available in sufficient numbers. Earlier work had identified that tussock-forming grasses (e.g. *Dactylis glomerata* L. and *Holcus lanatus* L.) provided appropriate and relatively stable conditions during the winter (Luff, 1966) and resulted in greater survival compared to other plants (D'Hulster and Desender, 1982). Such habitats typically occurred between field margins and hedgerows, and these were found to support high densities of overwintering generalist predators (Sotherton, 1984; 1985). These predators subsequently colonised the adjacent field in spring, but in large fields species which dispersed by walking took until June to reach field centres (Coombes and Sotherton, 1986). Unfortunately, between the 1950s and 1970s many hedgerows in the UK were removed, purportedly to increase agricultural productivity and efficiency. Many remaining hedgerows have become degraded, including the

hedgebase where herbicide and fertiliser drift destroys the complex plant community (Bealey *et al.*, 2009). The creation of 'island habitats' across fields was devised as a way to replace these losses with a simple-to-manage habitat that would provide overwintering cover and encourage a more extensive and earlier coverage of the field with generalist predators (Thomas *et al.*, 1991).

Evaluating effectiveness

Initial studies confirmed that the banks were quickly colonised by very high densities (up to 1,500 per m^2) of overwintering beetles (Thomas *et al.*, 1992), which led to their being called 'beetle banks' (Box 19.3). The mean density (585 per m^2) across a number of later studies was lower (Table 19.1), and in these studies, densities were maintained for up to 10 years and were comparable to or even higher than those of field margins (Thomas, 2001; Collins *et al.*, 2003; Macleod *et al.*, 2004). Considerable variation was found between years and study sites and was attributed to the many different variables (e.g. soil type and landscape composition) and anthropogenic impacts (e.g. crop management practices) occurring in adjacent fields. Overall, the invertebrates found within the banks comprised Carabidae, Staphylinidae (mostly *Tachyporus* spp.) and Araneae (mostly Linyphiidae).

The effectiveness of beetle banks may be estimated by calculating to what extent they are likely to increase the number of predatory natural enemies within the adjacent field. If beetle banks are expected to enhance predators to 75 m on either side (this being the recommendation given above) and support on average 585 predators per m^2 or 2,180 per m^2 at the highest estimate (Collins *et al.*, 2003), then this would raise the number of predators by 3.9 per m^2 or 14.5 per m^2 respectively, assuming all emigrate from the beetle bank during the summer. Actual predator densities within a cereal field without a beetle bank varied between 29.3 per m^2 on 3 June declining to 11 per m^2 by 13 July (Holland *et al.*, 2004). Thus, depending on predator densities within the beetle banks and fields (i.e., predators not originating from the beetle bank), beetle banks can supplement existing predator densities by over 50% (Figure 19.4).

Two studies examined whether the beetle banks led to more even predation across fields. Predation of artificial prey still occurred at the maximum distance used

Box 19.3 How to create beetle banks

The banks are created by ploughing two furrows together to create a raised bank 40 cm high and approximately 2 m wide. The ends of the beetle bank remained separated from the field margin by a spray boom width to minimise the disruption to agricultural operations.

To create optimal overwintering conditions for beetles the banks are best sown with tussock-forming grass species. A range of different grass species and mixtures were compared to see if this affected colonisation rates, but the most aggressive grass, *Dactylis glomerata* (Cock's Foot), quickly dominated although it does provide appropriate conditions (Thomas *et al.*, 1991; Collins *et al.*, 2003) and consequently most farmers only sowed this species (Thomas, 2000).

Photo: J.M. Holland

These grasses were found to maintain their structure for at least a decade and flowering plants also started to colonise the strips (Thomas *et al.*, 2001). Based upon the distance that beetles were dispersing from the banks, it was advised that fields larger than 15 ha should be divided by multiple beetle banks spaced no further than 150 m apart (Thomas *et al.*, 1991).

Photo: J.M. Holland

Photo: P. Thompson

Table 19.1 Densities of predatory natural enemies within beetle banks in three studies.

Reference	Year	Carabidae	Densities (per m²)		
			Staphylinidae	Araneae	Total predators
Macleod *et al.*, 2004	1987	11	1	6	17.9
	1988	111	44	22	177.1
	1989	20	39	26	85.3
	1990	14	28	43	84.6
	1991	53	84	48	185.2
	1992	72	125	45	242.6
	1993	45	91	25	160.9
Collins *et al.*, 2003	1994	80	377	136	593
	1995	301	857	89	1247
	1996	423	1550	207	2180
	1997	79	351	84	514
Thomas, 2001	1997	200	340	380	920
	1998	250	480	470	1200
	Mean				585.2

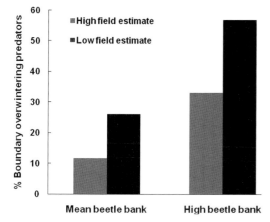

Figure 19.4 Proportion of predators originating from beetle banks for low and high estimates of predator densities within fields (see text for details).

(60 m), although predation was highest on the bank itself (Thomas, 1990). The impact on naturally occurring aphid infestations of excluding ground-dispersing predators was evaluated within enclosed plots established at 8, 33, 58 and 83 m from a beetle bank (Collins *et al.*, 2002). The mean number of aphids and aphid peak populations were reduced up to 58 m, but reductions were greatest at 8 m. The generalist predators probably had a lower than anticipated impact because the aphids invaded the crop relatively late in the season and increased rapidly, outstripping their control as found in other exclusion studies (Holland *et al.*, 1996).

The distribution of generalist predators may be affected by the presence of a beetle bank. A brief wave of emigration of generalist predators from the banks was detected in April or May (Thomas *et al.*, 1991; 2000; Collins *et al.*, 2002) followed by a period in which there was an even spread of boundary-overwintering predators across the adjacent field (Thomas *et al.*, 2000). More recent studies of insect spatial distribution within fields surrounded by hedgerows found that numbers of boundary-overwintering species were highest within 60 m of the boundary (Holland *et al.*, 2009) with a peak in their numbers during May, followed by a decline during June, also found by Thomas *et al.* (2000). This explains why in a small field (2.2 ha) the more mobile boundary-overwintering predators (e.g. *Tachyporus* spp. and linyphiid spiders) showed no association with margins (Holland *et al.*, 1999). The impact of beetle banks may be reduced if predators remain within them all year round. The density of Carabidae decreased by two-thirds between winter and spring with a slight increase in summer compared to spring (Thomas, 2001). Between winter and spring, Staphylinidae densities declined by three-quarters and Araneae densities declined by a half, indicating a similar emigration. Some losses may be ascribed to overwinter mortality rather than emigration.

Uptake of beetle banks

Despite widespread promotion of the concept since the early 1990s by the Game Conservancy Trust and Game and Wildlife Trust, uptake in the UK is poor, even after the approach was supported under English agri-environment schemes (AES). In the Countryside Stewardship scheme and its successor, the Environmental Stewardship (ES) scheme, payments were the equivalent of approximately £12 per 100 m length per year. The Entry Level Scheme which forms the first tier of ES has been adopted by 70% of farms in England, but only 1.4% had established beetle banks by 2009 (Boatman et al., 2007). In an earlier farmer survey it was highlighted that more information on the biocontrol potential of beetle banks was needed, and this may partly explain the poor uptake (Thomas, 2000). Other likely reasons are that the threat from cereal aphids is diminishing and insecticides are relatively cheap and can easily be added to the fungicide spray programme. In most cases, the farmers were establishing beetle banks for their wider biodiversity benefits, for game management, or to demarcate areas to aid farming operations. The banks support other invertebrate taxa including grasshoppers, some butterflies and invertebrates important as food for bird chicks, and thus they are used as nesting habitat, especially by grey partridge *Perdix perdix* L. (Thomas et al., 2001) and harvest mice (*Micromys minutus* Pallas).

The beetle bank concept has also been tested across the world, for example in Denmark (Reidel, 1992), Sweden (Chiverton, 1989), Finland (Helenius, 1995), the USA (Carmona and Landis, 1999) and New Zealand (Berry, 1997), but in many countries grassy perennial field margins provide a similar function and are the commonest way of dividing fields, although a raised bank does give some extra benefit on heavier soils by creating drier conditions (Sotherton, 1985). In the UK, grass margins (buffer zones) have been widely established (73,000 ha by 2009) using ES funding, and these may likewise support overwintering generalist predators and supply alternative prey in the summer for a range of predators and parasitoids (Meek et al., 2002). Landscape-scale evaluations confirmed that aphid control was related to the proportion of area occupied by grass margins (Holland et al., 2008a). However, exclusion studies that isolated the impact of ground and flying predators revealed that more than 90% control was achieved by flying predators compared to a maximum of 40% by ground-based predators alone (Holland et al., 2008a). This level of control was only achieved when fields were surrounded by 6 m-wide grass margins, and control occurred more slowly (Holland et al., 2008b). These studies were, however, conducted in June and July when populations of boundary-overwintering predators were declining. This decline may be due to natural mortality and predators retreating back to the margins, although summer sampling indicated this was not occurring (Thomas, 2001). Alternatively, the high level of control may be the consequence of predation by the larger field-overwintering species (e.g. the carabids, *Pterostichus* spp.). These species increased rapidly from early June and far outweighed those originating from the boundaries (Holland et al., 2009).

Economics

The economic benefits of beetle banks have been estimated based only upon the cost of establishment and income foregone for the land occupied by the bank rather than any measure of reductions in insecticide use or yield gain. In 2002, the establishment costs were £975 per ha with subsequent costs of £2 per ha for income foregone from the land occupied (Collins et al., 2002). Thus the agri-environment scheme payments of £600 per ha would cover these costs within two years and be more profitable in following years. The cost of an insecticide was between £3 and £12 per ha without application costs, but aphicides are typically added to a fungicide programme. Therefore, beetle banks are not economically feasible without the AES payments based upon savings in insecticide costs alone. However, there are other benefits such as enhancing natural enemy populations that may operate farm-wide, and the enhancement of wildlife.

Summary

It is unlikely that beetle banks will ever be widely used in cereal crops whilst cheap and efficient insecticides remain available, but for organic producers and those producing horticultural crops in Europe where the range of insecticides is dwindling owing to revisions to pesticide legislation (Council directive 91/414/EEC) then they may contribute to an IPM programme. Other habitats that are already present on farmland, such as grassy strips between fields or associated with fence lines, ditch edges, shelterbelts and the base of hedgerows, can support high densities of overwintering generalist predators (Griffiths *et al.*, 2008). These areas may also be suitable for breeding, aestivation and foraging, providing a source of alternative prey for periods when pests are insufficient. Such habitats may also be supplemented with flowers and hedgerows as these can provide a source of pollen and nectar for their natural enemies such as hoverflies (Cowgill *et al.*, 1993; Hickman and Wratten, 1996), parasitoids (Berndt *et al.*, 2006; Vollhardt *et al.*, 2010), lacewings (Robinson *et al.*, 2008) and predatory Heteroptera (Coll, 1998), whilst also diversifying the range of habitats for natural enemies. Floral resources can boost parasitoid movement, longevity and fecundity (Berndt and Wratten, 2005); however, flower species must be carefully chosen so that they are exploited primarily by the natural enemies rather than the pests (Baggen *et al.*, 1999; Lavandero *et al.*, 2006; Winkler *et al.*, 2009) and that natural enemy parasitism is not enhanced (Jonsson *et al.*, 2009). Combining the two habitats is not recommended because the tussock-forming grasses usually outcompete the flowering species. However, providing both habitats within fields may create synergistic advantages. Such farm and landscape diversification can enhance natural enemy activity (Bianchi *et al.*, 2006). Diversifying the type and number of predator guilds such as ground- and crop-active, day- and night-active, at the farm and landscape scale may also improve the robustness of biological control (Weibull *et al.*, 2003).

CONCLUSIONS

Many variables determine whether biodiversity-based habitat manipulation approaches can be successfully implemented in agroecosystems, including the value of the crop, the nature of the pest damage and the susceptibility of key pests to suppression by natural enemies. The three case studies outlined in this chapter illustrate both the broad applicability of habitat manipulation to enhance biological control, and the challenges to implementing it. We have demonstrated how the ecological services of beneficial arthropods can be integrated into pest management programmes for basic grains, fresh produce and fibre crops in temperate, Mediterranean and subtropical growing environments. Habitat manipulation schemes for provisioning habitat, floral resources, and prey to natural enemies of targeted pests must be designed according to the specific needs of the natural enemy complex in time and space. Habitat manipulation must also fit easily into the grower's way of doing things, and growers will not be persuaded to take up these conservation biocontrol technologies without evidence that they reduce the incidence of pests and are cost effective. Demonstrating that habitats incorporated into agricultural systems for enhancing natural enemies have additional benefits beyond biocontrol can improve their uptake as they may appeal to a grower's interests or philosophy. For example, beetle banks can be adopted to also improve nesting and feeding habitats for game birds. The additional environmental benefits created by the adoption of conservation biocontrol such as reduced use of pesticides and therefore a reduction in off-farm pollution or encouragement of biodiversity, may create financial savings elsewhere (e.g. reduced costs of removing pesticides from drinking water), which should be recognised, and some of this could be returned to the farmers through incentive schemes. Even where financial support is available, farmers may still not adopt conservation biocontrol when insecticides remain effective and cheap. Growers may seek alternative technologies only when faced with pest resistance (e.g. as has occurred in protected cops), pressure from retailers (e.g. crop assurance schemes) or statutory restrictions (e.g. Denmark, organic certification). Nevertheless, each of the examples described in this chapter shows how growers and researchers have worked together to gather practical information on biodiversity-based approaches to suppress insect pests by providing habitat, resources, and biological diversity. As restrictions on insecticide use increase in North America, Europe and other parts of the world, these models of collaboration between growers and researchers to suppress pests while reducing insecticide use may become more common.

REFERENCES

Altieri, M.A. and Whitcomb, W.H. (1979) The potential use of weeds in the manipulation of beneficial insects. *HortScience*, 14, 12–18.

Baggen, L.R., Gurr, G.M. and Meats, A. (1999) Flowers in tri-trophic systems: mechanisms allowing selective exploitation by insect natural enemies for conservation biological control. *Entomologia Experimentalis et Applicata*, 91, 155–161.

Bealey, C., Ledder, E., Robertson, H. and Wolton, R. (2009) Hedgerows – their wildlife, current state and management needs. *British Wildlife*, 20, 323–329.

Berndt, L.A. and Wratten, S.D. (2005) Effects of alyssum flowers on the longevity, fecundity, and sex ratio of the leafroller parasitoid *Dolichogenidea tasmanica*. *Biological Control*, 32, 65–69.

Berndt, L.A., Wratten, S.D. and Scarratt, S.L. (2006) The influence of floral resource subsidies on parasitism rates of leafrollers (Lepidoptera: Tortricidae) in New Zealand vineyards. *Biological Control*, 37, 50–55.

Berry, N.A. (1997) *Abundance, activity and distribution of soil-surface arthropod predators in arable farmland*. Doctor of Philosophy thesis, Lincoln University, New Zealand.

Bianchi, F.J.J.A., Booij, C.J.H. and Tscharntke, T. (2006) Sustainable pest regulation in agricultural landscapes: a review on landscape composition, biodiversity and natural pest control. *Proceedings of the Royal Society B*, 273, 1715–1727.

Blackman, R.L. and Eastop, V.F. (2000) *Aphids on the world's crops: an identification and information guide*, 2nd edn, John Wiley & Sons, Ltd, Chichester.

Boatman, N.D., Jones, N.E., Garthwaite, D. and Pietravalle, S. (2007) Option uptake in entry level scheme agreements in England. *Aspects of Applied Biology*, 81, 309–316.

Blumberg, A.Y. and Crossley, D.A. (1982) Comparisons of soil surface arthropod populations in conventional tillage, no-tillage and old field systems. *Agro-Ecosystems*, 8, 247–253.

Bugg, R.L., Wäckers, F.L., Brunson, K.E., Dutcher, J.D. and Phatak, S.C. (1991) Cool-season cover crops relay intercropped with cantaloupe: influence on a generalist predator, *Geocoris punctipes* (Hemiptera: Lygaeidae). *Journal of Economic Entomology*, 84, 408–416.

Bugg, R.L., Chaney, W.E., Colfer, R.G., Cannon, J.A. and Smith, H.A. (2008) *Flower flies (Diptera: Syrphidae) and other important allies in controlling pests of California vegetable crops*. University of California, Division of Agriculture and Natural Resources, Publication 8285, University of California Press, Davis, CA.

Carmona, D.M. and Landis, D.A. (1999) Influence of refuge habitats and cover crops on seasonal activity- density of ground beetles (Coleoptera: Carabidae) in field crops. *Environmental Entomology*, 28, 1145–1153.

CDFA (2008) Agricultural Overview, California Department of Food and Agriculture, URL: http://www.cdfa.ca.gov/statistics/PDFs/AgResourceDirectory2008/1_2008_OverviewSection.pdf

Chambers, R.J. (1988) Syrphidae. *World crop pests: aphids, their biology, natural enemies and control*, vol. B (eds A.K. Minks and P. Harrewijn), Elsevier, Amsterdam, pp. 259–270.

Chandler, A.E.F. (1968a) Some host–plant factors affecting oviposition by aphidophagous Syrphidae. *Annals of Applied Biology*, 61, 415–423.

Chandler, A.E.F. (1968b) The relationship between aphid infestations and oviposition by aphidophagous Syrphidae. *Annals of Applied Biology*, 61, 425–434.

Chaney, W.E. (1998) Biological control of aphids in lettuce using in-field insectaries, in *Enhancing biological control: habitat management to promote natural enemies of arthropod pests* (eds C.H. Pickett and R.L. Bugg), University of California Press, Berkeley, pp. 73–83.

Chaney, W.E. (1999) Lettuce aphid update. *Monterey County Crop Notes*. University of California Cooperative Extension, Salinas.

Chaney, W.E. and Smith, H.A. (2005) Controlling lettuce aphids naturally. *Coastal Grower*, Winter, 32–36.

Chaney, W.E., Smith, H.A. and Dlott, F.K. (2006) *Insectary crops to enhance syrphid fly activity in organic lettuce on California's Central Coast* (poster). Entomological Society of America Annual meeting, 10–13 Dec. 2006, Indianapolis, URL: http://esa.confex.com/esa/2006/techprogram/paper_26341.htm

Chiverton, P.A. (1986) *Predator density manipulations and its effects on population of* Rhopalosiphum padi (Homoptera: Aphididae) in spring barley. *Annals of Applied Biology*, 106, 49–60.

Chiverton P.A. (1989) The creation of within-field overwintering sites for natural enemies of cereal aphids. *Proceedings of the 1989 Brighton crop protection conference – Weeds*, 3, 1093–1096.

Colfer, R. (2004) Using habitat management to improve biological control on commercial organic farms in California, in *Fourth California Conference on Biological Control* (ed. M.S. Hoddle), July 13–15, 2004, Berkeley. University of California Press, Berkeley, pp. 55–62.

Coll, M. (1998) Living and feeding on plants in predatory Heteroptera, in *Predatory Heteroptera* (eds M. Coll and J.R. Ruberson), Entomological Society of America, Lanham, pp. 89–129.

Collins, K.L., Boatman, N.D., Wilcox, A. and Holland, J.M. (2002) The influence of beetle banks on cereal aphid population predation in winter wheat. *Agriculture, Ecosystems and Environment*, 93, 337–350.

Collins, K.L., Boatman, N.D., Wilcox, A.W. and Holland, J.M. (2003) A 5-year comparison of overwintering polyphagous predator densities within a beetle bank and two conventional hedgebanks. *Annals of Applied Biology*, 143, 63–71.

Conservation Technology Information Center (2010) 2008 *Amendment to the National Crop Residue Management Survey Summary*. URL: http://www.ctic.purdue.edu/media/pdf/National%20Summary%202008%20(Amendment).pdf

Coombes, D.S. and Sotherton, N.W. (1986) The dispersal and distribution of predatory Coleoptera in cereals. *Annals of Applied Biology*, 108, 461–474.

Corbett, A. and Rosenheim, J.A. (1996) Impact of natural enemy overwintering refuge and its interaction with the surrounding landscape. *Ecological Entomology*, 21, 155–164.

Cowgill, S.E., Wratten, S.D., and Sotherton, N.W. (1993) The selective use of floral resources by the hoverfly *Episyrphus balteatus* (Diptera: Syrphidae) on farmland. *Annals of Applied Biology*, 122, 499–515.

De Lima J.O.G. and Leigh, T.F. (1984) Effect of cotton genotypes on the western bigeyed bug (Heteroptera: Miridae). *Journal of Economic Entomology*, 77, 898–902.

D'Hulster, M. and Desender, K. (1982) Ecological and faunal studies on Coleoptera in agricultural land III. Seasonal abundance and hibernation in the grassy edge of a pasture. *Pedobiologia*, 23, 403–414.

Edwards, C.A, Sunderland, K.D. and George, K.S. (1979) Studies of polyphagous predators of cereal aphids. *Journal of Applied Ecology*, 16, 811–823.

EJF (2007) *The deadly chemicals in cotton*, Environmental Justice Foundation in collaboration with Pesticide Action Network OK, London.

ERS (2008) *Data sets, organic production*, Economic Research Service, URL: http://www.ers.usda.gov/Data/Organic/

Eveleens, K.G., van den Bosch, R. and Ehler, L.E. (1973) Secondary outbreak induction of beet armyworm by experimental insecticide applications in cotton in California. *Environmental Entomology*, 2, 497–503.

FAO (2010) *The importance of cover crops in conservation agriculture*. Food and Agricultural Organization of the United States, URL: http://www.ctic.purdue.edu/media/pdf/National%20Summary%202008%20(Amendment).pdf

Griffiths, G.J.K., Holland, J.M., Bailey, A. and Thomas, M.B. (2008) Efficacy and economics of shelter habitats for conservation biological control. *Biological Control*, 45, 200–209.

Guthrie, D., Hutchinson, B., Denton, P. *et al.* (1993) Conservation tillage. *Cotton Physiology Today*, 4(9), 1–4.

Heimpel, G.E. and Jervis, M.A. (2005) Does nectar improve biological control by parasitoids?, in *Plant-provided food for carnivorous insects: a protective mutualism and its applications* (eds F.L. Wäckers, P.C.J. van Rijn and J. Bruin), Cambridge University Press, Cambridge, pp. 267–304.

Helenius, J. (1995) Enhancement of predation through within-field diversification, in *Enhancing biological control: habitat management to promote natural enemies of arthropod pests* (eds C.H. Pickett and R.L. Bugg), University of California Press, Berkeley, pp. 121–160.

Hickman, J.M. and Wratten, S.D. (1996) Use of *Phacelia tanacetifolia* strips to enhance biological control of aphids by hoverfly larvae in cereal fields. *Journal of Economic Entomology*, 89, 832–840.

Hokkanen, H. and Pimentel, D. (1989) New associations in biological-control – theory and practice. *Canadian Entomologist*, 121, 829–840.

Holland, J.M., Thomas, S.R. and Hewitt, A. (1996) Some effects of polyphagous predators on an outbreak of cereal aphid (*Sitobion aveanae* F.) and orange wheat blossom midge (*Sitodiplosis mosellana* Géhin). *Agriculture, Ecosystems and the Environment*, 59, 181–190.

Holland, J.M., Perry, J.N. and Winder, L. (1999) The within-field spatial and temporal distribution of arthropods within winter wheat. *Bulletin of Entomological Research*, 89, 499–513.

Holland, J.M., Winder, L., Woolley, C., Alexander, C.J. and Perry, J.N. (2004) The spatial dynamics of crop and ground active predatory arthropods and their aphid prey in winter wheat. *Bulletin of Entomological Research*, 94, 419–431.

Holland, J.M., Oaten, H., Moreby, S. and Southway, S. (2008a) The impact of agri-environment schemes on cereal aphid control, in *Landscape management for functional biodiversity* (eds W.A.H. Rossing, H-M. Poehling and M. van Helden), IOBC/WPRS Bulletin, 34, 33–36.

Holland, J.M., Oaten, H., Southway, S. and Moreby, S. (2008b) The effectiveness of field margin enhancement for cereal aphid control by different natural enemy guilds. *Biological Control*, 47, 71–76.

Holland, J.M., Birkett, T. and Southway, S. (2009) Contrasting the farm-scale spatio-temporal dynamics of boundary and field overwintering predatory beetles in arable crops. *BioControl*, 54, 19–33.

Jonsson, M., Wratten, S.D., Robinson, K.A., and Sam, S.A. (2009) The impact of floral resources and omnivory on a four trophic level food web. *Bulletin of Entomological Research*, 99, 275–285.

Landis, D.A., Wratten, S.D. and Gurr, G.M. (2000) Habitat management to conserve natural enemies of arthropod pests in agriculture. *Annual Review of Entomology*, 45, 175–201.

Lavandero, I., Wratten, S.D., Didham, R.K. and Gurr, G. (2006) Increasing floral diversity for selective enhancement of biological control agents: A double-edged sword? *Basic and Applied Ecology*, 7, 236–243.

Leidner, J. and Kidwell, B. (2000) Bringing back the bobwhite: hardcore ag with a soft edge gives quail a chance on today's farms and ranches. *Progressive Farmer*, Sept. 2000, 22–25.

Lewis, W.J., Haney, P.B. and Phatak, S. (1996) Continued studies of insect population dynamics in crimson clover and refugia/cotton systems. Part I: Sweep and whole plant sampling. *Proceedings of the 1996 Beltwide Cotton Conferences*, 2, 1108–1116.

Liu, Y.B. (2004) Distribution and population development of *Nasonovia ribisnigri* in iceberg lettuce. *Journal of Economic Entomology*, 97, 883–890.

Luff, M.L. (1966) The abundance and diversity of the beetle fauna of grass tussocks. *Journal of Animal Ecology*, 35, 189–208.

Maan, A. and Beam, S. (2009) Unapproved feeding of cotton plant by-products, including gin trash or cotton stalks. *California Dairy Newsletter*, 1(2), 7.

Macleod, A., Wratten, S.D., Sotherton, N.W. and Thomas, M.B. (2004) 'Beetle banks' as refuges for beneficial arthropods in farmland: long-term changes in predator communities and habitat. *Agricultural and Forest Entomology*, 6, 147–154.

Mann, B.P., Wratten, S.D., Poehling, M. *et al.* (1991) The economics of reduced-rate insecticide applications to control aphids in winter-wheat. *Annals of Applied Biology*, 119, 451–464.

McPherson, R.M., Smith, J.C. and Allen, W.A. (1982) Incidence of arthropod predators in different soybean cropping systems. *Environmental Entomology*, 11, 685–689.

Meek, B., Loxton, D., Sparks, T., Pywell, R., Pickett, H. and Nowakowski, M. (2002) The effect of arable field margin composition on invertebrate biodiversity. *Biological Conservation*, 106, 259–271.

Monterey County (2009) *Crop report*. URL: http://www.co. monterey.ca.us/ag/pdfs/CropReport2009.pdf

NASS (2010) *Vegetables 2009 summary*, National Agricultural Statistics Service, URL: http://www.nass.usda.gov/ Publications/Ag_Statistics/2010/index.asp, pp. 21–23.

NCC (2011) *US mill consumption report*. National Cotton Council, URL: http://www.cotton.org/econ/textiles/mill-consumption.cfm

Nicholls, C.I., Parrella, M. and Altieri, M.A. (2001) The effects of a vegetational corridor on the abundance and dispersal of insect biodiversity within a northern California organic vineyard. *Landscape Ecology*, 16, 133–146.

OTA (2010) *Organic cotton facts*. Organic Trade Association. URL: http://www.ota.com/organic/mt/organic_cotton.html

Poehling, H-M., Freier, B. and Klüken, A.M. (2007) IPM Case Studies: Grain, in *Aphids as crop pests* (eds H.F van Emden and R. Harrington), CAB International, pp. 597–612.

Riedel, W. (1992) Linear biotopes in cereal fields to enhance polyphagous predators. *Tidsskrift for Planteavls Specialserie*, 86, 175–184.

Rijk Zwaan (2010) *Lettuce: worldwide range of successful varieties*. URL: http://www.rijkzwaanexport.com/RZZ/Export/ siteexp.nsf/webpages/ 580CCD1D94CD35E8C1257248003DACA8

Robinson, K.A., Jonsson, M., Wratten, S.D., Wade, M.R. and Buckley, H.L. (2008) Implications of floral resources for predation by an omnivorous lacewing. *Basic and Applied Ecology*, 9, 172–181.

Ruberson, J.R., Lewis, W.J., Waters, D.J., Stapel, O. and Haney, P.B. (1995) Dynamics of insect populations in a reduced-tillage, crimson clover/cotton system. Part I: Pests and beneficial on plants. *Proceedings of the 1995 Beltwide Cotton Conferences*, 2, 814–817.

Ruberson, J.R., Phatak, S.C. and Lewis, W.J. (1997) Insect populations in a crimson clover/strip tillage system. *Proceedings of the 1997 Beltwide Cotton Conferences*, 2, 1121–1125.

SAN (2007) *Transitioning to organic production*. Sustainable Agriculture Network. URL: http://www.sare.org/ publications/organic/organic.pdf

Shonouda M.L., Bombosch S., Shalaby A.M. and Osman S.I. (1998) Biological and chemical characterization of a kairomone excreted by the bean aphids, *Aphis faltae* Scop. (Horn., Aphididae) and its effect on the predator *Metasyrphus corollae* Fabr. 1. Isolation, identification and bioassay of aphid-kairomone. *Journal of Applied Entomology*, 122, 15–23.

Smith, H.A. and Chaney, W.E. (2007) A survey of syrphid predators of *Nasonovia ribisnigri* in organic lettuce on the Central Coast of California. *Journal of Economic Entomology*, 100, 39–48.

Smith, H.A., Chaney, W.E. and Bensen, T.A. (2008) Role of syrphid larvae and other predators in suppressing aphid infestations in organic lettuce on California's Central Coast. *Journal of Economic Entomology*, 101, 1526–1532.

Smith, R.I., Klonsky, K.M. and De Moura, R.L. (2009) *Sample costs to produce romaine hearts – leaf lettuce*. Central Coast region, Monterey and Santa Cruz Counties. University of California Cooperative Extension Publication LT-CC-09-1. URL: http://coststudies.ucdavis.edu/current.php

Sotherton, N.W. (1984) The distribution and abundance of predatory arthropods overwintering on farmland. *Annals of Applied Biology*, 105, 423–429.

Sotherton, N.W. (1985) The distribution and abundance of predatory Coleoptera overwintering in field boundaries. *Annals of Applied Biology*, 106, 17–21.

Sprague, M.A. and Triplett, G.B. (1986) *No tillage and surface-tillage agriculture: the tillage revolution*, John Wiley & Sons, Inc., New York.

SWCS (2006) *Environmental benefits of conservation on cropland: the status of our knowledge*, Soil and Water Conservation Society, Ankeny.

Thead, L.G, Pitre, H.N. and Kellogg, T.F. (1985) Feeding behaviour of adult *Geocoris punctipes* (Say) (Hemiptera: Lygaeidae) on nectaried and nectariless cotton. *Environmental Entomology*, 14, 134–137.

Thies, C. and Tscharntke, T. (1999) Landscape structure and biological control in agroecosystems. *Science*, 285, 893–895.

Thomas, M.B. (1990) The role of man-made grassy habitats in enhancing carabid populations in arable land, in *The role of ground beetles in ecological and environmental studies* (ed. N.E. Stork), Intercept Ltd, Andover, pp. 77–85.

Thomas, S.R. (2000) Progress on beetle banks in UK arable farming. *Pesticide Outlook*, April, 51–53.

Thomas, S.R. (2001) *Assessing the value of beetle banks for enhancing farmland biodiversity*. Doctor of Philosophy thesis, University of Southampton.

Thomas, M.B., Wratten, S.D. and Sotherton, N.W. (1991) Creation of 'island' habitats in farmland to manipulate populations of beneficial arthropods: densities and emigration. *Journal of Applied Ecology*, 28, 906–917.

Thomas, M.B., Mitchell, H.J. and Wratten, S.D. (1992) Abiotic and biotic factors influencing the winter distribution of predatory insects. *Oecologia*, 89, 78–84.

Thomas, S.R., Goulson, D. and Holland, J.M. (2000) Spatial and temporal distributions of predatory Carabidae in a winter wheat field. *Aspects of Applied Biology*, 62, 55–60.

Thomas, S.R., Goulson, D. and Holland, J.M. (2001) Resource provision for farmland gamebirds: the value of beetle banks. *Annals of Applied Biology*, 139, 111–118.

Tillman, P.G. (2006) Sorghum as a trap crop for *Nezara viridula* (L.) (Heteroptera: Pentatomidae) in cotton. *Environmental Entomology*, 35, 771–783.

Tillman, G., Schomberg, H., Phatak, S. *et al.* (2004) Influence of cover crops on insect pests and predators in conservation-tillage cotton. *Journal of Economic Entomology*, 97, 1217–1232.

Tillman, G., Lamb, M. and Mullinix, B. Jr. (2008) Pest insects and natural enemies in transitional organic cotton in Georgia. *Journal of Entomological Science*, 44, 11–23.

Tillman, P.G., Northfield, T.D., Mizell, R.F. and Riddle, T.C. (2009) Spatiotemporal patterns and dispersal of stink bugs (Heteroptera: Pentatomidae) in peanut-cotton farmscapes. *Environmental Entomology*, 38, 1038–1052.

Tourte, L., Smith, R.I., Klonsky, K.M. and De Moura, R.L. (2009) *Sample costs to produce organic leaf lettuce*. Central Coast Region, Santa Cruz and Monterey Counties. University of California Cooperative Extension Publication LT-CC-09-0. URL: http://coststudies.ucdavis.edu/files/lettuceleaforganiccc09.pdf

Triplett, G.B. (1986) Crop management practices for surface-tillage systems, in *No tillage and surface-tillage agriculture: the tillage revolution* (eds M.A. Sprague and G.B. Triplett), Wiley & Sons, Inc., New York, pp. 149–182.

Triplett, G.B. Jr. and Dick, W.A. (2008) No-tillage crop production: a revolution in agriculture. *Agronomy Journal*, 100, 153–165.

UCSC (2010) *Amigo Bob Cantisano, organic farming advisor, co-Founder, eco-farm conference*. University of California Santa Cruz, University of California Press. URL: http://digitalcollections.ucsc.edu/cdm4/document.php?CISOROOT=/p15130coll2andCISOPTR=81andCISOSHOW=77

Van Helden, M., van Heest, H.P.N.F., van Beek, T.A. and Tjallinghi, W.F. (1995) Development of a bioassay to test phloem sap samples from lettuce for resistance to *Nasonovia ribisnigri*. *Journal of Chemical Ecology*, 21, 761–774.

Verheggen, F.J., Ludovic, A., Bartram, S., Gohy, M. and Haubruge, E. (2008) Aphid and plant volatiles induce oviposition in an aphidophagous hoverfly. *Journal of Chemical Ecology*, 34, 301–307.

Vollhardt, I.M.G., Bianchi, F.J.J.A., Wäckers, F.L., Thies, C. and Tscharntke, T. (2010) Spatial distribution of flower vs. honeydew resources in cereal fields may affect aphid parasitism. *Biological Control*, 53, 204–213.

Weibull, A.C., Ostman, O. and Granqvist, A. (2003) Species richness in agroecosystems: the effect of landscape, habitat and farm management. *Biodiversity and Conservation*, 12, 1335–1355.

Winkler, K., Wäckers, F.L., Kaufman, L.V., Larraz, V. and van Lenteren, J.C. (2009) Nectar exploitation by herbivores and their parasitoids is a function of flower species and relative humidity. *Biological Control*, 50, 299–306.

Williams, M.R. (2011) *Cotton insect losses 2010*. URL: http://www.entomology.msstate.edu/resources/tips/cotton-losses/data/2010

Synthesis

Chapter 20

CONCLUSION: BIODIVERSITY AS AN ASSET RATHER THAN A BURDEN

Geoff M. Gurr, William E. Snyder, Steve D. Wratten and Donna M.Y. Read

Biodiversity and Insect Pests: Key Issues for Sustainable Management, First Edition. Edited by Geoff M. Gurr, Steve D. Wratten, William E. Snyder, Donna M.Y. Read.

INTRODUCTION

Too often over recent decades, biodiversity and associated environmental protection have been portrayed as a burden to society. 'Green' initiatives are often seen to impede economic development and force human enterprises (such as farming) to make compromises just to conserve some esoteric environmental entity. We hope that this book will help persuade even the most hard-nosed farmer or policy-maker that biodiversity can be an asset with great economic value.

The contributed chapters in this book illustrate the various ways in which biodiversity is being harnessed to improve pest management. Importantly, many of the chapters go beyond what critics might term 'theoretical hand-waving', providing concrete examples of biodiversity at work. We present examples and case studies that range as far afield as Europe (Tillman *et al.*), Africa (Khan *et al.*), the Americas (Landis *et al.*), Australasia (Landis *et al.*) and Asia (Gurr *et al.*; Lu and Puyan). Naturally the bulk of the book deals with agricultural systems and pests but, as chapter 18 by Shrewsbury and Leather shows, biodiversity is a powerful force in ornamental and amenity systems in urban areas too. Many of the chapters (e.g. chapter 2, Snyder and Tylianakis) show that basic ecological principles have been used to develop practical, effective pest management strategies that are achieving wide adoption by growers. It is now clear that biodiversity is the opposite of a 'millstone' that retards farm profitability. Rather, biodiversity is a valuable resource in its own right that provides free (or inexpensive) pest control and other ecosystem services of great monetary value.

SHADES OF GREEN – DIFFERING CONCEPTIONS OF BIODIVERSITY

Pale green biodiversity

A general theme from the contributed chapters is that not only is there a great range of ways in which biodiversity can support pest management but, more profoundly, there are very significant differences in the scales at which biodiversity can be found and exploited. At the smallest scale, authors such as Koul (dealing with the important topic of identifying plant compounds with insecticidal properties, chapter 6) and Horgan (dealing with plant breeding, chapter 15) illustrate the view of biodiversity as a resource from which useful genes or compounds can be isolated. This is most starkly apparent in the conclusion to Koul's chapter in which strategies such as the cloning and widespread intensive growth of plants with especially useful compounds are recommended. Such measures are far removed from a utopian vision of returning modern agricultural vistas to natural landscapes. They do, however, represent a valid aspect of the applied use of biodiversity, albeit one that may be considered 'pale green' rather than the deeper shade of green exemplified by other strategies (Plate 20.1). Certainly the intensive production of botanical insecticides is preferable to the production of conventional synthetic insecticides from non-renewable resources. Already the production of some botanical insecticides is a very significant industry. For example, the island state of Tasmania (Australia) has become one of the world's largest suppliers of pyrethrum from extensive cultivation of the pyrethrum daisy, *Tanacetum cinerariifolium* (Trevir.) Sch Bip. (Figure 20.1). These fields of flowers are favourably perceived by the general public to the extent that they have aesthetic value as tourist attraction (e.g. www.welcome2australia.com.au/tasmania/ulverstone/). On the other hand, this form of 'input substitution' – where naturally produced toxins simply replace synthetically produced ones within a traditional spray schedule – is not without its risks. The most well studied of these is toxicity to natural enemies of, for example, neem (*Azadirachta indica* A. Juss) (Schmutterer, 1997) and Meliaceae (Peveling and Ely, 2006).

Chapter 11 by James *et al.* on chemical ecology deals with compounds from plants but, rather than using these as input substitution toxins, posits the use of herbivore-induced plant volatile compounds (HIPVs) to trigger natural enemy attraction to plants under attack. This is generally conceived to involve spraying HIPVS onto plants but, unlike the application of botanical or other insecticides, the mode of action is truly bio-rational in that it triggers plant defences and allows the natural ecological process of top-down control to operate (albeit at an artificially elevated level). As knowledge of the chemical ecology of insect–plant interactions develops further it will be possible to more readily develop new 'push–pull' and similar strategies that do not depend upon exogenous application of HIPVs or other compounds; rather they depend on identifying and using plants themselves that are particularly effective at attracting natural enemies or at deterring or being trap plants

Figure 20.1 Large-scale cultivation of irrigated pyrethrum daisy (*Tanacetum cinerariifolium*) in Tasmania, Australia (reproduced with permission of Botanical Resources Australia Pty Ltd: http://www.ishs.org/news/?p=1126).

for pest species. The suggestion of Khan *et al.* to develop plants with improved capacity to exert these effects is another example of biodiversity as a source of genes (or gene-mediated traits) that will allow such future developments.

Intermediate green biodiversity

Other chapters in this book represent a less pale shade of green (Plate 20.1: centre). Here biodiversity is used not as a resource for compounds or genes (entities from biodiversity but not themselves living), but species and guilds of living organisms that directly or indirectly provide the ecosystem service of pest suppression. Chapter 9 by Wäckers and van Rijn constitutes a detailed resource for workers seeking to use plants to provide natural enemies with critical plant-derived foods such as pollen and nectar. A key message from this analysis is that the commonly held notion that 'all diversity is good' is misleading. Poorly chosen resource plants will be ineffective and may even provide sustenance for pests rather than natural enemies. Thus, plant choice is critical when seeking to use plant biodiversity strategically to promote natural pest control. The second broad class of using biodiversity in the form of species of living organisms is exemplified by chapter 3 by Welch *et al.* on generalist predators, where many species might be expected to act in concert to suppress pests. Often schemes to conserve generalists

have focused on sowing flowering plants, which provide alternative foods, or tussock grasses, which provide overwintering habitat and non-pest prey. These plantings are often managed like traditional crops, albeit with the goal of providing food for beneficial arthropods rather than humans or livestock.

Deep green biodiversity

Ultimately, several chapters in this volume reflect a deeper shade of green in which the use of biodiversity is less exploitative and designed to restore to the farm aspects of, or facsimiles for, the natural environment that existed before industrial agriculture. Examples of this include areas of planted woodland and hedgerows or even 'beetle banks'. Chapter 17 by Landis *et al.*, which deals with use of native plants, is a good example. Most of the work on the use of plant diversity to support natural enemies has focused on a few plant species such as buckwheat, alyssum or phacelia, which are exotic to most countries where they are used. Landis *et al.* focus instead upon the vast number of native plants available in each country that can support natural enemies. In such cases, the restoration of native plant communities and effective biological control do not need to be traded off from one another; instead, they work in tandem. Native plantings have the obvious potential to become entirely self-supporting, reducing costs associated with cultivated exotic annual plants that benefit natural enemies.

It is clear from the chapters by Gámez-Virués *et al.* and Scherber *et al.* (chapters 7 and 8) that the ecological processes involved in pest outbreaks and their suppression by natural enemies can operate at scales greater than individual fields. Highly mobile insects and 'ballooning' spiders often readily cross kilometres of landscape, such that broad regions must be considered. This raises the intriguing, although practically daunting, idea that entire landscapes might be manipulated to suppress pest outbreaks. Recent work highlighting the importance of habitat connectivity (in contrast to the relatively well-known effects of habitat composition) in cotton landscapes (Perovic *et al.*, 2010) suggests how the concept of selectivity – best known in the context of selecting nectar plants that support natural enemies whilst denying benefit to pests (Baggen and Gurr, 1998) – might be extended to the design of landscapes that promote beneficial over pest insects. The 'Revegetation by Design' case study presented in chapter 17 by Landis *et al.* suggests additional possibilities for new pest management tactics at the landscape scale.

Other than the direct role of native plants in biodiversity-based pest management, their use will also maximise the value of vegetation to wildlife. This is significant because already 40% of the world's 150,000,000 km² total land area is used for agriculture and only 13% is in protected areas such as reserves where biodiversity is afforded some level of protection. Furthermore, the area of land suitable for productive agriculture is declining as a result of urbanisation and degradation. This makes it imperative to meet the needs of an expanding human population whilst maintaining biodiversity wherever possible, including agricultural lands.

CONSERVING BIODIVERSITY AND FEEDING NINE BILLION PEOPLE: MUTUALLY EXCLUSIVE OR COMPLEMENTARY GOALS?

In addressing whether ecological agriculture might rise to the challenge of feeding the nine billion people that are expected to share our planet by the middle of the century, Pretty (2011) provides a wakeup call. We are failing to address the need to feed the *current* population of around seven billion. Around one-third of these people are undernourished. Moreover, current levels of agricultural production are not environmentally sustainable and contribute to eroding the capacity of land to produce food in future years.

Estimates of the size of the 'ecological footprint' of humans – the area of land required to support our demands from nature for food, clean water and air, etc., if the entire world population was to have a living standard that developed countries take for granted – is at least five Planet Earths! Clearly we live in a more constrained situation and the rapidly expanding world population will need to feed itself on just our single planet. Put bluntly, there is very little new land that can be pressed into agricultural production. Much of the *potential* land (with at least reasonable fertility and water supply) is urbanised or used for biodiversity conservation. So, can the answer to producing more food be found in pursuing the 'dream' of the Green Revolution; with technological advance in genetics (plant breeding) and especially chemistry (pesticides and syn-

thetic fertilisers)? Global data on pesticide production and the use of other inputs (Figure 20.2) suggests that this is the default strategy, the one that will be played out unless a dramatic change occurs. Trends for use of inputs ranging from nitrogen, phosphorous, water and pesticides have increased in a more or less linear manner since the end of the Second World War and show no sign of abating (Tilman *et al.*, 2002). Annual pesticide production currently exceeds 3 million tons (Figure 20.2). We believe that a better route to a more sustainable future is to embrace ecology. This will involve a shift to an agricultural paradigm that is less reliant on non-renewable resources and makes more effective use of biodiversity for it constitutes the ultimate 'tool box'. The irony is that agriculture has been a leading cause of global biodiversity *loss* (Green *et al.*, 2005) (including loss of genetic diversity, Horgan, chapter 15 of this volume) so has been steadily eroding its very foundation.

A 'business as usual' approach will fail to meet the food, fuel and fibre needs of the expanding human population and further deplete the viability of the planet's other species and ecosystems. The link between biodiversity conservation and capacity to meet important Millennium Development Goals, particularly the eradication of extreme hunger and poverty, is well recognised (Sachs *et al.*, 2009). An alternative to 'business as usual' is to develop agricultural systems that are 'ecologically engineered' to maximise natural processes such as fixing nitrogen from the atmosphere by vigorous symbiotic bacterial activity on crop plant roots, encouraging natural enemies of pests to suppress outbreaks and pollinating insects to pollinate crops. These are examples of ecosystem services that can be provided by biodiversity (Gillespie and Wratten, chapter 4 of this volume; Figure 20.3). But, because industrialised agriculture tends to greatly deplete biodiversity in farm landscapes, technological replacements have been developed. These include synthetic fertilisers, insecticides and domesticated bees, but these technologies are associated with problems such as reliance on non-renewable resources, pollution, cost and sustainability. An ecological engineering approach involves encouraging biodiversity in farm landscapes to restore ecosystem function whilst simultaneously providing land area for flora and fauna conservation that is in addition to that in protected reserves.

Chapter 12 by Escalada and Heong provides guidance on how the adoption of biodiversity-based pest

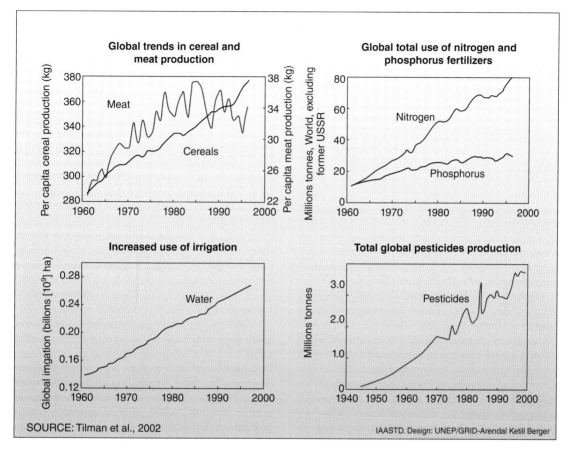

Figure 20.2 Global trends in agricultural inputs and production. From *Agriculture at a crossroads: Global Report*, by the IAASTD. Copyright © 2009 IAASTD. Reproduced by permission of Island Press, Washington, DC.

management may be promoted but also highlights the fact that achieving this can be difficult. A key aspect that tends to promote adoption and expansion appears to be effective communication of the benefits of the new technology to those who might use it. The power of this message is increased when the technology simultaneously provides multiple benefits. This was evident at a meeting of a small community of poor farmers attended by one of the authors (GMG) in central Vietnam. The farmers were asked (through an interpreter) why they had so enthusiastically adopted new, ecologically based approaches for pest management. One of the answers was simply 'our chickens have stopped dying'. Promoting parasitoid activity with nectar plants had lessened the need for insecti-

cide use to the extent that spraying and its attendant off-target impacts had become uncommon. Less anecdotally, chapter 16 by Khan *et al.* includes a long list of benefits that derive from the use of biodiversity in the 'push–pull' system. These include suppression of parasitic striga weeds and the production of fodder for cattle. These additional benefits have proven important in the widespread adoption of this approach in Africa.

Our view of the benefits of harnessing biodiversity is consistent with a recent, comprehensive assessment of the state of world agriculture that pointed to the need for sustainable practices based on ecological approaches (McIntyre *et al.*, 2009, p. 3). The basis for that conclusion was that:

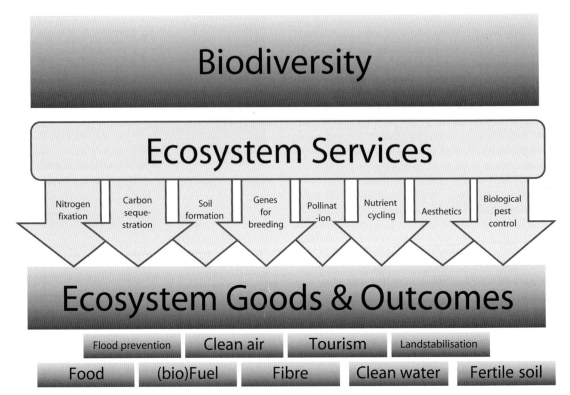

Figure 20.3 The case for on-farm biodiversity; not a burden but a resource that, via ecosystem services, provides food and other ecosystem goods and outcomes.

A range of fundamental natural resources ...[including biological diversity] provide the indispensable base for the production of essential goods and services upon which human survival depends, including those related to agricultural ecosystems.

So, agricultural systems are not only embedded with other ecosystems (natural, semi-natural and urban); there are also intricate linkages between natural resource use and the economic activities of society and the physical environment. Understanding the complexities of these systems and their linkages – the job of ecology – is key to the development of practices that will build, rather than erode, the resilience of agriculture.

De Schutter (2010) wrote in a report for the United Nations Environment Programme (UNEP) that ecologically based farming practices could double food production in a decade. However, De Schutter went on to stress the importance of ambitious public policies in order to take it beyond the experimental stage. It seems that agroecology science, if not its application, has come of age. The food demands of an increasing human population are clashing with the fast-growing realisation that high-input agriculture, which depends on fossil fuels with associated environmental and human health problems, can no longer be relied upon to service this demand. Agroecology is increasingly considered to be the most likely basis of 'future farming'.

The potential for ecologically based technologies to actually deliver beneficial outcomes is apparent in data from developing countries that ranges across diverse farming systems including smallholder irrigated and wetland rice, coastal artesian and urban-based (Pretty *et al.*, 2006). The sustainable technologies introduced included many that are detailed in this book, including growing secondary crops such as vegetables on the bunds around rice fields, incorporating agroforestry, using locally appropriate crop varieties and practising integrated pest management (IPM). The average yield increase following adoption of sustainable technologies in such cropping systems ranged from 22% to 146%, with an overall increase for all projects of 80%. Collectively, these ecologically sound methods that make optimal use of natural and human resources and are based on locally available or inexpensive technologies are referred to as 'sustainable intensification' (Pretty, 2011).

Despite the ongoing growth in manufacture and use of pesticides, there are indications that it would be unwise to continue with a high level of dependency on synthetic toxicant-based pest management. The discovery of new molecules and registration of new products is at a very low rate, compounds are being withdrawn from the market for economic and legislative reasons (especially in 'developed' countries), and this is compounded by emerging resistance to many pesticides. In Ireland recently, growers facing extensive herbicide resistance in weeds in brassicas were told by a senior agro-chemical executive that rescue by 'the cavalry appearing from over the hill' was unlikely; companies had nothing new to offer.

Manipulating crop and non-crop habitat to mitigate pest problems is a recurring theme in many of the chapters in this book and it is a foundation of the agroecology discipline. We are reminded of work by pioneers such as Price, Root, Pimentel and van Emden in this context. The last-named began by putting containers of cut flowers in a crop to attract pests' natural enemies. Gradually came a more detailed understanding of how the ecological fitness and practical impact of natural enemies could be improved through the provision of nectar for energy and pollen for protein. The fact that the fly family Syrphidae are called 'flowerflies' in the USA points to the fact that the association between adults of this group for floral resources has long been recognised in a natural history sense, although not previously exploited as a practical agroecological option to improve biological control. Advances in knowledge now reveal and exploit complexities in flower architecture, nectar sugar ratios, food-web dynamics, trophic cascades and 'attract and reward'. Below ground, too, bioactive fungi and bacteria can promote crop plant defences (Altieri *et al.*, chapter 5 of this volume) so, in fact, the armoury of theory, tools and techniques available for novel pest

management has never been greater. The range of methods available to researchers continues to grow (see Symondson, chapter 10 of this volume). As De Schutter's 2010 UNEP Report suggested, we still need to break away from purely experimental evidence and put these truly sustainable practices into the commercial arena. However, the stage is set for this to happen as the chapters in the 'Application' section of this book and meta studies such as Pretty *et al.* (2006) demonstrate. Changes in policy and practice are needed, and this book, we hope, will be a convincing step in effecting those vital changes in how we produce the world's food.

FUTURE DIRECTIONS

The chapters of this book provide encouraging evidence that strategic use of biodiversity can provide great benefits for agriculture. However, important work remains to be done. A key need is to provide more examples of biodiversity at work to solve pest problems, where economic benefits that exceed management costs can be clearly documented. Beyond this, we identify a few vexing questions that await further clarification. These range from fundamental issues related specifically to biodiversity and ecosystem function through to the global food system:

1. Biodiversity aspects beyond richness. Throughout most chapters of this book, and indeed in much recent research, the term 'biodiversity' is equated with the number of species (or other genetic units). However, ecologists have long thought that evenness, or the relative abundances of species, is also important to ecosystem functioning. Here, ecological communities with equally abundant species are considered healthiest, similar to the idea of a 'balance of nature' that is familiar enough to have become a household term. Some recent work (e.g. Crowder *et al.*, 2010) suggests that balancing communities can provide benefits for pest control similar to those provided by greater species richness; challenges remain, however, in understanding how to manage agroecosystems to encourage greater evenness. Greater evenness might provide benefits to natural pest control at both the field and landscape levels, although this topic has received little attention.

2. The development of multi-faceted schemes. The successful conservation of natural enemies is a complex process with many steps: natural enemies must be attracted, conserved, and encouraged to attack target pests. This complexity has perhaps contributed to the relative dearth of success stories in this area. One promising route towards increasing the success rate, is the development of strategies that bring together several different conservation approaches. For example, chapter 11 of this volume by James *et al.* presents the example of 'attract and reward' strategies, where natural enemies are first attracted into a crop through the initiation of HIPVs (the attraction, tactic 1) where they then are rewarded through the provision of plants that provide nectar resources (the reward, tactic 2). Similarly, Khan *et al.* (chapter 16) present the strategy known as 'push–pull', where herbivores are deterred from entering a crop through the planting of plants that give off disagreeable chemical cues (the push, tactic 1) and further drawn into plants other than the crop that is being protected which have an agreeable scent (the pull, tactic 2). This synergistic joining of two different approaches may have many other uses; for example, the provisioning of native plants might be made within landscapes designed to encourage natural enemies but deter pests.

3. Critically important in the development of these biodiversity-based approaches is the availability of funding. Generally, private investment will not support the development or adoption of these types of approaches because it is difficult for them to be patented or make a revenue stream. This places a great responsibility on researchers in universities and publicly funded agencies where public good research can be conducted and, via liaison with organisations such as those in the Consultative Group on International Agricultural Research, work with end-users to develop practicable technologies.

4. Increasing numbers of landholders value the presence of flora and fauna for aesthetic reasons but a pragmatic farmer is more likely to embrace change if offered benefits such as a field margin flower strip that is also a lucrative secondary crop, tree strips that provide shelter for crops and livestock or groundcovers that fix nitrogen and prevent soil erosion. This has been termed 'multi-function agricultural biodiversity' (Gurr *et al.*, 2003). The types of approaches set out in this book will be further promoted in countries with agri-environmental schemes that make payments to farmers for being good stewards of biodiversity and as schemes are developed to encourage on-farm carbon sequestration (Wade *et al.*, 2008). Ultimately these approaches place agriculture as a 'part or nature' rather than 'apart from nature'.

5. Finally, notwithstanding the fundamental importance of biodiversity to agriculture, policy change relating to social and economic factors will be important. Pretty *et al.* (2010) include in their 'top 100 questions of importance to the future of global agriculture' 41 questions relating to social capital, development, governance, markets and consumption patterns. Reflecting the importance of social rather than technological issues, Altieri and Rosset (2002, p. 2) say:

> The world today produces more food per inhabitant than ever before . . . The real causes of hunger are poverty, inequality and lack of access to food and land.

Politicians and economists will need to rise to the challenges of their respective domains if the global community is to fully realise the promise of biodiversity to deliver better pest suppression and other ecosystem services for food security.

CONCLUSION

Even a cursory glance through the chapters in this book reveals input from fields as diverse as economics and communication theory, from plant breeding to natural product chemistry as well as the community, landscape and chemical branches of ecology. It seems clear from the now detailed (if not yet complete) understanding of theory and the successful instances of applied use presented in this book, that biodiversity – in its various shades of green – is a powerful force for placing pest management on a more sustainable footing in the twenty-first century. Further uptake of biodiversity for pest management will depend on researchers from a broad range of disciplines working together. It will be necessary to integrate these strategies into wider schemes that provide multiple ecosystem benefits, and this task will require changes in the policy arena (De Schutter, 2010).

A crucial factor in success will be participation and appropriate educational opportunities for farmers, many of whom will be risk-averse and reluctant to change practice. Attitudes do change, however. An anecdotal illustration of this is the contrast between two quotations. A saying used by farmers in the USA in the twentieth century: 'Even as a single hair casts a shadow does a weed steal profit from the harvest' promoted the notion of 'no room' for any non-crop diversity in farming. A much more sophisticated and knowledge-intensive attitude to farming is evident in a recent quotation from a pioneering Swedish farmer, who said 'I am a photosynthesis manager and an ecosystem-service provider' (Peter Edlin, personal communication).

Those who embrace on-farm biodiversity as a key resource for providing ecosystem goods and outcomes will reap benefits both for themselves and for society more generally.

ACKNOWLEDGEMENTS

Part of this chapter is based on a talk by Geoff M. Gurr at the 2011 Consultative Group on International Agricultural Research (CGIAR) Science Forum 2011, Beijing, China.

REFERENCES

Altieri, M. and Rosset, P. (2002) Ten reasons why biotechnology will not ensure food security, protect the environment or reduce poverty in the developing world, in *Ethical issues in biotechnology* (eds R. Sherlock and J. Morey), Rowan & Littlefield, 175–182.

Baggen, L.R. and Gurr, G.M. (1998) The influence of food on *Copidosoma koehleri*, and the use of flowering plants as a habitat management tool to enhance biological control of potato moth, *Phthorimaea operculella. Biological Control*, 11, 9–17.

Crowder, D.W., Northfield, T.D., Strand, M.R. and Snyder, W.E. (2010) Organic agriculture promotes evenness and natural pest control. *Nature*, 466, 109–112.

De Schutter, O. (2010) Report submitted by the Special Rapporteur on the right to food, Oliver De Schutter. United Nations General Assembly.

Green, R.E., Cornell, S.J., Scharlemann, J.P. *et al.* (2005) Farming and the fate of wild nature. *Science*, 307, 550–555.

Gurr, G.M., Wratten, S.D. and Luna, J.M. (2003) Multifunction agricultural biodiversity: pest management and other benefits. *Basic and Applied Ecology*, 4, 107–116.

McIntyre, B.D., Herren, H.H., Wakhungu, J. and Watson, R.T. (2009) *Agriculture at a Crossroads. Global Report*. International Assessment of Agricultural Knowledge, Science and Technology for Development. Island Press, Washington, DC.

Perovic, D.J., Gurr, G.M., Raman, A. and Nicol, H.I. (2010) Effect of landscape composition and arrangement on biological control agents in a simplified agricultural system: a cost-distance approach. *Biological Control*, 52, 263–270.

Peveling, R. and Ely, S.O. (2006) Side-effects of botanical insecticides derived from Meliaceae on coccinellid predators of the date palm scale. *Crop Protection*, 25, 1253–1258.

Pretty, J. (2011) Can ecological agriculture feed nine billion people? *Monthly Review*, 61, unpaginated.

Pretty, J., Noble, A.D., Bossio, D. *et al.* (2006) Resource-conserving agriculture increases yields in developing countries. *Environmental Science and Technology*, 4, 1114–1119.

Pretty, J., Sutherland, W.J., Ashby, J. *et al.* (2010). The top 100 questions of importance to the future of global agriculture. *International Journal of Agricultural Sustainability*, 8, 219–236.

Sachs, J.D., Baillie, J.E.M., Sutherland, W.J. *et al.* (2009) Biodiversity conservation and the millennium development goals. *Science*, 325, 1502–1503.

Schmutterer, H. (1997) Side-effects of neem (*Azadirachta indica*) products on insect pathogens and natural enemies of spider mites and insects. *Journal of Applied Entomology*, 121, 121–128.

Tilman, D., Cassman, K.G., Matsons, P.A., Naylor, R. and Polasky, S. (2002) Agricultural sustainability and intensive production practices. *Nature*, 418, 671–677.

Wade, M.R., Gurr, G.M. and Wratten, S.D. (2008) Ecological restoration of farmland: progress and prospects. *Philosophical Transactions of the Royal Society B*, 363, 831–847.

INDEX

Page numbers in *italics* refer to figures and boxes, those in **bold** refer to tables. In subheadings, NE means 'natural enemies'.

Biodiversity and Insect Pests: Key Issues for Sustainable Management, First Edition. Edited by Geoff M. Gurr, Steve D. Wratten, William E. Snyder, Donna M.Y. Read.
© 2012 John Wiley & Sons, Ltd. Published 2012 by John Wiley & Sons, Ltd.